PHYSICS OF THE EARTH AND THE SOLAR SYSTEM

GEOPHYSICS AND ASTROPHYSICS MONOGRAPHS

Editor

B. M. McCORMAC, *Lockheed Palo Alto Research Laboratory, Palo Alto, Calif., U.S.A.*

Editorial Board

R. GRANT ATHAY, *High Altitude Observatory, Boulder, Colo., U.S.A.*
P. BANKS, *Radioscience Laboratory, Stanford University, Stanford, Calif., U.S.A.*
D. M. HUNTEN, *Department of Planetary Science, University of Arizona, Tucson, Ariz., U.S.A.*
M. KUPERUS, *Sonnenborgh Observatory, University of Utrecht, The Netherlands*
R. LÜST, *European Space Agency, Paris, France*
R. ROSNER, *Department of Astronomy and Astrophysics, University of Chicago, Ill., U.S.A.*
G. WEILL, *GDTA, Toulouse, France*

Volume 31

PHYSICS OF THE EARTH AND THE SOLAR SYSTEM

Dynamics and Evolution, Space Navigation, Space-Time Structure

by

BRUNO BERTOTTI

Department of Nuclear and Theoretical Physics, University of Pavia, Italy

and

PAOLO FARINELLA

Department of Mathematics, University of Pisa, Italy

KLUWER ACADEMIC PUBLISHERS

DORDRECHT / BOSTON / LONDON

Library of Congress Cataloging in Publication Data

Bertotti, B., 1930-
Physics of the earth and the solar system / Bruno Bertotti, Paolo Farinella.
p. cm. -- (Geophysics and astrophysics monographs ; v. 31)
ISBN 0-7923-0535-3
1. Solar system. 2. Geophysics. 3. Astrophysics. I. Farinella, Paolo. II. Title. III. Series.
QB501.B48 1990
523.2--dc20 89-27782

ISBN 0-7923-0535-3

Published by Kluwer Academic Publishers,
P.O. Box 17, 3300 AA Dordrecht, The Netherlands.

Kluwer Academic Publishers incorporates
the publishing programmes of
D. Reidel, Martinus Nijhoff, Dr W. Junk and MTP Press.

Sold and distributed in the U.S.A. and Canada
by Kluwer Academic Publishers,
101 Philip Drive, Norwell, MA 02061, U.S.A.

In all other countries, sold and distributed
by Kluwer Academic Publishers Group,
P.O. Box 322, 3300 AH Dordrecht, The Netherlands.

Printed on acid-free paper

Printed in the Netherlands

CONTENTS

INTRODUCTION

The purpose of this book is to give an outline of our current rational understanding, on the basis of mathematical models and quantitative estimates, of the structure, the properties and the evolution of the solar system in our space age. The descriptive and the experimental aspects are not stressed; rather, we concentrate upon the physical principles and processes which govern their behaviour. Particular attention is given to the orders of magnitude of the physical quantities and to the approximations used; exercises are provided at the end of each chapter to help the development of this rational understanding.

The subject matter is developed starting from concrete objects; the relevant physical principles in part are discussed in Chapter 1 and then recalled or developed as the need arises: mathematical rigour does not have a priority. We assume that the reader is acquainted with undergraduate physics, in particular with dynamics, electromagnetism, special relativity, thermodynamics and statistical physics; from this basis he will hopefully be led up to the threshold of current research. In choosing the sequence of topics we have tried, when possible, to respect the criterion of increasing difficulty; in many cases a substantial effort was put in to simplify and clarify the material. As one sees from the index, the range of our topics touches geophysics on one side and space physics and astronomy on the other; we do not dwell much in the border topics. In between, we range from the earth, especially as seen in its relationships with the solar system, to interplanetary space, the planets and the other constituents. We devote particular attention to two important topics which have witnessed outstanding advances during the last years, especially with the use of spacecraft: precise measurements in space and the evolution of the solar system. The physics of the magnetospheres, ionospheres and interplanetary plasma, about which a very large and complex body of knowledge has accumulated, is amply discussed in other books; in this work we choose to confine ourselves to the basic mechanisms at work.

In going through this immense variety of phenomena and processes one is struck by the surprising amount of common elements needed for a rational understanding; in particular, *gravitation* has the role of a protagonist: being the main binding force for large bodies and their dynamics, it provides a unifying common theme. Although the newtonian theory of gravitation is most of the times sufficient, the recent, striking advances in space physics and radio science have made the effects due to general relativity an essential ingredient of the dynamics of planets and satellites. It is therefore essential that the reader be introduced to the idea and the concrete use of curved space-time; we do so (Ch. 17) in the appropriate approximation, with little advanced mathematical formalism.

Before the second world war our subject developed mainly through intelligent, painstaking and separate collection and interpretation of data in different fields: terrestrial measurements of gravity and magnetic anomalies; planetary observations with optical telescopes; observation of sunspots and solar activity; and so on. Progress occurred through slow accumulation and change, within the framework of long-established paradigms, according to the well known patterns of a "normal" science. This favoured an extreme specialization and the growth of closed scientific communities, whose main purpose was the application and the development of a particular experimental technique. There was not enough information nor adequate theoretical tools to provide a unified picture of the solar system.

After the second world war, as a result also of the great progress in military technology, new and much more expensive experimental techniques were developed, in particular radio science and space navigation. They have changed our subject drastically, providing us with direct information about the magnetosphere, interplanetary space, the dynamics, the atmosphere and the interior of planets and satellites. The accuracy with which the relevant quantities have been measured has increased several tenfolds. For example, the Very Long Baseline Interferometry (VLBI) technique provides angular measurements with an accuracy better than 10^{-3} arcsec, compared with the error ≈ 1 arcsec in the angular position of celestial sources with optical ground telescopes. The direct measurement of distances in the solar system is a completely new achievement: e. g., now we know the distance of the moon to better than 10 cm. With the help of coherent radio beams, well stabilized in frequency by means of atomic standards, relative velocities are measured to accuracies of μm/s. The theoretical paradigms also changed; in particular, we now have much clearer ideas about the origin and the evolution of the solar system.

These new instruments are generally characterized by their much greater size, complexity and economic value; in this, as in other fields, we had a qualitative transition from a "little" to a "big" science, where the experimental programs require a very long preparation and large investments of energy and money. This change in character has not come about by chance. Military projects, in particular the development of radar and rockets, had a great influence on the new techniques currently used in space research. The great engineering accomplishments needed to bring a payload into space and to track artificial satellites are, of course, of paramount military interest, so that civilian and military activities are inextricably connected. It must be recognized, however, that, as far as we know, present military satellites do not carry weapons: by tacit restraint space activities of all kinds are accepted and respected by all nations. This state of affairs would of course drastically change if weapons capable of damaging or destroying spacecraft were deployed and space would be used as a base for warfare. This will endanger all civilian space activities as well and negate the concept of outer space as a common property of mankind.

One must also recognize the great impact of space exploration, which has shifted the boundary of unknown lands to the natural satellites, the planets and interplanetary space. Now our machines can actually go out there and report back; the magic spell of exploration, so well described in the story of Ulysses in Dante's *Commedia* (Hell, Canto 26), is still with us. This fascination produces a convergence of wills, creates and perpetuates well organized scientific communities; it truly generates a scientific power, anchored to extensive economic interests.

We use throughout the cgs system of units; Maxwell's equations are written in esu, not rationalized units (see eq. (1.59)). Bold symbols denote cartesian vectors. Sometimes, in a diadic notation, a bold, italic and capital letter denotes a tensor of second rank (e.g., eq. (1.38)). Partial derivatives (i. e., with respect to x) are indicated either with $\partial/\partial x$ or with ∂_x.

To help the reader, we have indicated with a star the more difficult sections, which can be left out in a first reading. The problems at the end of each chapter are in no particular order; they are also starred to denote difficulty. At the end of each chapter some books and review articles are listed for further reading. Here we quote a few outstanding general textbooks:

C.W. Allen, *Astrophysical Quantities*, Athlone Press (1964);

V.M. Blanco and S.W. McCuskey, *Basic Physics of the Solar System*, Addison-Wesley (1961);

W.K. Hartmann, *Moons and Planets*, Wadsworth (1983);

W.M. Kaula, *An Introduction to Planetary Physics*, Wiley (1968);

A.E. Roy, *Orbital Motion*, Hilger (1978);

F.D. Stacey, *Physics of the Earth*, Wiley (1977).

We had discussions with many people; here we mention in particular L. Anselmo, J.A. Armstrong, P.L. Bernacca, M. Carpino, A. Cellino, D.R. Davis, R. Greenberg, E. Melchioni, A. Milani, F. Mignard, A.M. Nobili, P. Paolicchi, A. Rossi, V. Vanzani and V. Zappala', who provided useful suggestions and comments.

B. Bertotti wishes to dedicate this book to Annamaria; P. Farinella dedicates it to his parents, for their love and support.

USEFUL PHYSICAL QUANTITIES

arcsec = 1" = $4.848 \cdot 10^{-6}$ rad.

AU = $1.496 \cdot 10^{13}$ cm = $1.496 \cdot 10^{11}$ m
Astronomical Unit (semimajor axis of the earth's orbit)

c = $2.998 \cdot 10^{10}$ cm/s = $2.998 \cdot 10^{8}$ m/s
velocity of light

e = $4.803 \cdot 10^{-10}$ esu = $1.601 \cdot 10^{-19}$ Coulomb
magnitude of electron charge

eV = $1.602 \cdot 10^{-12}$ erg = $1.602 \cdot 10^{-19}$ Joule = 11,600 K
energy and temperature associated with 1 electron volt

G = $6.673 \cdot 10^{-8}$ dyne cm^2/g^2 = $6.673 \cdot 10^{-11}$ Nm^2/kg^2
gravitational constant

h = $6.625 \cdot 10^{-27}$ erg s = $6.625 \cdot 10^{-34}$ J s
Planck constant

H = 55 Km/s$\cdot$Mpc
Hubble constant

J_2 = $(C - A)/(M_\oplus R_\oplus^2)$ = 0.001082
quadrupole coefficient of the earth

k = $1.380 \cdot 10^{-16}$ erg/K = $1.380 \cdot 10^{-23}$ J/K
Boltmann constant

ly = $9.46 \cdot 10^{17}$ cm = $9.46 \cdot 10^{15}$ m
light year

$L_\odot$ = $3.90 \cdot 10^{33}$ erg/s = $3.90 \cdot 10^{26}$ J/s
solar luminosity

m = $-2.5 \cdot \log_{10} \Phi$ + const
apparent magnitude of a star with a flux Φ; the sun at 10pc would have magnitude 4.72

m_e = $9.109 \cdot 10^{-28}$ g = $9.109 \cdot 10^{-31}$ kg
electron mass

m_p = $1.673 \cdot 10^{-24}$ g = $1.673 \cdot 10^{-27}$ kg
proton mass

$M_\odot$ = $1.99 \cdot 10^{33}$ g = $1.99 \cdot 10^{30}$ kg
solar mass

$m_\odot$ = $1.48 \cdot 10^{5}$ cm = $1.48 \cdot 10^{3}$ m
solar gravitational radius

$M_\oplus$ = $5.98 \cdot 10^{27}$ g = $5.98 \cdot 10^{24}$ kg
earth mass

$m_\oplus$ = 0.44 cm = $4.4 \cdot 10^{-3}$ m
earth gravitational radius

M_J = 318 $M_\oplus$ = $M_\odot/1047$
Jupiter mass

mas = 0.001"; it corresponds to 3.1 cm on the surface of the earth

$n_\oplus$ = $1.991 \cdot 10^{-7}$ rad/s
mean motion of the earth's orbit

pc = $3.086 \cdot 10^{18}$ cm = $3.086 \cdot 10^{16}$ m
parsec

$R_\oplus$ = $6.37 \cdot 10^{8}$ cm = $6.37 \cdot 10^{6}$ m
mean earth radius

$R_\odot$ = $6.96 \cdot 10^{10}$ cm = $6.96 \cdot 10^{8}$ m
mean solar radius

v_{esc} = $\sqrt{(2GM_\oplus/R_\oplus)}$ = 11.2 km/s
escape velocity from the earth

y = $3.15 \cdot 10^{7}$ s
a year

ϵ = $23^\circ\ 27'$
obliquity of the ecliptic

λ_D = $\sqrt{(kT/4\pi ne^2)} = 743\ \sqrt{[(kT/eV)(n\ cm^3)]}\ cm =$
$= 7.43 \cdot 10^{-3}\ \sqrt{[(kT/eV)(n\ m^3)]} m$
Debye length

σ = $2\pi^5 k^4/15c^2 h^3 = 5.67 \cdot 10^{-5}\ erg/(cm^2\, sK^4) =$
$= 5.67 \cdot 10^{-8}\ W/(m^2 K^4)$
Stefan-Boltzmann constant

Φ_0 = $1.38 \cdot 10^6\ erg/(cm^2\ s) = 1.38 \cdot 10^3\ J/(m^2\ s)$
solar constant (at 1 AU)

ω_p = $\sqrt{(4\pi ne^2/m_e)} = 5.64 \cdot 10^4\ \sqrt{(n\ cm^3)}\ rad/s = 56.4\ \sqrt{(n\ m^3)}\ rad/s$
plasma frequency

Ω_c = $eB/m_e\ c = 1.76 \cdot 10^7\ (B/gauss)\ rad/s =$
$= 1.76 \cdot 10^{11}\ (B\ m^2/weber)\ rad/s$
electron cyclotron frequency

$\omega_\oplus$ = $(7.29 \cdot 10^{-5} - 1.90 \cdot 10^{-11}\ t/y)\ rad/s = 2\pi/LOD$
sidereal angular velocity of the earth and its secular decrease; t is measured from the year 1900.0

Ω_{pr} = $50.26\ arcsec/y = 7.72 \cdot 10^{-12}\ rad/s$
lunisolar precession constant

1. DYNAMICAL PRINCIPLES

In this chapter we lay down the dynamical equations that govern the motion and determine the structure of a planetary body and its environment. The basic model is that of a continuum, whose local properties are described by the matter density ρ, the flow velocity $\mathbf{v}$ and the temperature T. If the fluid is electrically conductive, we must add the magnetic field $\mathbf{B}$. These quantities obey a set of partial differential equations which will be discussed in the relevant approximations, with examples and illustrations taken from the physics of the solar system. Gravitation plays an essential role here: a long range force, it becomes more and more important with increasing size. It determines the shape, the size and the structure of the planets, the stars, the galaxies and the universe itself. It is balanced primarily by the thermal pressure and secondarily by centrifugal forces: this balance is responsible for the usually spherical shape of cosmic bodies. When a body is sufficiently small and cold, however, it solidifies; its structure is then determined by elastic forces. In cosmic bodies different transport processes play important roles; in particular, the transport of momentum determines viscosity and the transport of energy is the basis for heat conduction. In an electrically conductive medium we have a large variety of electric and magnetic processes, like the generation of large scale magnetic fields in rotating bodies, acceleration of particles, propagation of different kinds of electromagnetic waves. In an ionized gas, when collisions are unimportant, the appropriate description requires the velocity space and gives rise to a much greater variety of phenomena.

1.1. GRAVITATIONAL EQUILIBRIUM

The structure of a big cosmic body is usually determined by the balance between the gravitational pull and the internal pressure; this pressure, in turn, depends upon the state of matter, the composition, the heat flow, the magnetic field, and so on. These factors are influenced by the way the body was formed and its previous history and their determination requires the laws of microscopic physics: the evaluation of the state of pressure, determined by the pressure tensor, for example, constitutes the necessary link between the microscopic state of each part and its overall structure and shape. If – as is the case in a fluid – this microscopic state has no privileged direction, in particular has no stratification, the pressure tensor is isotropic and is determined by a single scalar, the pressure. Then, if there are no external reasons for asymmetry, like rotation, a configuration of spherical equilibrium is possible: the pressure is a function of the radial distance only and will be able to compensate the radial gravitational pull. A deviation from the spherical equilibrium will occur, for instance, if the body, or part of it, is a solid, capable of supporting a force tangential to its surface; this situation will be described by a strain and a stress tensor which is not isotropic.

We now derive the condition of equilibrium of a spherical mass of density ρ(r) and pressure P(r). The total mass within the distance r from the centre is

$$M(r) = 4\pi\int_0^r dr' r'^2 \rho(r') . \tag{1.1}$$

The symbol M (without any argument) will denote the total mass and R the radius. The gravitational acceleration g(r) at a given distance r from the centre of the body has a radial, inward direction and is equal to the one produced by a point mass M(r) placed at the centre:

$$g(r) = GM(r)/r^2 . \tag{1.2}$$

One proves this theorem using the gravitational analogue of Gauss' theorem in electrostatic. The gravitational acceleration **g** of a point mass is obtained from the electrostatic field produced by charge q by replacing q with – Gm; the mathematical properties of the two quantities are the same and, therefore, the gravitational analogue of Gauss' theorem is:

$$\int_S d\mathbf{S}\mathbf{n} \cdot \mathbf{g} = 4\pi GM, \tag{1.3}$$

where S is a closed surface with outgoing normal **n** and M is the total mass inside. (There is also a fundamental correspondence between the electrostatic force $qq'\mathbf{r}/r^3$ between two charges and the gravitational force $-Gmm'\mathbf{r}/r^3$ between two masses: one obtains the latter by replacing q with $m\sqrt{(-G)}$; this correspondence embodies the whole physics of both fields of force. Gravitation is electrostatics with an imaginary charge; for this reason, for example, electrostatic plasma oscillations correspond to the unstable gravitational instability.) Let us apply (1.3) to a sphere of radius r inside our body with mass M(r). $\mathbf{g} = -g\mathbf{r}/r$ has, by symmetry, a radial direction and points inward; hence the left–hand side of eq.

(1.3) is $-4\pi r^2 g$ and the theorem (1.2) follows at once. The elegance of this argument is to be contrasted with the complicated three–dimensional integration which one must perform with the naive, direct method of summing the vector contributions to **g** from all constituent elements of the body; but the argument works only when the body is spherically symmetric.

The gravitational potential energy per unit mass U, defined by $\mathbf{g} = -\nabla U$, is, of course, $-GM/r$ outside the body; but inside it is *not* equal to $-GM(r)/r$ and a theorem similar to the one proved for the acceleration does not hold. To evaluate the potential energy, note that inside a thin spherical shell of mass dM and radius r' the gravitational pull vanishes; hence the gravitational potential energy is constant and equal to its value just outside, $-GdM/r'$. The total value of U(r) is then the sum of the contributions from the shells outside r and those inside; the latter amounts to $-GM(r)/r$. Therefore

$$U(r) = -G\left[4\pi\int_r^R dr' r'\rho(r') + \frac{M(r)}{r}\right] . \qquad (1.4)$$

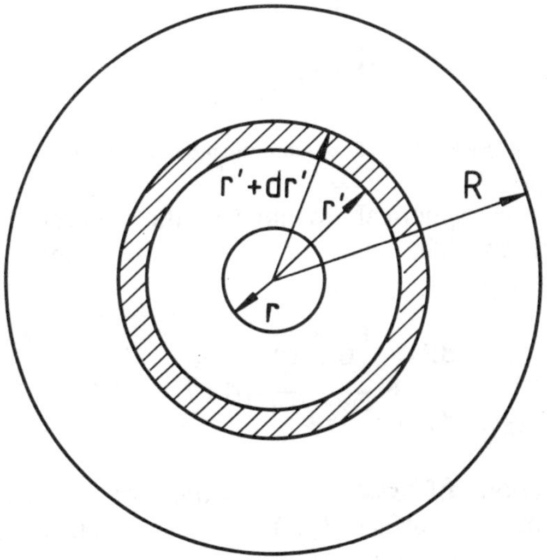

Fig. 1.1. The calculation of the gravitational potential energy inside a spherical body. The contribution from the outer layers is calculated by summing the terms due to each spherical shell dM; for the inner layers we use the standard result from Gauss' theorem.

For a sphere of uniform density

$$U = -\frac{GM}{2R}\left[3 - \frac{r^2}{R^2}\right], \tag{1.5}$$

which reduces to $-$ GM/R on the surface, as it should.

An infinitesimal element between r and (r + dr), with a base dS is subject to three forces:

– the pressure from below, dSP(r) (positive);

– the pressure from above, $-$ dSP(r + dr) = $-$ dSP(r) $-$ dSdrdP/dr (negative);

– the gravitational pull, df = $dSdrG\rho(r)/r^2$.

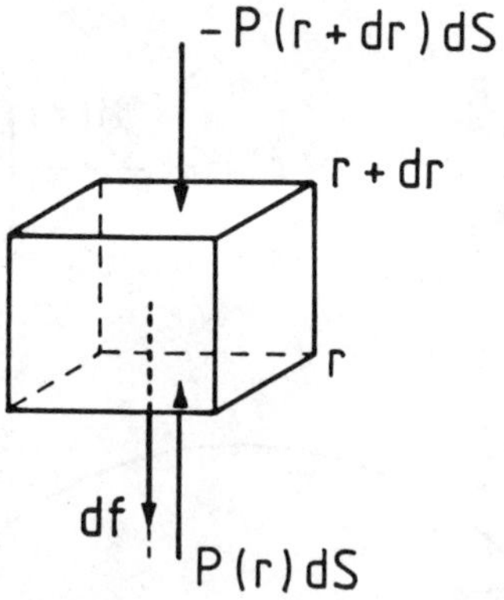

Fig. 1.2. The condition of equilibrium of a spherical body under gravity.

Because of the spherical symmetry the lateral forces give a net vanishing contribution. The sum of these three forces must sum up to zero in a static situation:

$$\frac{dP}{dr} + \frac{GM(r)}{r^2}\rho(r) = 0. \tag{1.6}$$

This is the equation of *static equilibrium* of a body with spherical symmetry. Since at the surface P(R) = 0, we can compute the pressure at the centre with a direct integration:

$$P(0) = G\int_0^R dr\rho(r)\frac{M(r)}{r^2}. \tag{1.7}$$

For example, if the density is uniform, we get

$$P(0) = 2\pi G\rho^2R^2/3 = 3GM^2/8\pi R^4. \qquad (1.8)$$

For the earth this gives $P(0) = 1.7\cdot10^{12}$ dyne/cm^2, a value several orders of magnitude greater than the pressure accessible in laboratory experiments; as a consequence, our knowledge of the state of the earth's interior (in particular the transport and the melting coefficients) is indirect and subject to great uncertainties. It is important, in particular, to evaluate the internal temperature; this requires a detailed understanding of the energy production mechanisms and the energy transport processes (in particular, conduction, convection and radiative transport). It often happens that the temperature distribution depends upon the past history, in particular, for a planet, the radiative energy received from the sun.

A basic quantity one needs in order to understand the energetics of a gravitationally bound body is its total gravitational energy

$$W = \tfrac{1}{2}G\int\int dm(\mathbf{r})dm(\mathbf{r}')/|\mathbf{r} - \mathbf{r}'| = 2\pi\int_0^R dr\ r^2\rho(r)U(r); \qquad (1.9)$$

for a uniform distribution (eq. (1.5)) this is

$$W = -\ 3GM^2/5R\ , \qquad (1.10)$$

equal to $2.3\cdot10^{39}$ erg for a homogeneous earth. If the gravitational pull is balanced by thermal pressure the thermal and gravitational energies are of the same order of magnitude and the temperature can easily be estimated. If the mean molecular number is A, there are in the body M/Am_p molecules (m_p is the proton mass, here assumed to be equal to the neutron mass); the total thermal energy is therefore of order kTM/Am_p, so that

$$kT \approx GMAm_p/R. \qquad (1.11)$$

This condition holds in the sun and other normal stars. In the earth we would get $kT \approx 0.013$ A erg = 0.8A eV (1 eV corresponds to about 11,000 K); however, the earth's binding energy is mainly provided by intermolecular forces and eq. (1.11) gives only an upper bound to the temperature.

A similar estimate can be obtained from the equilibrium equation (1.6). We evaluate the pressure gradient dP/dr by P/R, where P is a typical pressure in the interior and R is the size of the body; similarly, we estimate $M(r)/r^2$ by M/R^2. Therefore eq. (1.6) gives

$$P/R \approx GM\rho/R^2 \approx GMAm_pn/R^2,$$

where $n = \rho/Am_p$ is the number density of the molecules of the body. For the interior of the earth (assuming A = 50) this gives a pressure $P \approx 3 \cdot 10^{12}$ dyne/cm². If we now assume an equation of state for a perfect gas (the simplest assumption we can make to evaluate the pressure),

$$P = nkT = \rho kT/m, \tag{1.12}$$

we get the previous estimate (1.11), m being the mean molecular mass.

In a mixture the pressure is just the sum of the partial pressures and is still proportional to the total number density n of the particles. For two gases of molecular masses m_1 and m_2, for example, we have

$$n = \frac{\rho}{(1 - \alpha)m_1 + \alpha m_2} \quad (\alpha = n_2/(n_1 + n_2)). \tag{1.13}$$

Here α is the *concentration* of the second species by number.

In order to describe completely the structure of a static, spherical cosmic body we need the pressure P, the matter density ρ, the temperature T and, possibly, the composition as function of r. Leaving aside, for the time being, variations of composition, we have, to determine these three quantities: the equilibrium condition (1.6); an equation of state of the type (1.12) connecting in finite terms pressure, density and temperature; and a heat transport equation to describe the energy balance, to be discussed later. They have to be supplemented with the appropriate boundary conditions (finite values at the centre and vanishing mass density and pressure at the surface r = R). Their solution is an important task of cosmic physics and requires a clear understanding and appropriate modelling of the processes inside a body.

The main reason for a violation of the spherical symmetry of a cosmic body is rotation. It induces a deformation which is generally axially symmetric around the rotation axis; the ellipticity (assumed to be small)

$$\epsilon = \frac{1}{2}\left(1 - \frac{R_p^2}{R_e^2}\right) \simeq \frac{R_e - R_p}{R}, \tag{1.14}$$

measured in terms of the equatorial and the polar radii, is a good parameter to measure this deformation. In this case ϵ is of the order of the ratio μ between the rotational energy and the gravitational binding energy of the body or, in terms of its angular velocity ω,

$$\epsilon \approx \mu = R^3\omega^2/GM \tag{1.14'}$$

(see Sec. 3.3). This deformation is of fundamental importance in cosmic physics also because it is enhanced by collapse or by compression. If a body is not coupled to the environment or to other bodies, its angular momentum L, of the order of magnitude $MR^2\omega$, remains constant; hence ω is proportional to $1/R^2$ and the ellipticity to 1/R. For example, when a star like the sun (with radius $R = 7\cdot10^{11}$ cm) collapses to a neutron star with a radius of 20 km, the angular velocity increases by a factor $1.2\cdot10^9$ and the ellipticity by a factor $3.5\cdot10^4$.

Sometimes it is useful to measure the mass of a body in terms of its *gravitational radius*

$$m = GM/c^2 , \tag{1.15}$$

equal to 0.43 cm for the earth and 1.5 km for the sun. This is the distance from a point of mass M at which the newtonian energy of a particle falling from infinity equals its rest energy, making relativistic corrections essential. In ordinary bodies m is always much less than the radius, so that relativistic effects, of order m/R, are very small. They are discussed in Ch. 17.

The angular momentum per unit mass of a body L/M also determines a length

$$a = L/cM = GL/mc^3 ; \tag{1.16}$$

it is equal to 0.28 km for the sun and to 3.2 m for the earth. We show in Ch. 17 that this length characterizes a peculiar correction to the gravitational force (the *gravitomagnetic force*), corresponding to an effective potential energy per unit mass of order ma/r^2.

1.2 DYNAMICAL EQUATIONS

When there is motion, we have a dynamic equilibrium and eq. (1.6) must be supplemented with the inertial force. In spherical symmetry a fluid element of mass $\rho drdS$, which at time t has a radial velocity v(r, t), at time (t + dt) has a velocity v(r + dr, t + dt); here dr = v(r, t)dt is the radial displacement undergone by the fluid element. Its acceleration must be computed by adding the change of the velocity field at a given place and the change due to the fact that the fluid element moves:

$$\frac{v(r + vdt, t + dt) - v(r,t)}{dt} \Rightarrow \frac{dv}{dt} = \frac{\partial v}{\partial t} + v\frac{\partial v}{\partial r} . \tag{1.17}$$

The first term on the right-hand side is the ordinary, *eulerian* derivative and the whole expression defines the *lagrangean* derivative. More generally, in a fluid motion the lagrangean derivative defines the rate of change suffered when one moves with the fluid:

$$d/dt = \partial/\partial t + \mathbf{v} \cdot \nabla. \tag{1.18}$$

An index t will always denote ordinary, eulerian derivatives.

We have therefore, in place of eq. (1.6) and neglecting viscosity,

$$\rho\left[\frac{\partial v}{\partial t} + v\,\frac{\partial v}{\partial r}\right] + \frac{\partial P}{\partial r} + \frac{GM(r)}{r^2}\,\rho = 0. \tag{1.19}$$

If the flow is stationary, v is a function of r alone and this reads

$$\rho v\,\frac{dv}{dr} + \frac{dP}{dr} + \frac{GM(r)}{r^2}\,\rho = 0. \tag{1.20}$$

Radial flows are important for cosmic bodies; in particular, a radial wind of charged particles is emitted by the sun and produces striking phenomena in interplanetary space (Ch. 13).

When there is motion *mass conservation* must be taken into account. A shell between r and r + dr contains the mass $4\pi r^2 \rho dr$. In an infinitesimal time dt a particle at r moves by v(r, t)dt, so that

$$r^2 \to r^2 + 2rvdt.$$

The shell thickness dr and the mass density ρ change according to

$$dr \to dr + dr\partial_r v dt, \qquad \rho \to \rho + (\partial_t \rho + \partial_r\, \rho v)dt.$$

The total change in the quantity $r^2 \rho dr$ must vanish; hence

$$2rv\rho + r^2\left[\frac{\partial \rho}{\partial t} + v\frac{\partial \rho}{\partial r}\right] + r^2\rho\frac{\partial v}{\partial r} = \frac{d\rho}{dt} + \rho\left[\frac{\partial v}{\partial r} + \frac{2v}{r}\right] = 0. \tag{1.21}$$

The last bracketed quantity is just the divergence of the radial flow $\mathbf{v} = v\mathbf{r}/r$; in the form

$$\partial_t \rho + \mathbf{v} \cdot \nabla\rho + \rho\nabla \cdot \mathbf{v} = \partial_t \rho + \nabla \cdot (\rho\mathbf{v}) = 0 \tag{1.21'}$$

the equation is valid also in the general case, when spherical symmetry is not assumed. It can be easily interpreted by recalling that, if in a time dt a fluid element is displaced from **r** to (**r** + **v**dt), an infinitesimal volume dV associated with it changes into dV(1 + dt$\nabla \cdot$ **v**) (see the Problem 1.1). Since the total mass of the volume remains constant, the change $d\rho$ in mass density fulfills

$$\rho dV = (\rho + d\rho)dV(1 + \nabla \cdot \mathbf{v}dt) ,$$

Thus

$$d\rho/dt + \rho\nabla \cdot \mathbf{v} = 0, \qquad (1.21'')$$

which is eq. (1.21') in lagrangean form. In an incompressible flow (ρ = const) the velocity is free of divergence.

To describe a generic motion of a fluid, we need the appropriate generalization of (1.19) to account for the conservation of momentum:

$$\rho(\partial_t \mathbf{v} + \mathbf{v} \cdot \nabla\mathbf{v}) + \nabla P + \rho\nabla U = 0 . \qquad (1.22)$$

U is the gravitational potential energy per unit mass, determined by the matter distribution through *Poisson's equation*

$$\nabla^2 U = 4\pi G\rho . \qquad (1.23)$$

This differential equation must be supplemented with the condition of vanishing at infinity; then it is equivalent to the integral form (*Poisson's integral*)

$$U(\mathbf{r}) = - G\int d^3 r'\rho(\mathbf{r}')/|\mathbf{r} - \mathbf{r}'| . \qquad (1.24)$$

For a point mass M this gives the monopole solution U = – GM/r. The static equilibrium of a fluid body of arbitrary shape is determined by (from eq. (1.22))

$$\nabla P + \rho\nabla U = 0; \qquad (1.25)$$

its surface, where P = 0, is an equipotential surface. This is physically realized by the surface of the oceans and is called the *geoid*. At an altitude z from the ground we can approximately write U = gz; g, the gravity acceleration, varies little from place to place. The direction of the vertical, along the z axis, also has small variations produced by the inhomogeneities of the earth.

1.3 DYNAMICS OF SOLID BODIES

In Sec. 1.1 we have shown that the radius of a cosmic body of given mass, if bound by gravitational forces, is determined by the inner temperature. Therefore its size is determined by the previous history, the active energy sources (e. g., radioactive materials) and the energy transport processes (see Sec. 1.5). The discussion was based on the assumption of a fluid in which each molecule is free to move about and the equation of state (1.12) holds, at least approximately. When the temperature is below the freezing point of the material (i. e., when the thermal energy is smaller than the binding energy of a molecule to the neighbouring ones), the body bevomes solid and quite a different analysis is called for.

Solidity means that each element, in absence of external forces, has a reference position **r**. The state of motion of a solid body is described by a small *displacement* field **s**(**r**, t) of an element from the reference position. The dynamical effects of the displacement, unaffected by a translation, must depend upon the derivative tensor $\partial_i s_j$, which is the sum of a symmetrical and antisymmetrical part:

$$\partial_i s_j = \tfrac{1}{2}(\partial_i s_j + \partial_j s_i) + \tfrac{1}{2}(\partial_i s_j - \partial_j s_i) \ .$$

It can be shown that the antisymmetrical part has no dynamical effects: being proportional to the components of the curl of the vector field **s**, it describes a rotation of an infinitesimal element of the body and produces no stress. The symmetrical part

$$e_{ij} = \tfrac{1}{2}(\partial_i s_j + \partial_j s_i) \tag{1.26}$$

– the *strain tensor* – can be split into a diagonal part with trace

$$\theta = e_{ii} = \partial_i s_i = \nabla \cdot \mathbf{s} \tag{1.27}$$

and in a left over, trace-free part

$$e_{ij} - \delta_{ij}\theta/3 = -(\partial_i s_j + \partial_j s_i)/2 - \delta_{ij}\theta/3. \tag{1.28}$$

$\theta = \nabla \cdot \mathbf{s}$ measures the *dilatation* of an infinitesimal volume element produced by the deformation; in other words, a volume element dV becomes dV(1 + θ). The trace-free part (1.28) conserves the volume.

The effect of the strain upon the dynamics is described by the *stress tensor* P_{ij}. If we remove the material on one side of a surface element dS with normal **n**, to keep the body on the same state of motion (or of rest) we must exert from the empty side (where the normal lies) a force $P_{ij}n_j dS$. The stress tensor has therefore the dimension of a pressure, or energy density. In a fluid this force is orthogonal to the surface element, corresponding to an isotropic

$$P_{ij} = -P\delta_{ij}; \tag{1.29}$$

the scalar *P* is the *pressure*. In general there is also a force parallel to the surface element (*shear stress*). It can be shown that the stress tensor is symmetric, with six independent components for every point.

A solid under a small deformation behaves *elastically*, namely the elastic stress is a linear function of the strain (*Hooke's law*). This is expressed by a fourth rank (*Hooke's*) tensor as follows:

$$P_{ij} = C_{ijkh} e_{kh} \ . \tag{1.30}$$

It can be shown that the energy density of the deformation is given by the quadratic form

$$w = \tfrac{1}{2} C_{ijkh} e_{ij} e_{kh} \ . \tag{1.31}$$

Once this is given we can calculate the stress directly:

$$P_{ij} = \partial w / \partial e_{ij} \ . \tag{1.32}$$

In a generic material Hooke's tensor has 21 independent components; but in an isotropic material the elastic properties, and hence the energy (1.31), must be determined entirely by scalars. It turns out that there are only two quadratic invariants which can be constructed with a symmetric tensor like e_{ij}: θ^2 and

$$e_{ij} e_{ij} = e_{11}{}^2 + e_{22}{}^2 + e_{33}{}^2 + 2e_{12}{}^2 + 2e_{23}{}^2 + 2e_{31}{}^2 .$$

In this case the elastic energy is of the form

$$w = \tfrac{1}{2}\lambda\theta + \mu e_{ij} e_{ij} \ ; \tag{1.32}$$

the scalars λ and μ are *Lamé's constants*, with the dimensions of a pressure. From eq. (1.32) we get Hooke's law for an isotropic medium

$$P_{ij} = \lambda\theta\delta_{ij} + 2\mu e_{ij} \ . \tag{1.33}$$

In planetary bodies like the earth λ is about equal to μ.

Two particular examples are useful to show the physical meaning of Lamé's constant. A one-dimensional compression, with $e_{11} = \partial_x s_x$ only different from zero, produces a stress

$$P_{11} = (\lambda + 2\mu)\partial_x s_x . \tag{1.34}$$

A shear deformation, with only the component $e_{12} = \partial_y s_x/2$, produces a shear stress

$$P_{12} = \mu \partial_y s_x . \tag{1.35}$$

μ is the *rigidity* and expresses the resistance to shear. Under a three–dimensional compression, with $e_{ij} = \delta_{ij}\theta/3$, a scalar stress arises

$$P_{ij} = \delta_{ij}(\lambda + 2\mu/3)\theta = \delta_{ij}k\theta, \qquad (1.36)$$

which can be balanced by an additional outside hydrostatic pressure $-k\theta$. This is also the change in the pressure which produces the compression from a volume 1 to $1 + \theta$; hence

$$dP/d\rho = k/\rho \qquad (1.37)$$

is the *compressibility*. Here k is the *bulk modulus*, equal to about $5\mu/3$ in the earth's interior.

The total stress force exerted upon a volume element V bounded by a surface S by the surrounding medium is given by

$$\int_S ds n_i P_{ij} ,$$

by Gauss' theorem, equal to

$$\int_V dV \partial_j P_{ij} .$$

The inertial force for an infinitesimal volume dV is proportional to the second lagrangean derivative of the displacement **s** ; it, and not the fluid velocity, is the appropriate field variable. Equating this term to the sum of the stress force and the gravitational force we get :

$$\rho \frac{d^2 \mathbf{s}}{dt^2} - \nabla \cdot \boldsymbol{P} + \rho\nabla U = 0 . \qquad (1.38)$$

To keep the synthetic vector notation we denote by $\boldsymbol{P}$ the stress tensor. If λ and μ are constant, the momentum equation reads

$$\rho \frac{d^2 \mathbf{s}}{dt^2} = \mu\nabla^2 \mathbf{s} + (\lambda + \mu)\nabla\theta - \rho\nabla U . \qquad (1.39)$$

This medium is capable of propagating both transversal and longitudinal waves with velocities determined by Lamé's constants (Ch. 5).

The elastic behaviour depends also upon the time–scale of the external forces; it is possible that a material, elastic at high frequencies, behaves plastically under a prolonged stress and shows flow phenomena. This is indeed the case of the interior of the earth.

1.4. TRANSPORT

Neither eq. (1.22) nor (1.39) take into account viscosity. Viscosity arises when the mean free path ℓ_C of the particles which constitute the body is not negligible with respect to its characteristic scale of spatial variation; then momentum is microscopically transferred from one fluid element to the other, with the result that the velocity field tends to become more uniform. Although the description of transport processes requires a complicated analysis belonging to the kinetic theory, a simple argument suffices to get the relevant orders of magnitude.

Assume that a mean quantity Q associated to each particle changes in the z-direction only. For heat conduction, Q is the mean molecular kinetic energy; for viscosity it is the mean molecular momentum; for diffusion the mean concentration of a solute. The number of molecules crossing the unit surface orthogonal to z in the unit of time is approximatively nv_T, where v_T is the thermal speed and n their number density; of course they carry from one side to the other of the surface the value of Q they had at the last collision and give it up, so to

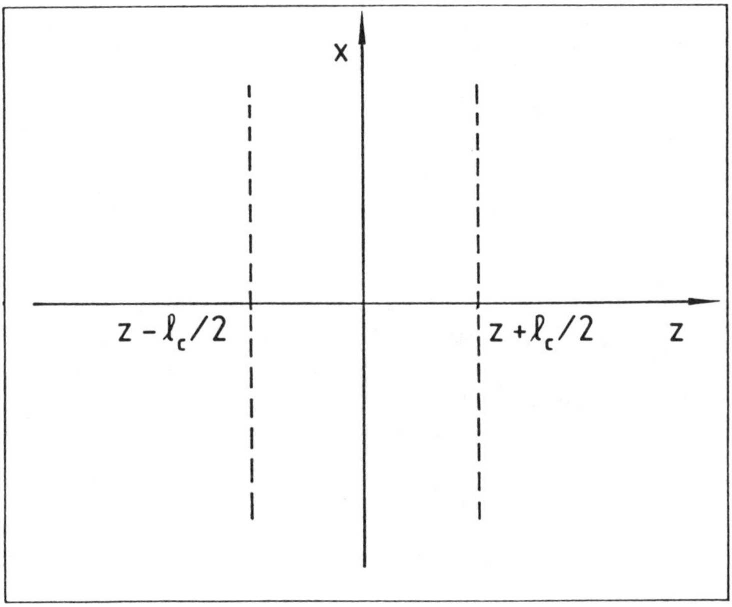

Fig. 1.3. A physical quantity is transported across a surface z = cost. by the molecules which travel freely from $z - \ell_C/2$ to $z + \ell_C/2$ or vice-versa.

speak, when they collide again. If z is the coordinate of the surface, we can say that these two collisions occur approximately at $(z - \ell_c/2)$ and $(z + \ell_c/2)$: each molecule then carries from one side to the other an amount of the quantity Q equal to $[Q(z - \ell_c/2) - Q(z + \ell_c/2)]$. If Q does not change much in a mean free path, its overall flux is

$$F_Q = nv_T[Q(z - \ell_c/2) - Q(z + \ell_c/2)] = - nv_T\ell_c dQ/dz. \quad (1.40)$$

This flux is proportional to the gradient of Q and occurs in the direction in which Q decreases, thus trying to make the medium more uniform. The coefficient of ndQ/dz is the *diffusion coefficient.* A more refined calculation based on kinetic theory shows that its correct value is $\ell_c v_T/3$. To avoid unnecessary complications, here ℓ_c is defined only in order of magnitude.

The mean momentum $Q = mv$ is transported across a surface $z =$ const. at the rate

$$F(mv) = - nmv_T\ell_c v'(z) .$$

As a consequence, the total momentum ρdVv in a volume $dV = dxdydz$ changes if the momentum transported through z is different from the one transported through $(z + dz)$. This change per unit time is therefore given by

$$- nmv_T\ell_c[v'(z) - v'(z + dz)]dxdy = \rho v_T\ell_c dVv''(z)$$

and produces a force per unit volume equal, in order of magnitude, to $\rho v_T\ell_c v''(z)$. It tends to make the flow more uniform.

$$\eta = \rho v_T\ell_c \quad (1.41)$$

is the *viscosity coefficient,* with dimensions $[ML^{-1} T^{-1}]$.

A general expression of the viscous force can be derived from first principles similarly to the stress force in a solid. A fluid element of volume V is acted upon by the surrounding elements through the velocity gradients which can transport momentum through the bounding surface S. We introduce the *viscous stress tensor* q_{ij} to define the force

$$q_{ij}n_j dS$$

exerted on the element dS, with an outer normal **n**. The corresponding force per unit volume is $\partial_j q_{ij}$.

Like the stress tensor, q_{ij} is symmetric; moreover, for small velocity gradients, it must be linear in the derivative tensor $\partial_i v_j$. The antisymmetric part of this tensor corresponds to a rotational velocity of the fluid element and can be shown to have no dynamical effects; what counts is its symmetric part. Just like for eq. (1.33), when the microscopic properties of the fluid have no privileged directions, the only admissible form of the viscous stress tensor is

$$q_{ij} = \eta(\partial_i v_j + \partial_j v_i) - \frac{2}{3}(\eta - \zeta)\delta_{ij}\partial_k v_k \ . \qquad (1.42)$$

If the scalar coefficients η and ζ are constant, we have the viscous force per unit volume

$$\partial_j q_{ij} = \eta\nabla^2 v_i + \frac{1}{3}(\eta + 2\zeta)\nabla_i(\nabla \cdot \mathbf{v}) \ . \qquad (1.43)$$

Kinetic gas theory shows that the second coefficient ζ vanishes or is very small. We are therefore able to complete eq. (1.22) as follows:

$$\rho(\partial_t \mathbf{v} + \mathbf{v} \cdot \nabla\mathbf{v}) + \nabla P + \rho\nabla U = \eta\nabla^2\mathbf{v} + \frac{1}{3}\eta\nabla(\nabla \cdot \mathbf{v}) . \qquad (1.44)$$

For an incompressible flow we have only the first term on the right-hand side:

$$\rho(\partial_t \mathbf{v} + \mathbf{v} \cdot \nabla\mathbf{v}) + \nabla P + \rho\nabla U = \eta\nabla^2\mathbf{v} \ . \qquad (1.45)$$

This is the *Navier-Stokes* equation. The quantity η/ρ is called *kinematic viscosity*, with dimensions $[L^2/T]$. Eq. (1.22) is time-reversible; in other words, if $\mathbf{v}(\mathbf{r}, t)$ is a solution, so is $\mathbf{v}(\mathbf{r}, -t)$. In agreement with the irreversible character of transport processes, the addition of viscosity destroys this property.

Viscosity also increases the order of the partial differential equation; therefore the determination of the solution requires more boundary conditions than the inviscid flow. It can be expected, therefore, that the transition $\eta \to 0$ is catastrophic, in the sense that it collapses the set of solutions into one of fewer dimensions. If in the transition the correct, full boundary conditions are kept, clearly the solution cannot tend everywhere to an inviscid solution, for which some of those conditions cannot be fulfilled. In this way the important phenomena of the *boundary layer* arises: the viscosity is everywhere negligible, except in very thin regions (i. e., near solid bodies surrounded by the flow), where the velocity gradient is very large. A very important boundary layer situation occurs in the atmosphere near the ground; there, besides the condition of vanishing velocity at the surface we have also to take into account complex heat transfer processes. Mathematically, boundary layers must be treated with the methods of *singular perturbation theory*.

The relevance of viscosity is generally described by the dimensionless *Reynolds' number*

$$Re = \rho Lv/\eta \ , \qquad (1.46)$$

L is a characteristic macroscopic length and v a typical velocity of the flow. Boundary layer phenomena arise when Re is much larger than unity; in this limit the flow may also become unstable and turbulent.

While for vector quantities, like the momentum, arguments more precise than eq. (1.40) are needed to get the exact conservation law, for scalars we can reason more directly as follows. Let $\int QndV$ the amount of the scalar quantity Q contained in the volume V; its change is given by (using Gauss' theorem)

$$\frac{d}{dt}\int_V dV(Qn) = -\int_S ds(\mathbf{n}\cdot\mathbf{F}) = -\int_V dV(\nabla\cdot\mathbf{F}) \ ,$$

leading to the differential conservation law

$$n\frac{dQ}{dt} = \nabla\cdot(nv_T\ell_c\nabla Q) \ . \qquad (1.47)$$

We have used the fact that the number density n is conserved, so that the lagrangean derivative of ndV vanishes.

For example, let $Q = fkT/2$ be the mean energy of each particle, with f degrees of freedom. When there is only translatory motion $f = 3$ and $Q = 3/2\ kT$. If only the temperature T varies, eq. (1.47) leads to the *heat equation*

$$\partial T/\partial t = v_T\ell_c\nabla^2 T = \chi\nabla^2 T \ . \qquad (1.48)$$

The quantity $\chi = v_T\ell_c$ is the *heat transport coefficient.* The main feature of eq. (1.48) is that if at the beginning we have a temperature spike, after a time t it spreads to a distance of order $\sqrt{(\chi t)}$.

The analysis of the different ways of energy transport is very important in the physics of cosmic bodies, in particular for their atmospheres (see Ch. 7). They include, besides conventional heat conduction, described by eq. (1.48), radiative transport and convection. The problem is further complicated by phase transitions and the corresponding latent heats. The determination of the temperature distribution is necessary also because it enters the momentum conservation through the equation of state (1.12); but there is an important case in which this can be avoided. In general the equation of state expresses the fluid pressure $P(\rho, s)$ as a function of the matter density and the *entropy per unit mass* s; the partial derivative

$$\partial P(\rho,\ s)/\partial \rho = c_S^2 \qquad (1.49)$$

is the square of the *sound speed* c_S.

If the fluid element in its motion does not exchange thermal energy with the surrounding, its entropy $s\rho dV$ is constant; we say, its motion is *adiabatic*. Generally speaking, this occurs when the motion is so fast to prevent heat exchange. Another relevant approximation is when all fluid elements have at the beginning the same specific entropy s; together with the first assumption, this implies that the entropy s is independent of space and time. In this case the states at any given moment of two fluid elements are related by an *adiabatic transformation* and the equation of state gives the pressure $P(\rho)$ as a function of the mass density alone (*barotropic* regime): there is no need of a detailed description of the energy balance.

For a perfect gas this relation can be expressed in a differential way as follows:

$$d(\ln P) = \gamma d(\ln \rho), \qquad (1.50)$$

where $\gamma = c_p/c_v$, the *polytropic index*, is the ratio of the specific heats at constant pressure and at constant volume. γ is related to the number of degrees of freedom f by

$$\gamma = (f + 2)/f \qquad (1.51)$$

and equals 5/3 for point molecules (f = 3).

$$c_S = \surd(\gamma/P\rho) \qquad (1.52)$$

is the sound speed. Eq. (1.50), integrated, gives the usual relation

$$P/\rho^\gamma = \text{const.} \qquad (1.53)$$

Another important quantity is the *internal energy per unit mass*, equal to f/m times the energy for each degree of freedom, $\frac{1}{2}kT$:

$$c_V T = \frac{1}{2m} fkT = \frac{fP}{2\rho} = \frac{P}{\rho(\gamma - 1)}\ . \qquad (1.54)$$

This define also the specific heat at constant volume

$$c_V = fk/2m = k/m(\gamma - 1). \qquad (1.54')$$

This is the *polytropic regime.*

When the gas is not perfect eq. (1.50) still holds, but now γ is a function of the density and eq. (1.53) does not follow. Similar relations tie pressure and temperature and density and temperature:

$$d(\ln P) + \gamma_2/(1 - \gamma_2)d(\ln T) = 0 \tag{1.55}$$

$$d(\ln T) - (\gamma_3 - 1)d(\ln\rho) = 0 . \tag{1.56}$$

For a perfect gas $\gamma_2 = \gamma_3 = \gamma$ is constant.

Eq. (1.53) includes, as a particular case, the isothermal equation of state for a perfect gas, corresponding to $\gamma = 1$. A linear relation between pressure and density holds also for another case, which, however, is of little interest in the physics of the solar system. A gas of photons or, in general, of massless particles, is characterized by an energy-momentum tensor with a vanishing trace; this means

$$P = \rho c^2/3 . \tag{1.57}$$

In this case the sound speed is $c/\sqrt{3}$. The same statement can be made for a gas of ultrarelativistic particles, that is to say, particles whose momentum is much larger than the rest mass multiplied by c.

1.5. MAGNETOHYDRODYNAMICS

At a sufficiently high temperature and low density and possibly under a strong ultraviolet radiation (from the sun!) a gas may become partially or totally ionized (Ch. 8). Unless one deals with very high frequencies and/or very small wavelengths, such a gas remains neutral, i. e., to a good approximation the electron density is compensated by the density of the ions. In this case one can still describe it as a single, neutral fluid; however, the relative motion of electrons and ions produce electric currents and magnetic fields. A charge q moving with velocity $\mathbf{u}$ in a magnetic field $\mathbf{B}$ suffers the Lorentz force $q\mathbf{u} \times \mathbf{B}/c$. Summing these elementary forces for all the charged particles in a volume element dV we get

$$\Sigma q\mathbf{u} \times \mathbf{B} = (\mathbf{J} \times \mathbf{B})dV,$$

where $\mathbf{J} = \Sigma q\mathbf{u}$ is the current density. Therefore the force per unit volume $(\mathbf{J} \times \mathbf{B}/c)$ must be added to momentum conservation law, (e. g., eq. (1.22)):

$$\rho(\partial_t + \mathbf{v} \cdot \nabla)\mathbf{v} + \nabla P + \rho\nabla U = \mathbf{J} \times \mathbf{B}/c . \tag{1.58}$$

We have also Maxwell's equations:

$$\nabla \cdot \mathbf{B} = 0 \ , \qquad (1.59_1)$$

$$c\nabla \times \mathbf{B} = 4\pi\mathbf{J} \ , \qquad (1.59_2)$$

$$c\nabla \times \mathbf{E} + \partial\mathbf{B}/\partial t = 0 \ . \qquad (1.59_3)$$

In our condition of (almost) electrical neutrality the remaining Maxwell equation defines the net charge density $\nabla \cdot \mathbf{E}/4\pi$, which must be much smaller than the ion charge density. In the second equation we have neglected the displacement current $\partial\mathbf{E}/\partial t$. We can do so, for example, when the electric field has an order of magnitude given by Ohm's law for infinite conductivity (1.65), namely vB/c; then the first term in eq. (1.59_2) is larger than the neglected term by $0(c^2/v^2)$. In this approximation electromagnetic waves, for which E and B are of the same order of magnitude, are excluded; moreover, the ratio of the space and time characteristic scales is just v.

The current density **J** is determined, in turn, by the electric and magnetic fields through *Ohm's law.* This law says, essentially, that currents in a medium at rest flow at a rate proportional to the electric field:

$$\mathbf{J} = \sigma\mathbf{E} \ . \qquad (1.60)$$

σ, the *electrical conductivity*, can be expressed in terms of microscopic quantities as follows. A current arises when the electric fluid has a relative velocity **u** with respect to the medium; this creates a friction, which in steady state balances the electric force. The frictional force upon a single electron has the form $-m_e\nu_c\mathbf{u}$, where ν_c is the relevant collision frequency. When this is balanced by the electric force we get

$$\mathbf{u} = -\ e\mathbf{E}/m_e\nu_c \ ,$$

or, in terms of the current density,

$$\mathbf{J} = -\ ne\mathbf{u} = (ne^2/m_e\nu_c)\mathbf{E} \ . \qquad (1.61)$$

This gives, in order of magnitude, the conductivity σ (it has the dimension of a frequency).

In a conductor moving with a velocity **v**, Ohm's law must be modified. If a wire segment crosses a magnetic field, an electric field ($\mathbf{v} \times \mathbf{B}/c$) arises (Lenz's law). Therefore eq. (1.60) becomes

$$\mathbf{J} = \sigma(\mathbf{E} + \mathbf{v} \times \mathbf{B}/c) \ . \qquad (1.62)$$

Since, as said earlier, in this approximation the charge density is neglected, the current must be free of divergence. This leads to the constraint equation

$$\nabla \cdot \mathbf{J} = \nabla \cdot (\mathbf{E} + \mathbf{v} \times \mathbf{B}/c) = 0, \qquad (1.62')$$

which gives the (small) amount of charge needed to generate the given electric field. Eq. (1.62) is only a (good) approximation; corrections to Ohm's law (1.62) are often needed with a better treatment based upon kinetic theory. Note also that when there is an external magnetic field, or in presence of other factors which destroy the isotropy of the fluid, the conductivity is, in general, a tensor quantity and eq. (1.62) must be appropriately generalized.

The magnetohydrodynamical equations carry forward in time the velocity eq. (1.58), the density (eq. (1.21)) and the magnetic field (1.59_3). (1.59_1) is an auxiliary initial condition, consistent with (1.59_3); (1.59_2) defines the current density, (1.62) the electric field and (1.62') the charge. If the fluid is not barotropic, we need, of course, an energy equation. Using the relation

$$(\nabla \times \mathbf{B}) \times \mathbf{B} = (\mathbf{B} \cdot \nabla)\mathbf{B} - \tfrac{1}{2}\nabla B^2$$

we can write the magnetic force per unit volume as

$$\mathbf{J} \times \mathbf{B}/c = (\nabla \times \mathbf{B}) \times \mathbf{B}/4\pi = -\nabla(B^2/8\pi) + (\mathbf{B} \cdot \nabla)\mathbf{B}/4\pi;$$

or, in components, using the divergence condition (1.59_1),

$$-\partial_i(B^2/8\pi) + \partial_j(B_j B_i/4\pi) \; .$$

This force corresponds to a *magnetic stress tensor*

$$P_{(mag)ij} = \delta_{ij}B^2/8\pi - B_i B_j/4\pi \; . \qquad (1.63)$$

The first term on the right–hand side is a pressure to be added to the fluid pressure P: we can have a *magnetic confinement*, in which the internal pressure of a fluid is balanced by an external magnetic pressure. This is an important way to maintain a plasma confined in a laboratory machine; it has also an essential role in the interaction between the earth dipole field and the solar wind (Ch. 9). The second contribution has off–diagonal components and generates shear stresses. The force exerted upon a surface element dS whose normal makes an angle θ with the magnetic field is $dSBB\cos\theta/4\pi$. It has the same expression as in medium made up of elastic wires along the magnetic force, whose tension per unit orthogonal surface element is $B^2/4\pi$; it has the tendency to make the lines of force as straight as possible and it can generate waves (see Sec. 8.3.)

By means of Ohm's law (1.62) we can eliminate the electric field from eq. (1.59_3). When the conductivity σ is constant we get

$$\frac{\partial \mathbf{B}}{\partial t} = \nabla \times (\mathbf{v} \times \mathbf{B}) + \frac{c^2}{4\pi\sigma} \nabla^2 \mathbf{B} \ . \tag{1.64}$$

In a medium at rest ($\mathbf{v} = 0$) this is a heat–like equation; the inverse conductivity acts like a diffusion coefficient, spreading and dissipating an initial magnetic field; more precisely, after a time t an initial spike spreads to a distance of the order $c\sqrt{(t/4\pi\sigma)}$. In the case of the dipole magnetic field of the earth, this points to the need of an active regenerating mechanism (see Ch. 6).

Finally, note that in a perfectly conductive fluid ($\sigma = \infty$) the relation

$$\mathbf{E} + \mathbf{v} \times \mathbf{B}/c = 0 \tag{1.65}$$

must hold. Contrary to newtonian dynamics, the electromagnetic field determines the component of the velocity orthogonal to the lines of force, and not its time derivative:

$$\mathbf{v}_\perp = c\mathbf{E} \times \mathbf{B}/B^2 \ ; \tag{1.66}$$

the velocity along **B**, however, has to be determined on the basis of the momentum law (1.58). Eq. (1.65) shows also that the electric and magnetic fields are orthogonal; any initial electric field along the lines of force is quickly neutralized by the free motion of charges.

Eqs. (1.65, 62) are of great importance in the physics of the motion of a conductive body through the plasma in the magnetosphere or interplanetary plasma. For example, Io, the satellite of Jupiter, moving through the dipole magnetic field of the planet, develops a potential difference of about 500 kV and carries a current of 5 MA across its body or its ionosphere. This fact has great consequences for the whole magnetosphere of Jupiter (See Ch. 13). Artificial, earth–bound satellites also show a similar effect.

1.6 CONSERVATION OF MAGNETISM AND VORTICITY

In this Section we deal with the problem of the conservation of fluxes in fluid motion. We need first the rate of change of a surface element $d\mathbf{S}$ dragged along by a velocity field $\mathbf{v}$. A line element $d\mathbf{r}$ following the flow changes according to

$$d(d\mathbf{r})/dt = (d\mathbf{r} \cdot \nabla)\mathbf{v} \, . \qquad (1.67)$$

A volume element dV can be expressed as the scalar product of a surface element $d\mathbf{S} = \mathbf{n}dS$ and a vector dr: $dV = d\mathbf{S} \cdot d\mathbf{r}$. Taking in to account the rate of change of dV (eq. (1.21")) and noting that **dr** is arbitrary we find

$$d(dS_i)/dt = dS_i \, \nabla \cdot \mathbf{v} - dS_j \, \partial v^j/\partial r^i \, . \qquad (1.68)$$

The lagrangean change of the magnetic field is easily obtained from eq. (1.64):

$$\frac{d\mathbf{B}}{dt} = \frac{\partial \mathbf{B}}{\partial t} + (\mathbf{v} \cdot \nabla)\mathbf{B} = (\mathbf{B} \cdot \nabla)\mathbf{v} - \mathbf{B} \, \mathrm{div} \, \mathbf{v} + \frac{c^2}{4\pi\sigma}\nabla^2\mathbf{B} . \quad (1.69)$$

Combining the last two equations we finally get the change in the magnetic flux through dS:

$$\frac{d(\mathbf{B} \cdot d\mathbf{S})}{dt} = \frac{c^2}{4\pi\sigma} \, d\mathbf{S} \cdot \nabla^2\mathbf{B} \quad . \qquad (1.70)$$

When the conductivity is infinite, the flux through a surface attached to the fluid is constant: one can say, the magnetic field is *frozen in*. This is important when the fluid motion compresses the material across the lines of force, thereby increasing the field intensity B. In the collapse of a cosmic body of a size R, B increases as $1/R^2$, a process which can produce an enormous amplification. A similar property holds when a charged particle moves in a slowly varying magnetic field: the flux embraced in a Larmor gyration remains almost constant (see Sec. 9.1).

It is also interesting to calculate the time change along the flow of the vector product (dr × **B**). Combining eqs. (1.69) and (1.67) we obtain

$$d(d\mathbf{r} \times \mathbf{B})dt = -(d\mathbf{r} \times \mathbf{B})\nabla \cdot \mathbf{v} + d\mathbf{r} \times (\mathbf{B} \cdot \nabla)\mathbf{v} + $$
$$+ (d\mathbf{r} \cdot \nabla)\mathbf{v} \times \mathbf{B} + c^2 d\mathbf{r} \times \nabla^2\mathbf{B}/4\pi\sigma . \qquad (1.71)$$

Let us now consider two points **r** and (**r** + dr) initially lying on the same line of force, so that (dr × **B**) = 0. Because of the antisymmetry of the vector product, the second and the third terms on the right–hand side kill each other: in an infinitely conductive plasma the condition (dr × **B** = 0) holds for ever. This suggests that a line of force can be identified by the fluid elements which sit on it. This is the *freezing in* theorem, of great importance in cosmic physics, in particular for the solar wind.

A result similar to eq. (1.70) can be deduced for the *vorticity*

$$\mathbf{w} = \nabla \times \mathbf{v} \tag{1.72}$$

of a flow which represents the rotational velocity of a fluid element with respect to its centre of gravity. To obtain the rate of change of its flux ($\mathbf{w} \cdot d\mathbf{S}$) we take the curl of eq. (1.45)/ρ, noting that

$$v_j \partial_j v_i = v_j(\partial_j v_i - \partial_i v_j) + \tfrac{1}{2}\partial_i v^2 \ .$$

The first term on the right-hand side is ($\mathbf{v} \cdot \mathbf{w}$); the last term gives no contribution. In eq. (1.45)/ρ, $\nabla P/\rho$ is also killed by the curl operator when the flow is incompressible or polytropic (i. e., when P is a function of ρ only). We therefore end up with

$$\frac{\partial \mathbf{w}}{\partial t} = \nabla \times (\mathbf{v} \times \mathbf{w}) + \nabla \times \frac{\eta \nabla^2 \mathbf{v}}{\rho} \ . \tag{1.73}$$

This equation has the same structure as eq. (1.64); in addition, the vorticity $\mathbf{w}$ is divergenceless, as $\mathbf{B}$. Hence, when the viscosity is negligible ($\eta \to 0$) the vorticity fulfills the same theorems as the magnetic field; first,

$$d(\mathbf{w} \cdot d\mathbf{S})/dt = 0 \ ; \tag{1.74}$$

that is to say, its flux through a surface dragged along by the flow is constant; secondly, two neighbouring fluid elements lying on a vorticity line will stay on it. The vorticity lines in a polytropic, inviscid fluid, are "anchored" to the matter.

1.7. KINETIC THEORY

It is now time to question the main assumption used so far, that the flow can be described completely with the mean quantities ρ, $\mathbf{v}$ and T (in addition, if needed, to the electromagnetic quantities). This assumption is based upon the hypothesis that the material is in *local thermodynamical equilibrium*; that is to say, that collisions have had the time and the space to produce a local Boltzmann distribution of velocities, in spite of external perturbing factors.

A set of N point particles in a box is described by their 6N coordinates in phase space, which evolve in time according their interactions and the external forces, including those exerted by the wall. If N is very large the random collisions will eventually destroy any initial order and bring the gas in the state of maximum probability consistent with the constants of motion of the system, in particular the energy. The final state of *thermodynamical equilibrium* is a maxwellian distribution in velocity space **u**, uniform throughout the box:

$$f = n\left(\frac{m}{2\pi kT}\right)^{3/2} \exp\left[-\frac{mu^2}{2kT}\right] . \qquad (1.75)$$

$f(\mathbf{r}, \mathbf{u}, t)d^3r d^3v$ is the number of particles present in the volume $d^3r d^3v$ around the point $(\mathbf{r}, \mathbf{u})$ in phase space. In general the five fluid quantities n (number density), **v** (mean velocity) and T (effective temperature) are obtained from f by averaging:

$$n(\mathbf{r}, t) = \int f(\mathbf{r}, \mathbf{u}, t)d^3u, \qquad (1.76_1)$$

$$n(\mathbf{r}, t)\mathbf{v}(\mathbf{r}, t) = \int \mathbf{u}\, f(\mathbf{r}, \mathbf{u}, t)d^3u, \qquad (1.76_2)$$

$$3n(\mathbf{r}, t)T(\mathbf{r}, t) = m\int u^2\, f(\mathbf{r}, \mathbf{u}, t)d^3u = mv_T^2, \qquad (1.76_3)$$

where v_T is the *thermal speed.* Kinetic theory shows that thermodynamical equilibrium is attained in two stages. First, after a few collision times the fluid is described by a single *distribution function* f in velocity space; however, f depends on space and time only through the mean quantities n, **v** and T, which now are, in general, functions of space and time:

$$f = n\left(\frac{m}{2\pi kT}\right)^{3/2} \exp\left[-\frac{m|\mathbf{u} - \mathbf{v}|^2}{2kT}\right] . \qquad (1.75')$$

It is easy to check that the relationships (1.76) are fulfilled. The gas is now in *local thermodynamical equilibrium* and is described by a fluid model and its conservation laws (mass, momentum and energy), which provide the time evolution for n, **v** and T. If the system is closed and isolated, it will be slowly brought into a uniform state (eq. (1.75) by diffusion and dissipative processes. For radiation, we also have a state of local thermal equilibrium, to be discussed in Ch. 7.

Local thermodynamical equilibrium holds only if the typical scales for variations in time and in space are, respectively, much larger than the collision time $1/\nu_c$ and the mean free path $\ell_c = v_T/\nu_c$; the collision frequency ν_c is obtained by averaging the relevant cross section σ_c with the velocity distribution f:

$$\nu_c = n \langle u\sigma_c \rangle \ . \tag{1.77}$$

When the two above conditions are not fulfilled, or when we wish to study general deviations of the velocity distribution from the Maxwell's distribution (1.75), instead of the fluid equations a dynamical description in phase space must be adopted, using a differential equation for the particle distribution function f(**r**, **u**, t) (*Boltzmann's equation*). It describes two different kinds of phenomena: the free dynamics of the particles under the external field of force and also, in the case of a plasma, the mean electromagnetic field generated by them; and the close collisions, which tend to bring about the maxwellian distribution. We therefore have two extreme models of a gas, both of interest in the solar system: the *free molecular flow*, when collisions are unimportant (*Knudsen regime*) and the fluid approximation (1.75), when collisions dominate. In the Knudsen flow of a neutral gas each particle moves independently of the other ones and the use of a distribution function is futile. In a plasma and in a many body gravitating system, *collective interactions* at long range produce important global dynamics effects even when collisions are unimportant. In both these cases the interparticle forces, proportional to $1/r^2$, are felt at very large distances, albeit at a weak level. A particle, so to speak, "collides" all the time with many particles simultaneously and is slowly, randomly and weakly deflected by their field (see Sec. 16.4).

A gas of particles of charge Ze (nuclei) and – e (electrons) with overall neutrality is called a *plasma* if the number of particles in a sphere of radius

$$\lambda_D = \left[\frac{kT}{4\pi e^2 n}\right]^{\frac{1}{2}} \tag{1.78}$$

is very large. This characteristic length, named after *Debye*, depends upon the temperature T and the electron density n (while the nuclei have a mean density n/Z). It describes the *screening* property of the plasma: the potential of a charge in a plasma is screened beyond a distance λ_D (Problem 1.8). In the table we give typical values of the relevant parameters in the solar wind.

Electron density n (cm^{-3})	Temperature T (^{0}K)	Electron mean free path ℓ_c(cm)	Debye length λ_D(cm)	$n\lambda_D{}^3$
5	$2 \cdot 10^5$	10^{13}	1400	$1.4 \cdot 10^{10}$

Table 1.1. Typical plasma parameters in the solar wind at 1 AU.

If a sphere of radius λ_D is entirely devoid of positive charges, its electrostatic potential is of the order

$$en\lambda_D{}^3/\lambda_D \approx kT/e;$$

hence in this case the electrostatic energy of an electron is of the same order as its thermal energy. It is therefore reasonable to expect that statistical fluctuations – which produce energy fluctuations at most of order kT – can produce substantial charge voids only within regions of size smaller than λ_D. We can say, therefore, that at distances smaller than λ_D we have random interactions of charges – either the true charges or the "statistical" charges appearing and disappearing all the time – through their Coulomb field. This is a disordering process and it can be shown it has the same features as the collisions of single molecules in the regime of the Boltzmann's equation; it tends to restore the local maxwellian distribution (1.75) for each species.

But now the situation is complicated by the fact that in a dynamical situation and possibly under external fields, also particles at a distance greater than λ_D interact. Small deviations from charge neutrality and also electric currents may occur, which make themselves felt at large distances through their electric and magnetic fields. These fields are not due to statistical fluctuations, and determine reversible processes. It can be shown that in this case, and neglecting collisions, the distribution function f of each species is governed by *Vlasov's equation*

$$\frac{\partial f}{\partial t} + \mathbf{u} \cdot \frac{\partial f}{\partial \mathbf{r}} + \mathbf{A} \cdot \frac{\partial f}{\partial \mathbf{u}} = 0\ , \tag{1.79}$$

where **A** is the total electromagnetic force produced by the particles themselves (through their current and charge) and any external field.
The main feature of this equation is its lack of linearity. This can generate wave couplings and turbulence in a collisionless plasma flow (i. e., in the bow shock of the magnetosphere). Morcover, functions of

space coordinates (like particles' density) exhibit a peculiar irreversible behaviour due to phase mixing in velocity space (*Landau damping*).

Vlasov's equation is the main tool to study a plasma far from thermal equilibrium; in particular, wave propagation and instability in phase space; temperature differences between electrons and nuclei; deviations from isotropy in the velocity distribution; high-velocity streams and so on. All these features are important in the solar wind and the magnetosphere. The effects of collisions are described by a suitable right-hand side to Vlasov's equation.

PROBLEMS

1.1. Prove the relation

$$d(\ln dV)/dt = \nabla \cdot \mathbf{v}$$

used in the derivation of the mass conservation law (eq. (1.21')).

*1.2. Model a cosmic body with a core. The density profile

$$\rho(r) = \rho(0)[1 - (r/R)^n]$$

reaches its median value $\rho(0)/2$ at $R/2^{1/n} = r_c$. We may call this value the radius of the core. If $n \ll 1$ the core is very small; if $n \gg 1$ the density is practically constant and there is no core. Given the total mass and the radius, compute, as a function of n, the gravitational energy and the central pressure.

1.3. Calculate the ratio a/m for the main planets and the sun (eqs. (1.14, 15)). This ratio is important in determining the final black hole state of a collapsing body.

1.4. Estimate the ellipticity of the planets and the sun.

1.5. Determine the distribution of stress and the lengthening of a freely hanging rod subject to its own weight.

1.6. Find the velocity profile of an incompressible, viscous flow along a flat surface.

1.7. What is the electric potential difference across a spacecraft in a low earth orbit?

*1.8. Find the potential of a spherical conductor in an isothermal electron plasma. Such a plasma can be simply described by the equation of hydrostatic equilibrium for the electron fluid (with pressure P_e)

$$-\nabla P_e + ne\nabla V = 0 .$$

Thus the electron density is proportional to exp(eV/kT). When the exponent is small, Poisson's equation can easily be solved for the electrostatic potential V. For a simpler version of this problem, consider an infinite, plane conductor.

1.9. Find the velocity field corresponding to a straight vortex line (defined by a vorticity **w** which vanishes everywhere, except at the origin, with a constant flux).

1.10. Using Gauss' theorem, find the gravitational field of an infinite, circular cylinder and an infinite plane slab.

*1.11. Assuming that the third order moments of the distribution function f(**r**, **u**, t) vanish, derive the differential equations for the number density n and the mean velocity **v** (eqs. ($1.76_{1,2}$)). The second order moments are described by a general, anisotropic stress tensor.

1.12. Estimate the viscosity and Reynolds' number of the atmosphere near the ground.

2. *THE GRAVITATIONAL FIELD OF A PLANET*

Deviations of the shape of a cosmic body from spherical symmetry show up in its gravitational field; conversely, corrections to the *monopole* gravitational field GM/r provide important information about the interior. The main deviation is due to rotation and corresponds to a correction in the gravitational potential proportional to $1/r^3$ (the *quadrupole* term). Smaller disturbances in the mass distribution produce on the surface smaller corrections which decrease with a higher power of the distance from the centre. The gravitational field of a generic, non–spherical body is appropriately described by a powerful mathematical tool, the set of *spherical harmonic* functions. This chapter introduces this tool and applies it to the earth. The physics of gravity anomalies of small size is also discussed.

2.1. SPHERICAL HARMONICS

The theorem (1.2) shows that the external gravitational field of a spherically symmetric body is determined only by its mass and does not bear any trace of its density distribution. Outside, the internal structure manifests itself only when the spherical symmetry is violated; in this case the full Poisson's integral (1.24) must needs be used. The appropriate mathematical tool to compute this integral and to describe the gravitational field of an isolated body of arbitrary shape is the expansion of the potential energy U in *spherical harmonics*; we now briefly recall the main properties of this expansion, of fundamental importance in many fields of mathematical physics.

If $P_\ell(\mathbf{r})$ is a homogeneous polynomial of degree ℓ, solution of Laplace's equation, the function $U_\ell(\mathbf{r}) = P_\ell(\mathbf{r})/r^{2\ell+1}$ is also harmonic. This result can easily be proved by direct calculation, remembering the property of the homogeneous polynomials : $\mathbf{r} \cdot \nabla P_\ell = \ell P_\ell$. The functions $Y_\ell = P_\ell/r^\ell$ depend only on the direction $\mathbf{r}/r$ and are termed *spherical harmonics of degree* ℓ. It can be shown that there are just $(2\ell + 1)$ independent, harmonic and homogeneous polynomials of degree ℓ, hence just $(2\ell + 1)$ independent spherical harmonics $Y_{\ell m}$, functions of the colatitude θ and the longitude φ in a polar coordinate system. The index m, ranging over $(2\ell + 1)$ values, denotes the *order*.

P_0 is a constant and so is Y_{00}. P_1 is an arbitrary linear combination of x, y and z; hence we can choose as a basis for Y_1 the three functions

$$x/r = \sin\theta\cos\varphi,\ y/r = \sin\theta\sin\varphi,\ z/r = \cos\theta.$$

For Y_2 we can choose, for example, the five–dimensional basis

$$xy/r^2 = \sin^2\theta\cos\varphi\sin\varphi,\ xz/r^2 = \sin\theta\cos\theta\cos\varphi,$$

$$yz/r^2 = \sin\theta\cos\theta\sin\varphi,\ (x^2 - y^2)/r^2 = \sin^2\theta(\cos^2\varphi - \sin^2\varphi),$$

$$(2z^2 - x^2 - y^2)/r^2 = 3\cos^2\theta - 1.$$

The last function is noteworthy because it is independent of φ.

Before discussing the expansion in spherical harmonics we introduce the *Legendre polynomials* $P_\ell(u)$ of order ℓ. Their importance in the physics of gravitation stems from their definition. $P_\ell(u)$ is the coefficient of α^ℓ in the power expansion of the function

$$(1 - 2\alpha u + u^2)^{-\frac{1}{2}} = \Sigma_\ell P_\ell(u)\alpha^\ell \ ; \tag{2.1}$$

the summation ranges from 0 to ∞. This expansion shows up in the gravitational potential energy produced at a point **r** by a mass point placed at **r'**:

$$\frac{1}{|\mathbf{r} - \mathbf{r}'|} = \frac{1}{r}\left[1 - 2\frac{r'}{r}\cos\psi + \frac{r'^2}{r^2}\right]^{-\frac{1}{2}} = \Sigma_\ell P_\ell(\cos\psi)\frac{r'^\ell}{r^{\ell+1}}. \tag{2.2}$$

Here $\mathbf{r}\cdot\mathbf{r}' = rr'\cos\psi$. The first four polynomials read

$$P_0 = 1,\ P_1 = u,\ P_2 = \tfrac{1}{2}(3u^2 - 1),\ P_3 = \tfrac{1}{2}(5u^3 - 3u). \tag{2.3}$$

We also define the *associated Legendre functions* by

$$P_{\ell m}(u) = (1 - u^2)^{m/2}\frac{d^m P_\ell(u)}{du^m} \qquad (m \le \ell). \tag{2.4}$$

They reduce to P_ℓ when m = 0. It can be shown that $P_{\ell m}(u)$ (including m = 0) has $(\ell - m)$ zero's in the relevant interval (–1, 1). For other mathematical details we refer the reader to appropriate textbooks.

The spherical harmonics are the generalization to the unit sphere of the functions $\exp(im\varphi)$ on the unit circle; indeed, the polynomials of n–th order $r^m\exp(im\varphi)$ are harmonic in the equatorial plane; the expansion in spherical harmonics is analogous to the Fourier expansion on the unit circle. In our case, too, it is convenient to exploit the formal power of complex numbers to define a complex basis:

$$Y_{\ell m}(\theta, \varphi) = N_{\ell m} P_{\ell m}(\cos\theta)\exp(im\varphi). \tag{2.5}$$

The *order* m ranges from $-\ell$ to ℓ; the *degree* ℓ from 0 to ∞. $N_{\ell m} = N_{\ell,-m}$ are dimensionless, normalization coefficients and

$$P_{\ell,-m}(\cos\theta) = P_{\ell m}(\cos\theta). \tag{2.4'}$$

The functions (2.5) fulfil

$$Y_{\ell m}(\theta, \varphi) = Y_{\ell,-m}^*(\theta, \varphi); \tag{2.5'}$$

moreover, they are "orthogonal" with respect to the scalar product

$$\int d\Omega\ Y_1\ Y_2^* = 0; \tag{2.6}$$

the integral is performed on the unit sphere. A suitable choice of the normalization coefficients completes the orthonormality conditions

$$\int d\Omega\ Y_{\ell m}(\theta,\varphi)\ Y_{\ell' m'}^*(\theta,\varphi) = 4\pi\delta_{\ell\ell'}\delta_{mm'}. \tag{2.7}$$

This requires

$$N_{\ell 0} = \sqrt{(2\ell + 1)}\ , \tag{2.8$_1$}$$

$$N_{\ell m} = \sqrt{(2\ell+1)}\left[\frac{(\ell - m)!}{(\ell + m)!}\right]^{\frac{1}{2}}. \tag{2.8$_2$}$$

In the language of complex vector spaces, we can say that the set (2.5) (with $-\ell \le m \le \ell$) is a *unitary basis* in a $(2\ell + 1)$-dimensional space with respect to the scalar product defined by the average on the unit sphere. It can be shown also that the set (2.5) is complete; in other words, any regular function f on the unit sphere can be expressed as an expansion in spherical harmonics:

$$f(\theta, \varphi) = \Sigma_\ell \Sigma_m f_{\ell m} Y_{\ell m}(\theta, \varphi)\ . \tag{2.9}$$

We shall not hereinafter write explicitely the summation limits. The expansion coefficients are obtained by means of eq. (2.7):

$$f_{\ell m} = (1/4\pi) \int d\Omega f(\theta,\varphi) Y_{\ell m}^*(\theta,\varphi). \qquad (2.10)$$

If f is real,

$$f_{\ell m} = f_{\ell,-m}^* . \qquad (2.10')$$

An arbitrary rotation R determines a map $P(\theta, \varphi) \rightarrow P'(\theta', \varphi')$ of the unit sphere onto itself and transforms $Y_{\ell m}(P)$ into the new function $Y_{\ell m}'(P) \equiv Y_{\ell m}(P')$. The defining properties of the harmonic functions (the order of the polynomial $r^\ell Y_{\ell m}$ and the harmonic property) are not affected by this mapping; hence $Y_{\ell m}'$ is also a spherical harmonic of order ℓ and must be expressible as a linear combination of the spherical harmonics of the same order:

$$Y_{\ell m}'(P) = \Sigma_{m'} D^{(\ell)}_{mm'}(R)\ Y_{\ell m'}(P) . \qquad (2.11)$$

The set of square matrices $D^{(\ell)}$ (one for each rotation R), of order $(2\ell + 1)$, constitutes a representation of the *rotation group* (they obey its multiplication table) and reduces to the set of ordinary rotations for $\ell = 1$. It can be shown that this representation is unitary with respect to the scalar product (2.9). This group-theoretical argument provides the justification for the choice (2.5) for the basis, against the "naive", and in geophysics more usual, real set (2.13). It can be shown that these representations are *irreducible*, that is to say, they do not leave invariant any subspace of the space S_ℓ of $(2\ell + 1)$ dimensions over which they operate. Moreover, every single–valued matrix representation of the rotation group is a linear combination of the set $D^{(\ell)}$; in other words, the spherical harmonics as a whole provide the basis vectors whose transformations under the rotation group generate every single-value matrix representation of the group itself. There are also two–value representations in which two different matrices correspond to each rotation R. They are labelled by half integer values of the index ℓ and describe the spin of elementary particles. These important group–theoretical properties provide powerful tools to deal with the spherical harmonics expansions.

We list here the complex, normalized spherical harmonics of the first three orders:

$$Y_{00} = 1, \qquad (2.12_1)$$

$$Y_{10} = \sqrt{3} z/r = \sqrt{3}\cos\theta, \quad Y_{11} = (x + iy)/r = e^{i\varphi}\sin\theta, \qquad (2.12_2)$$

$$Y_{20} = \sqrt{5}(3\cos^2\theta - 1)/2, \quad Y_{21} = \sqrt{3}\cos\theta \sin\theta \exp(i\varphi),$$
$$Y_{22} = \sin^2\theta \exp(2i\varphi). \tag{2.12$_3$}$$

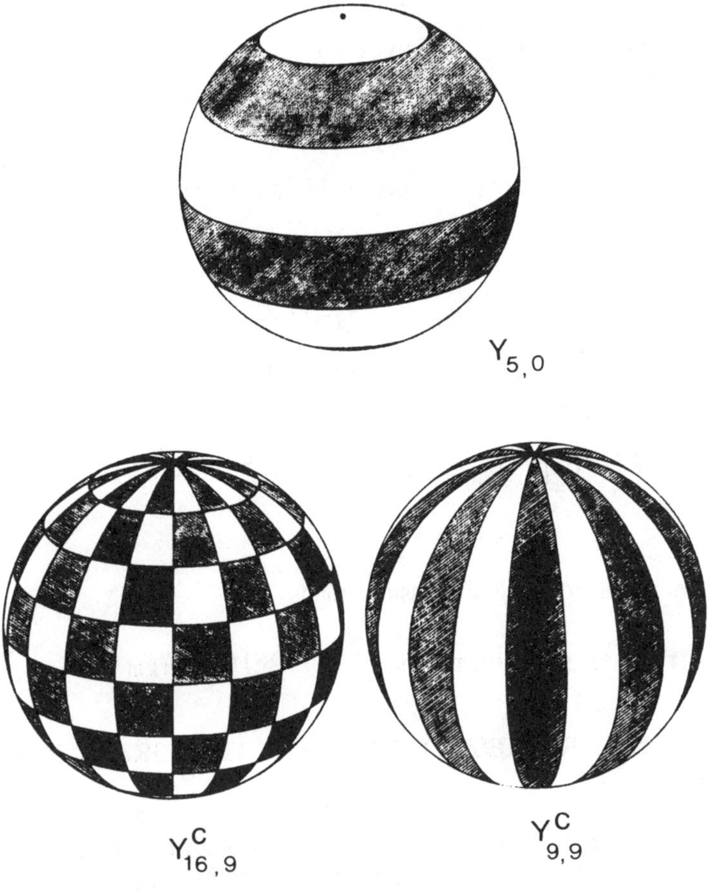

Fig. 2.1. The qualitative behaviour of the spherical harmonics is indicated by the meridians and the parallels on which they vanish; here they are drawn for P_{50}, $P_{16,9}\cos(9\varphi)$ and $P_{9,9}\cos(9\varphi)$. In general the harmonics have (ℓ − m) zero's along half a meridian and m zero's along half a parallel.

In place of the complex basis (2.5) the real (and orthonormal) set

$$N_{\ell 0} P_\ell(\cos\theta),$$
$$\sqrt{2} N_{\ell m} P_{\ell m}(\cos\theta) \cos m\varphi \ (m = 1, 2, \ldots \ell), \tag{2.13}$$
$$\sqrt{2} N_{\ell m} P_{\ell m}(\cos\theta) \sin m\varphi \ (m = 1, 2, \ldots \ell).$$

is more commonly used. The corresponding expansion

$$f(\theta, \varphi) = \Sigma_\ell \{C_{\ell 0} N_{\ell 0} P_\ell(\cos\theta) +$$

$$+ \sqrt{2}\Sigma_m N_{\ell m} P_{\ell m}(\cos\theta)[C_{\ell m}\cos m\varphi + S_{\ell m}\sin m\varphi]\} , \quad (2.14)$$

when compared with eq. (2.9), gives the new coefficients

$$C_{\ell 0} = f_{\ell 0},$$

$$C_{\ell m} = (f_{\ell m} + f_{\ell m}^*)/\sqrt{2},\ S_{\ell m} = i(f_{\ell m} - f_{\ell m}^*)/\sqrt{2} \quad (2.15)$$

and the old ones

$$f_{\ell 0} = C_{\ell 0} ,\quad f_{\ell m} = (C_{\ell m} - iS_{\ell m})/\sqrt{2} . \quad (2.16)$$

Sometimes it is convenient to describe each component (ℓ, $m \neq 0$) with an amplitude $J_{\ell m}$ and a phase $\varphi_{\ell m}$:

$$f_{\ell m} = 1/\sqrt{2}J_{\ell m}\exp(i\varphi_m),\ J_{\ell m} = \sqrt{(C_{\ell m}^2 + S_{\ell m}^2)}. \quad (2.17)$$

$$U_{\ell 0} = J_{\ell 0} . \quad (2.18)$$

is real. The corresponding expansion reads

$$f(\theta, \varphi) = \Sigma_\ell \Sigma_{m\geq 0} J_{\ell m} N_{\ell m} P_{\ell m}(\cos\theta)\ \cos(\varphi - \varphi_{\ell m}). \quad (2.19)$$

2.2. HARMONIC REPRESENTATION OF THE GRAVITY FIELD OF A PLANET

The spherical harmonics provide a representation of the gravitational potential energy per unit mass produced by an isolated body:

$$U(\mathbf{r}) = -\frac{GM}{r} - \frac{GM}{R}\Sigma_\ell \Sigma_m U_{\ell m} Y_{\ell m}(\theta, \varphi)\left(\frac{R}{r}\right)^{\ell+1} = -\frac{GM}{r} + \delta U; \quad (2.20)$$

here

$$U_{\ell m}^* = U_{\ell,-m}. \quad (2.20')$$

With the introduction of its mass M and mean radius R we have made the coefficients $U_{\ell m}$ dimensionless and set $U_{00} = 1$, fixing the ratio

GM/R. It can be shown that an arbitrary harmonic function which vanishes at infinity has the form (2.20). Each ℓ corresponds to a different kind of deviation from spherical symmetry, the larger ℓ giving rise to a smaller gravitational field at great distances: the gravitational field corresponding to the order ℓ is $O(r^{\ell+1})$.

With the spherical harmonics we can construct another class of harmonic functions which diverge at infinity:

$$V(\mathbf{r}) = \Sigma_\ell \, \Sigma_m \, V_{\ell m} \, r^\ell \, Y_{\ell m}(\theta, \varphi). \qquad (2.21)$$

They are useful to study the gravitational field inside a body and, together with (2.20), cover all the harmonic functions.

The $\ell = 0$ term in the expansion (2.20) is the main, monopole term GM/r. Note that so far the origin of the coordinates is undetermined. Under a translation $\mathbf{r} = \mathbf{r}' + \mathbf{a}$

$$\frac{1}{r} = \frac{1}{|\mathbf{r}' + \mathbf{a}|} = \frac{1}{r'}\left[1 + \frac{\mathbf{r}'\cdot\mathbf{a}}{r'^2} + O\left(\frac{a^2}{r'^2}\right)\right];$$

hence a displacement of the origin produces in a harmonic function U an additional dipole term (and other terms of higher order as well); therefore the three coefficients U_{1m} can always be taken equal to be nil, and we shall do so. The first relevant deviation from spherical symmetry is the *quadrupole* term with $\ell = 2$, described by five arbitrary coefficients; however, if the body is axially symmetric around the z–axis, $U_{22} = U_{21} = 0$ and only U_{20} survives; in this case we have only one free parameter. One usually writes this approximation as

$$U(\mathbf{r}) = -\frac{GM}{r}\left[1 - J_2\left(\frac{R}{r}\right)^2 \frac{3\cos^2\theta - 1}{2} + \ldots\right]. \qquad (2.22)$$

J_2 (which includes the factor $\surd 5$ of eq. (2.12_3)) is the dimensionless measure of the quadrupole coefficient, related to polar flattening.

Let us now compute J_2 in terms of the mass distribution. The integral form of eq. (1.2) (Poisson's integral) reads

$$U(\mathbf{r}) = -G\int dV'\rho(\mathbf{r}')/\surd(r^2 + r'^2 - 2\mathbf{r}\cdot\mathbf{r}'). \qquad (2.23)$$

Expanding the denominator with respect to r'/r we get

$$U = -\frac{G}{r}\int dV'\rho(\mathbf{r}')\left[1 + \frac{\mathbf{r}\cdot\mathbf{r}'}{r^2} + \frac{3(\mathbf{r}\cdot\mathbf{r}')^2 - r^2r'^2}{2r^4} + O\left(\frac{r'^3}{r^3}\right)\right].$$

The first term is the monopole contribution $-$ GM/r; the second term is the dipole contribution and vanishes when the origin of the coordinates is placed at the centre of gravity:

$$U_{1m} = \int dV \mathbf{r} \rho(\mathbf{r}) = 0. \tag{2.24}$$

The third term reads

$$U_2 = -\frac{G}{r^5}\,\mathbf{rr}:\tfrac{1}{2}\int dV'\rho(\mathbf{r}')(3\mathbf{r}'\mathbf{r}' - \mathbf{I}r'^2) = -\frac{G}{r^5}\,\mathbf{rr}\cdot\mathcal{Q} \tag{2.25}$$

and, being proportional to $1/r^3$, must be the quadrupole term (A bold character denotes here a second rank tensor.) The quadrupole tensor

$$\mathcal{Q} = \tfrac{1}{2}\int dV\rho(\mathbf{r})(3\mathbf{rr} - \mathbf{I}r^2) \tag{2.26}$$

is symmetric and trace–free; hence U_2 can be written as P_2/r^5, where $P_2 = \mathbf{rr} \cdot \mathcal{Q}$ is a quadratic harmonic form. Thus U_2 depends on five arbitrary coefficients and is of the form

$$U_2 = -\frac{GM}{R}\sum U_{2m}Y_{2m}(\theta, \varphi)\left(\frac{R}{r}\right)^3.$$

When the body has an axial symmetry around the z-axis, the x and y variables can be interchanged; only the diagonal part of $\mathcal{Q}$ does not vanish and depends upon a single parameter:

$$Q_{zz} = \tfrac{1}{2}\int dV\rho(\mathbf{r})(2z^2 - x^2 - y^2) =$$

$$= \int dV\rho(\mathbf{r})(x^2 + z^2 - x^2 - y^2) = A - C; \tag{2.27$_1$}$$

$$Q_{xx} = Q_{yy} = \tfrac{1}{2}\int dV\rho(\mathbf{r})(2x^2 - y^2 - z^2) =$$

$$= \tfrac{1}{2}\int dV\rho(x^2 + y^2 - x^2 - z^2) = \tfrac{1}{2}(C - A). \tag{2.27$_2$}$$

$$A = \int dV\rho(\mathbf{r})(y^2 + z^2)$$

and its cyclic permutations are the moments of inertia relative to the principal axes; in our case A = B. Then we have

$$U_2 = -\frac{G}{r^3}\left[(A-C)\cos^2\theta + \frac{(C-A)\sin^2\theta}{2}\right] = \frac{G(C-A)}{2r^3}(3\cos^2\theta - 1).$$

Comparing this with eq. (2.22), we see that J_2 is a dimensionless measure of the deviation from spherical symmetry of the body:

$$J_2 = (C - A)/MR^2 = 0.00108 \qquad (2.28)$$

for the earth. It is of the same order of magnitude as the ellipticity (1.14) and provides important information about the mass distribution in the interior. It can be measured by studying the perturbations induced in the orbits of natural and artificial satellites.

We now generalize eq. (2.27) and express $U_{\ell m}$ in terms of the spherical harmonic representation of the mass density

$$\rho(\mathbf{r}) = \Sigma_\ell \Sigma_m \rho_{\ell m}(r) Y_{\ell m}(P) \qquad (2.29)$$

(P is a point on the unit sphere). To do this we use in eq. (2.23) the expansion (2.3) and the *addition theorem for spherical harmonics*

$$P_\ell(\cos\Psi) = \Sigma_m Y_{\ell m}(P) Y_{\ell m}^*(P')/(2\ell + 1) \ . \qquad (2.30)$$

Here Ψ is the angle between the unit vectors OP and OP'. Using the orthonormality properties (2.7) we find for the coefficients $U_{\ell m}$ in eq. (2.20)

$$U_{\ell m} = \frac{4\pi}{(2\ell + 1)MR^\ell} \int dr\, r^{\ell+2} \rho_{\ell m}(r) \ . \qquad (2.31)$$

The radial integral is limited above by the largest distance from the origin at which there is matter.

2.3. LOW ORDER GRAVITATIONAL POTENTIAL

Consider a planet rotating with an angular velocity ω around the z–axis; the total potential energy U_T includes also the centrifugal potential:

$$U_T = U - \tfrac{1}{2}\omega^2 r^2 \sin^2\theta +$$

$$+ (GM/r)[1 - \tfrac{1}{2}J_2(R/r)^2(3\cos^2\theta - 1)] + O(R/r)^3 \ . \qquad (2.32)$$

If the oblateness, measured by J_2, is due to rotation (see Sec. 3.3), the two terms in this equation responsible for the deviation from spherical symmetry must be, on the ground, of the same order of magnitude; this justifies the previous estimate (1.14'). The higher harmonic coefficients describe more complicated deviations from axial symmetry.

	m = 0	m = 1	m = 2	m = 3	m = 4
ℓ = 2	– 484.165	0	2.439		
		0	–1.400		
ℓ = 3	0.957	2.030	0.904	0.721	
		0.250	–0.620	1.413	
ℓ = 4	0.539	–0.533	0.347	0.991	–0.190
		–0.475	0.664	–0.201	0.308

Table 2.1. GEM T-1 values for the earth of $C_{\ell m}$ and $S_{\ell m}$ in units of 10^{-6} (eq. (2.14)), with respect to the normalized harmonics (eqs. (2.5,8)); in particular, $C_{2\,0} = J_2/\sqrt{5}$.

One expects the coefficients of the geopotential to change appreciably only over the time scale of the evolution of the earth due, for example, to the readjustment of the crust after geological events. In particular, when the last glaciation ended about 20,000 years ago, the thick ice cover over the polar caps melted into the oceans, thereby removing weight from them. The crust then slowly started taking up a new shape, appropriate to the new mass distribution. The time scale of this relaxation is determined by viscosity; the process is still under way. As a result, we should expect to observe a decrease in the oblateness parameter J_2. It is most interesting that this prediction has been tested through the measurement of the motion of the node of the satellite LAGEOS; the value

$$dJ_2/dt = -3\cdot 10^{-11}/y \tag{2.33}$$

was found, in rough agreement with the expectation (see Ch. 20 and C.F. Yoder et al., Nature, **303**, 757 (1983)). The geopotential changes also because tidal effects (see Sec. 4.3); there are also small (mainly seasonal) changes due to the motion of the oceans and the atmosphere.

The *geoid* is an equipotential surface which describes the surface of an ideal ocean, with no waves or currents, hence at constant pressure (eq. (1.25)). Since the gravitational potential on the ground differs from $-GM/r$ by terms of order 10^{-3} or smaller, it is reasonable to expect that the shape of the geoid will differ by the same order from a sphere; thus in computing the small geopotential corrections we can use the spherical approximation. Therefore from eq. (2.20) we get for the radial equation of the geoid

$$R + \delta r(\theta,\varphi) = \mathrm{const} + R\,\Sigma_{\ell\geq 2}\Sigma_m U_{\ell m} Y_{\ell m}(\theta,\,\varphi)\ . \tag{2.34}$$

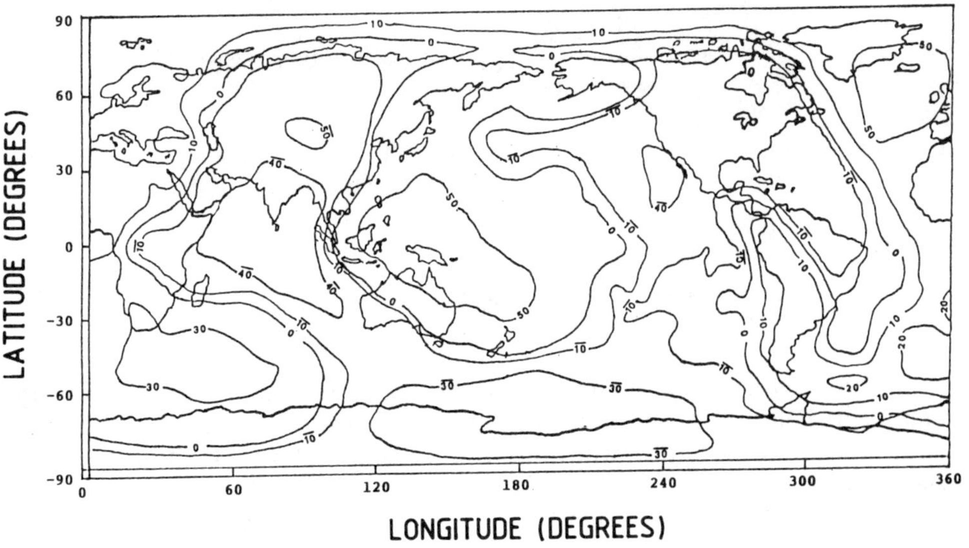

Fig. 2.2. Isolevel lines of the geoid. The numbers give height in meters over the mean ellipsoid. The geoid features indicated in this map are due to large scale and deep lying density inhomogeneities below the crust.

In the next chapter we shall discuss the main, $\ell = 2$ contribution, which must include also the centrifugal potential energy; here we show, as an application, how the coefficient U_{22} changes the shape of the equatorial geoid. On the equator $\theta = \pi/2$; U_{21} and U_{20} do not affect this shape and, from eq. (2.16) and Table 2.1, we get:

$$\delta_2 r(\pi/2, \varphi) = \text{const} + R[U_{22}\exp(2i\varphi) + U_{22}^{*}\exp(-2i\varphi)] =$$

$$= \text{const} + 25.4^{\circ}\cos 2(\varphi - 14.9^{\circ}) \ . \qquad (2.35)$$

The equatorial section of the geoid is elliptic, with the major axis directed along 15° W longitude. The term in eq. (2.31) corresponding to $\ell = 3$, $m = 0$ is proportional to $[5\cos(3\theta) - 3\cos\theta]$ and therefore describes a north–south asymmetry. The construction of an accurate geoid is an important task of geophysics (Fig. 2.2).

Another important effect of the deviations of the gravity field of the earth from spherical symmetry is the change in the local value of the acceleration of gravity g (*gravity anomaly*). The vector ∇U_T defines the local vertical, while its modulus gives g. In computing small changes of g we can neglect the difference between the vertical and the radial direction; we also set r = R, which is good enough on the geoid. From eq. (2.20) we find first

$$g = \frac{GM}{r^2} + \frac{GM}{R^2} \Sigma_{\ell>0} \Sigma_m U_{\ell m} Y_{\ell m}(\theta,\varphi)(\ell + 1) \ .$$

On the geoid, however, we cannot neglect the correction of the main term due to the difference between r and R (eq. (2.34)); this gives

$$g = \frac{GM}{R^2} (1 + \Sigma_{\ell>0} \Sigma_m U_{\ell m} Y_{\ell m}(\ell - 1)) \ . \qquad (2.36)$$

The measurement of the gravity anomaly on the ground is a very important field of geophysics with practical applications (like search for mineral and oil deposits, which have a density different from that of the surrounding rocks). There are also periodic changes in the gravity anomaly due to the body tides of the earth (Sec. 4.4). Field instruments (*gravimeters*) have an accuracy of 10 to 100 microgals (1 gal = 10^{-6} cm/s^2). Laboratory instruments based on the measurement of the free–fall of a small mass provide an absolute accuracy for the gravity acceleration of one or two microgals. We shall not discuss further these topics of strictly geophysical character.

2.4. FINE STRUCTURE OF THE GRAVITY FIELD OF THE EARTH

Since $P_\ell(u)$ is a polynomial of order ℓ, its derivative is of order ℓP_ℓ; consequently, the derivatives of $Y_{\ell m}(\theta,\varphi)$ are of the order $\ell Y_{\ell m}$ or smaller. On the ground, the spherical harmonics of order ℓ correspond to a contribution to the geopotential with a characteristic horizontal scale R/ℓ. To get the vertical scale in the case $\ell \gg 1$, note that in eq. (2.20) we can set h = (r – R) (h $\ll$ R) and get

$$\begin{aligned} (R/r)^{\ell+1} &= (1 - h/R)^\ell = \exp[\ell \ln(1 - h/R)] = \\ &= \exp\{ -\ell h/R + O[\ell (h/R)^2]\} . \end{aligned} \qquad (2.37)$$

Therefore the components of the order ℓ ($\gg$ 1) decrease exponentially with height with the same scale R/ℓ. Similarly, from eq. (2.31) we can find the thickness of the layer which is responsible for these components: introducing the depth d = (R – r) ($\ll$ R) we have for $\ell \gg 1$

$$\begin{aligned} (r/R)^{\ell+2} &= (1 - d/R)^\ell = \\ &= \exp[\ell \ln(1 - d/R)] = \exp(-\ell d/R) . \end{aligned} \qquad (2.38)$$

Therefore the contribution of the matter distribution decreases exponentially with depth with the same scale R/ℓ. Because of the high geophysical interest of the crust, where the continental drift is taking place, it is important to map the gravitational field with a resolution better than the thickness of the crust, say, 30 km; this requires the measurement of the spherical harmonic coefficients $U_{\ell m}$ up to $\ell = R/(30\text{ km}) \cong 200$. The number of these real coefficients is the sum of all the terms $(2\ell' + 1)$, where ℓ' is any integer between 0 and ℓ; this number, $(\ell^2 + 2\ell)$, runs into the tens of thousands. This shows the complexity of the task also from the point of view of data analysis,

At large ℓ the general trend of the geopotential is given by the mean

$$U_\ell = \left[\frac{1}{2\ell + 1} \Sigma_m |U_{\ell m}|^2\right]^{\frac{1}{2}} , \qquad (2.39)$$

analogous to the energy spectrum for a two-dimensional scalar field. It has been noticed that the empirical law (*Kaula's rule*)

$$U_\ell \approx 10^{-5}/\ell^2 \quad (\ell \gg 1) \qquad (2.40)$$

is followed fairly well, not only for the earth, but also for some planets. It is not known whether this law has a theoretical justification.

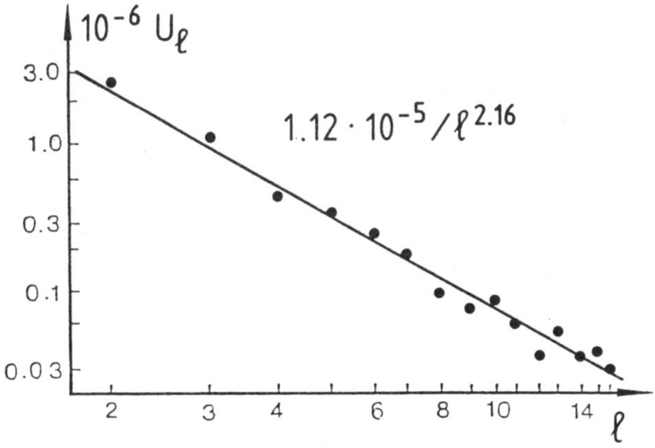

Fig. 2.3. The mean values (eq. (2.39)) of the geopotential coefficients as a function of their order ℓ. The line gives the empirical approximation (2.40).

With this approximation we can estimate the order of magnitude of the contribution to the geopotential which has a characteristic scale R/ℓ.

Consider in the expansion (2.20) the partial sum

$$\delta_\ell U(\mathbf{r}) = -(GM/R)\,\Sigma_{\ell'\geq\ell}\Sigma_m U_{\ell' m} Y_{\ell' m} \qquad (2.41)$$

of the components of order greater or equal to ℓ. Because of the fast decrease of the amplitude with ℓ (eq. (2.38)), we can expect this to be dominated by terms of order ℓ' not much greater than ℓ. The order of magnitude of $\delta_\ell U(\mathbf{r})$ is obtained by averaging its square over the sphere. Using the orthonomality relations (2.8) and (2.37) we get

$$<(\delta_\ell U)^2> = \int d\Omega[\delta_\ell U(\mathbf{r})]^2/4\pi =$$

$$= (GM/R)^2\,\Sigma_{\ell'\geq\ell}(2\ell' + 1)U_{\ell'}{}^2. \qquad (2.42)$$

With eq. (2.40) this gives

$$\delta_\ell U = (GM/R)\cdot 10^{-5}/\ell = gR\cdot 10^{-5}/\ell\ . \qquad (2.43)$$

Dividing by the characteristic scale ℓ/R we get the order of magnitude of the corresponding gravity anomaly, independent of ℓ:

$$\delta_\ell g = (GM/R^2)\cdot 10^{-5} = 0.01\ \mathrm{cm/s^2}\ . \qquad (2.44)$$

At large ℓ the curvature of the earth may be neglected and a simpler representation of the gravity field is useful. Consider a "flat" earth with an inhomogeneous mass distribution in the half space $z < 0$. For a given z (> 0) consider the Fourier expansions with respect to x and y of the gravitational potential and the matter distribution, functions of z and a two–dimensional Fourier vector $\mathbf{k} = (k_x, k_y)$ (to be understood); they fulfil Poisson's equation (1.23), namely

$$\frac{d^2U(z)}{dz^2} - k^2U(z) = 4\pi G\rho(z) \quad (k^2 = k_x{}^2 + k_y{}^2). \qquad (2.45)$$

The solution vanishing when z is very large reads

$$U(z) = -\frac{2\pi G}{k}\int_{-\infty}^{0} dz'\rho(z')\exp[k(z' - z)]. \qquad (2.46)$$

This shows that, for a given wave number $\mathbf{k}$, the potential is determined by the density inhomogeneities within a layer of depth approximately equal to $1/k$; and that the potential decreases exponentially with height with the same characteristic scale. Therefore k can be approximately identified with ℓ/R.

A promising way to get a global mapping of the gravity field of the earth at a very high resolution is by tracking a low flying spacecraft. Consider, for simplicity, a spacecraft at a constant altitude h, flying over a "flat earth" along the x direction, with an unperturbed velocity v_0. The conservation of energy means

$$v_0 \delta v_x = \delta U = \Sigma_k U(h) \exp(i k_x v_0 t). \qquad (2.47)$$

Therefore the Fourier coefficient of the velocity along the track gives directly the Fourier component of U for $k_x = \omega/v_0$, $k_y = 0$.

Eqs. (2.46) and (2.37) show clearly the main difficulty we have to face in a measurement of this kind: the potential perturbation decreases exponentially with height with a characteristic scale $1/k = R/\ell$. The component with $\ell = 200$, for example, at 180 km is attenuated by the factor $\exp(-180 \cdot 200/6400) = 0.0036$. The ground value (2.43) of the potential, 32,000 $(cm/s)^2$, becomes at that altitude 115 $(cm/s)^2$. The perturbation in the velocity of a satellite at 180 km is about 1 μm/s (eq. (2.47)). Because of the atmospheric drag it is practically impossible to fly a space vehicle in a lower earth orbit; we must therefore use a very sensitive instrument to measure velocities. A device of this kind is provided by a microwave phase–locked loop which exploits the Doppler effect between two satellites or between the ground and a satellite (Sec. 19.4); the velocity change quoted above corresponds to a fractional frequency change of the order of 10^{-14}, which is quite accessible to commercial frequency standards (Sec. 3.1).

PROBLEMS

2.1. The second degree harmonics (2.13_3) generate a 5 by 5 representation of the rotation group. Construct the matrices corresponding to the rotations around the coordinate axes.

*2.2. A symmetric tensor of the second rank with vanishing trace has five independent components. Show that it transforms under rotations according to the representation $D^{(2)}$ defined by eq. (2.11).

2.3. Calculate the potential perturbation and the gravity anomaly on a flat earth due to (i) a uniform sphere and (ii) a uniform infinite cylinder buried under the surface and with a different density. Hint: use Gauss' theorem.

2.4. Estimate the velocity change of a spacecraft in a low orbit flying across a mountain range. Hint: model the range as a cylinder placed on the surface of a flat earth.

2.5. Calculate the spherical harmonics of order 3. From Y_{30} find the north–south asymmetry of the profile of the geoid on a meridian.

2.6. The term U_{22} in the gravitational potential of the earth produces a modulation on the potential profile on the equatorial plane as a function of longitude. Obtain the amplitude and the phase of this modulation, important for geosynchronous satellites (see Sec. 18.4).

2.7. Using the harmonic character of the harmonic functions, find the *Legendre equation*, that is, the ordinary differential equation fulfilled by the associate Legendre functions.

2.8. Construct a basis for the harmonic functions of third order.

*2.7. Estimate the relaxation time τ_r of the earth after the last ice age, which ended about 20,000 y ago, from the measured value of dJ_2/dt and the ice load on the polar caps (about $3 \cdot 10^{22}$ g each).

*2.8. Express the average of the square of the geopotential and the gravity anomaly in terms of the harmonic coefficients and discuss the convergence properties of Kaula's rule (2.40).

2.9. Show that the moment of inertia of a spherically symmetric body is given by

$$A = B = C = (8\pi/3)\int dr\ r^4\ \rho(r).$$

For the earth it is 0.3335 R^2M; compare the coefficient with the one for a uniform density.

FURTHER READING

A good and classical textbook on geodesy and theoretical gravimetry is W.A. Heiskanen and H. Moritz, *Physical Geodesy*, Freeman (1967). On an easier level and with greater concern for applications, is G.D. Garland, *The Earth's Shape and Gravity*, Pergamon Press (1965). We also quote W.M. Kaula, *Theory of Satellite Geodesy*, Blaisdell, Waltham, Mass. (1966) and M. Caputo, *The Gravity Field of the Earth*, Academic Press (1967).

3. *THE ROTATION OF THE EARTH AND OTHER PLANETS*

This and the following chapter are devoted to a complex subject which ties together three different fields: astrometry, geophysics and celestial mechanics. The earth provides the basic frame of reference with respect to which most of our celestial observations are made: the determination of its (variable!) rotation with respect to the stars and the planets is a prerequisite of astrometry, the discipline that determines the position and the motion of celestial objects. This is of course connected with the fundamental problem of absolute rotation in mechanics. This chapter is also an appropriate place to introduce the problem of measurement of time and distance in space physics. Rotation produces an oblateness and at the same time is affected by the internal motion of the planet, an important chapter of geophysics. The presence of other bodies in the solar system also deeply influences the structure and the rotation of the earth through tidal phenomena.

In this chapter we shall develop two idealized models which are of great help in understanding the rotational dynamics of an isolated body. We can compute exactly the effect of a uniform and slow rotation upon a liquid mass; at the opposite extreme the rotation of a rigid body shows precession, and possibly nutation. The effects of outside gravitating bodies, in particular, tides, is considered in the following chapter.

3.1. MEASUREMENTS OF TIME AND DISTANCE

In traditional metrology we have two distinct units for length and time. The centimeter is defined through interferometric measurements as a given multiple of the wavelength of a stable optical line. The definition of the second of time uses very accurate frequency standards provided by radio resonances in atomic and nuclear systems: a clock is obtained by counting the number νT of periods elapsed in a time interval T. Obviously the fractional accuracy in T is equal to the fractional accuracy in the frequency ν. The best frequency standard currently available is based upon the hyperfine splitting of the ground energy level of atomic hydrogen; this splitting is due to the interaction between the spin of the electron and the spin of the nucleus and corresponds to the resonant frequency ν = 1,420,405,751.68 Hz. The corresponding frequency standard, called *hydrogen maser*, is a microwave cavity tuned to that line; its frequency can be stabilized to better than a part in 10^{15} for averaging times of the order of one hour. It has been found that different atomic frequency standards agree to within their present accuracy; the *atomic time* t is the fundamental time variable.

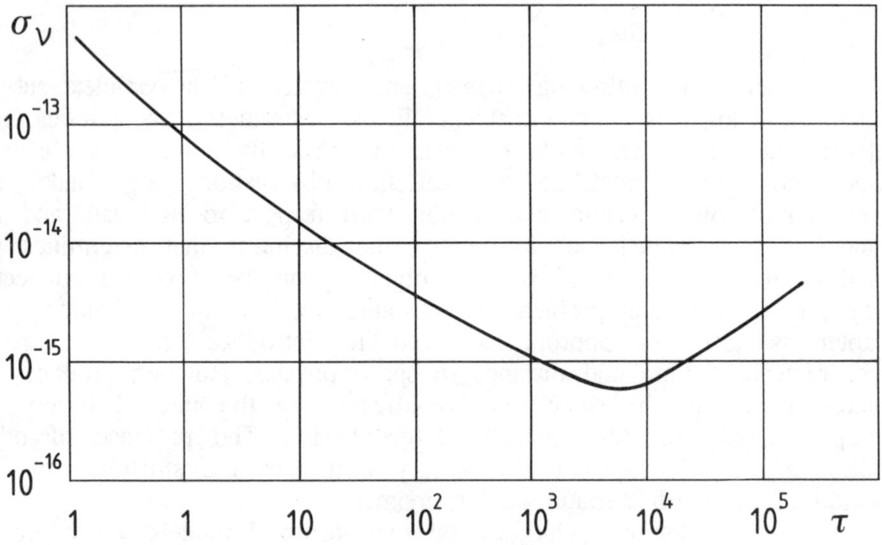

Fig. 3.1. Typical stability curve of a commercial hydrogen maser. The diagram reports, on logarithmic scales, the root mean square of the fractional frequency change, averaged over a time τ ("Allan's variance") (Sec. 19.4):

$$\sigma_\nu(\tau) = \tfrac{1}{2}\langle(\nu_\tau(t+\tau) - \nu_\tau(t))^2\rangle,$$

where $\nu_\tau(t)$ is the average of the frequency between the times t and t + τ.

Another important time variable is the phase $\varphi(t)$ of the earth (*hour angle*). The conventional origin of this phase is the intersection of the equatorial plane with the plane of the ecliptic at a fixed epoch (γ *point*, see Fig. 3.3). Instead of the angle φ one often uses the *universal time*

$$UT(t) = \varphi(t)/\omega_0, \tag{3.1}$$

obtained with a nominal earth rotation frequency ω_0.

The units of time and distance so defined are independent. By transferring the frequency standard from the microwave to the optical band it is possible to measure the frequency of the line whose wavelength determines the standard of length and therefore to determine their product, the velocity of light, with the dimensions cm/s. This transfer over a frequency range of about 10 decades, however, is subject to errors; moreover, the stability of the lasers used in interferometric techniques is much less than the stability of atomic frequency standards. To overcome these practical difficulties, another point of view has emerged and gained acceptance, i.e., to forego the interferometric standard as a primary one and to consider the velocity of light as a conventionally fixed quantity. If it is taken to be unity, one measures lengths in units of light seconds. The "poor" accuracy of interferometric instruments will then result in errors in lengths, and not in the definition of an independent standard of length.

This point of view is quite appropriate to space physics. There distances are usually measured with the transit times of light or radio pulses, timed with atomic clocks; or by means of the Doppler effect. The latter method uses a radio frequency standard and determines the

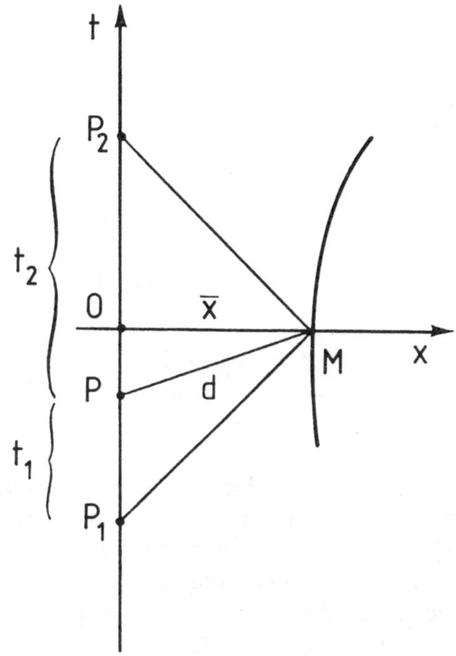

Fig. 3.2. The chronometric measure of distance. In the frame (t, x), we have $M \equiv (0, \bar{x})$, $P \equiv (t_1 - \bar{x}, 0)$. In special relativity the proper length of the hypotenuse MP is the square root of the *difference* of the squares of the time-like and the space-like sides:

$$d = \sqrt{[(\bar{x} - t_1)^2 - \bar{x}^2]} = \sqrt{[t_1(2\bar{x} - t_1)]} = \sqrt{[t_1 t_2]}.$$

relative velocity of the source and the receiver; by integration, the distance is obtained. In both cases, only the time standard is used. (Interferometric measurements in space have been planned, but not performed yet.)

The theory of relativity provides another theoretical argument in support of this view. There we have only one fundamental geometrical quantity, the *proper interval.* According to the *Equivalence Principle,* proper time intervals are measured by atomic time. Distances and time intervals are given by the proper interval in the space-like and the time-like case, respectively. The velocity of light is conventionally set equal to one.

It is interesting to see how one can formulate the measurement of distance in a relativistic context. The distance of an object in space can be measured with a clock in the following way (see Fig. 3.2). A source ℓ at an event P_1, moving uniformly, sends a short electromagnetic pulse to a moving mirror M. At M the pulse is reflected back to P_2 on ℓ. The distance of M from an event P on ℓ separated from P_1 and P_2 by the time intervals t_1 and t_2 is given by

$$d = \sqrt{(t_1 t_2)}\,. \tag{3.2}$$

When $t_1 = t_2$ we get the usual relation that the distance is half the round-trip light time; eq. (3.2) accounts for the relativity corrections due to the lack of simultaneity.

Time and frequency can be transferred from one point to another by electromagnetic signals. The observed frequency depends on the velocity of the source $\mathbf{v}_1$ and the velocity of the observer $\mathbf{v}_2$; in flat space-time the frequency shift is given by (Sec. 17.5):

$$\frac{\nu_2}{\nu_1} = \left[\frac{1 - v_1^{\,2}}{1 - v_2^{\,2}}\right]^{\frac{1}{2}} \frac{1 - \mathbf{v}_2 \cdot \mathbf{k}}{1 - \mathbf{v}_1 \cdot \mathbf{k}}\,. \tag{3.3}$$

Here $\mathbf{k}$ is the wave vector of the light. The terms linear in v give the ordinary *Doppler effect*; the square root term is the relativistic *transversal Doppler effect*, which gives a non-vanishing shift also when the motion is orthogonal to the ray. For a typical planetary velocity of 30 km/s the relativistic corrections are about a part in 10^8 and easily observable. Of course, in that case we have also a gravitational shift, of the same order of magnitude as, and indistinguishable from, the transversal Doppler effect.

With microwave links it is now possible to measure the distance of an object in the solar system with an accuracy of a few meters. This has been done, in particular, with the *Viking* spacecraft on Mars for about six years. Since the forces acting on a planet are much better known than those on artificial satellites, that are strongly affected by non-gravitational forces (see Sec. 18.1), it was possible to determine the semimajor axis of Mars with a fractional accuracy of about 10^{-11}. For earth satellites (namely, the moon and LAGEOS, see Sec. 20.2), laser tracking is used, with an accuracy of about 5 cm for the moon and 1 cm for LAGEOS, and a fractional accuracy of about 10^{-9}.

This extraordinary precision can be compared with the time when only optical observations and, therefore, angles between celestial objects were available; their distance D could be deduced from the known value of the earth radius $R_\oplus$ if their parallax $R_\oplus/D$ was measured. The distance of the moon, which has a parallax of about 1 degree, was known in antiquity; but for the sun, with a parallax of about 9 arcsec, and the planets, things were much more difficult. To achieve the required accuracy it was necessary to know well the law of atmospheric refraction, to use telescopes of good optical and mechanical qualities and to be able to locate precisely the image in the focal plane with a micrometric screw.

The measurement of the parallax of Mars and, indirectly, of the sun, was done for the first time in 1672 by Flamsteed in England and by Richter – an assistant to Cassini in Paris – in Cayenne. The object of the measurement was the displacement of the image of Mars in a few hours due to the rotation of the earth (see Problem 3.1). By Kepler's third law, the value of GM for the sun was also obtained in this way.

3.2. ROTATING FRAMES

Since most measurements are done from the earth, the basic frame of reference is earth-bound. The origin is taken at its centre of mass, defined as the point with respect to which the dipole component of the gravitational potential of the earth vanishes (eq. (2.24)). Were the surface of the earth rigid, it would be straightforward to define with respect to it three orthogonal directions; when a high accuracy is desired the relative plate motion must be taken into account and an "ideal" crust must be defined by averaging out this motion (Sec. 5.5).

In order to be able to use the observations relative to this frame, we must know its absolute acceleration and absolute rotation. The adjective "absolute" historically dates back to the newtonian conception of dynamics; but in an operational sense, it refers to the fact that the equations of motion in a rotating and accelerated frame contain apparent forces (see eq. (3.29). A cartesian frame of reference is *absolute*, or *inertial* if it has no apparent forces. As E. Mach has pointed out in his seminal work *Die Mechanik in ihrer Entwicklung historisch-kritisch dargestellt* (1893; English translation published by Open Court, 1960 as

The Science of Mechanics: A Critical and Historical Account of its Development), this formulation is epistemologically inconsistent, since it requires an arbitrary splitting of the force into a "real" part and an "apparent" part; in other words, one should not formulate dynamical laws and at the same time use them to determine the inertial frames. We need a precise, kinematical rule, independent of dynamical considerations, to define "absolute" rotation and "absolute" acceleration. There are no absolute kinematical quantities: motion can be described only relatively to other bodies, for example, distant matter in the universe. If this is not done, the formulation of dynamical laws is, according to Mach, incomplete.

Therefore, we should say that a cartesian frame of reference is inertial if its acceleration and its rotation with respect to the most distant matter in the universe vanishes. This definition raises a theoretical and an experimental problem. Theoretically, the prescription "with respect to the distant matter" should be made more precise, stating, for example, how the reference objects are weighed according to their distance. We know, for example, that stars do not provide a good reference because they share the rotation of the galaxy. Its period ($\approx 10^8$ y) gives an angular speed of about 0.01 arcsec/y, which has observable dynamical consequences. In practice, the best kinematical determination of the absolute rotation vector of the earth is done with respect to distant quasars by means of Very Long Baseline Interferometry (VLBI) (see Sec. 20.5).

The experimental problem is, how accurately do we need to measure rotation? This is determined by the accuracy with which we can measure in a local dynamical system the apparent forces produced by an "absolute" rotation. These measurements can be done most precisely with the moon or with the satellite LAGEOS (see Sec. 20.2); it turns out that the smallest detectable Coriolis force corresponds at present to an angular velocity of about 2 mas/y $\cong 10^{-8}$ rad/y. The accuracy of the VLBI frame is slightly better. It is interesting to note that this limit is much larger than Hubble's constant $H \approx 10^{-10}$/y; even if the universe had a differential rotation rate with respect to the local dynamical frame of the same order of magnitude as its expansion rate H, it would still go undetected. Indeed, from the isotropy of the cosmic background radiation it can be concluded that the upper limit to the differential rotation of the universe is much less than H.

We do not have relevant measurements of the linear acceleration of the earth with respect to the universe (We can measure only the relative radial velocity with respect to galaxies, not its time derivative.) However, the effect of an absolute acceleration – independent of position – upon local dynamics, is nil; only tidal forces, arising from acceleration gradients, have an effect.

The absolute rotation of the earth-bound frame is determined by the instantaneous rotation vector $\boldsymbol{\omega}(t)$; $2\pi/\omega(t)$ is the *length of the day* (LOD). The angular velocity of the earth suffers a secular deceleration due to the torque exerted by the moon (see Ch. 4)

$$d\omega/dt = -\,6\cdot10^{-22}\ \mathrm{rad/s^2}\,, \qquad (3.4)$$

corresponding to a quadratic correction in the phase of the earth φ

$$\delta\varphi = -\,3\cdot10^{-22}\ \mathrm{rad}(t/s)^2\ . \qquad (3.4')$$

Because of this effect, the length of the day increases in a year by about $2\cdot10^{-5}$ s, an amount which is not negligible if its accumulation over millennia is considered. In addition, we have changes in the length of the day over shorter time scales due to moving masses within and on the earth, in particular the oceans and the atmosphere.

To describe the motion of the instantaneous axis of rotation we must introduce the astronomical frame of reference on the celestial

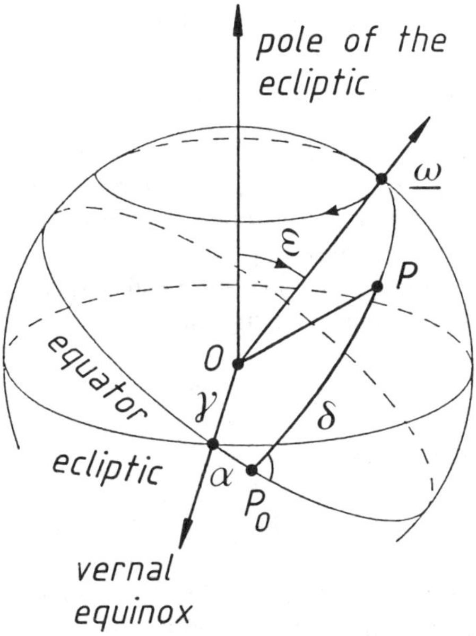

Fig. 3.3. The equatorial frame of reference on the celestial sphere: the right ascension α and the declination δ of a direction OP are defined with respect to the instantaneous equator at a given, conventional epoch. The precession moves the axis of rotation on a circular cone whose axis is the pole of the ecliptic, and whose aperture is specified by the obliquity angle ϵ.

sphere. The equator and the plane of the ecliptic (defined as the instantaneous orbital plane of the earth) trace two great circles on the sky. The position of these circles at a given epoch (say, 1950) defines the celestial coordinates of any astronomical object P (Fig. 3.3). The right ascension is the angle between the projection P_0 of P on the equator and the *vernal equinox* (or γ *point*), one of the two intersections between the equator and the ecliptic. The declination is the angle PP_0.

The main motion of the pole of rotation in the sky is the *lunisolar precession*, a slow rotational motion of the axis around the pole of the ecliptic, with a period of about 26,000 y. As a consequence, the equinox moves westward and the equatorial coordinates of a fixed star change. This motion is due to the torque exerted by the moon and the sun upon the oblate earth (Sec. 4.1). The angular speed of the rotation axis (*constant of precession*) has the value

$$\Omega_{pr} = 50.29''/\mathrm{y}.$$

There is also a secular change in the angle ϵ between the two great circles (*obliquity of the ecliptic*), due to planetary perturbations:

$$d\epsilon/dt = -\,0.468''/\mathrm{y}\ .$$

In addition to this secular motion, there are smaller and faster variations (*nutations*).

Even in the absence of external torques the angular velocity of the earth is not constant, due to to its oblateness. What is constant is its angular momentum **L**; $\boldsymbol{\omega}$ moves around **L** with an approximate period of 14 months (*Chandler's wobble*; see Secs. 3.5 and 3.6). This motion is best described in a frame tied to the ideal, rigid earth, whose axes $\mathbf{e}_x$, $\mathbf{e}_y$, $\mathbf{e}_z$ rotate with the angular velocity $\boldsymbol{\omega}(t)$:

$$d\mathbf{e}_x/dt = \boldsymbol{\omega} \times \mathbf{e}_x,\ d\mathbf{e}_y/dt = \boldsymbol{\omega} \times \mathbf{e}_y,\ d\mathbf{e}_z/dt = \boldsymbol{\omega} \times \mathbf{e}_z.$$

The intrinsic components are

$$\omega_x = \boldsymbol{\omega} \cdot \mathbf{e}_x,\quad \omega_y = \boldsymbol{\omega} \cdot \mathbf{e}_y,\quad \omega_z = \boldsymbol{\omega} \cdot \mathbf{e}_z\ .$$

Since $\boldsymbol{\omega}(t)$ differs very little from a constant vector $\boldsymbol{\omega}_0$, the unit vectors **e** are the sum of a main part rotating at the frequency ω_0 and a perturbation $\delta\mathbf{e}$ fulfilling

$$d\delta\mathbf{e}/dt - \boldsymbol{\omega}_0 \times \delta\mathbf{e} = \delta\boldsymbol{\omega} \times \mathbf{e}_0\ . \qquad (3.5)$$

A component of $\delta\boldsymbol{\omega}$ sinusoidal at the frequency σ_n in the right-hand side of eq. (3.5) drives $\delta\mathbf{e}$ at the frequency ($\omega_0 \pm \sigma_n$); hence the intrinsic components have the form

$$\omega_x(t) = \Sigma_n \omega_{xn} \exp[i(\omega_0 \pm \sigma_n)t] \ , \text{ etc.}$$

In the inertial frame we have therefore a set of eigenfrequencies very near the day frequency ω_0, within a band of the order of Chandler's frequency $\sigma = 2\pi/(14$ months); conversely, a very low frequency component of the inertial angular velocity results in a body-fixed component very near ω_0.

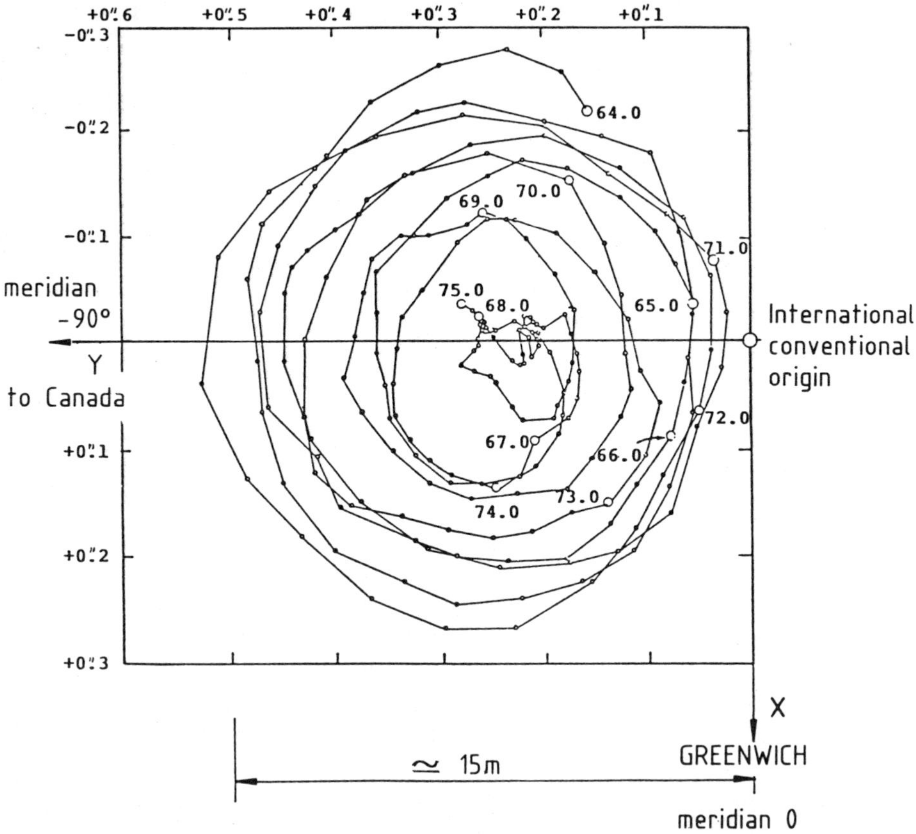

Fig. 3.4. The motion of the axis of instantaneous rotation near the north pole. The graph is constructed with 22 points per year, but data at 5 day intervals and, recently, even more frequently are available.

Traditionally the length of the day was measured with accurate timings of the passage of a star through a given direction. The orientation of the polar axis with respect to the earth determines the colatitude of a given place (the angle with the local vertical); to find the latitude instruments were used, like the *zenith tube,* capable of determining the angle between the zenith and a given star. The mean of its values at an interval of 12 hours is the latitude. To get the position of the pole several stations are used.

The determination of the polar vector with respect to an inertial frame is a branch of astrometry and requires a precise positioning of stars and planets. Planetary astrometry determines with respect to the planetary frame the equator and the ecliptic as follows. The planets move around the sun in keplerian orbits, whose perturbations can be accurately computed with newtonian dynamics in an inertial frame. If their angular positions in the sky were determined relative to a rotating (stellar) frame, discrepancies would arise with the predictions. We can therefore determine *dynamically* the rotation vector of the earth with a best fit of the observations to the theory. This method has provided in historical times an accurate determination of the constant of precession.

The stars in the solar neighbourhood exhibit the superposition of two motions. Random proper motions are present, but they can be averaged out locally if sufficient stars are available, and reduced by considering stars further away. We have also a collective rotational motion around an axis orthogonal to the galactic equator, due to the (differential) rotation of the galaxy with a period of about 10^8 y. Therefore any determination of the rotation of the earth relative to the stars in fact is not absolute. However, note that the galactic equator is known (to within about 1 mrad); if it is fixed, the two components of the earth rotation vector orthogonal to the galactic axis can be determined absolutely.

Another dynamical method uses precise measurements of distance or velocity of earth-bound satellites (natural and artificial). In absence of perturbations the normal to the orbital plane of an earth satellite and its perigee determine an absolute cartesian triad. To determine this frame we must accurately know the pertubations and determine the rotational position of the earth with respect to the actual triad. This can be done with a precision comparable to VLBI with the satellite LAGEOS (see Sec. 20.2).

Finally, as said earlier, a direct, kinematical determination of polar motion and LOD is currently done with VLBI (see Sec. 20.5). The refinement and the comparison between the results of these four methods is an important current field of research, with many applications, also of practical nature.

3.3. SLOW ROTATION OF A FLUID BODY

We would like to determine the shape and the gravitational potential of a rotating liquid body, characterized by a scalar pressure. When the rotation is uniform, the problem is best set in the rotating frame, where it is reduced to the effect of the centrifugal potential energy (per unit mass)

$$U_C = -\tfrac{1}{2}\omega^2 r^2 \sin^2\theta , \qquad (3.6)$$

to be added to the gravitational potential energy U (eq. (2.20)). θ is the colatitude. Of course, for a dynamical theory of motion with respect to the rotating earth, Coriolis' force must be taken into account as well; we shall consider in Ch. 7 its effects on atmospheric and oceanic dynamics. Here we confine ourselves to the equilibrium distortion when the rotation is slow; in other words, we look for a perturbation in the gravitational potential energy linear in the forcing energy U_c, corresponding to small deformations. This approximation is made rigorous by means of the small parameter

$$\mu = \omega^2 R^3 / GM , \qquad (3.7)$$

of the order of the ratio of rotational to gravitational energy.

U_c can be written also in the form

$$U_C = \frac{\mu GM}{3R}\left(\frac{r}{R}\right)^2\left(1 - \frac{3\cos^2\theta - 1}{2}\right) = \frac{\mu GM}{3R}\left(\frac{r}{R}\right)^2 (1 - Y); \qquad (3.6')$$

that is to say, its expansion in spherical harmonics has only components in the vector spaces S_0 and S_2, belonging to $\ell = 0$ and $\ell = 2$. Because of the axial symmetry, the $\ell = 2$ component contains only the $m = 0$ term

$$Y = Y_{20} = (3\cos^2\theta - 1)/2. \qquad (3.8)$$

The perturbation δU in the gravitational potential energy induced by U_c can also be expanded, for a fixed r, in spherical harmonics. Which components of this expansion differ from nought? To address this question, consider a generalization in which the external forcing potential is an arbitrary quadratic function of the coordinates, corresponding to a vector in S_0 and a vector in S_2. The perturbation δU in the self-potential produced by the displacement of masses is, outside the body, a harmonic function determined by a $(2\ell' + 1)$-dimensional vector for each $S_{\ell'}$ space. With our assumption of a small perturbation we have linear mappings from S_0 onto $S_{\ell'}$ and from S_2 onto $S_{\ell'}$. These mappings are solely determined by the intrinsic properties of the body in the unperturbed state; they do not contain any privileged direction, but depend only upon scalar quantities (like the elastic coefficients). Such mappings and the matrices which determine them must therefore be invariant, that is to say, their numerical values be independent of the coordinate system. Now it can be shown that a linear invariant mapping from S_ℓ onto $S_{\ell'}$ exists only if $\ell = \ell'$; and then it is a simple proportionality relation. This is a particular case of the *Clebsch-Gordon theorem* and hinges upon the crucial fact that the spaces S_ℓ are irreducible under the rotation group. We therefore conclude that δU also has components in S_0 and S_2 only and that they are proportional to the corresponding components of U_c.

Outside the body U is harmonic and axially symmetric. Therefore it must be of the form (see eq. (2.22):

$$U = -\frac{GM}{r}\left[1 - J_2\left(\frac{R}{r}\right)^2 Y\right], \tag{3.9}$$

with the constant J_2 to be determined as a function of μ. The quadrupole correction is sufficient to describe the effect and no higher order terms are present. The total external potential energy per unit mass is

$$U_T = U + U_C = -\frac{GM}{r}\left[1 - J_2\left(\frac{R}{r}\right)^2 Y\right] - \frac{\mu GM}{3R}\left(\frac{r}{R}\right)^2 (1 - Y). \tag{3.10}$$

The gravity acceleration is almost radial and reads (in modulus)

$$g = \frac{\partial U_T}{\partial r} = \frac{GM}{r^2} - 3J_2\frac{GMR^2}{r^4} Y - \frac{2\mu}{3}\frac{GM}{R^3} r(1 - Y).$$

On the surface of the body, deformed from $r = R$ to $r = R + \delta r(\theta)$, there is a small change in g:

$$\delta g(\theta)/g = - 2\delta r(\theta)/R - 3J_2 Y - 2\mu(1 - Y)/3, \tag{3.11}$$

where we have neglected all the terms quadratic in small parameters. The surface of the body is determined by the condition of being an equipotential for U_T (eq. (1.25)); neglecting non-linear terms we have

$$\delta r/R + J_2 Y - \mu(1 - Y)/3 = \text{const}. \tag{3.12}$$

The constant on the right-hand side is not determined yet, because no precise definition was given for the unperturbed radius R. We supplement it by requiring that R to be the mean radius; in other words, the average of δr over the sphere vanishes. Since Y also has a vanishing average, the constant is $-\mu/3$ and the condition becomes

$$\delta r/R + (J_2 + \mu/3)Y = 0. \tag{3.12'}$$

Combining (3.11) and (3.12') we can eliminate J_2 and obtain a condition which refers only to quantities measurable on the ground:

$$\delta g/g - \delta r/R = \mu(5Y - 2)/3. \tag{3.13}$$

Taking the difference between the poles and the equator, we get the ellipticity (1.14') as a function of the gravity anomaly:

$$(g_p - g_e)/g + (R_e - R_p)/R = 5\mu/2. \tag{3.14}$$

This relation is due to *Clairaut.*

To go further and to determine J_2, the internal density must be specified. With a uniform density (as for a liquid) δU must be harmonic also inside the body, since the unperturbed potential (1.5) takes care of the internal mass; but then the known angular dependence determines the radial dependence. We cannot have a function of the type (2.20), which diverges at the origin; we must use instead the class (2.21) of harmonic functions. Therefore, for $r < R$,

$$U = -\frac{GM}{R}\left[\frac{3R^2 - r^2}{2R^2} - a\left(\frac{r}{R}\right)^2 Y + b\right].$$

The requirement of continuity at $r = R$ requires $a = J_2$ and $b = 0$:

$$U = -\frac{GM}{R}\left[\frac{3R^2 - r^2}{2R^2} - J_2\left(\frac{r}{R}\right)^2 Y\right]. \tag{3.15}$$

Let us note an important consequence of this relation, discovered by Newton. From eq. (3.10) we see that the gravitational potential energy inside the body is a quadratic function of the cartesian coordinates (plus a constant); hence the gravity acceleration is linear in these coordinates. Along the coordinate vectors to the poles and the equator, respectively, we have, therefore, in terms of the surface values g_p and g_e,

$$g = g_p r/R_p, \qquad g = g_e r/R_e;$$

and the work to be performed to bring a unit mass from the centre to the surface in the two cases is, respectively, $\frac{1}{2}g_pR_p$ and $\frac{1}{2}g_eR_e$. On the other hand, the surface being an equipotential, they must be equal:

$$g_eR_e = g_pR_p. \tag{3.16}$$

Therefore the ellipticity (1.14') is determined by the gravity anomaly:

$$\epsilon = 1 - R_p/R_e = 1 - g_e/g_p = 5\mu/4. \tag{3.17}$$

We have made use of Clairaut's relation (3.14). Using eq. (3.12') we get, finally

$$J_2 = \mu/2. \tag{3.18}$$

The centrifugal potential energy U_c (3.6') (or, rather, its part with vanishing mean) generates a deformation potential energy δU (eq. (3.9)); their ratio on the surface is a constant (eq. (2.28))

$$\kappa_S = 3J_2/\mu = 3G(C - A)/\omega^2R^5, \tag{3.19}$$

called *secular Love number*. We have just shown that this number is 3/2 for a liquid earth.

The separate measurements of the three independent quantities ϵ, J_2 and $(1 - g_e/g_p)$ and the comparison with their ideal values as functions

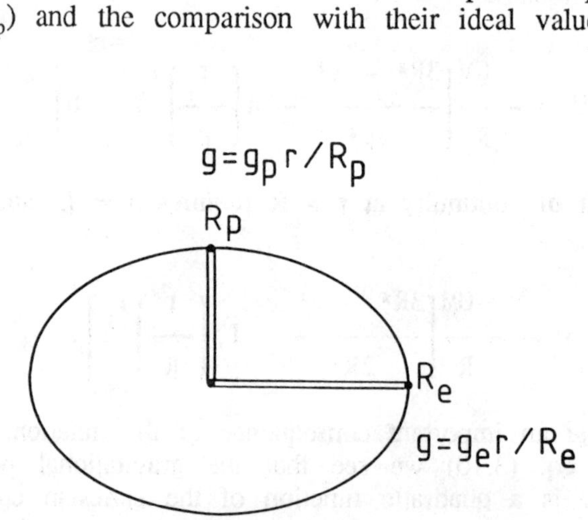

Fig. 3.5. The "canals" method due to I. Newton. One computes the work needed to extract from the centre of the planet a test body along two canals drilled from the pole and from the equator. The work done is the same in the two cases; hence the gravity accelerations are inversely proportional to the radii (eq. (3.16)).

of μ yield valuable information about the internal structure of a cosmic body. A core, always denser than the mantle, is expected to give, for a given μ, a smaller J_2. The decrease in gravity near the equator combines three different effects of the same sign (eq. (3.11)): a larger distance from the centre, the centrifugal force and a smaller amount of mass near a given point on the equator. A denser core will mainly attenuate the third effect, so that there a smaller gravity anomaly is expected. Finally, the ellipticity is a superficial phenomenon and should not be greatly affected by the core. The experimental and the theoretical values of these parameters are shown below for the earth, for which $\mu = 1/290 = 0.00345$.

	J_2	ϵ	$1 - g_e/g_p$
experimental	0.00108	0.00337	0.00529
theoretical	0.00173	0.00431	0.00431

The agreement is remarkable; in particular, Clairaut's relation (3.14) is well verified, an indirect evidence that the differential rotation is small.

*3.4. EQUILIBRIUM SHAPES OF BODIES IN RAPID ROTATION

What happens when the assumption of slow rotation is no longer valid, that is, when the dimensionless parameter μ (and, as a consequence, also ϵ and J_2) is not much smaller than unity? We restrict our analysis to the ellipsoidal shapes of equilibrium of fluid bodies with uniform density ρ (namely, self-gravitating incompressible "liquids"), but the qualitative results hold also in more complex cases. In this case, an obvious solution of Laplace's equation for the gravitational potential energy (per unit mass) U, which generalizes (3.15), is still a quadratic function of the coordinates, of the form:

$$U = - \pi G\rho(A_1 x^2 + A_2 y^2 + A_3 z^2). \qquad (3.20)$$

(x, y, z) are cartesian coordinates defined by the principal axes of inertia, with the origin at the centre of the ellipsoid; A_1, A_2, A_3 are dimensionless constants. Following a procedure similar to the one used in Sec. 3.3, it can be shown that this solution fits the continuity requirement for the potential energy at the surface provided that

$$A_1 = abc\int_0^\infty (a^2 + u)^{-3/2}(b^2 + u)^{-\frac{1}{2}}(c^2 + u)^{-\frac{1}{2}}du, \qquad (3.21_1)$$

$$A_2 = abc\int_0^\infty (a^2+u)^{-\frac{1}{2}}(b^2+u)^{-3/2}(c^2+u)^{-\frac{1}{2}}du, \qquad (3.21_2)$$

$$A_3 = abc\int_0^\infty (a^2+u)^{-\frac{1}{2}}(b^2+u)^{-\frac{1}{2}}(c^2+u)^{-3/2}du. \qquad (3.21_3)$$

a, b and c are the semiaxes of the ellipsoid. While in general these integrals can be expressed in terms of elliptic integrals, in particular cases they can be evaluated analytically. For instance, if $a = b = c$, $A_1 = A_2 = A_3 = 2/3$ corresponds to a non-rotating sphere (eq. (1.4)); if $a = b \succ c$ we have

$$A_1 = A_2 = \frac{\surd(1-e^2)}{e^3}\arcsin e - \frac{1-e^2}{e^2},$$

$$A_3 = \frac{2}{e^2} - \frac{2\surd(1-e^2)}{e^3}\arcsin e, \qquad (3.21')$$

where $e = \surd(1 - c^2/a^2)$ is the eccentricity of the oblate, axisymmetric ellipsoid. From eq. (3.20) the gravity components are still linear functions of the coordinates; for example, for the x-component

$$g_X = -\,\partial U/\partial x = -\,2\pi G\rho A_1 x\ . \qquad (3.22)$$

To derive the equilibrium shape we can apply again the method of Newton's canals. The work done against gravity to bring the unit mass from the centre to the surface along the three axes is $\pi G\rho A_1 a^2$, $\pi G\rho A_2 b^2$ and $\pi G\rho A_3 c^2$, respectively. If the ellipsoid is spinning about the shortest axis c, we must add in the two former cases the (negative) contribution of the centrifugal force; since the surface is equipotential, the work is the same in the three cases:

$$\pi G\rho A_1 a^2 - \omega^2 a^2/2 = \pi G\rho A_2 b^2 - \omega^2 b^2/2 = \pi G\rho A_3 c^2, \qquad (3.23)$$

For $a = b$, using eqs. (3.21') and $c = a\surd(1 - e^2)$, we obtain a relationship between ω^2 and e (or, equivalently, between ω^2 and c/a):

$$\frac{\omega^2}{\pi G\rho} = \frac{2\surd(1-e^2)}{e^3}(3-2e^2)\arcsin e - \frac{6}{e^2}(1-e^2). \qquad (3.24)$$

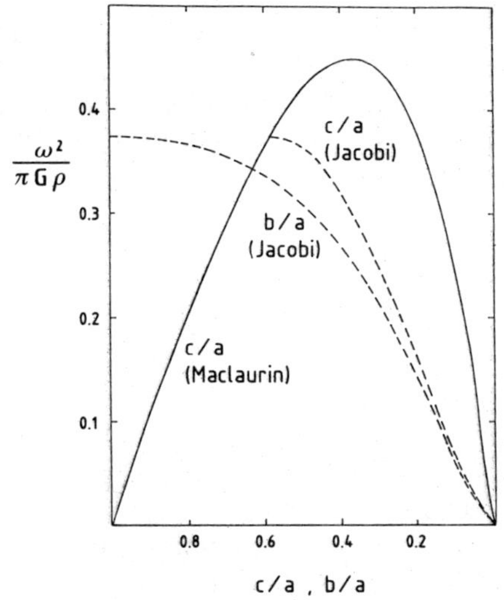

Fig. 3.6. The figure shows the variation of the axial ratios c/a and b/a as a function of the dimensionless parameter $\omega^2/\pi G\rho$ for Maclaurin spheroids (axisymmetrical, i.e. a = b ≻ c; solid line) and for Jacobi ellipsoids (triaxial, a ≻ b ≻ c; dashed lines). c is always assumed to be directed along the spin axis. These ellipsoidal shapes are consistent with equilibrium for a self-gravitating, spinning and "liquid" object.

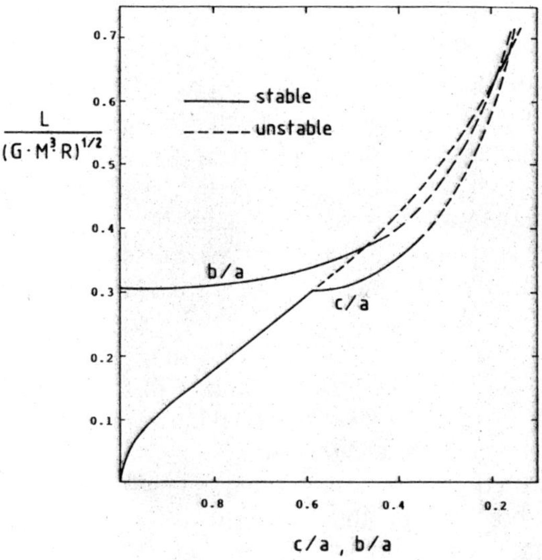

Fig. 3.7. A plot of the angular momentum of rotation L versus the axial ratios c/a and b/a for Maclaurin spheroids (for which b/a is always 1 and c/a ≻ 0.583) and for Jacobi ellipsoids (whose c/a curve bifurcates at c/a ≅ 0.583 from the Maclaurin curve. Dashed lines correspond here to unstable equilibrium figures of both types.

The dependence of ω^2 on c/a is illustrated in Fig. 3.6 (solid line), while Fig. 3.7 shows the dependence on c/a and b/a of the rotation angular momentum L. Notice that for slow rotation $\omega^2/\pi G\rho \cong 4\mu/3$, where μ is given by eq. (3.7), and thus it is proportional to the ratio of rotational to gravitational energies. When $\mu \ll 1$, we can also recover the slow-rotation formula (3.17).

L is given by

$$L = \sqrt{(GM^3 R)}\ \frac{\sqrt{3}}{5} \left(\frac{a}{R}\right)^2 \omega, \qquad (3.25)$$

where $R = (abc)^{1/3}$ is the radius of the sphere with the same volume as the spheroid. We notice that, although L increases monotonically with the flattening (owing to the rapidly growing moment of inertia), $\omega^2/\pi G\rho$ reaches the maximum value of 0.449 for c/a = 0.367 (namely e = 0.930), solution of the equation $d(\omega^2)/de = 0$, i. e.:

$$\frac{2}{e^3}(9 - 2e^2) - \frac{2(9 - 8e^2)}{e^4\sqrt{(1 - e^2)}} \arcsin e = 0. \qquad (3.26)$$

The fact that in the limit $\omega \Rightarrow 0$ we obtain not only a quasi–spherical equilibrium solution but also another, highly flattened one, can intuitively be understood by referring to Newton's method. In the flattened spheroidal solution, gravity is much stronger along the polar axis, and this compensates both for the fact that the equatorial axis is longer and for the centrifugal term. Anyway, it can be shown by linear stability analysis that this sequence of axisymmetrical solutions (called *Maclaurin spheroids*) becomes unstable at e = 0.813 (i. e., c/a = 0.583), corresponding to $\omega^2/\pi G\rho = 0.374$ and $L/\sqrt{(GM^3 R)} = 0.304$. The instability occurs with respect to small deformations which break the axial symmetry of the solutions. Since this instability lies on the ascending branch of the ω^2 versus e relationship, for each value of ω we have only one stable Maclaurin spheroid.

The fact that eq. (3.23) is symmetrical with respect to the exchange $a \leftrightarrow b$ does not mean that a = b is a necessary condition for equilibrium. In fact, as pointed out by Jacobi in 1834, if we assume $a \neq b \neq c$ we obtain from eqs. (3.23) and (3.21) the ω versus shape relationship:

$$\frac{\omega^2}{\pi G\rho} = 2\,\frac{A_1 a^2 - A_2 b^2}{a^2 - b^2}, \qquad (3.27)$$

together with a geometric condition, which yields b/a if c/a is known, or *vice versa*:

$$\frac{a^2 b^2}{a^2 - b^2}(A_2 - A_1) = c^2 A_3 . \qquad (3.28)$$

These equations represent a sequence of solutions with $a \geqslant b > c$ (the so-called *Jacobi ellipsoids*), which "bifurcates" from the Maclaurin sequence (a = b) just at the point where the latter becomes unstable with respect to non-axisymmetrical deformations. Figs. 3.6 and 3.7 show also how the axial ratios c/a and b/a vary as a function of ω^2 and L along the Jacobi sequence. The existence of triaxial equilibrium shapes can be understood in the same way as that of slowly rotating, very flattened ones: for a cigar-shaped body, gravity is much more intense along the b and c axes, compensating for the smaller centrifugal contribution. Along the Jacobi sequence, as the axial ratios decrease and the angular momentum grows, the spin rate becomes slower and slower, owing to the rapid increase of the moment of inertia.

The linear stability analysis can be applied also to Jacobi ellipsoids, for instance by exploring what happens when a small, "pear-shaped" deformation is introduced. As shown by Poincaré in 1885, such deformations cause instability when $b/a < 0.432$, $c/a < 0.345$, $\omega^2/\pi G\rho < 0.284$, $L/\sqrt{(GM^3 R)} > 0.390$. Thus, no stable ellipsoidal equilibrium shape is possible for a homogeneous, self-gravitating body when its angular momentum of rotation exceeds the threshold $0.390 \cdot \sqrt{(GM^3 R)}$; for higher angular momenta, only binary or multiple configurations, with a fraction of the angular momentum accounted by the orbital motion, may be consistent with equilibrium. Therefore, when a nearly homogeneous self-gravitating body receives an input of angular momentum (e. g., by an off-centre impact) such that the above-mentioned threshold is exceeded, fission into two or more components is the expected outcome. We shall discuss in Ch. 14 some applications of these results to the origin of some solar system bodies. For this purpose, it is important to note that the same qualitative features we have found by assuming uniform density (transition from axisymmetric to triaxial solutions, fission instability, etc.) are preserved when more general (and realistic) assumptions are used. Similar results have indeed been derived both when the density changes in such a way that the equidensity surfaces are still ellipsoids, with variable eccentricity, and when the internal structure is consistent with hydrostatic equilibrium and a polytropic (eq. (1.53)) relationship between pressure and density.

3.5. RIGID-BODY DYNAMICS: FREE PRECESSION

Were the earth perfectly rigid, its moments of inertia would be constant and the principal axes would provide a good coordinate system attached to the body. In reality their directions change slightly with time and depend upon the internal structure. For fine work (in particular, when the plate motion has to be taken into account), more complicated procedures are needed to define a body-fixed frame; the rigid earth approximation, however, is a fundamental and simple model, prerequisite to any further step. We confine ourselves to this model.

In the body-fixed frame Euler's equations for the angular velocity $\boldsymbol{\omega}$ read:

$$\begin{aligned} A\dot{\omega}_x + (C - B)\omega_y\omega_z &= \Gamma_x \\ B\dot{\omega}_y + (A - C)\omega_z\omega_x &= \Gamma_y \\ C\dot{\omega}_z + (B - A)\omega_x\omega_y &= \Gamma_z . \end{aligned} \tag{3.29}$$

The quadratic terms in the angular velocity are due to the Coriolis force. When the external torque $\boldsymbol{\Gamma}$ vanishes there are three fundamental solutions in which only one of the three components of the angular velocity differs from zero and is constant in time. This result, obtained by Maxwell, indicates that a body is free to rotate around a principal axis of inertia; however, the rotation around the axis corresponding to the middle moment of inertia is unstable.

When the body is axially symmetric ($A = B$) it is termed *oblate* if $C > A$ and *prolate* if $C < A$. In this case ω_z is a constant and the other two components fulfil

$$A d\omega_x/dt + (C - A)\omega_y\omega_z = 0 \tag{3.30$_1$}$$

$$A d\omega_y/dt + (C - A)\omega_x\omega_z = 0 . \tag{3.30$_2$}$$

One easily sees that both ω_x and ω_y are harmonically oscillating; it is convenient at this point to introduce the complex dimensionless variable

$$m = (\omega_x + i\omega_y)/\omega_z . \tag{3.31}$$

The solution of eqs.(3.30) is a uniform rotation with the *free precession frequency* σ:

$$m = m_0 \exp(i\sigma t), \ \sigma = \omega_z(C - A)/A. \tag{3.32}$$

Here m_0 is an arbitrary complex number, giving amplitude and phase of the oscillation. The vector (ω_x, ω_y) describes a uniform circular motion, with constant radius $\sqrt{(\omega_x^2 + \omega_y^2)}$; hence ω is a constant also. As a consequence, the axis of rotation describes a circular cone about the axis of symmetry; the motion is prograde in the usual case of an oblate body with $C > A$. In the case of the earth, the aperture of the cone is about 0.2 arcsec and m has an amplitude $\approx 10^{-7}$, corresponding to about 6 m in distance at the poles (see Sec. 3.2 and Fig. 3.4). From the expression (2.28) of J_2 and the numerical value

$$A = 0.329 \cdot MR^2 , \qquad (3.33)$$

we get

$$(C - A)/A = J_2/0.329 = 1/305 ,$$

corresponding to a precession period of 305 days. In reality the motion of the pole around the axis of symmetry is not quite circular and its main periodicity is about 430 days. These discrepancies are due to the fact that the earth is not perfectly rigid (see Sec. 3.6).

The motion of an axially symmetric body in an inertial frame can be interpreted with a geometrical construction due to Poinsot. Let $\mathbf{e}_x$, $\mathbf{e}_y$ and $\mathbf{e}_z$ be the unit vectors along the principal axes of the body ($\mathbf{e}_x$ and $\mathbf{e}_y$ are determined to within a rotation). The (constant) angular momentum

$$\mathbf{L} = A(\omega_x \mathbf{e}_x + \omega_y \mathbf{e}_y) + C\omega_z \mathbf{e}_z = A\boldsymbol{\omega} + (C - A)\omega_z \mathbf{e}_z \qquad (3.34)$$

is decomposed in two vectors of constant size, separated by a constant angle. Hence the vectors $A\boldsymbol{\omega}$ and $(C - A)\omega_z \mathbf{e}_z$ are rigidly connected to $\mathbf{L}$ and their loci are, respectively, two cones (called *a* and *b*) having $\mathbf{L}$ as axis and semi-apertures α and β determined by (see eq. (3.31))

$$\cos(\alpha + \beta) = \omega_z/\omega = (1 + |m_0|^2)^{-\frac{1}{2}} , \qquad (3.35)$$

$$\cos\beta = L_z/L = C\omega_z [A^2\omega^2 + (C^2 - A^2)\omega_z^2]^{-\frac{1}{2}} . \qquad (3.35')$$

If $C > A$, the vectors $\boldsymbol{\omega}$ and $\mathbf{e}_z$ lie on opposite sides of $\mathbf{L}$.

In its motion the body – on which the z-axis is fixed – rotates around the instantaneous axis of rotation $\boldsymbol{\omega}$. This motion is described by constructing the cone *c* having $\mathbf{e}_z$ as axis and containing $\boldsymbol{\omega}$; and by letting such cone roll without skidding around *a*. In fact, this motion maintains the required congruence of the triangle and consists indeed of successive infinitesimal rotations around the changing axis $\boldsymbol{\omega}$. This axis, the common generatrix of *a* and *c*, rotates around $\mathbf{L}$ in the same sense as the main rotation. Seen from the body, instead, the cone *c* is fixed and the cone *a* rolls on it without skidding with the angular velocity σ, dragging around the angular momentum and the rotation axis. In the usual case $C > A$ this motion occurs in the same sense as the main rotation; when $C < A$ (prolate body) the sense is opposite.

In the case of a slightly deformed oblate body, with

$$\delta = (C - A)/C \ll 1 , \qquad (3.36)$$

we see from eq. (3.35) that $\cos\beta$ and $\cos(\alpha + \beta)$ differ from each other by a small term of order δ; hence $\alpha = O(\delta)$, while β is finite. This shows that the rotations involved in the problem fall into two classes: the fast class, corresponding to the rolling of the big cone *c* on the small cone *a* (approximately the daily rotation); and the slow class, corresponding to the slow drift of the axis of the cone *a* due to its (fast) rolling on *c*.

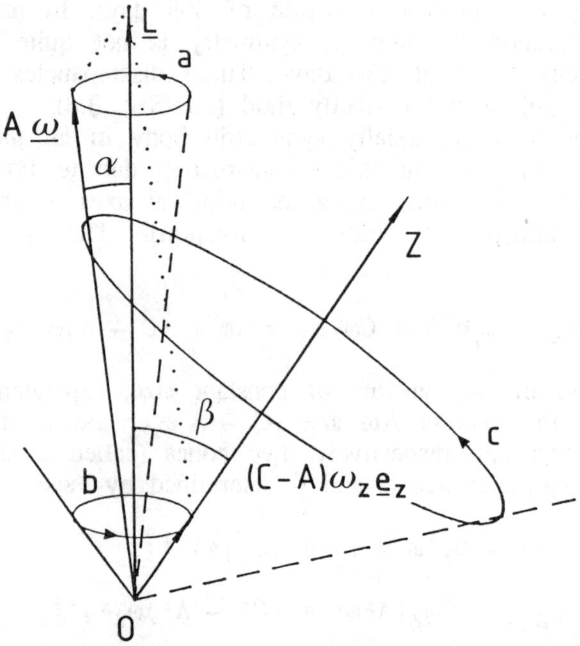

Fig. 3.8. The geometrical disposition of the symmetry axis, the angular momentum and the angular velocity for an oblate body. In an inertial frame the parallelogram with sides $A\underline{\omega}$ and $(C - A)\omega_z e_z$ rotates rigidly around its diagonal **L**, which is fixed in space. In a body-fixed frame the cone *c* rotates without skidding along the cone *a*, dragging around the principal axis of inertia.

Fig. 3.8 shows also the quantitative aspect of the motion. In the inertial frame the motion of the cone *c* is the superposition of two rotations: one with angular velocity $-\sigma = -\omega_z\delta$ along z, corresponding to the precessional motion in the body-fixed frame; and one with angular velocity vector $(\omega + \omega_z\delta e_z) = \mathbf{L}/A$. This is also the (large) angular velocity with which the figure rotates around **L**. In the body-fixed frame, the cone *a* has an instantaneous velocity $-\boldsymbol{\omega}$, sum of the velocity $\sigma = \omega_z\delta$ along z and $-$ **L**/A. The plane (L, ω) rotates around z with angular velocity σ.

Given δ and the arbitrary ratio ω_z/ω, equations (3.35) determine α and β. Let us solve them when δ is small. We have

$$\cos\beta = \frac{\omega_Z}{\omega}\left[1 + \delta\left(1 - \frac{\omega_Z{}^2}{\omega^2}\right) + O(\delta^2)\right]$$

and

$$\alpha = \frac{\omega_Z}{\omega}\left[1 - \frac{\omega_Z{}^2}{\omega^2}\right]^{\frac{1}{2}}\delta . \qquad (3.37)$$

In the case of the earth we have, in addition, $(\alpha + \beta) = 0.2$ arcsec $\cong 10^{-6} \ll 1$, so that the amplitude of the precession

$$\beta = \surd[2(1 - \omega_Z/\omega)] + O(\delta) = |m_0| + O(\delta) \qquad (3.38)$$

is small as well, and α is even smaller:

$$\alpha = \beta\delta = 3\cdot 10^{-9} \ll \beta . \qquad (3.38')$$

The axis of instantaneous rotation, as seen from an inertial observer, describes in about a day a small circle near the pole with radius $3\cdot 10^{-9}\cdot 6.4\cdot 10^{8}$ cm $\cong 2$ cm.

From Fig. 3.8 one can also derive the precessional velocity σ. Consider first the inertial frame. After a unit time the axis of symmetry S and the rotation axis R advance to S' and R' in such a way that, on the unit sphere, $SS'/\beta = RR'/\alpha = \omega'$, that is the common angular velocity with respect to **L**. Since the angles are small, we use the definition of angular velocity in a plane. The axis of symmetry S rotates around R with angular velocity $\omega = SS'/(\alpha + \beta)$. In the body-fixed frame, similarly, $RR'/(\alpha + \beta) = \sigma$ is the angular velocity of R with respect to S. Combining these two relations we obtain again

$$\sigma = \omega\alpha/\beta = \omega\delta. \qquad (3.39)$$

The angular velocity ω' of the rotation axis with respect to L is

$$\omega' = RR'/\alpha = \omega(1 + \delta). \qquad (3.39')$$

*3.6. THE ROTATION OF THE EARTH AND ITS INTERIOR STRUCTURE

The earth is not perfectly rigid and its rotational motion differs from the previously described mathematical model (see Fig. 3.4), the main difference being that the precession frequency σ is not $\omega\delta$, but about

$$\sigma = \omega\delta/f = \omega\delta/1.42 \ , \qquad (3.40)$$

corresponding to a period of 430 instead of 305 days. The motion of the pole is also somewhat irregular.

This increase in the period of Chandler's wobble is due to the fact that, owing to the (small) misalignment between the rotation axis and the axis of symmetry, there is a component of the centrifugal force perpendicular to the (mean) equator, reaching a maximum in the plane (ω, z). As a consequence, the oblateness swelling is slightly tilted in an S-shaped configuration; the centrifugal force, acting on the asymmetrical, displaced mass, produces a torque whose sign is in the direction of a diminishing amplitude β of the precession. The final effect of this gyroscopic torque is just to slow down the precession. The parameter $f = \sigma/\omega\delta$ can therefore provide information about the elastic coefficients of the earth.

To get a simple model of this effect, let us start with a general result of gyroscopic dynamics. Assume that a torque

$$\boldsymbol{\Gamma} = \Gamma(\omega_y, -\omega_x, 0)/\sqrt{(\omega_x^2 + \omega_y^2)} \quad (\Gamma > 0) \qquad (3.41)$$

acts on the body along a direction orthogonal to the (ω, z) plane in such sense as to decrease the angle β. We also assume $\beta \ll 1$, so that Euler's equations (3.29) read

$$A d\omega_x/dt + (C - A)\omega_y\omega = \Gamma\omega_y/\omega\beta \qquad (3.42_1)$$

$$A d\omega_y dt - (C - A)\omega_x\omega = -\Gamma\omega_x/\omega\beta \ . \qquad (3.42_2)$$

We see that the torque is formally equivalent to a decrease in the oblateness $\delta = (C - A)/C$, so that the new precession frequency is

$$\sigma = \omega\delta - \Gamma/A\omega\beta \ . \qquad (3.40')$$

Therefore from the value of f we can deduce

$$\Gamma = A\omega^2\beta\delta(1 - 1/f) \ . \qquad (3.43)$$

We now establish, in order of magnitude, the connection between the torque Γ and the elastic deformation and thereby estimate the rigidity of the earth. For simplicity, neglect all factors of proportionality of order unity. The centrifugal force $F_c = M\omega^2 R$ has a component βF_c normal to the equator and produces a shear stress $\beta F_c/R^2$ and a strain (eq. (1.33))

$$e = \beta F_c/\mu R^2 = \beta M\omega^2/\mu R \ , \tag{3.44}$$

where μ is the rigidity (see eq. (1.35)). This component is proportional to the cosine of the angular distance in the equatorial plane from the (ω, z) plane, being negative on the side of $\boldsymbol{\omega}$. Its average over a precession period vanishes; we are concerned here with the short-term response to this changing force. Over geological times, the earth can be regarded as fluid and will take up under the action of the mean centrifugal force, a figure of equilibrium of the kind described in Sec. 3.3, always lying in the (x, y) plane.

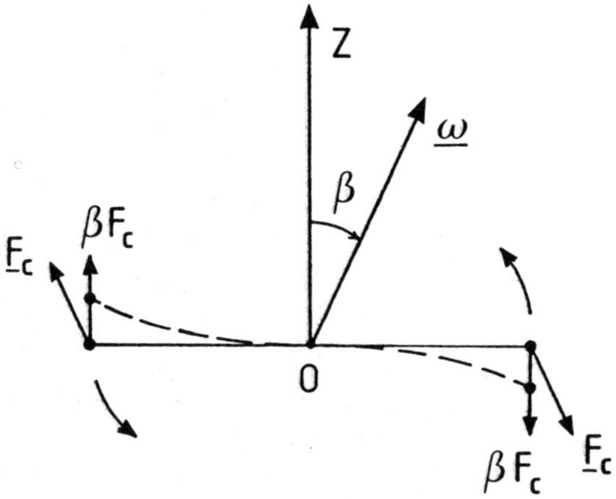

Fig. 3.9. The effect of finite rigidity on Chandler's wobble: the component of the centrifugal force F_c normal to the mean equator tilts the oblateness swelling in an S-shaped form.

Eq. (3.44) also gives the angular distortion of the equatorial swelling. This distortion produces on the surface a net redistribution of mass of order $Me\delta$, upon which the centrifugal force acts producing the torque (see Fig. 3.10)

$$\Gamma \approx Me\delta\omega^2 R^2 = M^2\omega^4 R\delta\beta/\mu \ . \tag{3.45}$$

From eq. (3.43) we find then

$$1 - 1/f = M\omega^2/R\mu. \tag{3.46}$$

It is interesting to note that the result does not depend either on the oblateness δ or the amplitude β. Our rough estimate for the rigidity is

$$\mu \approx fM\omega^2/R(f-1) \cong 1.5\cdot 10^{11} \text{ dyne/cm}^2 . \qquad (3.47)$$

As will be discussed in Ch. 5 (see Fig. 5.5), the earth is indeed fairly rigid above a distance of about 3500 km from the centre, with the rigidity coefficient of the order of 10^{12} dyne/cm^2. Considering the rough method, the result is not bad. Note also that from eqs. (3.44, 47), the angular displacement of the swelling is

$$e = \beta(f-1)/f \cong 3\cdot 10^{-6} .$$

The corresponding surface displacement is about $R\delta e \approx 6$ cm.

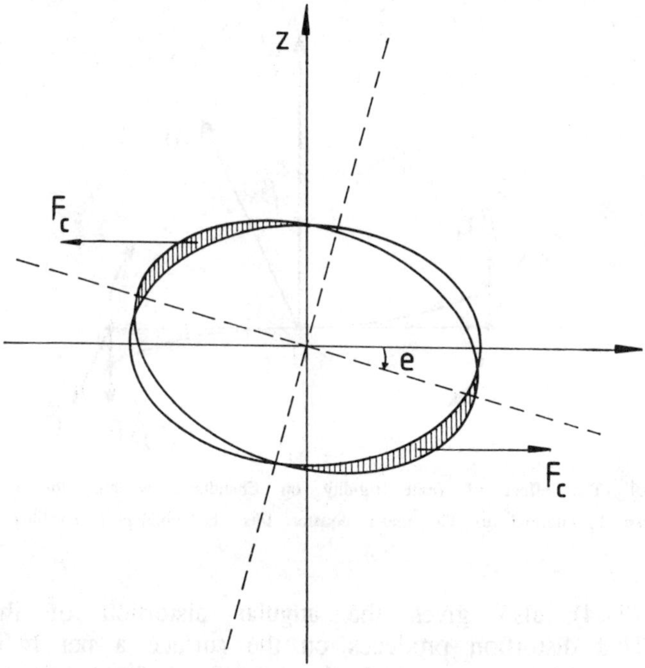

Fig. 3.10. The torque produced by the centrifugal force F_c acting on the tiny elastic deformation of the earth diminishes the free precession rate.

The polar motion is certainly affected also by any differential rotation within the earth; in particular, models have been developed in which the rigid crust is coupled to an inner, solid core with a smaller oblateness, leading also to an increase in the precession period.

The energy of a precessing body is higher than that of a purely rotating configuration. Since the earth has a non-vanishing viscosity, it can be expected that, in absence of energy sources, the precession will slowly dissipate away. It is therefore important to assess the power needed to maintain the observed amplitude β of Chandler's wobble. To estimate the wobble energy we consider, besides the real configuration with angular momentum (eq. (3.34))

$$L = C\omega(1 - \delta\beta^2) , \tag{3.48}$$

an axially symmetric configuration with the same angular momentum (along z) and therefore with angular velocity

$$\omega_0 = \omega(1 - \delta\beta^2) . \tag{3.49}$$

The internal dynamics conserves the angular momentum and therefore will tend to make the energy difference E_w between the two configurations as small as possible. We therefore appropriately call this difference the wobble energy:

$$E_W = \tfrac{1}{2}C\omega^2\beta^2\delta . \tag{3.50}$$

For the earth this is about $10^{-10}/300 \cong 3\cdot10^{-13}$ times the rotational energy

$$\tfrac{1}{2}C\omega^2 = 2.14\cdot10^{36}\ \text{erg}. \tag{3.51}$$

The dissipation of the wobble occurs because the equatorial swelling is deformed periodically with the precession frequency σ. A dissipative material, in absence of external forces, will damp a periodic oscillation in a time Q/σ; Q is the *quality factor* of that mode. Its evaluation is uncertain, but, typically, for solid bodies it ranges from $\approx$ 10 to 100. (As we shall see in Sec. 15.3, the earth's response to lunisolar tides is characterized by $Q \approx 20$.) The power required to sustain Chandler's wobble is therefore (eq. (3.50));

$$P = E_W\sigma/Q = \tfrac{1}{2}C\omega^3\beta^2\delta^2/fQ. \tag{3.52}$$

There are indications that earthquakes may provide the appropriate energy source, but this issue and the detailed spectrum of the wobble are still unsolved geophysical problems.

PROBLEMS

3.1. Discuss the parallactic motion of an object in the solar system due to the rotation of the earth (for example, consider an observation of 6 hours centred around midnight, when the source is at the zenith of the observer).

*3.2. Using vectorial notation, find out the position of the sun in the sky at a given latitude. Assume a spherical earth, a uniform rotation and a uniform revolution. This problem provides the basis for the construction of a sundial.

*3.3. Show that eq. (3.20) for the gravitational potential energy inside a homogeneous ellipsoid (with A_1, A_2, A_3 given by eq. (3.21) or (3.21')) reduces to eq. (3.15), with $J_2 = 0$, for a sphere.

*3.4. The prevailing right-hand circulation of cars and lorries creates a net angular momentum whose change slightly alters the angular velocity of the earth. Construct a rough model to evaluate this effect.

*3.5. As a counterpart of eq. (3.32), describe the motion of an isolated, axially symmetric and rigid body in an inertial frame.

*3.6. Find the position of an object which rotates with the angular velocity

$$\boldsymbol{\omega} = \omega(t)(m_x(t),\ m_y(t),\ 1),$$

in the approximation m_x, $m_y \ll 1$. This is a general description of polar motion. Hint: introduce the *hour angle* φ (such that $d\varphi/dt = \omega$) and work with the complex variable $m = (m_x + i m_y)$.

3.7. Find the expression of the relativistic Doppler effect (3.3) to within terms quadratic in the velocities (see eq. (17.33)).

3.8. A model of the 1964 earthquake in Alaska consists in the displacement by 22 m of a block with mass 10^{23} g. Find the change in LOD it produced if the displacement was along the meridian.

*3.9. Study the polar motion of a rigid earth with small deviations from axial symmetry: $(C - A) \gg (A - B) > 0$.

3.10 Show that eq. (3.24) reduces to (3.17) when $e \ll 1$.

FURTHER READINGS

The classical textbook on rigid rotation is E.J. Routh, *Advanced Dynamics of a System of Rigid Bodies*, Dover Editions (1955). On the earth's rotation an old, but still good book is W.H. Munk and G.J.F. Macdonald, *The Rotation of the Earth. A Geophysical Discussion*, Cambridge University Press (1960). More recent and complete is K. Lambeck, *The Earth's Variable Rotation: Geophysical Causes and Consequences*, Cambridge University Press (1980); see also *Reference Frames in Astronomy and Geophysics* (J. Kovalevsky, I.I. Mueller and B. Kolaczek, eds.), Kluwer (1989). The International Astronomical Union periodically sponsors Colloquia relevant to this topic: D.D. McCarthy and J.D. Pilkington, eds., *Time and the Earth Rotation*, Reidel (1979); E.M. Gaposchkin and B. Kolaczek, eds., *Reference Coordinate Systems for Earth Dynamics*, Reidel (1981); J. Kovalevsky, ed., *Reference Systems*, Reidel (1988). On the subject of Sec.3.4 (figures of rapidly rotating self-gravitating bodies), see S. Chandrasekhar, *Ellipsoidal Figures of Equilibrium*, Yale University Press (1969).

4. TIDAL EFFECTS

The relative acceleration between two nearby bodies produced by another gravitating, far away mass is determined by the gradient of the acceleration caused by the latter; this produces a *tidal acceleration*, proportional to their distance. Tidal processes are ubiquitous in the universe. They also appear when two bodies interact gravitationally and the size of at least one of them is not negligible with respect to their mutual distance. In this case gravity gradients give again rise to tidal forces and torques, that in turn cause deformations and changes in the rotational state of the affected bodies. In many cases tidal deformations are time-dependent, and thus dissipation of energy occurs, owing to violations of Hooke's law of elastic behaviour, viscosity in fluid regions or friction at interfaces. This dissipation is frequently the main source of long-term, accumulating changes in the rotational and orbital parameters of the celestial bodies. Although tidal physics has many astrophysical applications concerning stellar evolution and high energy processes occurring in close binary systems, as well as in galactic dynamics and morphology (close encounters between galaxies are a relatively frequent phenomenon), geophysics and planetary science provide the main arenas for tidal effects. We shall discuss some examples in this Chapter, leaving to Ch. 15 the analysis of tidal evolution processes in the solar system.

Tidal forces are a consequence of the equality between inertial and gravitational mass (see Sec. 17.1). Indeed, only in this case the relative acceleration between two near bodies in an external gravitational potential (per unit mass) U can be written in the "tidal" form, linear in their vectorial displacement:

$$\ddot{\mathbf{r}}_2 - \ddot{\mathbf{r}}_1 = - \nabla U(\mathbf{r}_2) + \nabla U(\mathbf{r}_1) = - (\mathbf{r}_2 - \mathbf{r}_1) \cdot \nabla\nabla U + \ldots$$

4.1. LUNISOLAR PRECESSION

As we have already anticipated in Sec. 3.1, the rotation rate vector $\boldsymbol{\omega}$ of the earth is variable at different time scales and due to disparate reasons. Here we discuss the physical mechanism of the *lunisolar precession*, which changes the direction of $\boldsymbol{\omega}$ in an inertial frame very significantly, but with a very long time scale. On the short run, the precession appears as a drift of $\boldsymbol{\omega}$ (i. e., of the *vernal equinox* or γ point, intersection between the celestial equator and the ecliptic) of about 50"/y – it is easily measured with the techniques of positional astronomy and can even be detected by naked eye observations, as confirmed by the fact that it was already known to ancient greek astronomers.

Let us model the earth as an oblate spheroid whose rotation is affected by the presence of a body (e. g., the sun) moving around it on a circular orbit of radius **R** lying on the ecliptic plane. The force due to the gravitational pull of the sun on a mass element dm of the earth at a geocentric position **r** is

$$d\mathbf{F} = GM_{\odot}dm \frac{\mathbf{R}-\mathbf{r}}{|\mathbf{R}-\mathbf{r}|^3} = GM_{\odot}dm \frac{\mathbf{R}-\mathbf{r}}{R^3}\left[1+\frac{r^2}{R^2}-\frac{2\mathbf{R}\cdot\mathbf{r}}{R^2}\right]^{-3/2}$$

$$= GM_{\odot}dm \frac{\mathbf{R}-\mathbf{r}}{R^3}\left[1+\frac{3\mathbf{R}\cdot\mathbf{r}}{R^2}+0\left[\frac{r^2}{R^2}\right]\right]; \qquad (4.1)$$

since $R \gg r$, we have used the multipole expansion of the force, dropping terms of order 2 and higher in r/R (see Fig. 4.1) (eq. (2.3)). The torque due to $d\mathbf{F}$ is $d\boldsymbol{\Gamma} = \mathbf{r} \times d\mathbf{F}$; integrating over the earth's volume V we obtain:

$$\boldsymbol{\Gamma} = \int_V \mathbf{r}\times d\mathbf{F} = \frac{GM_{\odot}}{R^3}\int_V \mathbf{r}\times\mathbf{R}dm + \frac{3GM_{\odot}}{R^5}\int_V (\mathbf{r}\cdot\mathbf{R})(\mathbf{r}\times\mathbf{R})dm, \quad (4.2)$$

where the first integral in the right-hand side is zero for symmetry reasons, since the origin is at the centre of mass. If we choose an equatorial reference system and call (x, y, z) and (X, Y, Z) the components of **r** and **R**, respectively, the second integral can be easily expressed in terms of the earth's principal moments of inertia (see eq. (2.27)), yielding for the components of $\boldsymbol{\Gamma}$:

$$\Gamma_X = \frac{3GM_{\odot}}{R^5}YZ(C-A), \quad \Gamma_Y = -\frac{3GM_{\odot}}{R^5}XZ(C-A), \quad \Gamma_Z = 0. \quad (4.3)$$

Again the computation has been very much simplified by the fact that the volume integrals with the products xy, xz and yz vanish for a

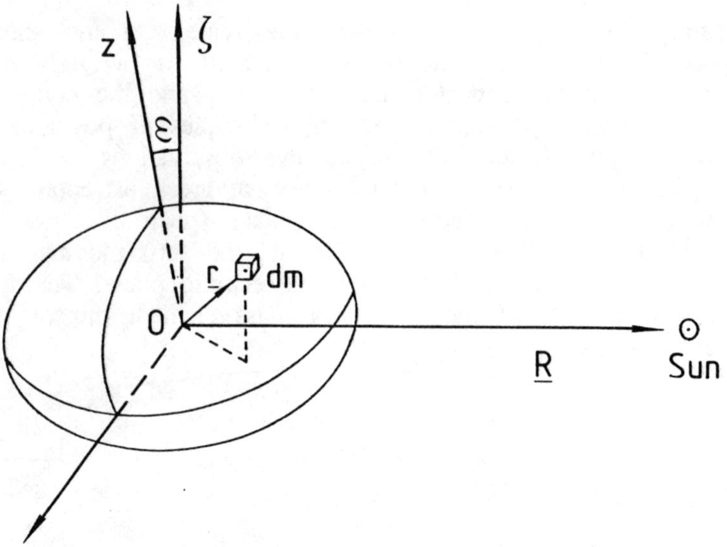

Fig. 4.1. Geometry of the solar torque on the flattened earth.

spheroidal earth. Now let us consider an ecliptic reference frame (ξ, η, ζ) whose ζ axis points to the north ecliptic pole, while the ξ axis points to the vernal equinox. In this frame the sun's position **R** is given in terms of the solar ecliptic longitude λ_0 by $\mathbf{R} = R(\cos\lambda_0, \sin\lambda_0, 0)$. The x and ξ axes coincide; the *obliquity of the ecliptic* ϵ is the inclination of the earth's polar axis with respect to the normal to its orbital plane (Sec. 3.2). Then the x-component of **R** is $X = R\cos\lambda_0$, while the other components are obtained by carrying out a rotation in the yz plane by an angle ϵ:

$$\begin{pmatrix} Y \\ Z \end{pmatrix} = \begin{pmatrix} \cos\epsilon & -\sin\epsilon \\ \sin\epsilon & \cos\epsilon \end{pmatrix} \begin{pmatrix} R\sin\lambda_0 \\ 0 \end{pmatrix} = \begin{pmatrix} R\sin\lambda_0\cos\epsilon \\ R\sin\lambda_0\sin\epsilon \end{pmatrix} . \qquad (4.4)$$

We are interested only in the secular precession, that is to say, on the effect of the torque (4.3) averaged over many years (revolutions of the earth); then from eq. (4.4) we see that $\langle\Gamma_y\rangle = 0$ and

$$\langle\Gamma_x\rangle = \langle\Gamma_\xi\rangle = \frac{3}{2}\frac{GM_\odot}{R^3}(C - A)\sin\epsilon\cos\epsilon . \qquad (4.5)$$

The angular momentum of the earth $\mathbf{L}_\oplus = C\boldsymbol{\omega}$ (along z) and the mean torque $\langle\Gamma\rangle$ (along $\xi \equiv x$) are orthogonal; hence $\mathbf{L}_\oplus$ changes secularly in direction only, and its mean velocity, parallel to $\langle\Gamma\rangle$, always lies in the ecliptic plane and is orthogonal to $\boldsymbol{\omega}$ itself. The dynamical equation

$$\langle dL_{\oplus}/dt\rangle = \langle\Gamma\rangle = \boldsymbol{\Omega}_{pr} \times \langle L_{\oplus}\rangle \qquad (4.6)$$

determines a mean precessional velocity $\boldsymbol{\Omega}_{pr}$ normal to the ecliptic and with modulus

$$\Omega_{pr} = \frac{\langle\Gamma\rangle}{L_{\oplus}\sin\epsilon} = \frac{3}{2}\,\frac{(C-A)}{C}\cos\epsilon\,\frac{n^2}{\omega}\,. \qquad (4.7)$$

Here n is the sun's mean motion. It corresponds to a period of about 82,000 y.

A torque of the same order of magnitude is produced by the gravity gradient of the moon. The same derivation applies, but now the angle ϵ is between the axis of the earth and the normal to the orbital plane of the moon. Neglecting for the moment the inclination of this plane to the ecliptic, from eq. (4.7) we get

$$\Omega_{pr,sun}/\Omega_{pr,moon} = M_{\odot}R^3_{moon}/M_{moon}R_{\oplus}^3 = 0.46 \qquad (4.8)$$

and this implies that the proximity of the moon more than compensates for its smaller mass. The proportionality to the mass of the perturbing body and to the inverse cube of its distance are typical of tidal effects, which are generated by gravity gradients. Both solar and lunar precessions are prograde and generate a total *lunisolar precession* with rate

$$\Omega_{pr,tot} = \Omega_{pr,sun} + \Omega_{pr,moon} = \Omega_{pr,sun}(1.46/0.46),$$

corresponding to a period of about 26,000 y (see eq. (3.4)).

In reality, both the sun and the moon perturb the rotational motion of the earth at the same time and in a full treatment of the problem their influence should be considered simultaneously. As the shapes and the relative orientation of the lunar and solar orbit with respect to the earth change, the torques on the earth vary with different time scales. Therefore, a number of small amplitude, but relatively rapid oscillations of the earth's spin axis are superimposed on the main long-term, precessional motion discussed above. For instance, the moon's orbit is inclined to the ecliptic by about 5 degrees, and since the lunar torque is greater than that of the sun, the instantaneous axis about which $\boldsymbol{\omega}$ precesses is in fact closer to the normal of the lunar orbit than to the pole of the ecliptic. However, the gravitational force of the sun causes the moon's nodal line (with respect to the ecliptic) to make a complete revolution in about 18.6 y, that is a period much shorter than the precessional period. Therefore, though on the average the precession occurs about the pole of the ecliptic, there is a slight nodding (i. e., a change in obliquity) in this motion with the same period of 18.6 y and an amplitude of 17.2 arcsec. This is the largest of many other such effects with similar causes, which have amplitudes ranging from about 0.01 to about 10 arcsec and periods ranging from fractions of a lunar orbital period (about 27.3 days) to several years. These effects are collectively called *nutations*.

As mentioned earlier, precession and nutation can easily be observed because they change the right ascension and the declination of celestial objects (Fig. 3.3). The ecliptic undergoes long term changes, with typical periodicities of $\approx 10^5$ y, due to the planetary perturbations on the earth's orbit. These changes affect the celestial coordinates and are traditionally combined with those arising from the motion in the axis of the earth to yield the *general precession* of the γ point.

4.2. TIDAL POTENTIAL

The fact that the earth lies (and rotates) in the gravity field of the sun and the moon, makes any point on the surface to feel a small and variable gravitational force of external origin. It is described by the *tidal potential* and causes of both the well known marine tides and small periodic deformations of the whole solid body of the planet.

Let us consider the potential U at a point P on the earth's surface caused by the moon's gravitational field. The moon's orbital motion is assumed to have an angular velocity (mean motion) n about an axis perpendicular to the lunar orbit through the centre of mass of the system.

$$U = -\frac{Gm}{|\mathbf{R} - \mathbf{r}|} - \frac{1}{2} n^2 b^2 , \tag{4.9}$$

where $\mathbf{r}$ and $\mathbf{R}$ are the same as in Sec. 4.1, m is the mass of the moon, b is the distance of P from the axis of orbital rotation (see Fig. 4.2). By Kepler's third law we get

$$n^2 R^3 = G(M_\oplus + m) . \tag{4.10}$$

Since, as in Sec. 4.1, $r \ll R$, we can expand $1/|\mathbf{R} - \mathbf{r}|$ (eq. (2.3)), retaining terms up to the second order in (r/R):

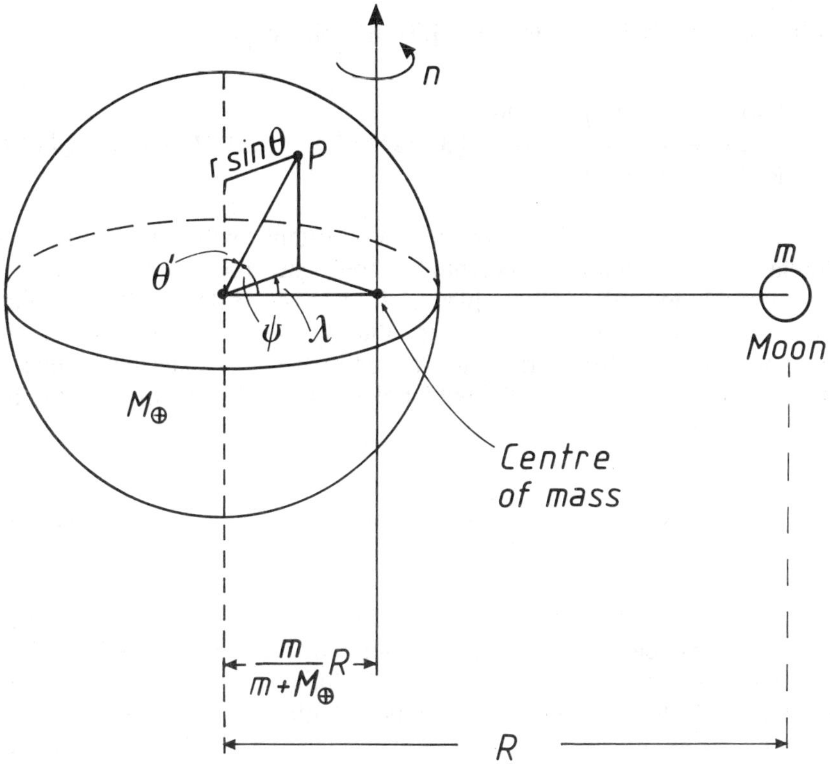

Fig. 4.2. Geometry of the tidal effect of the moon at a point P on the earth's surface.

$$\frac{1}{|\mathbf{R}-\mathbf{r}|} = \frac{1}{R}\left[1 + \frac{\mathbf{r}\cdot\mathbf{R}}{R^2} - \frac{1}{2}\frac{r^2}{R^2} + \frac{3}{2}\frac{(\mathbf{r}\cdot\mathbf{R})^2}{R^4} + 0\left(\frac{r}{R}\right)^3\right]. \quad (4.11)$$

Let us now call ψ the angle between **r** and **R** ($\mathbf{r}\cdot\mathbf{R} = rR\cos\psi$). By trigonometry (see again Fig. 4.2) we easily get

$$\cos\psi = \sin\theta'\cos\lambda \ , \quad (4.12)$$

where θ' is the colatitude of P with respect to the orbital plane and λ is its instantaneous longitude distance from the sublunar point. Moreover we have

$$b^2 = \left[\frac{mR}{m + M_\oplus}\right]^2 + (r\sin\theta')^2 - 2\left[\frac{mR}{m + M_\oplus}\right]\sin\theta'\cos\lambda =$$

$$= \left[\frac{mR}{m + M_\oplus}\right]^2 + r^2\sin^2\theta' - 2\left[\frac{mR}{m + M_\oplus}\right]r\cos\psi. \quad (4.13)$$

Substituting eqs. (4.10-13) in eq. (4.9) we obtain:

$$U = -\frac{Gm}{R}\left[1 + \frac{1}{2}\frac{m}{m+M}\right] - \frac{Gmr^2}{2R^3}(3\cos^2\psi - 1) - \tfrac{1}{2}n^2r^2\sin^2\theta' . \quad (4.14)$$

The first term in eq. (4.14) is just a constant, while the third can be interpreted as a rotational potential about an axis through the earth's centre normal to the orbital plane. It has an effect similar (although much smaller, since $n^2 \ll \omega^2$) to that of the diurnal rotation of the earth, that is, it gives rise to a permanent "equatorial bulge" and contributes to the $\ell = 2$ geopotential terms. The second term in eq. (4.14)

$$U_2 = -\frac{Gmr^2}{2R^3}(3\cos^2\psi - 1) \quad (4.15)$$

is the so-called *tidal potential*, proportional to m/R^3, like the torque of Sec. 4.1. It is clearly a second order zonal harmonic (see Sec. 2.1) which gives a deformation of the equipotential surface with the shape of a prolate, axisymmetric ellipsoid aligned with the position of the moon in the sky. This means that tidal effects have a dominant component of nearly semidiurnal periodicity (U_2 contains λ through $\cos 2\lambda$), exceeding 12^h by about 25^m because of the moon's orbital motion around the earth. An asymmetry between the two diurnal tides (*tidal inequality*) arises when the moon is displaced from the earth's equatorial plane, and at high latitudes this may even cause one of the two diurnal waves to disappear (see Fig. 4.3). As we shall see, the ellipticity of the lunar orbit and the superposition of solar tides (smaller but only for a factor ≈ 2, see Sec. 4.1) results in a complex wave pattern with many periods.

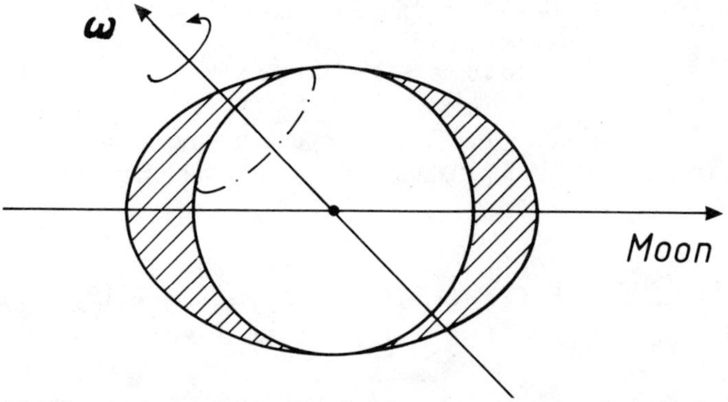

Fig. 4.3. The asymmetry in the diurnal tides when the moon lies off the equatorial plane can be seen by considering two points about 180 degrees apart along a parallel.

Let us now assume that the earth is rigid, so that its almost spherical mass distribution is not changed by the tidal potential. Then the fractional tidal variation in the gravity acceleration is

$$\frac{\delta g_R}{g} = \left(\frac{R_\oplus^2}{GM_\oplus}\right)\left(-\frac{\partial U_2}{\partial r}\right)_{r=R_\oplus} = -\frac{m}{M_\oplus}\left(\frac{R_\oplus}{R}\right)^3 (3\cos^2\psi - 1), \qquad (4.16)$$

that is of the order of 10^{-7} and detectable only with very sensitive gravity meters. The tidal deflection of the vertical $\delta\alpha_R$ is given by

$$\tan\delta\alpha_R = \left(\frac{1}{g}\right)\left(-\frac{1}{R_\oplus}\frac{\partial U_2}{\partial \psi}\right) = \frac{3m}{2M_\oplus}\left(\frac{R_\oplus}{R}\right)^3 \sin 2\psi , \qquad (4.17)$$

implying that $\delta\alpha_R$ is of the order of 0.01 arcsec. The variation in height of the equipotential surface is

$$\delta h_R = \frac{U_2}{g} = \frac{m}{M_\oplus}\left(\frac{R_\oplus}{R}\right)^3 R_\oplus\left(\frac{3}{2}\cos^2\psi - \frac{1}{2}\right) , \qquad (4.18)$$

that is of the order of one meter.

4.3. TIDES IN A NON-RIGID EARTH

We now abandon the assumption of rigidity and take into account the earth's deformation resulting from the tidal potential. Since the corresponding acceleration is small ($\approx 10^{-7}$ g from eq. (4.16)), it is reasonable to assume that the response of the solid earth is elastic, that is to say, linear in the external deforming potential (see Sec. 1.3). Moreover, the additional potential generated by the deformation itself can be taken to be also proportional to U_2 (like we assumed in Sec. 3.3 in studying the effect of a slow rotation potential). We thus define h as the ratio of the total tidal deformation δh_T of the surface to the displacement of the geoid δh_R given by (4.18), and κ as the ratio between the deformation potential U_T and U_2 at $r = R_\oplus$. The adimensional quantities h and κ are called *Love numbers*; a similar quantity appears in the theory of a rotating earth, see eq. (3.19). We have:

$$\delta h_T = hU_2/g , \qquad (4.19)$$

$$\delta g_T = -\frac{\partial(U_2 + U_T)}{\partial r} + \frac{\partial g}{\partial r}\,\delta h_T\,. \qquad (4.20)$$

Since $U_T(R_\oplus) = \kappa U_2(R_\oplus)$ is a second degree harmonic arising from the masses within $R_\oplus$, for $r > R_\oplus$ it must be proportional to $1/r^3$:

$$U_T(r) = \left[\frac{R_\oplus}{r}\right]^3 U_T(R_\oplus) = \left[\frac{R_\oplus}{r}\right]^3 \kappa U_2(R_\oplus) = \left[\frac{R_\oplus}{r}\right]^5 \kappa U_2(r) \qquad (4.21)$$

since U_2 is proportional to r^2 (see eq. (4.15)). Substituting eqs. (4.15, 19, 21) into eq. (4.20) we obtain

$$\delta g = -\left[1 + \frac{3}{2}\kappa + h\right]\left[\frac{\partial U_2}{\partial r}\right] = \left[1 + \frac{3}{2}\kappa + h\right]\delta g_R\,. \qquad (4.22)$$

Another measurable quantity is the variation in height δz of an equipotential with respect to the earth's surface (e.g., the surface of a pond with respect to its bottom). The tide would rise the equipotential by $(1 + \kappa)U_2/g$, but hU_2/g must be subtracted to account for the rise of the solid bottom, to yield

$$\delta z = (1 + \kappa - h)U_2/g. \qquad (4.23)$$

Finally, the tilt $\delta\alpha_T$ of the vertical is different from $\delta\alpha_R$ because of the enhancement by the factor $(1 + \kappa)$ of the potential and of a horizontal displacement of the earth's crust which is taken into account by a third coefficient l. In analogy with eq. (4.17), we have

$$\delta\alpha_T = (1 + \kappa - l)\left[-\frac{1}{gR_\oplus}\frac{\partial U_2}{\partial \psi}\right]. \qquad (4.24)$$

Measurements of these tidal effects on the earth's surface, as well as of the perturbations caused by U_T on the orbits of artificial satellites, have yielded for the earth's Love numbers the approximate values $h = 0.60$, $\kappa = 0.30$, $l = 0.08$.

This static equilibrium model, which assumes perfect elasticity, is approximate for two reasons. First, the earth includes fluid parts – the core and the oceans – in which the tidal response may be rate-dependent and cannot be represented by a single set of constants for all kinds of deformations. For instance, marine tides must be treated by a dynamical theory. They produce shallow water waves of wavelength much greater than the water depth H, with a propagation velocity of order $\sqrt{(gH)}$. Were this velocity larger than the speed $(\omega_\oplus - n)R_\oplus$ of the driving tidal force, we would have (like in forced oscillators) a quasi-equilibrium tidal wave in phase with the tidal potential; but since the opposite holds, we rather have out-of-phase *inverted tides*, namely, low tides where the equilibrium theory predicts high ones and *vice versa*. In reality the situation is much more complicated because of the complexity of sea floor geometry and of coastal lines, resulting in marine tides which depend very much on the location and can undergo strong enhancements with respect to the equilibrium amplitude in particular geographical scenarios.

The second reason why the equilibrium model is inadequate is friction, that is, dissipation of energy that always occurs in real materials and often is particularly large at the interfaces between solid and fluid regions. Like in forced oscillations with friction, the net result is a *phase lag* (again, because $\omega_\oplus \gg n$), implying that the maximum tidal bulge at some fixed position occurs after that point has passed the sublunar longitude. For the earth, this lag is of order of 3^0 (that is, about 12 minutes of time), the energy being dissipated in tidal currents in shallow seas. Although small, this dissipative lag has caused important, long-term changes in the dynamical state of the earth-moon system as well as of other planet-satellite systems (see Sec. 15.3).

4.4. TIDAL HARMONICS

The tidal potential U_2 given by eq. (4.15) and the related effects (both static and dynamical) discussed in Sec. 4.3 contain many different periodicities, since (i) the angle ψ depends on time through the geographical coordinates of the point P at **r**, which rotates with the earth, and the celestial coordinates of the moon and the sun; (ii) the $1/R^3$ factor also changes with time due to the orbital eccentricities of the moon and the sun and the perturbations on their orbits.

As for (i), it is useful to express by spherical trigonometry $\cos\psi$ as a function of the geographic colatitude θ of P, of the moon's declination δ (defined as its angular distance from the earth's equatorial plane) and of the local hour angle H(P) of the moon (i. e., its longitude distance from the plane of the meridian through P). We have

$$\cos\psi = \cos\theta \sin\delta + \sin\theta \cos\delta \cos H. \qquad (4.25)$$

Then, using trigonometric identities, in a frame of reference anchored to the earth we get from (4.15)

$$U_2 = \frac{GmR_{\odot}^2}{2R^3}\left[\sin^2\theta\cos^2\delta\cos 2H(P) + \sin 2\theta\sin 2\delta\cos H(P) + 3\left(\cos^2\theta - \frac{1}{3}\right)\left(\sin^2\delta - \frac{1}{3}\right)\right]. \tag{4.26}$$

At a given time the first term in eq. (4.26) is a sectorial spherical harmonic on the earth's surface, with the maximum at the equator, when the declination of the perturbing body is zero; the dependence on cos2H(P) implies that this term contains periodicities close to the semidiurnal period 12^h25^m. The corresponding spectral line is split into several components close to each other in frequency, because δ and R are also periodic functions of time, with longer periods. The second term in eq. (4.26) is a tesseral harmonic, giving rise to quasi-diurnal components which are largest at 45 degrees of latitude and when the declination of the perturbing body is highest, and vanish at the equator and at the poles (this term is the origin of the tidal inequality quoted in Sec. 4.2). The third term in eq. (4.26) depends on the latitude, but not on the longitude; it is a zonal harmonic vanishing at the two latitudes $\pm$ $35^{\circ}16'$. Its main periodicity comes from $\sin^2\delta$ and therefore is about 14 days for the moon and 6 months for the sun; the $R^{-3}(t)$ factor produces also monthly and yearly periodicities. Moreover, the constant part of this function causes a slight permanent flattening of the equipotential surface (by a few tens of cm) which adds up to the larger one due to the earth's rotation.

A quantitative analysis of the spectrum of U_2 can be made by expressing the coordinates of the moon and the sun as Fourier expansions with respect to time. Then U_2 can also be expressed as a Fourier series of six arguments which are (at least over a time span of the order of a century) linear functions of time. They are:

τ – Mean lunar time, that is, the time elapsed from the passage of the moon at the meridian. $2\pi/\dot{\tau}$ is the mean lunar day, equal to 1.03505 mean solar days (about 24^h50^m).

s – Mean longitude of the moon. $2\pi/\dot{s}$ is the orbital period of the moon, or *tropic month*, equal to 27.3216 days.

q – Mean longitude of the sun. $2\pi/\dot{q}$ is the orbital period of the earth, or *tropic year*, equal to 365.242 days.

p – Mean longitude of the lunar perigee. $2\pi/\dot{p}$ = 8.847 y is the *apsidal period* of the moon.

N – Mean longitude of the ascending lunar node. dN/dt is negative and $2\pi/|dN/dt| = 18.613$ y is the moon's nodal period.

p_s – Mean longitude of the perihelion. $2\pi/\dot{p}_s = 20{,}940$ y is the earth's apsidal period around the sun.

Origin	Symbol	Argument	Periodicity
Lunar principal wave	M_2	2τ	Semidiurnal
Solar principal wave	S_2	$2t = 2\tau + 2s - 2q$	Semidiurnal
Lunar eccentricity wave	N_2	$2\tau - s + p$	Semidiurnal
Lunar declination wave	K_2m	$2t' = 2(\tau + s)$	Semidiurnal
Lunar declination wave	K_1m	$t' = \tau + s$	Quasi-diurnal
Lunar principal wave	O_1	$\tau - s$	Quasi-diurnal
Solar principal wave	P_1	$\tau - q$	Quasi-diurnal
Solar declination wave	K_1s	$t' = t + q$	Quasi-diurnal
Lunar eccentricity wave	Q_1	$(\tau - s) - (s - p)$	Quasi-diurnal
Lunar declination wave	M_f	$2s$	Fortnightly
Lunar eccentricity wave	M_m	$s - p$	Monthly
Solar declination wave	S_{sa}	$2q$	Semiannual
Lunar constant flattening	M_0	–	Permanent deformation
Solar constant flattening	S_0	–	Permanent deformation

Table 4.1. The main Fourier components of the tidal potential.

All these angles are measured with respect to a conventional origin (the instantaneous vernal equinox). Combining them one obtains:

$t' = \tau + s$, the sidereal time;

$t = t' - q = \tau + s - q$, the mean solar time;

$2\pi/[d(s - N)/dt]$, the *draconitic month*, i. e., the period between two passages of the moon at a node, 27.2122 days;

$2\pi/[d(s - p)/dt]$, the *anomalistic month*, the period between two passages of the moon at perigee, 27.5546 days;

$2\pi/[d(s - q)/dt]$, the *synodic month*, the period of lunar phases, 29.5306 days;

$2\pi/[d(s - 2q + p)/dt] = 31.812$ days, the *evection* period;

$2\pi/[2d(s - q)/dt] = 14.765$ days, the *variation* period.

The two last periods are those of the main solar perturbations on the moon's eccentricity. It is worth noting that solar and lunar eclipses are periodic, with period of $\simeq$ 18 y and 11 d $\simeq$ 223 synodic months $\simeq$ 239 anomalistic months. This is *Saros period*, already known in antiquity.

The most important Fourier harmonics appearing in the spectrum of the tidal potential are listed in Table 4.1.

Of course the tidal effects discussed in Sec. 4.3 (for instance, the marine tides) will show the same basic periodicities as the forcing potential, but with amplitudes depending upon local features and possibly amplified by resonance phenomena.

*4.5. EQUILIBRIUM SHAPES OF SATELLITES

In Section *3.4 we have derived the equations for the ellipsoidal equilibrium shapes for isolated, spinning objects of constant density and having no internal strength. For satellites orbiting near their planet, provided their spin rate is synchronized with the orbital angular velocity (as we shall see in Ch. 15, this is in most circumstances the natural end-product of tidal evolution), the same kind of theory can be applied to derive the ellipsoidal equilibrium shapes corresponding to different values of the density, of the orbital distance (i. e., by Kepler's third law, of the orbital and rotational angular velocity) and of the satellite-to-planet mass ratio (which is usually much smaller than unity). We can qualitatively expect that satellites with a slow rotation, and consequently with a large orbital distance, are almost spherical, because tidal and centrifugal forces are negligible. On the other hand, satellites close to the planet will be distorted in a way inversely correlated with their density. As we shall see in Sec. 15.7, it is also easy to show that below some minimum orbital distance (the so-called *Roche* limit), no equilibrium shape is possible and a satellite with negligible tensile strength would be disrupted by tidal forces. Of course, all these results, including the concept of Roche limit, cannot be applied in a straightforward way when bodies having a finite material strength are considered. We shall now derive quantitatively the equilibrium shapes of homogeneous "fluid" satellites (which are called *Roche ellipsoids*) by applying the same type of argument we used in Sec. 3.4.

Let us consider a homogeneous ellipsoidal satellite of density ρ, semiaxes a, b and c (a being directed towards the planet's centre and c orthogonally to the orbital plane), orbital radius R and mass ratio $p = m/M$ with respect to the planet. Its angular velocity $\boldsymbol{\omega}$ is directed along c and, from the spin-orbit synchronization condition, its magnitude is given by

$$\omega^2 = G(M + m)/R^3 = (1 + p)n_0^2 , \tag{4.27}$$

where $n_0^2 = GM/R^3$. If we take a reference system with the origin at the satellite's centre and the axes (x, y, z) directed along (a, b, c) respectively, the centrifugal potential energy (per unit mass) due to the orbital motion of the satellite is

$$U_C = -\frac{1}{2}\omega^2\left[\left(x - \frac{MR}{M + m}\right)^2 + y^2\right] , \tag{4.28}$$

where we have corrected for the distance between the planet's centre and the centre of mass. As for the planet's gravitational potential energy U, we shall assume that the planet has a spherical shape (i. e., is not rapidly spinning and is not distorted by the satellite's gravitational field – this latter assumption being consistent only for small values of p and/or large orbital separations), and shall expand U about the satellite's centre in the following way:

$$U = -\frac{GM}{R}\left[1 + \frac{x}{R} + \frac{x^2 - \frac{1}{2}y^2 - \frac{1}{2}z^2}{R^2} + \ldots\right]. \tag{4.29}$$

Thus, if $U_T = U_c + U$, the satellite's gravity must include the tidal terms

$$-\frac{\partial U_T}{\partial x} = \frac{GM}{R^2} + \frac{2GM}{R^3}x + \omega^2\left(x - \frac{MR}{M + m}\right) = 2n_0^2 x + (1 + p)n_0^2 x,$$

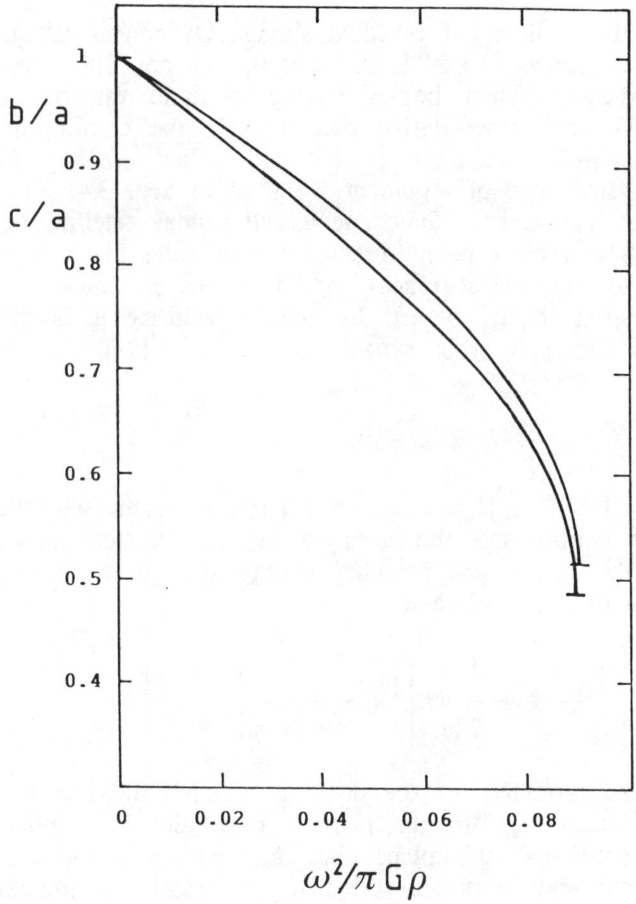

Fig. 4.4. The axial ratios c/a (lower curve) and b/a vs. $\omega^2/\pi G\rho$ for the sequence of p = 0 Roche ellipsoids, which ends up at the bars representing the so-called *Roche limit.*

$$-\frac{\partial U_T}{\partial y} = -\frac{GM}{R^3}y + \omega^2 y = -n_0^2 y + (1+p)n_0^2 y \ , \qquad (4.30)$$

$$-\frac{\partial U_T}{\partial z} = -\frac{GM}{R^3}z = -n_0^2 z \ .$$

Note that in the x-component the gravitational pull of the planet and the centrifugal force balance each other at the satellite's centre (we have used eq. (4.27)). Since these additional gravity terms are still linearly dependent upon the coordinates, we can consider, like in Sec. 3.4, three Newton's canals along the semiaxes a, b and c, and impose the condition that the work to bring a unit mass to the equipotential surface is the same in the three cases. We have

$$- 2\pi G\rho A_1 a^2 + (1 + p)n_0{}^2 a^2 + 2n_0{}^2 a^2 =$$

$$= - 2\pi G\rho A_2 b^2 + (1 + p)n_0{}^2 b^2 - n_0{}^2 b^2 =$$

$$= - 2\pi G\rho A_3 c^2 - n_0{}^2 c^2 , \qquad (4.31)$$

where A_1, A_2 and A_3 are the same integrals defined in Sec. 3.4, and depend on a, b and c. From this equation we can get the two relationships

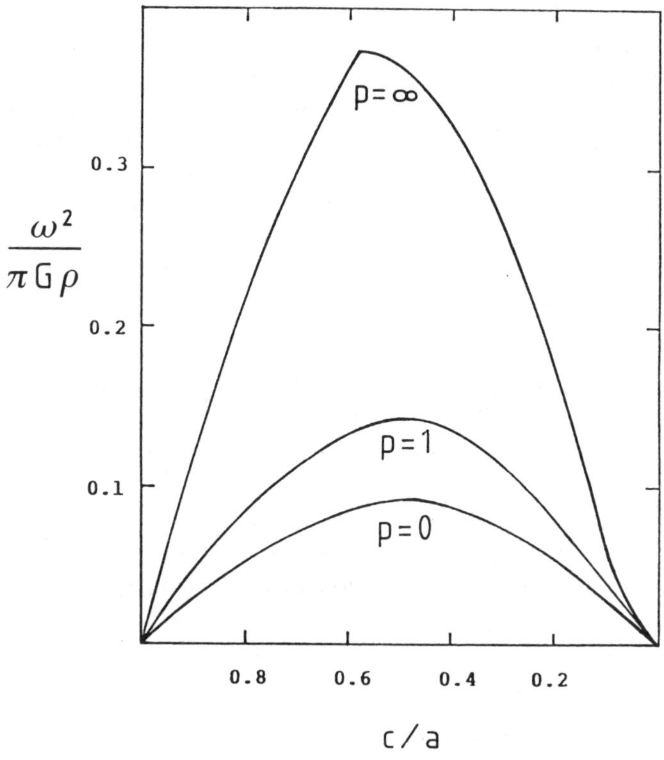

Fig. 4.5. The variation of $\omega^2/\pi G\rho$ as a function of c/a along the sequences of Roche ellipsoids corresponding to p = 0, p = 1 and p $\Rightarrow$ ∞. The maxima of the curves define here the Roche limit.

$$[(3 + p)a^2 + c^2]n_0^2 = 2\pi G\rho(A_1 a^2 - A_3 c^2) \tag{4.32$_1$}$$

$$(pb^2 + c^2)n_0^2 = 2\pi G\rho(A_2 b^2 - A_3 c^2) \ . \tag{4.32$_2$}$$

Given the mass ratio p, by inverting these equations numerically, it is possible to obtain the two axial ratios c/a and b/a as a function of n_0 (i. e., via eq. (4.27), of ω or R; see Fig.4.4).

Fig. 4.5 shows the behaviour of the sequences of Roche ellipsoids for $p = 0$, 1 and $p \Rightarrow \infty$ (the last case, provided $n_0^2 \Rightarrow 0$ and $n_0^2 p \Rightarrow \omega^2$, is coincident with that of a spinning, isolated body, and we find again the Maclaurin and Jacobi ellipsoids). In the case $p = 1$, the Roche figures are good approximations to the real equilibrium shapes of the two binary components only for $\omega^2/\pi G\rho \ll 1$, otherwise the mutual distortion effects cannot be neglected. From the figure it is apparent that the Roche ellipsoids behave qualitatively like we anticipated (the unexpected branch with decreasing values of ω^2 for increasing deformations can be shown to correspond to unstable equilibrium solutions). In particular, the fact that $\omega^2/\pi G\rho$ attains a maximum along the Roche sequences (depending on p) implies that for distances R less than that corresponding to this maximum no equilibrium solution is possible. This limit, called *Roche limit*, represents in other words the distance of closest approach between the two components consistent with equilibrium. Table 4.2 shows the properties of the "critical" Roche ellipsoids located at the Roche limit, for $p = 0$ and $p = 1$. We shall discuss some important applications of the Roche limit concept to planetary rings in Sec. 15.7.

p	$\omega^2{}_{max}/\pi G\rho$	$n_0{}^2{}_{max}/\pi G\rho$	$(c/a)_{cr}$	$(b/a)_{cr}$
0	0.09009	0.09009	0.4826	0.5113
1	0.14132	0.07066	0.4903	0.5408

Table 4.2. The properties of the Roche ellipsoids at the Roche limit.

PROBLEMS

4.1. Evaluate the Coriolis force on a low orbit satellite produced by an error in the lunisolar precession model of 2 mas/y during a mission lasting 3 y.

4.2. How high is the tide raised by the earth on the sun? And that raised on Saturn by its largest satellite, Titan? (use the data tabulated in Ch. 14).

4.3. Compute in order of magnitude the total energy fed into the earth's tidal bulges. Assuming that 10^{-3} of this energy is dissipated during one day, what is the corresponding power?

4.4. What is the critical water depth for having inverted marine tides, out of phase with the tidal potential?

4.5. Using the satellite data listed in Sec. 14.2 and looking at Figs. 4.4 and 4.5, find out the natural satellites of the solar system for which the expected tidal deformation exceeds 10^{-2}.

*4.6. Study the dynamics – equilibria and small oscillations – of two mass points connected by a short, rigid wire, and orbiting in a circular path around the earth.

FURTHER READINGS

P. Melchior, *The Tides of the Planet Earth.* Pergamon, Oxford (1983) provides the most comprehensive and systematic treatment of earth tides. On marine tides we quote only H. W. Schwiderski, *Ocean Tides,* Marine Geodesy, 3, 161 and 219 (1980).

5. THE INTERIOR OF THE EARTH

The interior of the earth (and of the other planets) is much less known than the surface of celestial bodies light-years away and also the interior of the sun. In fact, the physics and chemistry of solid and liquid phases, which are relevant for the planets, are much more complicated than those of nearly perfect gases (which form the stars). Moreover, the main source of information at our disposal – the propagation of seismic waves throughout the planet – is usually not under control of the geophysicists, but depends on the sudden occurrence of unpredictable events, the earthquakes. On the other hand, a planet is a complex machine whose internal processes continuously and deeply influence the morphology and the dynamics of the crustal layer, as well as the composition of the atmosphere. In this chapter, we describe how the seismic "probe" works, how quantitative models of the structure and of the thermal processes of the interior are built and, finally, the dynamical response of the surface layer to these processes, i.e., the so-called global tectonics.

5.1 SEISMIC PROPAGATION

As we have discussed briefly in Sec. 1.4, solid bodies respond *elastically* and linearly to deforming forces which are small enough and of small enough duration. In other words, the *strain* (measure of deformation) is proportional to the *stress* (force per unit area). Within the earth, transient forces arise mainly due to earthquakes: sudden fractures in the earth's solid crust can release as much as 10^{26} ergs and the corresponding disturbances can travel to large distances throughout the planet. Therefore, the propagation of seismic waves can be used as a powerful tool to investigate the physical properties of the earth's interior.

Neglecting the gravitational term in eq. (1.39) and taking the divergence, we obtain (when μ and λ are constant)

$$\rho \partial^2 \theta / \partial t^2 = (\lambda + 2\mu) \nabla^2 \theta \ , \qquad (5.1)$$

where $\theta = \nabla \cdot \mathbf{s}$. Since we are working in the linear approximation, the difference between eulerian and lagrangean derivatives is negligible. Alternatively, the curl of eq. (1.39) gives:

$$\rho \partial^2 (\nabla \times \mathbf{s})/\partial t^2 = \mu \nabla^2 (\nabla \times \mathbf{s}) \ . \tag{5.2}$$

Eqs. (5.1) and (5.2) are both D'Alembert's equations; therefore the divergence and the curl of **s** propagate as waves with the velocities $v_p = \surd[(\lambda + 2\mu)/\rho]$ and $v_S = \surd(\mu/\rho)$, respectively; they depend upon the Lame's constants of the material. The waves involving θ cause just dilatations and compressions and are called *S-waves*; the so-called *P-waves* are instead due to a rotational (or shear) deformation. In the earth, $\lambda \cong \mu$, so that the P waves are faster by about a factor $\surd 3$.

Let us consider now a wave front travelling through a material where ρ and the constants λ and μ change, as in the interior of a planet. If – as it normally happens – λ and μ increase with depth faster than ρ, the seismic velocities increase with depth also; therefore refraction makes the ray paths concave toward the planetary surface (recall Fermat's least time principle, Problem 5.2). The difference in travel times of the waves from a given source to different points on the surface (which can be directly measured) depends on the variation of velocity with depth, and hence on the physical properties of the planetary interior. In a spherically symmetric body it can easily be seen that the quantity (with the dimension of time)

$$p = r \sin(i)/v(r) \tag{5.3}$$

is constant along any given ray (here r is the distance from the centre, v is the propagation velocity and i is the angle of the ray with the radial direction; see Problem 5.1). If Λ is the angle at the centre measured along the ray and ds is the arc length element, we have $\sin(i) = r d\Lambda/ds$ and therefore

$$p = \frac{r^2}{v(r)} \frac{d\Lambda}{ds} \ . \tag{5.4}$$

Using the relationship $ds^2 = dr^2 + r^2 d\Lambda^2$ and defining R and r_{min} as the radius of the planet and the lowest radius reached by a given ray, from eq. (5.4) we can obtain the total angular distance ΔΛ and the travel time Δt :

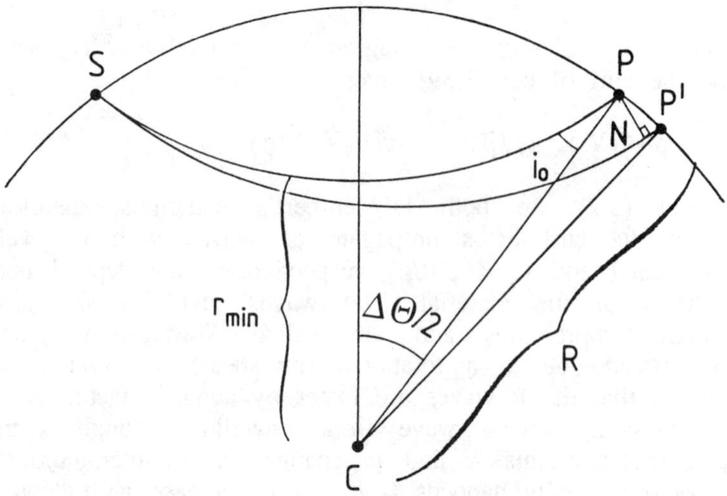

Fig. 5.1. Consider two neighbouring rays starting at the source S and emerging at P and P'. If at the surface $r = R$, $i = i_0$ and $v = v_0$, and we have $\sin(i_0) = NP'/PP' = [v_0\,\delta(\Delta t)/2]/[R\delta(\Delta\theta)/2] = (v_0/R)d\Delta t/d\Delta\theta$. Since $p = (r\sin(i))/v$ is constant along each ray, we obtain: $p = d\Delta t/d\Delta\varphi$.

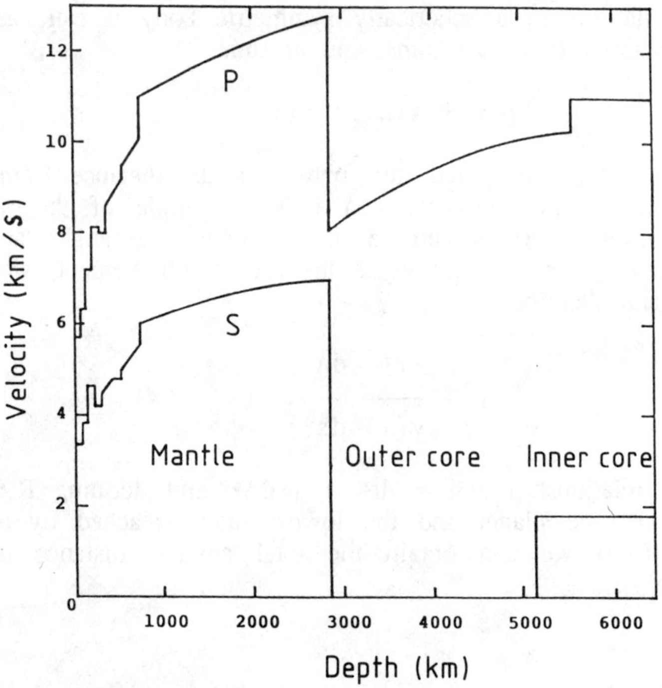

Fig. 5.2. The velocities v_S and v_P of S and P seismic waves as a function of depth in the earth as derived from earthquake data analysis.

$$\Delta\Lambda = 2p\int_{r_{min}}^{R} \frac{dr}{r\sqrt{[(r/v)^2 - p^2]}} , \tag{5.5}$$

$$\Delta t = 2\int_{r_{min}}^{R} \frac{ds}{v} = 2\int_{r_{min}}^{R} \frac{(r/v)^2 dr}{r\sqrt{[(r/v)^2 - p^2]}}$$

$$= p\Delta\Lambda + 2\int_{r_{min}}^{R} \frac{\sqrt{[(r/v)^2 - p^2]}dr}{r} . \tag{5.6}$$

From Fig. 5.1, comparing two neighbouring rays, we also see that $p = d\Delta t/d\Delta\Lambda$; this quantity, therefore, can be measured directly (if the location of the source of seismic waves is known) by comparing a set of values of $\Delta\Lambda$ and Δt provided by a suitable array of seismometers. The location of an earthquake is obviously not known *a priori*, but can be solved for if a large enough amount of data is available. Putting together the data obtained from many earthquakes and using eqs. (5.5) and (5.6), the function v(r) can be reconstructed (see Fig. 5.2).

*5.2. BOUNDARY EFFECTS

Discontinuities in the elastic properties of the material introduce reflection and refraction phenomena. In the case of electromagnetic waves, a boundary in general alters the mixture between different states of polarization; in our case, depending on the incidence angle, there is also a transfer of energy between S- and P-waves and, for the S-waves (which have two states of polarization), between the components parallel (SH) and normal (SV) to the surface of discontinuity. The letters H and V stand, respectively, for horizontal and vertical, corresponding to the usual case of a horizontal discontinuity. In the electromagnetic case we have also the phenomenon of total reflection, in which the disturbance does not quite vanish on the other side, but decreases exponentially, corresponding to an imaginary wave vector orthogonal to the discontinuity. Similarly, in the seismic case we also have to admit the possibility of imaginary orthogonal wave vectors. This leads to a new propagation mode, confined near the boundary (*Rayleigh waves*).

In order to describe boundary seismic phenomena, it is convenient to introduce the general decomposition of a vector field in an irrotational and a solenoidal component:

$$\mathbf{s} = \nabla u + \nabla \times \mathbf{a} \qquad (\nabla \cdot \mathbf{a} = 0) \; . \tag{5.7}$$

The scalar potential u is generic, but the vector potential **a** can, without loss of generality, be chosen to be solenoidal as well. From eqs. (5.1) and (5.2) they fulfil the equations

$$\nabla^2 \left[\frac{\partial^2 u}{\partial t^2} - v_P{}^2 \nabla^2 u\right] = 0 \tag{5.8}$$

$$\nabla^2 \left[\frac{\partial^2 \mathbf{a}}{\partial t^2} - v_S{}^2 \nabla^2 \mathbf{a}\right] = 0 \; . \tag{5.9}$$

Since we are looking for solutions of eqs. (5.8,9) regular in the whole space, we can use the fact that such a harmonic function is a constant and can be taken to be zero. Hence u and **a** satisfy d'Alembert equation for the propagation velocities v_P and v_S, respectively. We look for a solution across a surface of discontinuity at z = 0, corresponding to a frequency ω and a tangential wave vector k_x along x; then the wave vector along z, orthogonal to the boundary, is

$$k_Z = \pm \sqrt{(\omega^2/v_P{}^2 - k_X{}^2)} \tag{5.10$_1$}$$

for the P-wave and

$$k_Z = \pm \sqrt{(\omega^2/v_S{}^2 - k_X{}^2)} \tag{5.10$_2$}$$

for the S-wave. If k_z is real, the two signs correspond to impinging and reflected waves; the angle between the direction of propagation and the normal is $\pm$ arctg(k_x/k_z). If k_z is imaginary, the exponentially growing solution must be excluded. In this way any boundary problem is reduced to the fulfilment of the appropriate boundary conditions. They are determined by the physical nature of the interface, in particular, whether the two layers are free to slip with respect to each other. They boil down to two complex conditions which determine the stress vector at the boundary; because of the fact that every quantity is independent of y, its components reduce from three to two. A typical problem is the following: find out what happens to a plane wave impinging upon the discontinuity. We can have three cases: a longitudinal, P-wave; a transversal S-wave polarized horizontally (**a** along y) or vertically (**a** in the (x, z) plane). As a function of its complex amplitude, the two boundary conditions determine the complex amplitudes of the reflected and the transmitted waves. In this way, wave energy can be transferred from one mode to another.

We shall deal only with the simple, but important case of a free boundary, at which the stress vector

$$P_{i3} = \lambda\theta\delta_{i3} + 2\mu e_{i3} \qquad (5.11)$$

vanishes (eq. (1.33)). We also assume a P-wave and a horizontal, SH-wave, with

$$\mathbf{a} = a\mathbf{e}_y \ . \qquad (5.12)$$

Using eqs. (1.26), (1.27) and (5.7) one arrives at the following boundary conditions at z = 0:

$$2\frac{\partial^2 u}{\partial x \partial z} + \frac{\partial^2 a}{\partial x^2} - \frac{\partial^2 a}{\partial z^2} = 0 \ , \qquad (5.13_1)$$

$$2\mu\left[\frac{\partial^2 u}{\partial z^2} + \frac{\partial^2 a}{\partial x \partial z}\right] + \lambda\left[\frac{\partial^2 u}{\partial x^2} + \frac{\partial^2 u}{\partial z^2}\right] = 0 \ . \qquad (5.13_2)$$

As mentioned earlier, in the propagation case (k_z real) each mode is a superposition of a (+) mode with $k_z > 0$ and a (–) mode with $k_z < 0$; then eqs. (5.13) boil down to algebraic relations:

$$2k_x\sqrt{(\omega^2/v_p^2 - k_x^2)}(u_+ - u_-) + $$
$$+ (2k_x^2 - \omega^2/v_s^2)(a_+ + a_-) = 0 \qquad (5.14_1)$$

$$[2\mu(\omega^2/v_p^2 - k_x^2) + \lambda\omega^2/v_p^2](u_+ + u_-) + $$
$$+ 2\mu k_x\sqrt{(\omega^2/v_s^2 - k_x^2)}(a_+ - a_-) = 0. \qquad (5.14_2)$$

For example, given u_+ and $a_+ = 0$, one can compute the amplitude ratio $|a_-/u_+|$, whose square is the energy conversion factor from P into S-waves (see Problem 5.3).

With seismic waves it is possible to satisfy the free boundary conditions (5.13) with a superposition of S- and P-waves with *imaginary*

k_z, which confines them near the surface. This corresponds to the upper sign in eq. (5.10) (we take field quantities proportional to $\exp(-ik_z z)$); therefore we have to solve eqs. (5.14) with $a_- = u_- = 0$, which gives the "dispersion" relation for k_x. In computing it, note that the coefficient in the square bracket of eq. (5.14$_2$) can be expressed in terms of $v_S^2 = v_P^2\mu/(\lambda + 2\mu) \equiv q^2 v_P^2$:

$$(2\mu + \lambda)\omega^2/v_P^2 - 2\mu k_x^2 = \mu(\omega^2/v_S^2 - 2k_x^2) . \qquad (5.15)$$

Finally, the dispersion relation reads

$$4k_x^2\sqrt{(\omega^2/v_P^2 - k_x^2)}\sqrt{(\omega^2/v_S^2 - k_x^2)} =$$

$$= -(2k_x^2 - \omega^2/v_S^2)^2 , \qquad (5.16)$$

to be solved for k_x. Note, by the way, that the problem of a full conversion of P into S-modes or viceversa is solved by taking $u_+ = a_- = 0$ or $a_- = u_+ = 0$, which corresponds to changing the sign of the right hand side of eq. (5.16). In that case if $k_x < \omega/v_P$, so is $k_x < \omega/v_S$, corresponding to propagation for both modes. It can be shown that this solution occurs when $\lambda < 0.52(\lambda + \mu)$. But the novelty appears when the sign is negative, as indicated; then we could have a root with both $k_x > \omega/v_P$ and $k_x > \omega/v_S$, corresponding to a confined mode. A little algebra shows that the dispersion relation is a cubic equation for $p = (\omega/k_x v_S)^2$, with the velocity ratio q (< 1; eq. (5.15)) as a parameter;

$$f(p) = p^3 - 8p^2 + 16(3/2 - q^2)p - 16(1 - q^2) = 0 . \qquad (5.17)$$

Since $f(0) = -16(1 - q^2) < 0$ and $f(1) = 1 > 0$, there is always a real root p_0 in the interval (0, 1), corresponding to a real value of k_x. The velocity potentials have the form

$$u \propto \exp[i\omega t + z(\omega/v_S)\sqrt{(q^2 - 1/p_0)} - i\omega x/v_S\sqrt{p_0}] \qquad (5.18)$$

$$a \propto \exp[i\omega t + z(\omega/v_S)\sqrt{(1 - 1/p_0)} - i\omega x/v_S\sqrt{p_0}] . \qquad (5.19)$$

It is interesting to note that the phase velocity ω/k_x along the boundary is independent of frequency, so that a pulse preserves its form. It can be shown that the velocity of propagation of these *Rayleigh waves* always lies between $0.874v_S$ and $0.955v_S$. These waves are important in seismology because they can propagate to large distances. In fact, they suffer in general a smaller dampening than S and P waves for increasing distances from the source, since their energy spreads on a smaller volume. Thus, a typical seismogram appears like in Fig. 5.3.

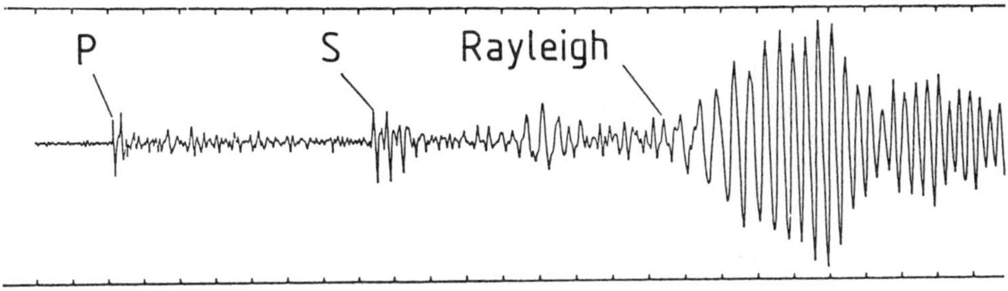

Fig. 5.3 Example of a seismogram, with signals produced by P, S and Rayleigh waves.

In the real earth, of course, the material properties and the velocities of propagation vary with depth. Since, as shown by eqs. (5.18), the Rayleigh waves penetrate the earth up to a depth of the order of their wavelength, when the latter is of the order or larger than ≈ 100 km (corresponding to periods ≈ 100 s or more), dispersion phenomena occurs: normally the velocity increases with wavelength, because longer wavelengths penetrate to deeper layers, where the velocity is higher. By obtaining from the data the velocity versus period relationships (the so-called *dispersion curves*) and comparing them with the predictions of different models, additional information is obtained on the properties of the crust and upper mantle layers.

The boundary conditions (5.14) cannot be satisfied when the P-wave is absent ($u = 0$); but pure SH surface waves can exist in a layered medium, e.g., when a discontinuity is present at some depth. One obtains in this scheme the so-called *Love waves*.

5.3 INTERNAL STRUCTURE OF THE EARTH

Our knowledge of the interior of the earth is almost entirely derived by the analysis of seismic wave propagation. Qualitatively, the main result is that, whereas P-waves propagate throughout the entire volume of the planet, S-waves cannot enter a spherical shell whose inner and outer surface can be located at $r \cong 0.20R$ and $r \cong 0.54R$, respectively. This is interpreted as being consistent with an overall structure where a solid mantle encloses a liquid outer core (liquids are, of course, unable to withstand shear deformations), which itself encloses a solid inner core. Another sharp, though less outstanding, discontinuity (the so-called *Moho*) separates the crust from the mantle, at a depth of about 10 km under the continents and a few km under the oceans. Considerable lateral variations have also been detected in the upper mantle layers (possibly as deep as 200 km), by analysis of both body waves – whose velocities

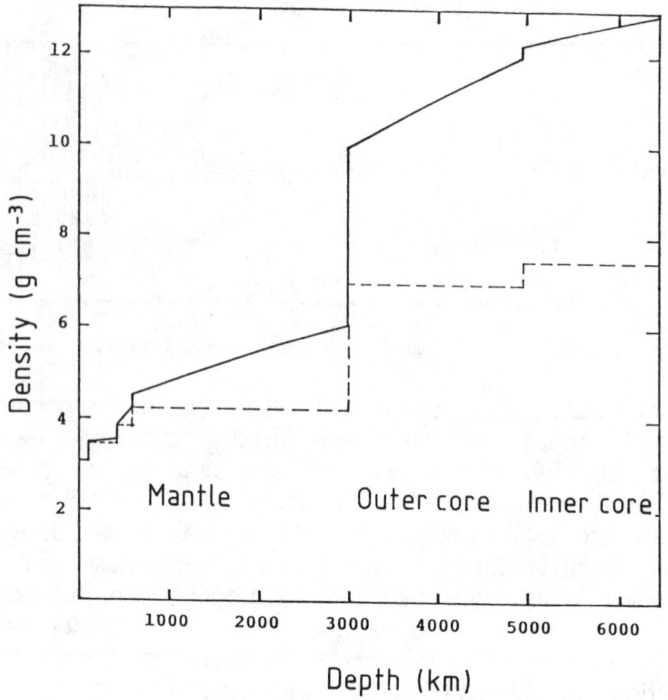

Fig. 5.4 Density profile for the earth's interior, derived from seismic wave data. The dashed line gives the corresponding "uncompressed" densities, extrapolated to zero pressure and room temperature.

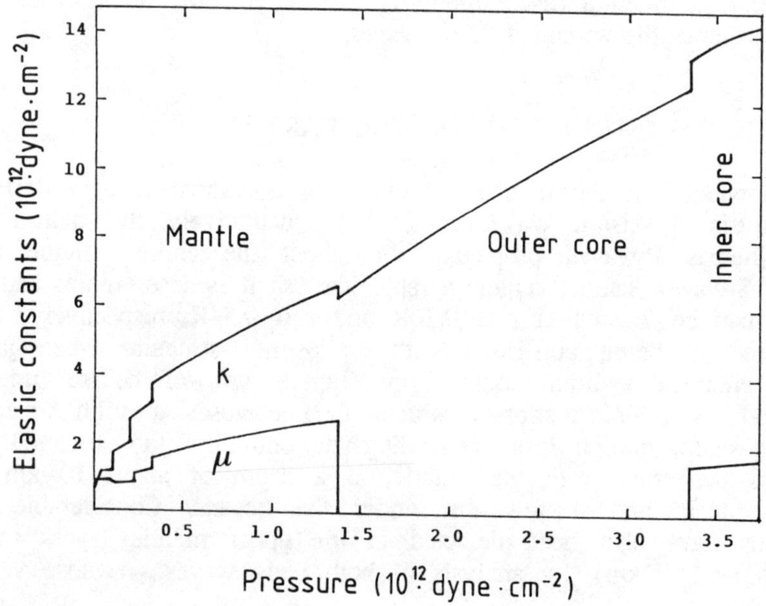

Fig. 5.5 Variations of elastic constants with pressure in the earth's interior, as derived from seismic wave data.

have a typical variability of ≈ 10% – and surface waves, whose dispersion curves are different for sub-continental and sub-oceanic regions.

From the quantitative point of view, starting from the inferred magnitudes of v_S and v_P as a function of depth, it is possible to derive immediately the ratios:

$$\frac{dP}{d\rho} = \frac{k}{\rho} = v_P^2 - \frac{4}{3}v_S^2, \qquad \frac{\mu}{\rho} = v_S^2. \tag{5.20}$$

Since the bulk modulus $k = \lambda + 2\mu/3$ (eqs. (1.36, 37) and μ are not known *a priori*, in order to get the function $\rho(r)$ we can use the the condition of hydrostatic equilibrium:

$$\frac{d\rho}{dr} = \frac{d\rho}{dP}\frac{dP}{dr} = -\frac{\rho^2}{k}\frac{GM(r)}{r^2}, \tag{5.21}$$

with M(r) given by eq. (1.1). Differentiating, this gives a second order differential equation for the unknown $\rho(r)$; inserting the inferred values of k/ρ as a fuction of r, it can be integrated numerically, using as initial conditions the values at the surface of $\rho(R) \cong 3.3$ g/cm³ and the derivative, provided by eq. (5.21). Thus the density profile is obtained and, from eq. (5.20), one gets $k(r)$, $\mu(r)$ and $P(r)$ (Figs. 5.4 and 5.5).

Another, independent method of obtaining information about the density profile uses the free oscillations of the earth as a whole. Large earthquakes cause shocks that make the planet vibrate, as a bell rings when struck by a hammer. There are many different modes which have been measured with increasing accuracy in the last 30 years. *Compressional*, or *spheroidal*, normal modes are classified by the expansion of the density perturbation in spherical harmonics; $\ell = 0$ corresponds to the fundamental mode of radial oscillations, which has been observed to have a period P_0 of 57^m, in good agreement with theoretical calculations based on earth's models. For low frequency modes the main restoring force is gravity; hence the order of magnitude of P_0, dictated by dimensional analysis, is given in terms of the mean density ρ by

$$2\pi/P_0 = \sqrt{(4\pi G\rho)}.$$

This is also Jeans' frequency (16.26) and, approximately, the mean motion of a low spacecraft. The next mode $\ell = 1$ does not exist because it would imply a displacement of the centre of mass; $\ell = 2$ produces in succession a prolate and an oblate deformation; and so on.

An *incompressible* mode is described by a solenoidal displacement field **s**. As shown, for example, in the Appendix III of S. Chandrasekhar's *Hydrodynamic and Hydromagnetic Stability*, Oxford (1961), any solenoidal vector field can be classified as *toroidal* when

$$\mathbf{s} = \mathbf{T} = \nabla \times (U(\mathbf{r})\mathbf{r});$$

and *poloidal* when:

$$\mathbf{s} = \mathbf{S} = \nabla \times [\nabla \times (V(\mathbf{r})\mathbf{r})];$$

in both cases they are determined by a single scalar, U or V. Again, the classification of these modes is based upon the expansion of U and V in spherical harmonics $Y_{\ell m}(\theta, \varphi)$ (Sec. 2.1). When the mode is excited by an earthquake at Q, it is convenient to place the pole $\theta = 0$ at Q. For a spherical and non-rotating earth, all the modes are degenerate with respect to the azimuthal number m. In the simple cases in which U or V are independent of φ, the flow lines are along the parallels for the toroidal case and in the meridian planes for the poloidal case. (See Sec. 6.4 for and application of these concepts to the dynamo theory.)

Especially for the low frequency modes, whose period is an appreciable fraction of a day, an accurate treatment should take into account the Coriolis force due to the earth's rotation. This removes the azimuthal degeneracy, corresponding to waves travelling in opposite directions with respect to the earth's rotation. This splitting has been actually observed in a few cases of very strong earthquakes. In view of the formal correspondence between Coriolis' force and the Lorentz force acting upon a charge in an external magnetic field, there is a close analogy between this splitting and the multiplets arising in atomic lines as a consequence of the Zeeman effect. Note, finally, that, as ℓ increases, only the layers near the surface are involved in the oscillation. Normal modes of oscillations are very important for all planetary bodies and for the sun; in the latter case they have recently contributed in an essential way to our knowledge of its internal structure.

If we aim at inferring something about composition from the density profile, we have to extrapolate densities to conditions observable in the laboratory, i. e. to a small pressure. For this purpose we cannot apply conventional elasticity theory, because, as shown by Fig. 5.5, the elastic constants, under pressures comparable to their value, change appreciably. A simple model can be set up by assuming a linear relationship, the so-called *Murnaghan's equation*:

$$k = k_0 + k_0' P , \tag{5.22}$$

where subscript 0 refers to values at zero pressure; we can see from

Fig. 5.5 that, at least for the mantle, this is a good approximation. Since $k = \rho dP/d\rho$, we have then

$$\frac{\rho}{\rho_0} = \left[1 + \frac{k_0'}{k_0} P\right]^{1/k_0'} = \left[\frac{k}{k_0}\right]^{1/k_0'} ; \qquad (5.23)$$

taking for the lower mantle $k_0 = 2.25 \cdot 10^{12}$ dyne/cm² and $k_0' = 3.35$ we obtain an "uncompressed" density ρ_0 of about 4 g/cm³ (the extrapolation to room temperature produces a much smaller correction). This is consistent with a substantially homogeneous chemical composition of the whole mantle, formed by a mixture of high pressure phases of iron, magnesium and silicon oxides. For the outer core the same method is less accurate, but the result (ρ_0 = 6.3 g/cm³) appears to imply that the main constituent is iron, which at the melting point (a reasonable estimate for the temperature there) has a density of about 7 g/cm³. The ≈ 10% difference between these densities implies that some lighter element is also present as a minor constituent: good candidates are sulfur, which alloys with iron even at low pressures and temperatures (the cosmic abundance of sulfur is ≈ 30% that of iron by mass, but it was a volatile in the preplanetary nebula); and oxygen, which can be incorporated into iron only at high pressures. The outer core is a very good candidate for the liquid, electrically conducing region where the earth's magnetic field is to be generated (see Ch. 6).

The interpretation of inner core data is more difficult. Solidity is suggested by the nonzero rigidity constant, but μ is abnormally low with respect to k for a close-packed crystalline solid. The composition is very probably a solid nickel-iron alloy, but what is the relevant phase of this solid is still an open question. It is interesting to notice that the boundary between the solid mantle and the liquid outer core coincides with a gross difference of composition; this is probably not the case for the boundary between inner and outer core, which more likely results just from the dependence of the melting point of ferrous materials on pressure. Finally, it is worth noting that the earth's core contains about 32% of the total mass of the planet, a fraction several times larger than that expected on the basis of the cosmic abundance of iron with respect to all other elements, except hydrogen and helium (see Table 13.1). This is a significant clue for the processes through which the material that subsequently formed the earth condensed from the primordial solar nebula.

5.4. HEAT GENERATION AND FLOW

Although the overall equilibrium of a planet does not involve explicitly its thermal state, the conditions prevailing in a planetary interior are strongly connected with the way heat is generated and transferred, with the resulting temperature distribution and with its evolution in time. If the planet is initially hotter than the surroundings, heat will be released at the surface, causing the temperature to increase with depth. (The energy input received from the sun, though very important in determining the state of the surface and of the atmosphere, makes a negligible contribution to the total heat content of a planet.) The detailed internal distribution at any time will depend on three factors:

(1) The initial state, where for "initial" we intend at the time the planetary accumulation process was completed, about $4.5 \cdot 10^9$ y ago. As we shall see, for a planet as big as the earth, the memory of the initial state is probably not yet lost. Note that if the heat content is proportional to the planet mass, since the dispersion of energy is proportional to the surface and the density of solid bodies does not vary strongly in the solar system, the time needed to "freeze" the interior (in absence of energy generation) is roughly proportional to the body size.

(2) The energy generation mechanism. The temperatures in the interior of large planetary bodies are of the order of $10^3 - 10^4$ K, too low for the onset of thermonuclear reactions; however, the planetary material contains small fractions of long-lived radioactive elements whose decay causes a continuous energy input.

(3) The energy transfer mechanism. Thermal energy can be transported by radiation, conduction and convection. At least the two latter mechanisms are important in the interior of planets.

As a "boundary condition", we can use the direct measurements of the heat flow through the earth's crust carried out on a global scale during the last few decades. The heat flux is typically 60 erg/(s cm^2), with a variability of a factor 2 or 3, depending on the type of crust and the intensity of geologic activity. However, this is not sufficient to deduce uniquely the thermal state of the interior, since a wide range of combinations of "primordial" heat and of radioactive heating may lead to the same surface conditions. For instance, the fact that the heat flux through the continental crust and through the ocean floors has the same order of magnitude is somewhat surprising, because in the former case the measured concentrations of radioactive elements in the rocks imply that a large fraction of the heat is generated in the crust itself, while this is not possible for the much thinner oceanic crust, where virtually all the heat flux must come from the mantle.

The decay of radioactive nuclei in a given volume of material occurs according to the equation

$$dN = -\alpha N dt, \tag{5.24}$$

where dN is the number of nuclei decayed in the time dt, N is their total number and α is a constant characteristic of every radioactive species. Integration of eq. (5.24) yields

$$N = N_0 \exp(-\alpha t) = N_0 \exp(-t \cdot \ln 2 / T_{\frac{1}{2}}) \ , \qquad (5.25)$$

where N_0 is the abundance at $t = 0$ and $T_{\frac{1}{2}}$ is the half-life of the species. The decay can involve the emission of massive particles, gamma rays and neutrinos. Apart from the latter, for which the earth is practically transparent, the energy of the emitted particles is rapidly absorbed by the surrounding material which is thereby heated. Four radioactive nuclei are of particular importance in the heating of planetary interiors: U^{235}, U^{238}, Th^{232} and K^{40}. Their properties are listed in Table 5.1 (including approximate abundances in natural rocks).

Isotope	Half-life (10^9 y)	Heat production (erg/(s·g))	Abundance in natural rocks (ppm)	
			Crust	Upper mantle
U^{235}	0.71	5.70	10^{-3} - 10^{-2}	10^{-4}
U^{238}	4.50	0.94	10^{-1} - 1	10^{-2}
Th^{232}	13.90	0.26	1 - 10	10^{-2} - 10^{-1}
K^{40}	1.25	0.29	10^{-1} - 10	10^{-2} - 10^{-1}

Table 5.1. Radioactive elements in the earth.

Taking the upper mantle abundances as representative for the whole earth, we obtain an energy production rate of the order of 10^{-7} erg $s^{-1} g^{-1}$, which is just the order of magnitude needed to give the observed average surface flux quoted earlier. As we said, the abundances of radioactive elements in crustal rocks are in general much higher, so that in the continents (where the crust is thicker) most radioactive heat probably comes from the crustal layers. On the other hand, the abundance values found for the upper mantle are of the same order as those of meteorites, and thus are probably typical of the primordial solid material out of which the planets accreted. The reason why the surface layers of the earth (and of the moon at the sites sampled in the Apollo missions) are enriched in radioactive material is not yet understood.

Coming to heat transfer mechanisms, we observe first that, contrasting with stars, radiative transfer inside a planet is not important, because the temperature is lower and, more important, the opacity to radiation is very large. On the other hand, conduction is often important. In fact, the geothermal flux at the surface is estimated by measuring the temperature gradient and the thermal conductivity of the material. In the interior of the earth we must add to the heat equation (1.48) a source term to account for radioactive energy generation:

$$\partial T/\partial t = \chi \nabla^2 T + h \ . \tag{5.26}$$

$c_p \rho h$, where c_p is the specific heat at constant pressure, is the heat produced per unit volume per unit time. According to eq. (5.26), the time for the heat to diffuse over a distance L will be $\tau \approx L^2/\chi$. Since for terrestrial-type materials the heat transport coefficient $\chi \approx 0.01$ cm²/s, for L = 100 km we get $\tau \approx 10^{16}$ s $\approx 3 \cdot 10^8$ y; τ is of the order of the age of the solar system for L ≈ 400 km, so that for larger bodies the initial heat is still locked inside the interior. For the earth, were the loss due to conduction alone, very little heat would have escaped from the interior of the planet. It is also interesting to note that due to ocean floor spreading (see Sec. 5.4), the residence time of oceanic lithosphere at the surface between formation at a ridge and sinking at a subduction zone is of the order of the thermal conduction time τ (since the lithosphere thickness is about 70 km and the drift velocity is a few cm/y). As we shall see, this is consistent with the hypothesis that the lithospheric motions are driven by convection currents in the mantle, since the oceanic lithosphere apparently remains at the surface just about the time needed to give up its residual heat.

A simple, but important result about the increase of temperature with depth in a medium subjected to gravity can be derived just from thermodynamical arguments. Consider a unit mass of material at the temperature T and pressure P and let a quantity dQ of heat enter its volume $V = 1/\rho$. From the first principle of thermodynamics we get

$$dQ = dU + PdV = dH - VdP = c_p dT + [(\partial H/\partial P)_T - V]dP, \tag{5.27}$$

where U is the internal energy, H = U + PV is the enthalpy and $c_p = (\partial H/\partial T)_P$ is the specific heat at constant pressure. Now from the second principle we have, in terms of the entropy S, dQ = TdS and since dH = dQ + VdP, using the free energy F = U − TS + PV (such that dF = − SdT + VdP), we have

$$\left(\frac{\partial H}{\partial P}\right)_T = T\left(\frac{\partial S}{\partial P}\right)_T + V = V + T\left[\frac{\partial}{\partial P}\left(-\frac{\partial F}{\partial T}\right)_P\right]_T =$$
$$= V - T\left[\frac{\partial}{\partial T}\left(\frac{\partial F}{\partial P}\right)_T\right]_P = V - T\left(\frac{\partial V}{\partial T}\right)_P . \tag{5.28}$$

Substituting eq. (5.28) in (5.27) we obtain

$$dQ = c_p dT - T(\partial V/\partial T)_p dP \,. \tag{5.29}$$

Of course dQ depends on the physical processes of heat transfer during the transformation. For an adiabatic compression dQ = 0, and eq. (5.29) becomes

$$\left(\frac{dT}{dP}\right)_S = \frac{T}{c_P}\left(\frac{\partial V}{\partial T}\right)_P = \frac{T}{c_P}\frac{\beta}{\rho} \,, \tag{5.30}$$

where we have used the definition of the volume expansion coefficient $\beta = (\partial V/\partial T)_p/V$. For a perfect gas $\beta = 1/T$; in this case, eq. (5.30) can be derived more simply using the equation of state (1.12).

Now, if $-z$ is the depth

$$-\left(\frac{dT}{dz}\right)_S = -\left(\frac{dT}{dP}\right)_S\frac{dP}{dz} = \left(\frac{T}{c_P}\frac{\beta}{\rho}\right)\rho g = \frac{g\beta}{c_P}T \tag{5.31}$$

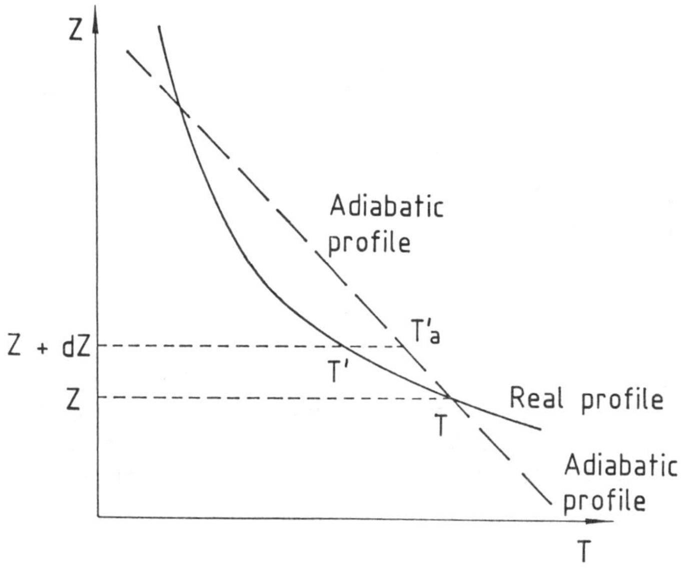

Fig. 5.6 The convective instability produced by a temperature gradient steeper than the adiabatic profile: if a fluid element is quickly moved from z to z + dz, its temperature is T_a', larger than the temperature T' of the surrounding, and therefore its motion continues.

This gradient is the *adiabatic temperature gradient.* Its importance lies in the fact that it is the largest gradient a fluid can support at rest in a stable situation. This can easily be seen as follows. If the temperature increases with depth faster than specified by eq. (5.31), an accidental upward displacement of a fluid element, if quick enough to be adiabatic, would bring it into a region where it is hotter than its new surroundings (see Fig. 5.6); its density would then be lower and the buoyancy force will tend to make it rise further. Its motion then would continue in an unstable way until, finally, heat is exchanged with the environment and the fluid element is brought back to the same temperature as the surrounding medium. The motion then stops, but as a result a net amount of heat is carried upward. A similar process occurs if the fluid element initially moves downward. If the temperature gradient is less than the adiabatic gradient any initial motion will be reversed by the buoyancy force, the corresponding oscillation frequency is Väisälä's frequency (see Sec.7.4). This particular way to transport heat is called *convection*, and occurs spontaneously whenever the gradient is larger than adiabatic; it is of great importance in cosmic bodies, where gravity is the main macroscopic force and there are often energy sources in the interiors.

The evaluation of the heat transport coefficient in a convective – and turbulent – regime is a difficult and still little understood problem. In many circumstances, however, the convective heat transport is so efficient that the temperature gradient is reduced to a very small relative excess over the adiabatic gradient (at which convection would cease); this arises from the very high efficiency of convective energy transport. Thus it is frequently a good approximation to assume in presence of convection an adiabatic temperature profile.

In the case of planetary interiors, a more convenient expression for the adiabatic gradient (5.31) can be obtained by defining the dimensionless *Grünesein's parameter*

$$\gamma_G = \beta k/\rho c_p \; , \tag{5.32}$$

whose value has been empirically found to be close to unity for virtually all geologically interesting materials, and almost constant for the range of temperatures and pressures found in the earth's interior (contrasting with its component parameters β, k and ρ). From eqs. (5.31, 32) we have then

$$\left(\frac{dT}{dz}\right)_S = -\frac{g\gamma_G\rho T}{k} \; ; \tag{5.33}$$

using the equilibrium law $dP = -g\rho dz$ and the definition (1.37), if γ_G

is assumed constant, we get

$$\ln T = \int \gamma_G dP/k = \gamma_G \ln \rho \ , \qquad (5.34)$$

simulating a polytropic gas with $\gamma = \gamma_G + 1$ (eq. (1.56)). This provides the temperature profile from $\rho(r)$.

What happens in the real earth? In the lithosphere the observed temperature gradient is on the average 15 K/km, consistently with a heat flow of $\approx$ 60 erg/(s cm^2) and a typical thermal conductivity $\approx 4 \cdot 10^5$ erg/(s cm K). (The conductivity is the product of specific heat, $\approx 10^7$ erg/g, the density and the heat transport coefficient χ, $\approx$ 0.01 cm^2/s.) Of course, in the (solid) litosphere no convection is possible and heat is transferred only by conduction.

In the mantle, on the contrary, convection plays a dominant role. At first glance this may seem surprising, since the propagation of shear elastic waves in it (see Sec. 6.1) guarantees that the material is not liquid; on the other hand, with the lithospheric temperature gradient, at the bottom of the lithosphere ($\approx$ 70 km in depth) the temperature becomes $\approx$ 1500 K, close to the melting point of most minerals. In fact, the mantle is somewhat intermediate between the fluid and the solid state. At temperatures somewhat below melting, crystalline substances under shear stresses respond in a way depending on the time scale of the shear: for very rapid deformations (such as seismic waves) they behave elastically, while for very long times they are "soft enough" to *creep*, i. e. undergo a very slow, steady deformation. In other words, rocky materials have a finite *viscosity* when their temperature approaches the melting point. In the earth's mantle the average melting temperature of the various mineral components increases with depth (because of the increasing pressure) at a rate of $\approx$ 0.5 K/km. This is only about a factor 2 higher than the adiabatic temperature gradient (5.31) ($T \approx 2000$ K, $c_p \approx 10^7$ erg/(g K), $\beta \approx 10^{-5}$ K^{-1}) or (5.33) (with $\gamma_G \approx 1$ and $k \approx 5 \cdot 10^{12}$ dyne/cm^2 – see Fig. 5.5). On the other hand, since the conductivity does not depend very much on temperature, the adiabatic gradient is too small to allow radioactively generated heat to be transferred by conduction only. As a consequence, very slow convective currents arise, compatible with creep deformations, and the temperature increases with depth at a rate which is close both to that of the melting temperature (allowing a low enough viscosity) and to that of the adiabatic gradient (making convection possible).

How was this situation established ? Even assuming that the earth started cold (an unlikely case, as we shall see in Ch. 16), the radioactive heat release probably sufficed to bring it close to melting in $\approx 10^8$ y (recall that this release is presently $\approx 10^{-7}$ erg/(s g), but it was higher in the past by about an order of magnitude – see Table 5.1; and that the specific heat is $\approx 10^7$ erg/g). Convection then set up and the situation became self-stabilized, since a slower convection would heat up the earth, but then the material would become "softer" and convection would be accelerated again. Convection stops only when the heat flux will become so low that conductive transfer along a quasi-adiabatic temperature gradient will match the flux itself.

For the liquid outer core, the situation is again different. Here, the material is fluid and the only condition required by convection is a strong enough energy generation. This condition is more demanding than for the mantle, because: (a) the abundance of radioactive elements in the iron core is uncertain, but probably lower than in the mantle for chemical reasons; (b) the thermal conductivity is higher than in the mantle, being dominated (like for the electrical conductivity of metals) by electrons in the conduction band. However, an alternative heat source is provided by latent heat and gravitational energy release due to the growth of the inner solid core. Gravitational energy becomes available because the liquid outer core contains ≈ 10% of light alloying constituents (like sulphur or oxygen – see Sec. 5.3), which are at least partially excluded from the solid phase and are displaced upwards as the inner core grows. The formation and growth of the inner core in the planetary centres occur essentially because for iron and iron-rich alloys the rate of increase of freezing temperature with pressure is higher than the adiabatic gradient $(\partial T/\partial P)_S$, which approximately holds in the liquid core. Moreover, in general, the addition of impurities depresses the freezing point of liquids (if the corresponding solid is purer); if the core starts as a liquid and then gradually cools, when some solid iron forms, the coexisting fluid is richer in the lighter constituents, and therefore less dense, than the overlying fluid (which has not yet undergone freezing) and therefore tends to rise and to start a compositionally driven convection. In any case, the onset of convection in the liquid core appears to be an essential prerequisite for the generation of a significant planetary-scale magnetic field.

5.5. TECTONIC MOTIONS

A real scientific revolution has affected geology during the 60's and the 70's, transforming this discipline from a mostly descriptive one, providing explanations only for small scale features and phenomena, to a mature science, based on a paradigmatic theory (the so-called *plate tectonics*) which explains on a global scale the most prominent characteristics and the evolution of the earth's outermost layers.

The evidence for extensive horizontal displacements of large sections of the earth's crust relative to each other comes from disparate types of data and techniques:

(1) Impressive similarities exist between continental margins now

separated by wide oceans, both in shape (an outstanding example are the Atlantic coast lines of Africa and South America) and in geological features, paleoclimatic histories, fossil records of ancient flora and fauna. In the first decades of our century this evidence was systematically collected and used by A. Wegener to support his theory of continental drift and of the origin of the present continents from a primordial, common land mass. This theory, though much debated, was rejected for a long time by the majority of geophysicists, mainly because no plausible physical mechanism was deemed capable of causing the proposed drift.

(2) The morphology of the oceanic floors, first globally mapped in the 50's, shows many prominent linear features associated with intense volcanic and seismic activity, which appear to divide the earth's crust into a discrete number of huge "plates", including the continents. These linear features are either long ridges, where the sea floor appears to be generated from below and then to diverge, or deep trenches, where the surface layer appears to plunge back into the mantle. They are frequently broken into many shorter sections somewhat displaced horizontally, and connected by *transform faults*, i. e., shear lines parallel to the direction of relative motion between the boundaries. Thus the continents do not "float" like rafts on the mantle, but move about as a consequence of the continuous generation and destruction of the oceanic crust.

The new data on oceanic floors showed also clearly that the bipartition of the earth's crust into oceans and continents is not just the result of the presence of large amounts of liquid water: the oceanic crust is thinner, younger, different in mineralogical composition (less acidic) and lies on the average some 4.6 km lower than the continental shelfs, with a comparatively steep slope in between. This dichotomy appears to be unique among terrestrial-type planets.

(3) The ideas about continental drift and sea-floor spreading received a detailed and quantitative support from paleomagnetic data, collected in the 50's and the 60's. As it will be discussed in Ch. 6, most crustal rocks retain a record of the local geomagnetic field at the time they cooled and solified or were deposited as sediments. In particular, the orientation of the geomagnetic poles relative to a given rock sample can be readily derived; and since the axis of the main (dipolar) geomagnetic field, when averaged over times $\gtrsim 10^4$ y, is believed to coincide with the rotational axis of the earth (see Sec. 6.3), pole positions relative to different continents and geologic epochs can be inferred. For rocks younger than about $2 \cdot 10^7$ y all the poles cluster about the present geographical pole; but for earlier epochs, substantial departures appear. These departures, derived independently for different continents, are not consistent with a coherent motion of the whole crust with respect to the rotational poles (as it happens, on a very small scale, for the polar motion phenomenon discussed in Sec. 3.5); on the

contrary, the poles relative to different continents can be reconciled only if large relative drifts have occurred on a time scale of $\approx 10^8$ y, in a way fully consistent with Wegener's picture, derived from entirely different data. As regards the oceanic crust, sea-floor spreading from ridges is confirmed by the observation of symmetric, linear stripes of alternately positive and negative magnetic anomalies on the two sides of the ridges, as it is expected from the known occurrence of reversals of the field polarity, recorded by cooling volcanic rocks. These data are consistent with spreading rates of 1 – 10 cm/y, and allow the determination of the relative velocity of neighbouring plates.

Other large scale crustal features can be interpreted in this framework, provided the mechanism for horizontal motions of the crust is recognized in their coupling with vertical motions in the underlying mantle, associated with convective currents (see Sec. 5.3). The basaltic ocean crust is then seen as a veneer on the *lithosphere*, the rigid layer $\approx$ 70 km thick and broken in different plates, moving and evolving due to convection in the deeper, plastic *asthenosphere*. At the ocean trenches (*subduction zones*), the lithosphere bends down forming a rigid, inclined slab which plunges into the mantle, producing earthquakes with intermediate and deep foci (contrasting with earthquakes associated with transform faults at ocean ridges, which are generated closer to the surface), until, at $\approx$ 700 km of depth, it is reassimilated in the mantle. But during this process the lighter material differentiates and emerges again as *andesitic magma*, resulting into extensive volcanism and forming island chains (e. g., Japan) or building up the edge of a continent (e. g., the Andes). On the other hand, when a collision between two continental blocks occurs, the continental crust is too light and thick for sinking into the mantle, and thus folds and forms extensive mountain systems (e. g., the Himalaya and the Alps).

The only type of volcanic and orogenetic activity which is unrelated to plate motions and to the upper mantle convection pattern is associated with the so-called *hot spots*. Some chains of volcanic islands (e. g., the Hawaii), lying far from the plate boundaries, show a progressive increase of age, which suggests plate motion across a steady volcanic source, possibly a narrow convective plume originating in the lower mantle. Some 120 hot spots, active in the last 10^7 y, have been recognized, both inside the continents (e. g., the volcanoes Nyamlagira and Nyiragongo in central Africa) and in the oceans, either superposed on ridges (e. g., Iceland) or well removed from them (e. g., Hawaii). A map of the main plates and hot spots is shown in Fig. 5.7.

The kinematics of plate movements can be summarized in a global model as follows. As shown by Euler in 1776, every displacement of a rigid plate on a spherical surface is a rotation. A rotation is determined by three parameters, two for the direction of the rotation axis and one for the rotation angle. Therefore, for two neighbouring plates A and

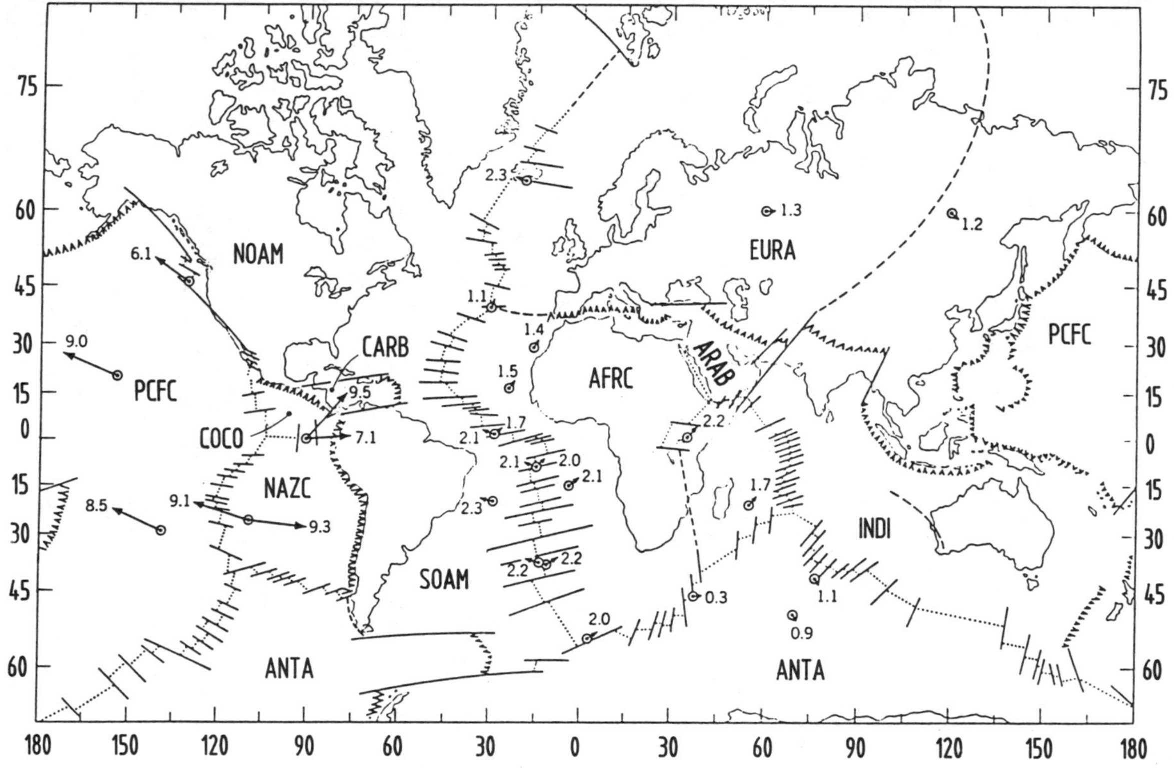

Fig. 5.7. A map of the main lithospheric plates, indicated by their four-letters international designations. Dotted lines are lines of divergence (ocean ridges or incipient oceans, like in the Red Sea and East Africa), while lines of convergence are marked by arrow heads showing the directions of motion of subducting lithospheric slabs; they are zones of very intense seismicity. A number of prominent "hot spots" are indicated by dotted circles, and the corresponding numbered arrows give motions of plates (in cm/y) relative to the hot spot, as derived from a global kinematical model. Single lines mark transform faults and their associated fractures; dashed lines indicate features of uncertain interpretation. (Adapted from Stacey.)

B, the instantaneous movement of A with respect to B can be represented by the magnitude of the angular velocity Ω_{AB} and the position (e.g., longitude φ_p and colatitude θ_p) of the instantaneous pole of rotation. The velocity relative to B of every point k (φ_k, θ_k) of A is then

$$V = \Omega_{AB} R \sin\Delta; \qquad (5.35)$$

here Δ is the angular separation between the chosen point k and the instantaneous pole of rotation, given by spherical trigonometry:

$$\cos\Delta = \sin\theta_p \sin\theta_k \cos(\varphi_p - \varphi_k) + \cos\theta_p \cos\theta_k . \qquad (5.36)$$

The position of the instantaneous pole of rotation can be found from the direction of the relative velocity vectors at two points along the border between the two plates: in fact, the great circles perpendicular to these velocity vectors meet at the pole of rotation. The direction of the velocity vectors can easily be determined by observing the orientation of the transform faults. On the other hand, the magnitude of the relative velocity (and hence of Ω_{AB}) can be derived at different points of the plate border from paleomagnetically determined ocean floor spreading rates; additional data on the directions of relative motion are obtained from the slip vectors at earthquakes. By choosing a set of main reference plates (usually eleven, shown in Fig. 5.7) and carrying out a least squares fit of all the available data (corresponding to several hundreds independent measurements) about their relative motions, a self-consistent global model can be obtained, yielding, for every pair of neighbouring plates, Ω_{AB}, φ_p and θ_p (see Table 5.1).

Of course, these parameters are not independent. Denoting by $\mathbf{\Omega}_A$ the absolute rotation velocity vector of the plate A, the relative rotation with respect to plate B is described by the vector

$$\mathbf{\Omega}_{AB} = \mathbf{\Omega}_A - \mathbf{\Omega}_B . \qquad (5.37)$$

For any triplet of plates these vectors fulfil the relationship

$$\mathbf{\Omega}_{AB} + \mathbf{\Omega}_{BC} + \mathbf{\Omega}_{CA} = 0 . \qquad (5.38)$$

Of course, this model is only approximate for two main reasons: first, the rotation rates and the poles are in fact averaged over a time span of at least a few million years; secondly, in the regions where the structure (and dynamics) of the crust is complex on a local scale and many micro-plates interact (e. g., in the Mediterranean area), global models provide inadequate representations. For these reasons it is very important that the plate velocities are now measured directly and independently. If the present motion is different from the motion averaged over geological times, the corresponding acceleration has important geophysical consequences. For example, the accumulation of stress energy produced by the interaction between neighbouring plates may suddenly be released with an earthquake and modify the resistance to the motion. Large research programs are now going on to measure

Plate pair	Latitude deg N	Longitude deg E	Rotation rate ($deg/10^6$ y)
NOAM-PCFC	48.77	– 73.91	0.852
COCO-PCFC	38.72	– 107.39	2.208
NAZC-PCFC	56.64	– 87.88	1.539
EURA-PCFC	60.64	– 78.92	0.977
INDI-PCFC	60.71	– 5.79	1.246
ANTA-PCFC	64.67	– 80.23	0.964
COCO-NOAM	29.80	– 121.28	1.489
AFRC-NOAM	80.43	56.36	0.258
EURA-NOAM	65.85	132.44	0.231
NOAM-CARB	– 33.83	– 70.48	0.219
COCO-CARB	23.6	– 115.55	1.543
NAZC-CARB	47.30	– 97.57	0.711
COCO-NAZC	5.63	– 124.40	0.972
NOAM-SOAM	25.57	– 53.82	0.167
CARB-SOAM	73.51	60.84	0.202
NAZC-SOAM	59.08	– 94.75	0.835
AFRC-SOAM	66.56	– 37.29	0.356
ANTA-SOAM	87.69	75.20	0.302
INDI-AFRC	17.27	46.02	0.644
ARAB-AFRC	30.82	6.43	0.260
AFRC-EURA	25.23	– 21.19	0.104
INDI-EURA	19.71	38.46	0.698
ARAB-EURA	29.82	– 1.64	0.357
INDI-ARAB	7.08	63.86	0.469
NAZC-ANTA	43.2	– 95.02	0.605
AFRC-ANTA	9.6	– 41.70	0.149
INDI-ANTA	18.67	32.74	0.673

Table 5.1. A global kinematical model for lithospheric plate motions, according to J.B. Minster and T.H. Jordan, J. Geophys. Res **83**, pp. 5331-5354 (1978). The instantaneous rotation rate vector is specified by the geographical coordinates of the corresponding pole and its magnitude. The first plate in every pair moves counterclockwise with respect to the second. The errors are usually of a few degrees for the pole coordinates, and of 0.1 - 0.01 $deg/10^6$ y for the rotation rates.

directly the present plate motions using extraterrestrial objects: the satellite LAGEOS (see Secs. 18.1 and 20.2), the set of Global Positioning Satellites (Sec. 18.1) and extragalactic radio sources with VLBI (Sec. 20.5). All these methods are capable to achieve the required accuracy of a few cm/y or better.

Of course, these global models by definition contain an arbitrary rotation of the whole system, since they can describe only relative motions. To describe the absolute motion of the plates with respect to the interior of the planet, we need to know the absolute motion of at least one plate or, alternatively, to identify a set of points on the surface which are stationary with respect to the deep interior. A "marker" role of this kind is played by the hot spots which, as said earlier, are probably linked to features in the lower mantle unrelated with upper mantle convection and lithopsheric motions. This is confirmed by the fact that hot spots remain in almost fixed relative positions. Moreover, the abundance and prominence of hot spot volcanism in regions like Africa and Eastern Asia (compared, for instance, with America) are consistent with the resulting, very small ($\lesssim 1$ cm/y) absolute velocities of the corresponding plates (see Fig. 5.7). In general, the absolute velocities of plates appear to correlate with the relative length of their subduction margins and to anticorrelate with the relative extensions of their continental parts; no clear correlation is apparent with the absolute size of the plates.

From the point of view of the energy budget, the mechanical energy needed to feed the observed seismic, volcanic and orogenetic activities in the lithosphere is of the order of $10^{18} - 10^{19}$ erg/s, some two orders of magnitude lower than the thermal energy flux coming from the earth's interior. However, the detailed causal relationship between the observed tectonic motions and the system of convective currents in the upper mantle is still largely unknown.

PROBLEMS

5.1. Prove that within a spherically symmetric planet, where the propagation velocity v of the wave is a function only of the radius r, along any given ray the quantity $p = r \sin(i)/v(r)$ is a constant (here i is the angle between the ray and the radial direction). Hint: consider three homogeneous spherical layers and apply Snell's law at the two boundaries.

*5.2. Deduce the constancy of p (previous problem) from Fermat's variational principle for geometrical optics. This principle says that the travel time

$$\int ds/v(\mathbf{r}(s))$$

between two points is a minimum. This entails for the path $\mathbf{r}(s)$ the differential equation:

$$\frac{d}{ds}\left(\frac{1}{v}\frac{d\mathbf{r}}{ds}\right) = \nabla\frac{1}{v}.$$

With spherical symmetry p is a first integral.

*5.3. Compute as a function of the incidence angle, the energy conversion factor from S- to P-waves and from P- to S-waves at a free boundary.

5.4. What is the heat flux needed to balance radioactive heat production in the earth?

5.5. Estimate the global kinetic energy of the plate motion.

5.6. On the basis of the model of Minster and Jordan for the plate motions (Table 5.1), compute the velocity of some important cities, i.e., New Delhi with respect to the Soviet Union.

*5.7. The parameters of Table 5.1 are not independent; the relative motion of 11 plates is determined by 30 parameters. Set up the appropriate formalism to determine by a least square fit the angular velocities of 10 plates with respect to the last.

FURTHER READINGS

Three useful books, exploring in more details the subject of this chapter, are F.D. Stacey, *Physics of the Earth*, Wiley (1977); W.M. Kaula, *An Introduction to Planetary Physics: the Terrestrial Planets*, Wiley (1968); Y.N. Zarkov, *Internal Structure of the Earth and the Planets*, MIR, Moscow (1986). The second part of the last book also contains up-to-date discussions about the interiors of the moon and the other planets. L. Brekhovskikh and V. Goncharov, *Mechanics of Continua and Wave Dynamics*, Springer-Verlag (1985) contains a clear discussion of wave propagation in a solid. For the model of plate motion quoted in Sec. 5.5, see J.B. Minster and T.H. Jordan, J. Geophys. Res., **83**, 5331-5354 (1978). On the normal modes, there is E.R. Lapwood and T. Usami, *Free Oscillations of the Earth*, Cambridge University Press (1981).

6. PLANETARY MAGNETISM

The fact that a small needle of lodestone suspended at its center of mass keeps an almost constant orientation with respect to the earth's body has been known for thousands of years, and widely used for navigation purposes. It was William Gilbert who in 1600 associated this effect with the magnetic properties of the earth itself. In his *De Magnete* – one of the earliest scientific treatises ever written – Gilbert stated that the earth behaves like a great permanent magnet, its magnetic force being closely similar to that of a uniformly magnetized sphere. In spite of these early practical and theoretical achievements, the study of the earth's magnetic field has progressed slowly. Its complex structure and rapid time variability, the fact that the overall field is produced by at least three different sources (currents in the conductive, rotating and convective core; magnetization of crustal rocks; plasma currents in the ionosphere and farther out) and, finally, the mathematical difficulties of any theory aiming at quantitative predictions concerning geomagnetism, can explain why this problem was once described by Einstein as one of the three most difficult, unsolved issues of physics. Today, thanks also to space missions, we have a much more detailed knowledge of the earth's field and also of the magnetic behaviour of other planets. Moreover, planetary magnetic fields provide important evidence about the structure of planetary interiors and, for the earth, about the motions of its crust, while their role has become apparent in determining the properties of the regions surrounding the planets and their interactions with the interplanetary medium.

6.1. THE MAIN DIPOLE FIELD

The magnetic field at the earth's surface $\mathbf{B}_0$ can be expressed in terms of its horizontal component H, its radial component A (positive downward) and the deflection angle of a magnetized needle with respect to the local geographical north (positive eastward), the so-called *magnetic declination* D. Also traditionally used is the deflection of the needle with respect to the horizontal plane I = arctan (H/A). Of course this is equivalent to using three rectangular components in the x (northward), y (eastward) and z (downward) directions:

$$\begin{pmatrix} B_0 x \\ B_0 y \\ B_0 z \end{pmatrix} = \begin{pmatrix} H \cdot \cos D \\ H \cdot \sin D \\ A \end{pmatrix} = B_0 \begin{pmatrix} \cos I \cdot \cos D \\ \cos I \cdot \sin D \\ \sin I \end{pmatrix} . \qquad (6.1)$$

The geomagnetic field strength is a fraction of one gauss (Γ). Each component has been globally measured to a relative accuracy of about 10^{-5} (e. g., by the MAGSAT satellite launched in 1979). The field strength changes by a factor of the order of 2 on the surface, and undergoes perceptible changes with time, in the order of 0.1% per year (see Fig. 6.1 and Sec. 6.3).

Since no free magnetic pole is known to exist ($\nabla \cdot \mathbf{B} = 0$), if one disallowes local currents and time variations (implying $\nabla \times \mathbf{B} = 0$), the field can be expressed as gradient of a scalar potential V_M obeying Laplace's equation $\nabla^2 V_M = 0$. As a consequence, at great distances from the region where the main part of the field is generated (i. e., the planetary core), its overall structure can be approximated by the field of a magnetic dipole **d** at the centre of the earth:

$$\mathbf{B} = -\nabla(\mathbf{d} \cdot \mathbf{r}/r^3) . \qquad (6.2)$$

In polar coordinates the components of **B** are

$$B_\theta = -\frac{1}{r}\frac{\partial V_M}{\partial \theta_M} = \frac{d}{r^3} \sin\theta_M \qquad (6.3_1)$$

$$B_r = -\frac{\partial V_M}{\partial \theta_M} = \frac{2d}{r^3} \cos\theta_M \qquad (6.3_2)$$

(θ_M, the *magnetic colatitude*, is the angle between **d** and **r**); on the surface, assumed to be a sphere of radius R, we have

$$B_{\theta_0} = -H = B_e \sin\theta_M ,$$
$$B_{r_0} = -A = 2B_e \cos\theta_M , \qquad (6.4)$$

where $B_e \equiv d/R^3$ is about 0.31 Γ (corresponding to $d = 7.9 \cdot 10^{25}$ Γ cm^3) and **d** is inclined at about 11^0 to the earth's polar axis.

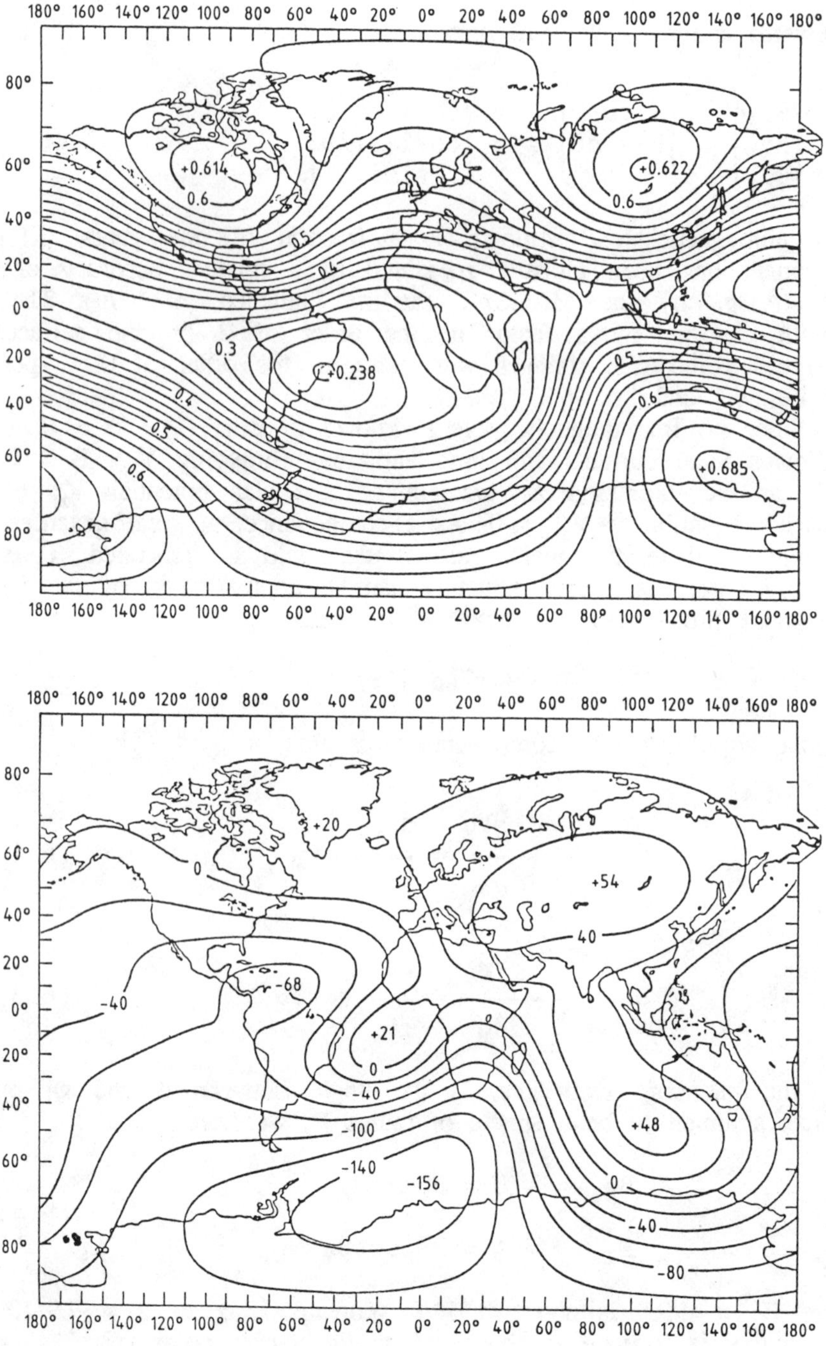

Fig. 6.1. The total intensity of the geomagnetic field at the surface (above, in Γ) and its secular change (in 10^{-5} Γ/y) for epoch 1965.0 [from Cain et al., J. Geophys. Res. **70**, 3647 (1965)].

From eq. (6.4) we get

$$B_0 = \sqrt{[(B_{\theta_0})^2 + (B_{r_0})^2]} = B_e\sqrt{(1 + 3\cos^2\theta_M)} \quad , \quad (6.5)$$

implying that the field is stronger near the magnetic poles (the polar value being twice the equatorial one) and that

$$\tan I = B_{r_0}/B_{\theta_0} = 2\cot g\theta_M \quad . \quad (6.6)$$

Since $B_r/B_\theta = dr/rd\theta_M$, the lines of force $r(\theta)$ are solutions of the differential equation

$$dr/rd\theta_M = 2\cot g\theta_M \quad , \quad (6.7)$$

readily integrated into

$$r = r_e\sin^2\theta_M \quad . \quad (6.8)$$

r_e is the radial distance at which the line of force crosses the (magnetic) equatorial plane. Although, as recognized by Gilbert, a dipole field can be generated by a uniformly magnetized sphere, the magnitude of d is so large that this kind of source is ruled out.

Other planets as well are known to have significant, dipole-like magnetic fields, which have been measured by spacecraft in planetocentric orbits for the moon, Mars and Venus and in fly-by's for Mercury, Jupiter, Saturn, Uranus and Neptune.

	R(km)	$B_e(\Gamma)$	$d(\Gamma cm^3) = B_eR^3$	Tilt to polar axis (°)
Mercury	2440	$2\cdot10^{-3}$	$2.7\cdot10^{22}$	?
Venus	6050	$<2\cdot10^{-5}$	$<4.3\cdot10^{21}$	—
Earth	6370	0.31	$8.1\cdot10^{25}$	11
Moon	1760	$<2\cdot10^{-6}$	$<10^{19}$	—
Mars	3390	$\leqslant3\cdot10^{-4}$	$<1.2\cdot10^{22}$	—
Jupiter	69800	4.2	$1.4\cdot10^{30}$	10
Saturn	58400	2.1	$4\cdot10^{29}$	$\leqslant1$
Uranus	25500	0.23	$3.8\cdot10^{27}$	60
Neptune	24750	0.1	$1.5\cdot10^{27}$	50

Table 6.1. The parameters of the main dipole field for the planets and the sun.

6.2. MAGNETIC HARMONICS AND ANOMALIES

Since the geomagnetic potential V_M is a harmonic function, it can be represented as a series of spherical harmonics. The coefficient of the terms of order ℓ are proportional to $1/r^{\ell+1}$ for an internal source (see eq. (2.20)) and to r^{ℓ} for an external source like the ionosphere (see eq. (2.21)). Therefore the ℓ-th harmonics are described by 4 coefficients for each value of the order $m \neq 0$ and by 2 coefficients for $m = 0$. Using the representation (2.1) of the spherical harmonics in terms of normalized, associated Legendre functions, we have

$$V_M = \frac{1}{R} \Sigma_{\ell m} P_{\ell m}(\cos\theta) \left\{ \left[c_{\ell m} \left(\frac{r}{R} \right)^{\ell} + (1 - c_{\ell m}) \left(\frac{R}{r} \right)^{\ell+1} \right] C_{\ell m} \cos m\varphi \right.$$

$$\left. + \left[s_{\ell m} \left(\frac{r}{R} \right)^{\ell} + (1 - s_{\ell m}) \left(\frac{R}{r} \right)^{\ell+1} \right] S_{\ell m} \sin m\varphi \right\} . \qquad (6.9)$$

Here θ and φ are the colatitude and the longitude; the coefficients $c_{\ell m}$ and $s_{\ell m}$ are numbers between 0 and 1 describing the fractional contributions of the internal sources. As in Ch. 2, the limits of summation with respect to ℓ (from 0 to ∞) and with respect to m (from 0 to ℓ) are understood. Note also that, since the monopole terms $\ell = 0$ are absent, a shift in the origin will change the quadrupole and higher terms, but not the dipole term.

Of course we do not measure V_M, but the magnetic field

$$B_\theta = \partial V_M / r\partial\theta, \quad B_\varphi = \partial V_M / r\sin\theta\partial\varphi, \quad B_r = \partial V_M / \partial r. \qquad (6.10)$$

With the traditional method, using magnetometers on the surface of the earth $r = R$, the four sets of coefficients c, s, C, S can, in principle, be measured (in fact B_θ and B_φ do not contain c and s, but B_r does). C. F. Gauss first applied this method in 1839, showing from the data available at that time that $c_{\ell m} = s_{\ell m} = 0$, i.e., the geomagnetic field is entirely internal. Today we know that this is only a good approximation, since a field up to $\approx B_e/100$ (with typical time scales ranging from milliseconds to 11 y, the solar cycle period) may be due to currents in the ionosphere or further out in the magnetosphere. These external field variations induce electric currents in the upper mantle and temporal variations of the internal field which are sensitively dependent upon the conductivity of the mantle. The latter quantity and its variations with depth can thus be estimated.

The internally generated field (from eq. (6.9)) is usually written as

$$V_M = R\Sigma_{\ell m}(R/r)^{\ell+1} P_{\ell m}(\cos\theta)(g_{\ell m}\cos m\varphi + h_{\ell m}\sin m\varphi),$$

where the *Gauss' coefficients* $g_{\ell m} = C_{\ell m}/R^2$, $h_{\ell m} = S_{\ell m}/R^2$ have the dimension of a magnetic field. The $\ell = 1$ terms give the dipole field: g_{10}, g_{11} and h_{11} are equal to R^3 times d_z, d_x and d_y, respectively (with the z-axis along the polar axis);

$$d = R^3 \sqrt{[(g_{01})^2 + (g_{11})^2 + (h_{11})^2]}$$

and the tilt angle is

$$\theta_M = \mathrm{arctg}\{\sqrt{[(g_{11})^2 + (h_{11})^2]}/g_{01}\}.$$

The quadrupole terms $\ell = 2$ (see Table 6.2) correspond in part to a displacement of the dipole from the centre of the earth by an amount of about 0.07 R (Problem 6.3). In general, the nondipole part of the field has a total intensity of the order of 10% of the dipole field.

ℓ	m	$g_{\ell m}(10^{-5}\Gamma)$	$h_{\ell m}(10^{-5}\Gamma)$
1	0	– 30339	0
1	1	– 2123	5758
2	0	– 1654	0
2	1	2994	– 2006
2	2	1567	130

Table 6.2. Gauss' coefficients for the dipole and the quadrupole magnetic field, for the same epoch as Fig. 6.1.

As regards the higher order coefficients, measurements at the earth's surface and by orbiting spacecraft have provided values up to $\ell \approx 25$. It is interesting to study the variation with ℓ of the quantity

$$R_\ell = (\ell + 1)\Sigma_m[(g_{\ell m})^2 + (h_{\ell m})^2], \qquad (6.11)$$

which represents the mean square value over the earth's surface of the magnetic field intensity produced by harmonics of ℓ-th degree. The factor $\ell + 1$ is a consequence of the appropriate normalization (2.7) for the spherical harmonics. We have come across a similar quantity (2.39) in discussing the spatial inhomogeneities of the gravitational field of the earth. As shown by Fig. 6.2, a knee in the (spatial) spectrum is apparent near degree 14: for $\ell \geqslant 14$, the spectrum is almost "white", with all harmonic degrees almost equally represented, while for $\ell \leqslant 14$ the relationship

$$R_\ell = 0.135 \cdot (3.70)^{-\ell} \cdot \Gamma^2 \qquad (6.11')$$

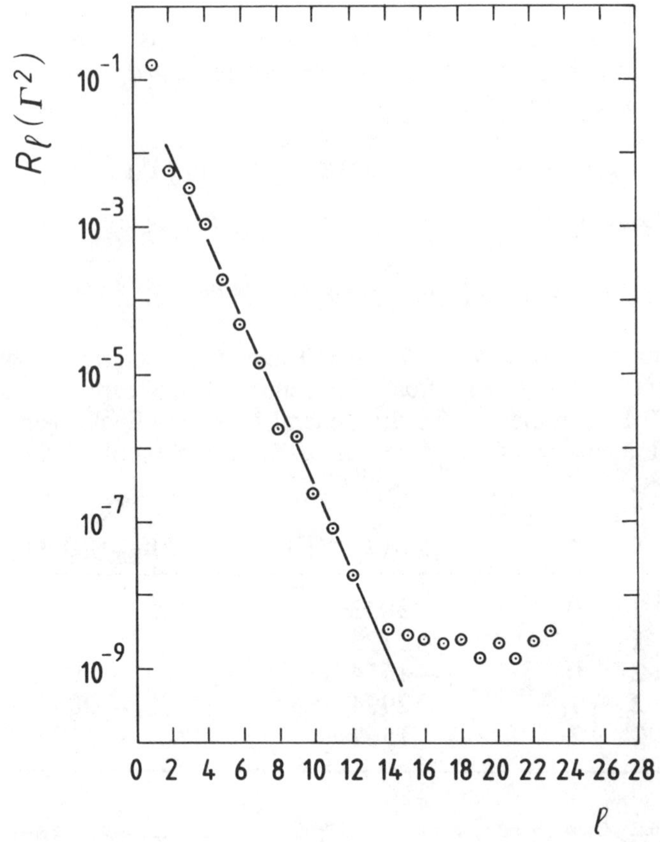

Fig. 6.2. The spatial spectrum of the geomagnetic field R_ℓ vs. ℓ, derived by the MAGSAT mission. Notice the knee at $\ell = 14$, where the best-fitting relationship (6.11') becomes no longer valid, and the spectrum becomes almost white (from Langel and Estes, Geophys. Res. Lett. **9**, 250, (1982)).

fits well the data. This behaviour is very different from the typical power spectra one has in random media, like Kaula's rule (Fig. 2.3). Here the data are approximated by a straight line in the (ℓ, $\log R_\ell$) plane, corresponding to the exponential decrease of a spherical harmonic of large order with height (see Sec. 2.4). It is therefore natural to attribute this dependence not to the intrinsic spectrum of the source, but to its distance from the surface. Since the field strength due to terms of degree ℓ varies with r as $(r/R)^{-(\ell+2)}$, extrapolating downward the spectrum (6.11') makes it progressively less dependent on ℓ until, at $r \approx R/\surd(3.70) \approx 0.52R \approx 3300$ km the spectrum loses its exponential dependence on ℓ. This occurs in the outer core, close to the seismically determined core-mantle boundary. This suggests that the low degree part of the geomagnetic field has its sources in the outer part of the core.

Thus the dominance of the dipole field, as seen at the surface, is not present at the depth where the field is generated, and is mostly a consequence of the significant distance between the source and the surface. For Jupiter and Saturn, and especially for Uranus and Neptune, the quadrupole and octupole terms are even larger with respect to the dipole than for the earth, possibly implying that the field-generating layers are closer to the planetary surfaces.

The high degree harmonics for $\ell > 14$, which show a white behaviour very close to the surface, are generated by magnetized crustal rocks lying within a depth of the order of R/14 (see Sec. 2.4); their amplitude is of the order of 10^{-4} that of the main field, and their wavelength ranges up to R/14 $\approx$ 400 km. There are hints that these high degree "anomalies", unlike the low degree field, are associated with large scale geological and tectonic features.

6.3. SECULAR CHANGES AND REVERSALS

Already in the 17th century it was discovered that the geomagnetic field is not steady, but varies in strength and orientation in a coherent way over large areas of the earth (see Fig. 6.1). These variations are much stronger than the transient changes due to phenomena in the ionosphere or the magnetosphere and probably occur on all time scales from ≈ 10 to $\approx 10^9$ y. Direct observations are available for the last few centuries, showing time scales up to $\approx 10^4$ y. The main observed secular effects are the following:

– The dipole field is decreasing in strength by about 0.05%/y ($1.5 \cdot 10^{-4}$ Γ/y) and is drifting westward in an almost precessional way at a rate of about $0^{\circ}.07$/y (corresponding to a period of $\approx$ 5000 y).

– The nondipole part of the field is changing more rapidly, by some $5 \cdot 10^{-4}$ Γ/y on the average, but with a large scatter. These variations give rise to an average westward drift of the nondipole features at about $0^{\circ}.2$/y (corresponding to an average rotation period of $\approx$ 1800 y with respect to the solid earth), but with a strong variability between different latitudes and different harmonic components. In fact, these features appear to deform, are displaced and disappear; they behave more like eddies in a fluid stream than like permanent or quasi permanent features.

Since the low-degree field is thought to be generated in the planetary core, this behaviour suggests that motions in the core do not occur in a steady, ordered way, but rather in a turbulent manner. It has also to be stressed that magnetic fields associated with other cosmic bodies (e. g., the sun and other stars) always show a high and somewhat erratic variability, probably a basic characteristic of their generating mechanism.

Paleomagnetism – the study of the ancient geomagnetic field by measurement of the remnant magnetization of rocks – provides information on previous field orientations and, to a lower accuracy, strengths, on much longer time scales, up to $\approx 10^9$ y. For instance, if we define a paleomagnetic pole as the axis of the dipole field that would give rise in the rock to a local field parallel to the measured magnetization, it can be determined by noticing that it must lie on the great circle defined by the ancient magnetic declination D, at an angular distance equal to the ancient magnetic colatitude θ_M, related by eq. (6.6) to the dip angle I measured in the rock. Due to their secular variations, the non-dipole components can be averaged out by taking many measurements from rock formations which cooled or were deposited over time spans of several thousand years. Many of these paleomagnetic poles have been determined for rocks up to $\approx 10^7$ y old, yielding the important result that the poles cluster about the earth's rotation axis. This leads to the so-called *axial dipole hypothesis*, that the dipole field, averaged throughout the history of the planet, is aligned with the rotation axis. This implies that, on very long time spans ($\succ 10^4$ y), the geomagnetic field can be approximated by an aligned, geocentric dipole, while the present 11° inclination is just a transient state. The axial dipole hypothesis on the one hand indicates that the earth's rotation is an essential ingredient of the mechanism generating the field, and on the other hand it enables geologists to reconstruct from paleomagnetic data the positions and motions of the continents relative to the rotational axis.

In spite of this regular, long-term behaviour of its orientation, the observed changes of the magnetic polarity in a series of lava flows or sedimentary layers show that the field underwent repeated reversals. Unlike the sun, for the earth this phenomenon displays no periodicity or regularity, but appears rather stochastic (see Fig. 6.3). The current reversal rate is a few times per million years (the last one occurred $7 \cdot 10^5$ y ago), but this rate has varied a great deal over geological times: two very long periods with no reversal, lasting 36 and 70 million years, began about 119 and 320 million years ago respectively. On the other hand, the actual reversals take place in ≈ 1000 y; on this time scales there are also many fluctuations, incomplete or temporary reversals, referred to as *paleomagnetic excursions*. Thus the zero-field

Fig. 6.3. The sequence of normal (black) and reversed (white) polarity states in the last 79 million years [from Heirtzler et al., JGR, **73**, 2119, (1968)].

state appears as an unstable state, which the system crosses rapidly before settling to a temporarily stable state with normal or reversed polarity. The time scale of 10^8 y in the reversal rate may, on the other hand, reflect changes in the pattern of convection currents in the fluid outer core.

*6.4. THE GENERATION OF PLANETARY MAGNETIC FIELDS

Under some conditions, which have been reviewed in Ch. 5, the heat content and the internal pressure of a planet are so high that its core behaves as a highly viscous fluid. When it contains metals, it is also an electric conductor and it is in principle possible that the kinetic energy of its internal motion is transformed, through magnetohydrodynamical effects, into magnetic energy. However, the origin of the main dipole field, more or less aligned with the angular momentum, is an important and little understood problem in planetary physics. Its mathematical basis is straightforward and deceptively simple: given a velocity field **v** (determined by the external forces and convection), solve the magnetic diffusion equation (1.64):

$$\partial \mathbf{B}/\partial t = \lambda \nabla^2 \mathbf{B} + \nabla \times (\mathbf{v} \times \mathbf{B}) \ . \qquad (6.12)$$

Here

$$\lambda = c^2/4\pi\sigma \qquad (6.13)$$

is the *magnetic diffusion coefficient* and σ is the electrical conductivity. In particular, we wish to characterize the velocity fields **v** which maintain and amplify a large scale, aligned dipole field, providing an effective *dynamo*. The effect of the last term in eq. (6.12) has been discussed in Sec. 1.6: it anchors the field lines to the fluid elements, dragging them along with the flow (*frozen-in field*).

In the interior of the earth λ is of order 10^4 cm^2/s; for a dynamo of size R, therefore, the magnetic decay time is

$$\tau_m = R^2/\lambda = (R/3\cdot 10^8\,\mathrm{cm})^2/(\lambda/10^4\,\mathrm{cm^2\,s^{-1}})\,3\cdot 10^5\,\mathrm{y}, \qquad (6.14)$$

much less than the age of the earth, implying that the field cannot be the residual of a primordial magnetization, but must be continuously regenerated. It is important to note that in terrestrial type materials the heat transport coefficient is $\approx 10^5$ times smaller, thus allowing the permanence over geological times of internal heat sources. In the enormous difference between these two diffusion processes lies the peculiarity of dynamo processes.

Magnetic diffusion processes are characterized by the *magnetic Reynolds number*

$$Re_m = Rv/\lambda \,, \qquad (6.15)$$

which gives the order of magnitude of the ratio between the last and the last but one term in eq. (6.12). We can estimate the velocity v in the earth's core (with a radius $R = 3 \cdot 10^8$ cm) from the known westward drift of 0.2^0 per year of the non-dipole field. At the radius R this corresponds to $v \approx 0.01$ cm/s, giving a magnetic Reynolds number of 300. Note that these velocities are several orders of magnitude higher than those of the crust ($\approx 10^{-7}$ cm/s). One can say, the core is characterized by a kinematical time scale

$$\tau_R = R/v = \tau_m/R\ell_m \approx 10^3 \text{ y}, \qquad (6.16)$$

much shorter than τ_m.

People have therefore investigated at length the possibility that the induction produced by a planetary velocity field will be able to compensate the much weaker diffusion effect and maintain an overall, axially symmetric field throughout the planet. This is the *laminar dynamo* problem. The general conclusion of this research of mathematical physics is that a high degree of symmetry is incompatible with the permanence of the configuration and that the magnetic decay eventually prevails. The reason for this is, roughly speaking, that, for a "reasonable" velocity field, the induction operator $\nabla \times (\mathbf{v} \times$ changes the symmetry character of the magnetic configuration, while the (scalar) diffusion operator does not: the two processes cannot compensate.

The studies of laminar magnetic configurations have clarified the transformation produced by the induction operator on two main types of symmetries: *poloidal* and *toroidal.* Both of them are axially symmetric (independent of longitude); the poloidal magnetic field (typically a dipole) has lines of force in the meridian planes and the toroidal magnetic field along the parallels. For example, it has been shown that the induction operator corresponding to differential rotation generates a toroidal component from a poloidal component. This is intuitively seen by noting that different parts of a line of force in a meridian plane at different distances from the axis of rotation are dragged around differently and in the interior, where the rotation is generally faster, develop a kink toward the eastward direction. The poloidal field so generated changes sign with latitude. This is the so-called ω-*effect* and has been studied by E.N. Parker.

Laminar dynamos are inadequate not only mathematically, but also factually: we know that generally convection induces in the core of the planet turbulent motions, whose elementary eddies may escape the "no go" theorems for symmetric dynamos. The view is now generally upheld that the overall, permanent magnetic field over a planetary scale is a by-product of the interaction and interference of elementary magnetic eddies driven by convection.

We can gain some understanding of the problem with the *mean field electrodynamics* (Steenbeck, 1966). The idea is to split the field quantities in a part $\mathbf{B}_0$, $\mathbf{v}_0$, which changes over the planetary scale R; and a fast varying part $\mathbf{B}'$, $\mathbf{v}'$, with a scale $\ell \ll R$. Similarly to what one does in the theory of turbulence, the primed quantities are interpreted as stochastic variables with zero mean and describe an ensemble of planets with the same large scale field, but different microfields. By averaging over the ensemble one is able to deduce statistical conclusions about the planetary field, irrespective of the particular realization of the turbulence. Indicating with angular brackets this averaging procedure, we obtain

$$\partial \mathbf{B}_0/\partial t = \lambda \nabla^2 \mathbf{B}_0 + \nabla \times (\mathbf{v}_0 \times \mathbf{B}_0) + \nabla \times \langle \mathbf{v}' \times \mathbf{B}' \rangle \,. \tag{6.17}$$

The turbulent field fulfils:

$$\partial \mathbf{B}'/\partial t = \lambda \nabla^2 \mathbf{B}' + \nabla \times (\mathbf{v}_0 \times \mathbf{B}') + \nabla \times (\mathbf{v}' \times \mathbf{B}_0) \,. \tag{6.18}$$

The first equation is the stochastic average of eq. (6.12) (note that $\langle \mathbf{B} \rangle = \mathbf{B}_0$ and $\langle \mathbf{v} \rangle = \mathbf{v}_0$); the second equation is (6.12) minus (6.17).

Leaving aside for the moment the macroscopic velocity $\mathbf{v}_0$, we must be in a situation in which the term $\nabla \times (\mathbf{v}' \times \mathbf{B}_0)$, which drives the turbulence, is of the same order of magnitude, or larger than, the diffusion term $\lambda \nabla^2 \mathbf{B}'$; that is to say, the magnetic Reynolds number of the turbulence $\ell v'/\lambda$ must not be smaller than B'/B_0. In principle we now must solve eq. (6.18) for $\mathbf{B}'$ in terms of the stochastic velocity field $\mathbf{v}'$, when $\mathbf{B}_0$ is given and slowly varying; and evaluate the effective electric field

$$\mathbf{E} = \langle \mathbf{v}' \times \mathbf{B}' \rangle / c \tag{6.19}$$

which drives eq. (6.17). Since $\mathbf{B}_0$ is essentially constant over the time scale of the microfield B', the relevant solution of eq. (6.18) is linear in $\mathbf{B}_0$, and so is **E**; moreover, since we are solving eq. (6.18) locally, over the scale ℓ, the solution depends only on the local value of $\mathbf{B}_0$ and not its spatial derivatives; therefore there is a dimensionless and slowly varying tensor α_{ij} which describes this relationship

$$E_i = \alpha_{ij} B_{0j} / c \,. \tag{6.19'}$$

This tensor is determined by the local statistical properties of the turbulent velocity field. The contribution from its skew part is of the same type as the velocity term in eq. (6.17), which corresponds to the electric field

$$\mathbf{v}_0 \times \mathbf{B}_0 / c$$

and can be neglected at this stage; what counts is the symmetric part. In an isotropic situation the tensor α_{ij} must be proportional to the unit matrix and eq. (6.19') reduces to

$$E_i = \alpha B_{0\,i} / c \ . \qquad (6.19")$$

On the basis of these simplifying, but reasonable assumptions we can understand the dynamo process.

Upon a reflection of the axes, the curl operator $\nabla \times$ does not change its sign; hence α goes into $-\alpha$ and is a *pseudoscalar*. This shows that the turbulent velocity field **v**' we need is rather peculiar, in that it must be possible to construct with it pseudoscalar quantities, which are invariant under rotations, but change sign under reflections. A typical quantity of this kind is the *kinematical helicity* $\mathbf{v}' \cdot \nabla \times \mathbf{v}'$, which occurs when the motion has a helical structure and it is not possible to find surfaces orthogonal to the velocity field. (The helicity vanishes when the velocity is proportional to the gradient of a scalar.) A configuration of this kind arises, for instance, when a convective cell rises in a rotating planet: when the rotation vector is directed towards the north pole the helicity is positive in the northern hemisphere and negative in the southern one (see Problem 6.6).

It is easy to see that the curl operator transforms the poloidal field into a toroidal field, and *vice versa.* (In fact, within the class of axially symmetric fields

$$\mathbf{B} = [B_r(r,\theta),\ \ B_\theta(r,\theta),\ \ B_\varphi(r,\theta)],$$

consider the flux Φ_m through any surface on a meridian plane and the flux Φ_p through any surface on a plane of parallels, orthogonal to the z-axis. A poloidal field ($B_\varphi = 0$) is characterized by the condition $\Phi_m = 0$; a toroidal field ($B_r = B_\theta = 0$) corresponds to $\Phi_p = 0$. Similarly, consider the circulation c_m along any closed circuit on a meridian plane, and the circulation c_p along any closed circuit on a plane of parallels; the poloidal and toroidal fields correspond, respectively, to $c_p = 0$ and $c_m = 0$. Since the circulation of the curl of a vector equals its flux, the theorem is immediately proved.)

The effective electric field (6.19") therefore will increase a toroidal field from a given poloidal field, and *vice versa.* This is the *α-effect* and can produce a dynamo in two ways. It can reproduce an original

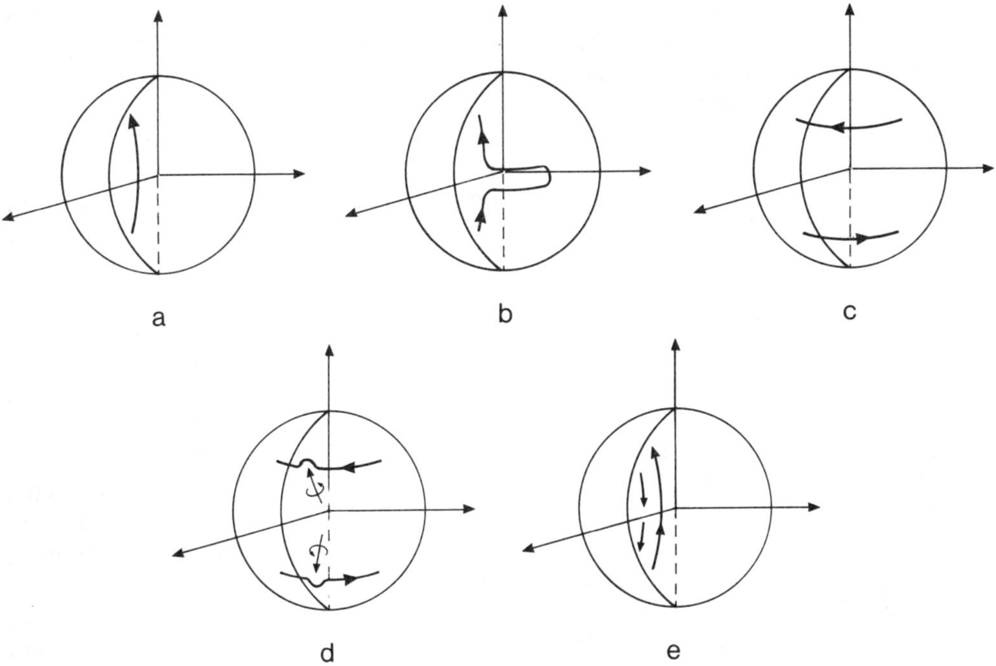

Fig. 6.4 A sequence of configurations leading to the enhancement of an initial poloidal field through the $\omega\alpha$ process.

dipole field which has been distorted by differential rotation into a toroidal configuration ($\alpha\omega$-*effect*); it can also be applied twice to restore the dissipation of an initial poloidal field (α^2 -*effect*).

Working out qualitatively a specific example of the $\alpha\omega$ effect on the basis of these abstract guidelines gives them an intuitive content and justification. Referring to Fig. 6.4, differential rotation produces a toroidal component in a line of force lying in a meridian plane (a, b, c). Rising convective cells create local kinks (d); the right sides of these kinks are displaced to deeper levels by rotation, therefore creating a poloidal, northward component. The left sides of the kinks are displaced to lower depths and generate an opposite poloidal component. The first, however, prevails because of the smaller volume available. As a result, the initial poloidal field is enhanced.

The actual conditions which make a dynamo possible in a given planetary interior; the mechanism for reversals and their frequency; the value of the dipole; the mathematics of dynamo models; and other problems are beyond our scope and not well understood.

The semiqualitative arguments discussed above can be applied to the magnetic fields of other planetary bodies as well (see Table 6.1). Among the terrestrial planets (including the moon), the earth's relatively strong, internally generated field is unique. This seems consistent with the fact that no other body in this class has both a rapid diurnal rotation (only Mars has a spin period comparable with that of the earth) and a large, fluid, convective and metallic (hence conducting) core. It must be emphasized that the role of the "rapid" rotation is uncertain, since no quantitative estimate is available of the minimum spin rate compatible with a dynamo; it is therefore possible that the small or absent global magnetic fields in Mercury, Venus and the moon are due to the properties of their cores (possibly unfavourable to convection) rather than to their slow rotations.

The high mean density of Mercury implies that its metallic core includes a large fraction of the planetary mass, but its spin rate has been slowed down by tidal effects to only $\approx 2\%$ of that of the earth; the small, but non-vanishing magnetic field detected by the *Mariner* 10 probe appears to require that some dynamo mechanism is effective in this planet. Another interesting case is that of the Moon: no global dipole has been detected by the Apollo orbiters, but small scale magnetic features and even some large scale anomalies are present; moreover, the returned lunar rock samples have significant remnant magnetizations. It is presently unclear whether these phenomena are related to "local" events (such as big impacts), or the moon once possessed a large dynamo generated field which magnetized the crust, but subsequently decayed and disappeared. This latter hypothesis requires both that the lunar crust was reoriented by a large polar wander, and that the moon has a small iron core (which, however, cannot account for a large mass fraction, since the moon's mean density is very close to that of the earth's mantle.)

For the giant planets the situation is different. On Jupiter, Saturn, Uranus and Neptune the *Pioneer* and *Voyager* probes detected strong global fields, with surface intensity comparable or higher than that of the earth, and extensive magnetospheres. These fields are presumed to be driven by dynamos arising in electrically conducting planetary interiors: in fact, all these planets have rotation periods smaller than one day (and possibly also a global differential rotation). On the other hand, hydrogen is known to convert to a metallic, conducting form at very high pressures and thus, at least in Jupiter and Saturn, it can play the role of iron in the earth. All the giant planets have non-dipole components relatively stronger than the earth, a fact that might be due to sources significantly nearer to their surfaces. As for the dipole orientations, the situation is puzzling: while Jupiter's inclination angle to the spin axis is similar to that of the earth, Saturn the dipole tilt is so small that only an upper limit of $\approx 1^{\circ}$ could be inferred from the data. On the other hand, for Uranus and Neptune the tilt angles are large; these planets are, therefore, *oblique rotators*, in which the rotation, dragging around the dipole field, creates in the magnetosphere a time dependent magnetic field and, hence, an electric field (similarly to what happens in the pulsars.) The latter has a component along the lines of force and, therefore, can accelerate particles (Problem 6.8). As for the origin of these skewed magnetic fields, it has been speculated that they may arise from the convection of conducting material in a thin shell near the surface.

PROBLEMS

6.1. Show that a uniformly magnetized sphere produces the same field of a suitable dipole situated at its centre: Hint: consider two spheres made of monopoles of opposing polarities, overlapping apart from a small shift.

6.2. Find the vector potential of which the dipolar field is the curl.

*6.3. Show that an arbitrary displacement of the origin adds to a magnetic dipole a particular magnetic quadrupole with three free components. From the measured value of the earth's magnetic quadrupole (Table 6.2), find out the displacement which minimizes the square of the quadrupole $Q_{ij}Q_{ij}$ (see eq. (2.25)). Only the component of the displacement along the dipole is determined, about 0.07 R.

6.4. Find the total energy of the dipole field of the earth and compare it with the rotational energy.

6.5. Suppose that the dipole field of the earth is produced by a uniformly magnetized sphere of radius r. If the magnetic energy energy is equal to the kinetic energy of convection, estimate its rms speed.

6.6. Calculate the helicity of the velocity field sum of a radial, uniform flow and an axial rotation with uniform angular speed.

6.7. Find, as a function of latitude and longitude, the declination D and the deflection from the horizontal plane I of a dipole field of given axis.

6.8. Estimate the energy gain of a proton in the magnetosphere of Uranus.

FURTHER READING

We suggest the general treatise by E.N. Parker, *Cosmical Magnetic Fields*, Clarendon Press, Oxford (1979). On dynamo processes and planetary magnetism in general there is a good book: H.K. Moffat, *Magnetic Field Generation in Electrically Conducting Fluids*. Cambridge University Press (1978). For a more recent review, see R.T. Merrill and M.W. McElhinny, *The Earth's Magnetic Field*, Academic Press (1983) and D.J. Stevenson, *Planetary Magnetic Fields*, Rep. Prog. Phys., **46**, 555-620 (1983). For the dynamo theory, see F. Krause and K.H. Rädler, *Mean-field Magnetohydrodynamics and Dynamo Theory*, Pergamon Press (1980) and F.H. Busse, *Recent Developments in the Dynamo Theory of Planetary Magnetism*, Ann. Rev. Earth Planet. Sci., **11**, 241 (1983).

7. *ATMOSPHERES*

A planetary atmosphere is essential for the development of life in the form we know. The radiation from the sun, the composition with easily combined elements and different chemical phases, the complicated dynamical and thermal structures which produce a great variety of conditions: all these conditions contribute to form a *milieu* favourable to the formation of complex organic molecules. Climatic conditions play here an essential role too. The atmosphere may also determine the outer appearance of a planet, in particular its albedo. Fluid dynamics of air and water is dominated by the effects of gravity and the Coriolis force, which affect local meteorological behaviour, global circulation and energy transfer. Over the whole planet, complicated dynamical patterns are established which depend upon many factors, like the external energy source – the solar flux –, the rotational velocity and the chemical composition and processes.

7.1. THE STRUCTURE OF THE ATMOSPHERE

In a gaseous star like the sun, the distinction between the interior and the atmosphere is not a sharp one: the density decreases gradually as the weight of the higher layers decreases and eventually merges with the interstellar gas density. In this case one can give only an optical definition of the atmosphere, as the layer through which we see by means of telescopes: its thickness is then approximatively equal to the mean free path of the photons. In a solid planet, instead, the atmosphere terminates at the surface and is created and maintained by the emission of gases from it (by sublimation, evaporation, leaking and deabsorption) and/or by the gravitational trapping of interplanetary gas. In principle, as the density decreases with altitude, there are two different physical regimes. At low altitudes, where the molecular mean free path is smaller than the characteristic spatial scale over which the density changes, the molecules have time and space enough to achieve local thermal equilibrium with the surrounding medium; in this regime we can describe the atmosphere as a fluid. At higher altitudes the mean free path increases and, eventually, becomes larger than the vertical scale of the density profile: we have a regime of *free molecular flow*, determined by collisions and radiative interactions. In this regime there is no gravitational confinement for the gas as a whole: a sufficiently fast molecule escapes in interplanetary space (see Sec. 8.4) and, conversely, some interplanetary molecules are trapped in the atmosphere. Were the planet surrounded by empty space and were no gas produced afresh from the ground, the atmosphere would eventually be completely lost. The presence and the composition of planetary atmospheres are therefore the result both of ongoing complicated processes and of the previous chemical history; great diversity is to be expected.

	Mean surface temperature (K)	Surface pressure (dyne/cm²)	Surface gravity (cm/s²)	Main constituents
Venus	750	$9 \cdot 10^4$	890	CO_2(96%), N_2(3%)
Earth	288	$1 \cdot 10^3$	980	N_2(78%), O_2(21%)
Mars	240	7	370	CO_2(95%), N_2(3%)
Jupiter	134*	$2 \cdot 10^3$ *	2500	H_2(89%), He(11%)

Table 7.1. The main properties of the atmospheres of four planets. The asterisks mean that for Jupiter data are not referred to the solid or liquid surface (which lies very deep in the planet and cannot be observed), but to some reference depth specified by the pressure itself. Abundances are given by number of molecules (or volume).

The density profile is determined by eq. (1.25) which, however, greatly simplifies when the thickness is much smaller than the radius R of the planet. Then the gravitational acceleration can be considered as constant; and a "flat earth" approximation, with the solid part in the half-space $z < 0$, is adequate. We then get

$$dP = - g\rho dz. \tag{7.1}$$

To solve the problem we need an additional relationship between pressure and density. In principle, this is provided by the equation of state (for a perfect gas, eq. (1.12)) and by an energy equation to govern the temperature. When the thermal conductivity is large enough to produce a uniform temperature, eqs. (7.1) and (1.13) give

$$P(z) = P(0)\exp(-z/H). \qquad (7.2)$$

$$H = kT/mg \qquad (7.3)$$

is the *scale height*, which for the earth, on the ground, varies between 6 km and 8.5 km. Here m is the mean molecular mass (see eq. (1.13)) and depends upon the composition. The ground values are:

$P(0) = 1.013 \cdot 10^6$ dyne cm^{-2}

$\rho(0) = 0.00129$ g cm^{-3}

$n(0) = 2.69 \cdot 10^{19}$ molecules cm^{-3}.

At the opposite extreme we can assume that the interchange of fluid elements between different layers can be done without heat exchange; the corresponding temperature profile T(z) is then *adiabatic* (see Sec. 5.3). From eq. (5.31), in the case of a perfect gas (usually a good approximation for an atmosphere, but of course not for a planetary interior), the *adiabatic temperature gradient* is (eq. (5.31))

$$|dT/dz|_a = g/c_p. \qquad (7.4)$$

For the particular case of a perfect gas, this result can be derived more simply than we did in Ch. 5, by noting that from the equation of state we get

$$Pd(1/\rho) + dP/\rho = kdT/m$$

and using eqs. (7.1) and (1.56) with $\gamma_3 = \gamma$,

$$k\gamma dT/m(\gamma - 1) = c_p dT = -gdz \qquad (7.5)$$

(see eq. (1.54')). If g is assumed to be constant, the temperature decreases linearly with height, with a gradient which for the earth at the ground level is about 10 degrees per km. We also get

$$d\rho/\rho = -dz/H_a = -dz/\gamma H. \qquad (7.6)$$

In other words, the density profile is still roughly exponential, but the adiabatic scale height H_a is larger than H by a factor γ.

The actual density profile of the atmosphere is indeed exponential up to about 120 km, but above this altitude the heat transport and absorption processes are much more complicated and efficient; as a result, the temperature is much higher and the density decay is much slower (see Fig. 8.5). The composition also changes because of complicated chemical reactions governed mainly by the solar radiation. There is also a remarkable ionization, origin of important electromagnetic phenomena, to be reviewed in Ch. 8.

The properties and the dynamics of planetary atmospheres are dominated by the main external energy source, the solar radiation. The sun emits, to a good approximation, electromagnetic radiation with an equilibrium spectrum corresponding to a temperature $T_\odot = 5800$ K, yielding a total luminosity

$$L_\odot = 4\pi R_\odot^2 \sigma T_\odot^4 \cong 3.9 \cdot 10^{33} \text{ erg/s}. \qquad (7.7)$$

Here $R_\odot \cong 7 \cdot 10^{10}$ cm is the solar radius, while

$$\sigma = \frac{2\pi^5 k^4}{15 h^3 c^2} = 5.67 \cdot 10^{-5} \text{erg cm}^{-2} \text{ s}^{-1} \text{ K}^{-4} \qquad (7.8)$$

is the well known *Stefan-Boltzmann constant.* At a distance D this produces a flux

$$\Phi = \frac{L_\odot}{4\pi D^2} = \left(\frac{R_\odot}{D}\right)^2 \sigma T_\odot^4 \;, \qquad (7.9)$$

which at 1 AU is the so-called *solar constant,* about $1.38 \cdot 10^6$ erg/(cm^2 s). The total amount of radiation impinging upon the earth is $1.7 \cdot 10^{14}$ kW, of which about 10^{14} kW are absorbed. This can be compared with the total energy consumption by human activities, about 10^{10} kW; of course, with a solar radiation flux of 1.4 GW/km^2, the local climate is necessarily affected near a big power station.

The subsequent history of a photon, upon reaching the earth, can be very different and complicated. It may be absorbed by the atmosphere and then reemitted; it may be reflected by the oceanic surfaces or diffused by the ground, to be partially intercepted by the clouds; it may also be absorbed by the ground. The *albedo* A of a surface is the fraction of the incoming solar radiation which is not absorbed; thus the fraction (1 – A) contributes to the thermal energy of the body. For the earth the mean albedo is about 0.3 greatly varying with latitude, longitude, metereological conditions and other factors (for instance, clouds and snow coverage increase very much the local albedo). In a steady state the power absorbed by a planet of radius R is $(1 - A)\pi R^2 \Phi$ and must be reemitted into space; with a uniform surface temperature T_s, it would be determined by

$$4\sigma T_S^4 = (1 - A)\Phi. \qquad (7.10)$$

Although a uniform temperature is possible only for infinite conductivity, this gives indeed a reasonable estimate of the surface temperature of the planets and suggests that it is roughly inversely proportional to the square root of the distance from the sun. A lack of uniformity in temperature between the day and the night hemispheres would decrease the effective emitting surface from $4\pi R^2$ to, say, $4\pi\zeta R^2$ ($\zeta < 1$); then the temperature of the hot part is increased by the factor $\zeta^{-1/4}$. Jupiter, however, is warmer than the heat balance equation predicts, an indication of an active internal heat source, either "fossile" thermal energy from the primordial accretion phase, or a slow collapse under the gravitational force. For the earth eq. (7.10) gives T_s = 255 K, lower than the actual average value, 288 K. This is due to the *greenhouse effect*, an important feature of the heat balance and the climate of planetary atmospheres. Its understanding requires an elementary analysis of the radiation transport in a stratified atmosphere (see later in this Section).

The climate of the earth in the geological past had important effecs on its biota, mainly through the ice ages. These changes are certainly in part due to the complex chemical evolution of the surface and to the tectonic motions of the crust; one wonders if changes in the solar luminosity and effects due to the dynamics of the solar system had a role also. According to the standard theory of stellar evolution the luminosity of the sun has increased by ≈ 30% during the last 4.5 billion years, the age of the solar system. We are interested, however, in changes over a shorter time scale. Note that, from eq. (7.13), a given fractional change in the luminosity produces - other things being equal - a fractional change in the surface temperature four times smaller. The SMM spacecraft, launched in 1980, did measure some rather small changes in the solar luminosity over time scales from minutes to a year, the largest being 0.15% over a time scale of 15 days. Changes over the solar cycle of 11 y are also likely, of course. It is uncertain whether these minute variations can have any climatic effects. For instance, there is some evidence of a much weaker solar activity - the so-called *Maunder minimum* - during the "little ice age" at the end of the 17th century (see Sec. 13.1).

The insolation of the earth at a given latitude may change also because of variations in its orbital motion, in particular its eccentricity and the longitude of the perihelion, produced by planetary perturbations. The obliquity of the ecliptic does also change for the same reason. For example, it was found that the obliquity changes approximately from 22° 30' to 24° 30', with a period of 41,000 y; this change produces a small unbalance in the temperature difference between north and south. Recently these characteristic climatological frequencies found a confirmation in the data obtained from the ice samples extracted from a drilling in Antarctica up to 2,083 m in depth. From the isotopic composition of these samples it was possible to deduce a mean temperature of the earth during the last 160,000 y and to correlate it with the changes in the insolation predicted on the basis of celestial mechanics. A report on these exceptional results from the ice "Vostok core" has appeared in *Nature* on October 1st, 1987. This coupling between climate and planetary motions, first suggested by E. Milankovitch, has recently found another important application from the discoveries about striking climate variations on the surface of Mars, with extensive evidence of ancient floods of liquid water and of long-period changes of the structure of the polar ice caps. Indeed, Mars has both a higher orbital eccentricity and larger long-term obliquity changes than the earth.

A radiative flux f, passing through an infinitesimal absorbing slab of thickness ds changes by

$$df = - f ds/\ell = - f\kappa\rho ds. \qquad (7.11)$$

Were the lenght ℓ constant, the flux would decrease exponentially. The absorbing coefficient $1/\ell$ is usually denoted with $\kappa\rho$ to stress the fact that it is often proportional to the material density ρ; the quantity κ, with dimension $[L^2/M]$, is the *opacity* of the material. In general it is a function of radiation frequency and position (in our case, height).

We now consider a stratified atmosphere, in which every quantity is a function of the height z alone. The quantity

$$\tau(z) = \int_z^\infty dz\kappa\rho \qquad (d\tau = - \kappa\rho dz) \qquad (7.12)$$

is the *optical depth.* This is a dimensionless quantity which vanishes on the top of the atmosphere and is determined by the amount of material above the height z. On the ground $\tau(0)$ gives the total optical thickness. A layer with $\tau \ll 1$ is transparent, or optically thin; optical thickness occurs in the opposite case $\tau \gg 1$. In general the opacity and the optical depth are functions of frequency.

Consider, for simplicity, the case of an atmosphere transparent to the solar radiation (no clouds!), but with significant optical thickness $\tau(z)$ for the infrared radiation emitted from the ground. This flux is determined by the Stefan-Boltzmann law

$$\Phi_r = \sigma T_s{}^4$$

in terms of the ground temperature T_s. Of this radiation only a part escapes, approximatively given by $\Phi_r \exp(- \tau(0))$; what is left is reemitted to the ground. To get an estimate of the ground temperature better than (7.10), consider a place whose zenith makes an angle θ with the sun. There we have the energy balance

$$(1 - A)\Phi_r\cos\theta = \Phi_r\exp[-\tau(0)] = \sigma T_s{}^4\exp[-\tau(0)]. \quad (7.13)$$

Therefore T_s is proportional to

$$(1 - A)^{\frac{1}{4}}\cos^{\frac{1}{4}}\theta \; \exp[\tau(0)/4].$$

This shows why the ground temperature decreases with height and how it is modulated – through the angle θ – with latitude, day and season.

For a quantitative break up of the energy balance of the atmosphere and the ground, consider the total mean flux absorbed by the earth'surface in the optical band, 237·1000 erg/(cm^2 s) (arising from 345 of mean impinging flux, less 108 reflected; in this paragraph we use the unit 1000 erg/(cm^2 s) = W/m^2.) Of this, only 169 reach the ground, while 68 are absorbed by the atmosphere. In the infrared, the earth emits 390 W/m^2, corresponding to the ground temperature of 288 K. Of this, only 237 escape in outer space; the rest, 153 W/m^2, is trapped in the atmosphere, producing the greenhouse effect. The atmosphere, being cooler than the ground, sends down a flux of 327 W/m^2, smaller than

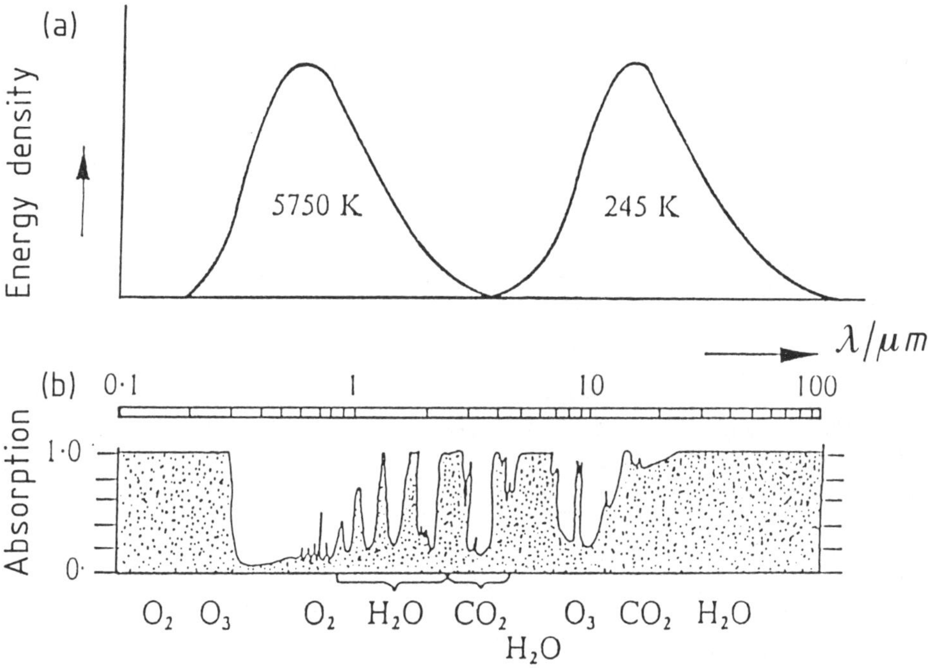

Fig. 7.1. The black body energy densities at T = 5750 K and 245 K (a) and the absorption coefficient {1 – exp(– τ(0))} of the earth's atmosphere in the visible and infrared bands. (From J.T. Houghton, *The Physics of Atmospheres.*)

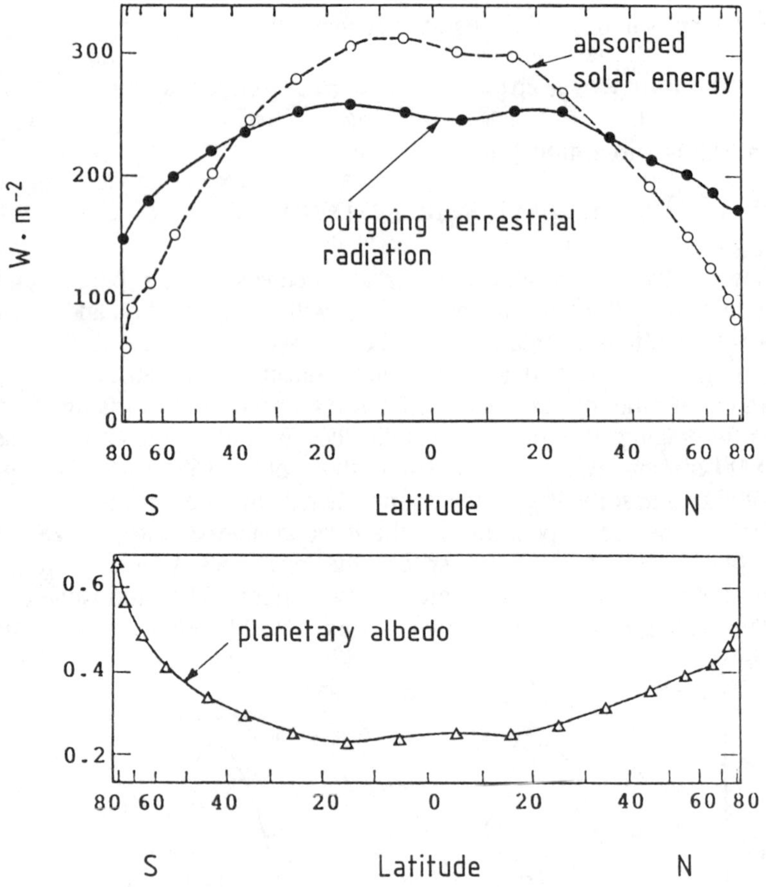

Fig. 7.2. As a function of latitude, the figure shows the earth's albedo (below) and the energy budget on top of the atmosphere (above). (From Houghton.)

the upward flux (390 W/m^2); therefore it loses energy at the rate (327 – 153 – 68 = 106)W/m^2. This loss is compensated by upward turbulent heat transfer and by evaporation heat.

The main sources of atmospheric infrared absorption are the rotation band of the water molecule, the water continuum between 8 and 13 μm, the ozone band at 9.6 μm and the CO_2 band at 15 μm (see Fig. 7.1). The latter is particularly important because it is affected by all combustion processes going on on the earth, in particular by petrol burning. As a consequence of the current increase in the CO_2 content (some % per year), it is widely believed that in the next 20 - 40 years the mean temperature of the earth will slightly increase and the global climate might be affected. An outstanding example of this greenhouse effect is provided by Venus' atmosphere, whose large surface density and

CO_2 abundance increase the surface temperature to ≈ 750 K, well above the value T_s resulting from (7.10).

It is also important to note that the earth's radiation budget depends on the latitude (see Fig. 7.2). At large latitudes, because of the cloud cover and the ice, the albedo is larger than near the equator; as a consequence, there is a net energy loss from the polar regions and an energy gain from the equatorial belt. This unbalance is compensated by atmospheric motions and ocean currents.

*7.2. RADIATIVE TRANSFER

The elementary analysis of radiative transfer in a plane stratified atmosphere that we have previously given can be replaced with a rigorous formulation of outstanding mathematical interest. Such a formulation is required for precise results and more complicated situations.

In local thermodynamical equilibrium (see Sec. 1.7) the spectrum of photons – the number of photons in the frequency interval $(\nu, \nu + d\nu)$ contained in a volume dV – is uniquely determined by the local temperature T in terms of *Planck's distribution function*:

$$n_\nu d\nu dV = \frac{8\pi\nu^2}{c^3} \frac{d\nu dV}{\exp(h\nu/kT) - 1} . \qquad (7.14)$$

The corresponding spectral energy density

$$u_\nu = h\nu n_\nu \qquad (7.15)$$

is proportional to ν^2 for frequencies much less than kT/h. The exponential in the denominator makes the total energy density

$$u = \int_0^\infty d\nu u_\nu = aT^4 \qquad (7.16)$$

(where $a = 4\sigma/c = 7.56 \cdot 10^{-15}$ erg cm^{-3} K^{-4}) finite (see eq. (7.8)). We denote by the subscript ν the spectral density of a given quantity.

Consider now a body in thermal equilibrium with the radiation around it. The exchange of photons between the body and the medium must be such as to maintain the equilibrium distribution (7.14); this implies an important condition for the emission and absorption coefficients of the body (*Kirchhoff's law*). Let us consider the simple case in which the body is thick enough to prevent transmission; in other words, an impinging photon, travelling at an angle θ from the normal, has a probability $\alpha(\nu, \theta)$ of being absorbed and a probability $[1 - \alpha(\nu, \theta)]$ of being reflected or diffused. Let $j_\nu(\theta)d\Omega d\nu$ be the intensity of emission from the unit surface in a solid angle $d\Omega$ and a frequency interval $(\nu, \nu + d\nu)$: such an emission must be equal to the corresponding absorbed flux of energy, that is, to the energy contained in a cylinder of volume $c \cdot \cos\theta$ and in a solid angle $d\Omega$ (see Fig. 7.3):

$$j_\nu(\theta)d\Omega d\nu = u_\nu d\nu d\Omega \cdot \cos\theta\alpha(\theta,\nu) \cdot c/4\pi \ .$$

This shows that in thermal equilibrium the ratio of the emission intensity to the absorption coefficient is a *universal function of the frequency*

$$j_\nu(\theta)/\alpha(\theta,\nu) = \cos\theta u_\nu \cdot c/4\pi \ . \qquad (7.17)$$

Provided the absorption coefficient α is independent of frequency and direction, we can find the total emission j by integrating this equation over half the unit sphere (namely, the angle θ ranging from 0 to $\pi/2$):

$$j = \int_0^\infty d\nu \int_{2\pi} d\Omega j_\nu(\theta) = \tfrac{1}{2}\alpha a T^4 \int_0^{\pi/2} d\theta \ \sin\theta\cos\theta = \alpha\sigma T^4 \ . \qquad (7.18)$$

For a completely opaque body ($\alpha = 1$) this is the usual emission law.

The basic quantity in the theory of radiation transfer is the *specific intensity* I_ν, related to the radiation energy in the frequency interval $(\nu, \nu + d\nu)$ which in the time dt crosses a surface element dS within a solid angle $d\Omega$:

$$dE_\nu = I_\nu \cos\theta \ d\Omega dt d\nu dS \ . \qquad (7.19)$$

This energy is, of course, proportional to the cosine of the angle θ between the normal to the surface element and the direction of the ray. The specific intensity is in general a function of frequency, position, time and direction.

To obtain the transfer equation when the radiation passes through matter, consider a cylinder of area dS and length ds parallel to the beam. The change in dE_ν from s to (s + ds) is the sum of four contributions: two losses, due to absorption by the cylinder and scattering away from the solid angle and the frequency interval we envisage; and two gains, due to emission by matter within the cylinder and scattering into the solid angle and the frequency interval. The total loss is proportional to the beam intensity and can formally be described by a generalized opacity coefficient (see eq. (7.11)); we write the total gain only in the particular and interesting case in which there is no scattering.

When conditions of local thermodynamical equilibrium apply, this gain is obtained from eq. (7.17);

$$j_\nu d\nu d\Omega dS dt = \alpha(\nu)\cos\theta \cdot u_\nu d\nu d\Omega dS dt/4\pi$$

is the corresponding contribution to the change in dE_ν and represents the part of the beam at (s + ds) which has been added by the matter inside the cylinder. α, the probability that a photon is absorbed by the cylinder, can be expressed in terms of the opacity (eq. (7.11)):

$$\alpha = ds/\ell = \kappa\rho ds. \qquad (7.20)$$

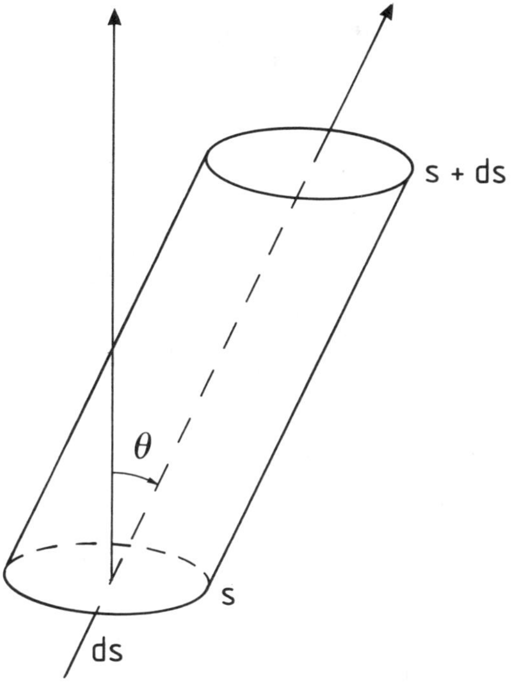

Fig. 7.3. The geometry of radiation transfer.

We have, finally:

$$\frac{dI_\nu}{ds} = -\kappa\rho\left[I_\nu - \frac{cu_\nu}{4\pi}\right] = -\kappa\rho(I_\nu - B_\nu), \qquad (7.21)$$

where

$$B_\nu = \frac{cu}{4\pi} = \frac{2h\nu^3}{c^2}\,\frac{1}{\exp(h\nu/kT) - 1} . \qquad (7.22)$$

In order to solve a radiative transfer problem, the temperature, the density and the opacity must be given; then eq. (7.21) propagates the specific intensity from an initial surface to the whole volume. In the simple case considered here the transfer equation (7.21) can be solved separately for each frequency; in more complicated cases (e. g., with *fluorescence*) the specific intensities at different frequencies are coupled together.

A thin planetary atmosphere is a *plane, parallelly stratified medium*, changing in a given direction θ with the variable $s = z/\cos\theta$; then it is convenient to measure s in terms of the optical depth τ (eq. (7.12)). On the ground $\tau(0) = \tau_0$. Eq. (7.21) reads now

$$\mu\, dI_\nu/d\tau = I_\nu - B_\nu \quad (\mu = \cos\theta). \qquad (7.23)$$

According to the common usage, angles are labelled with the variable $\mu \equiv \cos\theta$; the upward and the downward fluxes correspond, respectively, to positive and negative values of μ. Note that I_ν is a function of μ (besides ν and τ), but B_ν is not (the thermal emission is isotropic). The conservation of energy provides an integral of (7.23): the total flux is obtained by integrating $I_\nu\mu$ over angles and frequencies (eq. (7.19)),

$$\Phi = \int_{4\pi} d\Omega I\mu = 2\pi\int_{-1}^{1} d\mu\mu I \quad (I = \int_0^\infty d\nu I_\nu). \qquad (7.24)$$

independent of height in stationary conditions. Recalling that the global quantity corresponding to the spectral density B_ν is (eqs. (7.22, 14, 16))

$$B = acT^4/4\pi = \sigma T^4/\pi, \qquad (7.25)$$

we have

$$\mu dI/d\tau = I - B , \qquad (7.26)$$

and hence, as $d\Phi/d\tau = 0$, from eqs. (7.24, 26) we get

$$\mu dI/d\tau = I - \tfrac{1}{2}\int d\mu I = I - B . \qquad (7.26')$$

For simplicity, in the integrals with respect to μ we drop the limits of integration -1 and 1. This integrodifferential equation must be supplemented with the appropriate boundary conditions on the ground and at infinity ($\tau = 0$).

*7.3. GRAY ATMOSPHERES

When the opacity depends on the frequency, the ground, at which appropriate boundary conditions must be imposed, corresponds to different values of τ_0. The case of a *grey atmosphere*, in which the opacity and τ_0 are frequency independent, is easier and still illuminating. One first solves eq. (7.26') with the appropriate boundary conditions at $\tau = 0$ and $\tau = \tau_0$; then from eqs. (7.25, 26') deduces $B(\tau)$ and hence $T(z)$. In many cases this is the main information one needs. If the spectrum of the specific intensity is required, one must solve the linear, inhomogeneous equation (7.23) with B_ν given by (7.22). We shall be content here with the first step.

Our main physical interest is the *greenhouse effect*, due to the fact that the radiation emitted from the ground, in the infrared, may be absorbed by the atmosphere more than the solar radiation, in the visible Fig. 7.1a). Accordingly, we consider only the radiation balance for infrared radiation, assuming that a solar flux Φ is absorbed by the ground. The same flux must escape in the infrared at the top of the atmosphere (at $\tau = 0$):

$$\Phi = 2\pi\int d\mu\mu I(\tau) = 2\pi\int d\mu\mu I(0), \qquad (7.27)$$

with $I(0) = 0$ for $\mu > 0$. In the function I we have indicated only the argument τ. In stellar atmospheres we have a similar problem, but with no incoming flux. Φ is proportional to the first moment of the intensity; from eqs. (7.26, 27) we can also obtain the second moment of the intensity

$$K = \tfrac{1}{2}\int d\mu\mu^2 I \ : \qquad (7.28)$$

$$\frac{dK}{d\tau} = \frac{1}{4\pi}\Phi, \qquad K = \frac{1}{4\pi}\Phi\tau + \text{const}. \qquad (7.29)$$

Eq. (7.26) amounts to an integral equation for the function $B(\tau)$, whose exact solution requires sophisticated mathematical tools. However, it can be easily solved in a region of large optical opacity ($\tau \gg 1$). The larger the opacity, the nearer the radiation must be to thermal equilibrium, where its angular distribution is isotropic. Therefore in the

Taylor expansion around $\mu = 0$

$$I = I_0 + I_1\mu + \tfrac{1}{2}I_2\mu^2 + \ldots, \qquad (7.30)$$

we can assume $I_n \gg I_{n+1}$. From eqs. (7.26, 27, 28), to lowest order, we get

$$B = I_0, \qquad \Phi = 4\pi I_1/3, \qquad K = I_0/3\ ; \qquad (7.31)$$

and from eq. (7.29)

$$B = 3K = 3\Phi\tau/4\pi + \text{const}. \qquad (7.32)$$

The temperature grows proportionally to $\tau^{\frac{1}{4}}$. This fundamental solution is the starting point for a systematic approximation for the unknown function $q(\tau)$ (*Hopf function*) in

$$B = 3\Phi[\tau + q(\tau)]/4\pi\ . \qquad (7.32')$$

Its expression connects the deeper layers to the upper part of the atmosphere, where there is a strong deviation from isotropicity. Fortunately, it turns out that the function q is of the order of unity and increases monotonically from $q(0) = 0.58$ to $q(\infty) = 0.71$. Therefore the linear law (7.32) is a reasonable approximation in the lower part of thick atmospheres; and no great error is made if $q(\tau)$ is taken to be equal to $q_0 = 0.58$.

From the solution (7.32') one can easily integrate eq. (7.26) to get the angular dependence of the emerging radiation at $\tau = 0$:

$$I(0) = \int_0^\infty d\tau B(\tau)\exp(-\tau/\mu)/\mu. \qquad (7.33)$$

With $q = q_0$ (a constant) we have from eq. (7.32')

$$I(0) = 3\Phi(\mu + q_0)/4\pi\ . \qquad (7.34)$$

This shows an obvious and important feature of the radiation emitted by planets and stars; the centre of their disc, corresponding to $\mu = 1$, is brighter than the limb by the ratio $(1 + q_0)/q_0 \cong 2.7$ (*limb darkening*).

7.4. DYNAMICS OF THE ATMOSPHERE AND THE OCEANS.

The dynamics of the earth's atmosphere (and of the other planetary atmospheres) is the result of many complex factors beyond our scope:

the heat exchange with the ground, occurring in a thin boundary layer; the energy transfer in the troposphere and stratosphere, often by convection; phase changes and precipitations; chemical and photochemical reactions; the effects of human activity; and so on. In line with the priorities outlined in the introduction, we confine ourselves to two exemplary topics, geostrophic flow and gravity waves.

In a frame of reference rotating with the earth's spin rate there are four main forces: gravity, pressure, the centrifugal and Coriolis apparent forces. The centrifugal force is velocity independent and amounts to a slight change in the gravity acceleration g and the direction of the vertical $\mathbf{e}_z$. In a local problem it can be neglected.

Momentum conservation reads in this frame:

$$\frac{d\mathbf{v}}{dt} + 2\boldsymbol{\omega} \times \mathbf{v} = -\frac{1}{\rho}\nabla P + g\mathbf{e}_Z + \mathbf{a} = \mathbf{A}. \qquad (7.35)$$

For completeness, we have added a generic force **a**, in most cases due to friction. When considering dynamical effects over a scale smaller than the radius of the earth, it is appropriate to use a local cartesian frame of reference (see Fig. 7.4) and neglect the curvature of the earth. In this frame we split the velocity in a horizontal and a vertical component

$$\mathbf{v} = v_Z\mathbf{e}_Z + \mathbf{u}$$

and note that weather dynamics is essentially a horizontal phenomenon, with $v_z \ll u$ (say, 1 cm/s against 10^3 cm/s) and the horizontal scale ℓ much larger than the vertical scale H (say, 1000 km against 10 km). However, the horizontal motion is faster than the vertical:

$$u/\ell = u\nabla_\perp > v_Z/H \; . \qquad (7.36)$$

With these orders of magnitude, it is easy to see that the vertical momentum conservation reduces to the equilibrium eq. (7.1). In stationary conditions the horizontal momentum conservation reads

$$\mathbf{u} \cdot \nabla\mathbf{u} + \lambda\mathbf{e}_Z \times \mathbf{u} = -\frac{1}{\rho}\nabla_\perp P + \mathbf{a}_\perp = \mathbf{A}_\perp. \qquad (7.37)$$

Here

$$\lambda = 2\omega\cos\theta \qquad (7.38)$$

is the *Coriolis parameter* (θ is the angle between $\boldsymbol{\omega}$ and $\mathbf{e}_z$), assumed

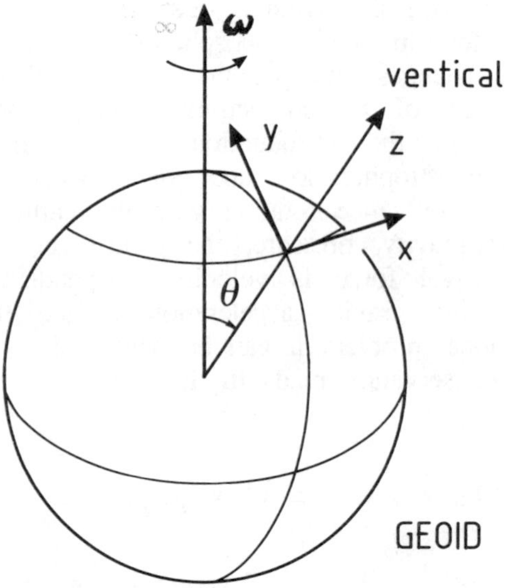

Fig. 7.4. The local frame of reference to describe geostrophic flow. Tha z axis is along the vertical, to wit, the normal to the geoid.

to be constant, in agreement with the "flat earth" approximation.

We see from eq. (7.18) that, leaving aside the frictional corrections, the pressure can be balanced not only, as usual, by the inertia term, but also by Coriolis force. Their ratio, of order

$$\mathrm{Ro} = u/(|\lambda|\ell), \tag{7.39}$$

is called *Rossby number*, and determines the winner in this competition. At mid latitudes Ro is typically of the order of 0.1, giving rise to the *geostrophic* approximation Ro $\ll$ 1. When this holds, it is convenient to replace eq. (7.37) with the equivalent form

$$\lambda \mathbf{u} = \mathbf{A} \times \mathbf{e}_z - \mathbf{u} \cdot \nabla \mathbf{u} \times \mathbf{e}_z \tag{7.37'}$$

and to solve it by iteration as follows:

$$\mathbf{u} = \frac{1}{\lambda}\,\mathbf{A} \times \mathbf{e}_z - \frac{1}{\lambda^3}\,(\mathbf{A} \times \mathbf{e}_z) \cdot \nabla(\mathbf{A} \times \mathbf{e}_z) \times \mathbf{e}_z + \ldots \tag{7.40}$$

The ratio of two successive terms is of the order of 1/Ro: this is effectively an expansion in powers of the Rossby number.

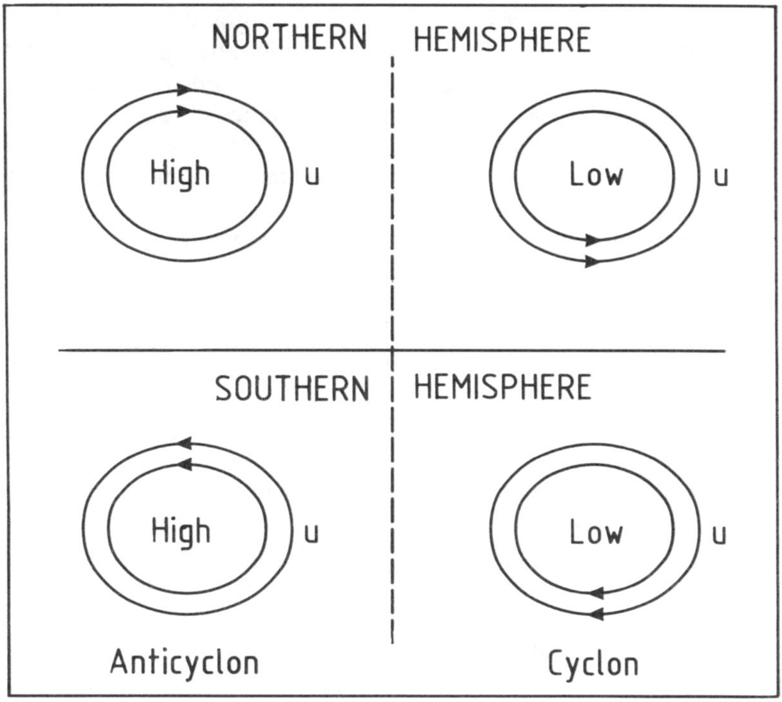

Fig. 7.5. The geometry of cyclonic and anticyclonic flows.

In the lowest approximation (neglecting for the moment the frictional term), we have

$$\mathbf{u} = -\nabla_{\perp}P \times \mathbf{e}_z/\lambda\rho \qquad (7.41)$$

and the flow follows the isobaric lines, orthogonal to the pressure gradient. In the northern hemisphere, where $\lambda > 0$, the flow occurs in a clockwise sense around a high-pressure region (*anticyclonic* regime) and in a counterclockwise sense around a pressure trough (*cyclonic* regime) (see Fig. 7.5). In the southern hemisphere, where $\lambda < 0$, the circulation sense is inverted. If we have also a frictional force (especially near the ground), proportional to the flow velocity

$$\mathbf{a}_{\perp} = -\nu_f \mathbf{u} \ , \qquad (7.42)$$

eq. (7.40) gives

$$\mathbf{u} = -\frac{1}{\lambda\rho}\nabla_{\perp}P \times \mathbf{e}_z - \frac{\nu_f}{\lambda^2\rho}\nabla_{\perp}P \qquad (7.41')$$

and there is now flow along the pressure gradient towards the low pressure side; air flows into the cyclonic troughs and out from anticyclones, producing a vertical circulation to balance the mass.

Near the equator, since the Coriolis parameter (7.38) is proportional to the sine of the latitude, the geostrophic approximation is invalid. Neglecting Coriolis force we can only balance the pressure forces with the centrifugal force. Consider, for example, a situation in which pressure and density depend only on the horizontal distance r from a given point; the pressure must increase outward to balance the centrifugal force produced by a uniformly rotating flow:

$$\rho u^2 = r dP/dr \ . \tag{7.43}$$

In the case of an axially symmetric and uniform flow it is easy to solve rigorously the momentum equation (Problem 7.6).

Among the large variety of waves which can propagate in the atmosphere we do not have to say much about acoustic waves, which propagate with the speed of sound (1.51). More interesting are instead those waves whose restoring force is gravity. They have been already implicitely mentioned in Sec. 5.4 in the context of the discussion about convective instability: indeed, in the opposite, stable situation in which the density gradient is less than adiabatic, an initially displaced fluid element will oscillate around its equilibrium position. The ancestor of all kinds of gravity waves is provided by a simple model in which a parallelly stratified atmosphere is in the equilibrium situation (7.1), but with an unspecified equation of state $P(\rho, s)$, where s is the entropy per unit mass. Following and refining the argument used in Sec. 5.4 to deal with convective instability, consider at z a small fluid element of given volume V and mass ρV. It is acted upon by two forces, its weight $-g\rho(z)V$ and Archimedes' force, equal and opposite to the weight of the fluid it displaces; of course in equilibrium they balance. If the element is displaced to $(z + \delta z)$, the weight is unchanged, but Archimedes' force is not, and an unbalance arises proportional to δz:

$$\delta F = -g\rho(z)V + g\rho(z + \delta z)(V + \delta V) .$$

As with the convective instability, we compute the change in volume δV by assuming an adiabatic displacement:

$$\delta V/V = -\delta\rho/\rho = -\delta P/\rho c_S{}^2 = -\delta z dP/\rho c_S{}^2 dz = g\delta z/c_S{}^2 .$$

We have used here the definition (1.49) of the speed of sound c_s (the partial derivative keeps the entropy constant !) and the equation of equilibrium (7.1). Finally, expanding in powers of δz,

$$\delta F = \rho V g(g/c_S{}^2 + d\rho/\rho dz)\delta z \ . \tag{7.44}$$

Since ρV is the mass of the fluid element, its motion is harmonic, with a frequency ω_V given by

$$\omega_V^2 = g(g/c_S^2 + \rho'/\rho) \ . \qquad (7.45)$$

(*Väisälä frequency*). To see its meaning, write the equilibrium condition (7.1) explicitely as

$$-\rho g = \frac{dP}{dz} = c_S^2\frac{d\rho}{dz} + b\ \frac{ds}{dz} \quad , \qquad (7.46)$$

where

$$b = \partial P(\rho, s)/\partial s \ . \qquad (7.47)$$

Then

$$\omega_V^2 = \frac{gb}{\rho c_S^2}\ \frac{ds}{dz} \qquad (7.45')$$

comes out to be negative, leading to convective instability, when the entropy s decreases with height; and positive, corresponding to oscillations, in the opposite case. Note that the negative sign corresponds to a density gradient stronger than adiabatic, as explained in Sec. 5.4.

In this simple model, sufficient to understand the basic mechanism of gravity waves, we have neglected propagation and other complications (see Problem 7.8). In the lower atmosphere Väisälä's frequency is of order of 0.05 Hz.

7.5. TROPOSPHERIC TURBULENCE

Near the ground the flow is strongly affected by friction and by transfer of moisture and heat. In normal conditions (kinematical viscosity $\eta/\rho \approx$ 0.1 cm^2/s) the Reynolds' number of the flow (1.46) is very large and we expect a *turbulent boundary layer* near the surface.

In a turbulent flow we expect that the velocity **v** is the sum of a mean value <**v**> and a fluctuating value **v**', similarly to what was done in the dynamo theory (Sec. 6.4):

$$\mathbf{v} = \langle\mathbf{v}\rangle + \mathbf{v}' \qquad (\langle\mathbf{v}'\rangle = 0) \ . \qquad (7.48)$$

While <**v**> is determined by the macroscopic conditions, **v**' changes for different realizations of the same macroscopic flow and can be studied only by statistical methods. In these methods the crucial quantity is the *spectral energy density* E(k) of the macroscopic flow. For an incompressible flow this amounts to a Fourier decomposition of the turbulent energy density

$$\tfrac{1}{2}\rho\langle v'^2\rangle = \int_0^\infty dk\ E(k). \qquad (7.49)$$

k is the wave number of a given component $\mathbf{v}'(\mathbf{k})$ of the turbulent velocity and its reciprocal measures the size of the corresponding eddy. The prediction of the spectral energy density E(k) and its change with time and/or with altitude is the main task – still partly unsolved – of the theory of turbulence.

An eddy of size 1/k has a characteristic time dependence $\tau_k = 1/[v'(k)k]$. It is important to compare it with the dissipation time

$$\tau_d = \rho/(\eta k^2). \qquad (7.50)$$

For $\eta/\rho = 0.1\ cm^2/s$ and v' = 1 cm/s the two times are equal when k = 10 cm^{-1}. Larger eddies will persist with an inviscid dynamics for a long time. Near the ground of the earth, energy is fed at a scale L of the order of the thickness of the boundary layer, generally much larger than the dissipation-dominated eddies; for all the wave numbers in between the energy can be transferred from one k to another only by the non-linear terms in the equations of motion. Indeed, the Fourier transform of the product of two functions with wave vectors $\mathbf{k}_1$ and $\mathbf{k}_2$ has a component at $(\mathbf{k}_1 + \mathbf{k}_2)$, generally at a larger distance from the origin in the **k**-space (Problem 7.13). For an incompressible flow the non-linearity comes from the convective term $\mathbf{v} \cdot \nabla\mathbf{v}$ in the equation for the momentum (1.22). In this regime we therefore have the following description of E(k). At wave numbers ≈ 1/L energy is fed into the turbulent eddies by an external power source; this energy is slowly transferred to larger wave numbers until, for the wave number k_d for which

$$\tau_k = \tau_d \qquad (7.51)$$

(or larger), dissipation prevails. When $k_d L \gg 1$, Kolmogorov and Heisenberg have noted that it is possible to predict the energy spectrum on the basis of dimensional considerations alone. Since in the range $k < k_d$ the eddies are not affected by dissipation, E(k) must be a universal expression given in terms of k and the power per unit mass ϵ (with dimensions $[L^2/T^3]$) fed into the system. One easily sees that this requirement leads to Kolmogorov's spectrum (see Problem (7.9))

$$E(k) = \alpha\epsilon^{2/3}k^{-5/3} \ , \tag{7.52}$$

where α is a constant without dimensions, usually of order unity. In this derivation we have neglected not only the compressibility, but also the vertical gradient; we have assumed an isotropic turbulence, whose spectrum depends only on the size of the wave number and not on its direction. It is pleasing to note that this power spectrum is indeed observed in the ground boundary layer and in other turbulent flows in the solar system, notably in the interplanetary solar wind. The same dimensional argument leads to an expression for the turbulent velocity

$$v'(k) = \alpha'(\epsilon/k)^{1/3} , \tag{7.53}$$

where α' is another dimensionless constant. In this way from (7.50,51) we obtain the value of the "dissipative" wave number

$$k_d = (\alpha'\rho/\eta)^{\frac{3}{4}}\epsilon^{\frac{1}{4}} . \tag{7.54}$$

An extension of Kolmogorov's spectrum to lower wave numbers, where energy is fed by large-scale instabilities, has been obtained by V. Canuto and his collaborators (Phys. Fluids **30**, 3391 (1987)).

The turbulence in the ground layer of the atmosphere has drastic consequences on its transport and optical properties. Again neglecting viscosity we can formally compute the average of the acceleration of a fluid element in eq. (1.22) by means of the decomposition (7.48):

$$\partial\langle \mathbf{v}\rangle/\partial t + \nabla \cdot (\langle \mathbf{v}\rangle \langle \mathbf{v}\rangle) + \nabla \cdot \langle \mathbf{v}'\mathbf{v}'\rangle \ .$$

The last term acts as a pressure tensor and may produce a shear which adds to the viscous shear. Consider, for example, the case in which the turbulent energy is drawn from the differential motion of a wind along the x-direction: $\langle \mathbf{v}\rangle = (v(z), 0, 0)$; then, when the corresponding Reynolds number

$$Re = v^2\rho/(dv/\eta dz) \tag{7.55}$$

is large, we expect a turbulent shear to develop proportional to dv/dz; in particular the term

$$\rho\langle v_x'v_z'\rangle = - \eta_t \ dv/dz \tag{7.56}$$

describes the vertical turbulent transfer of x-momentum (the minus sign indicates that such a transfer tends to destroy the velocity gradient). η_t

is the *coefficient of turbulent viscosity*. Its calculation requires a detailed theory, but with ordinary values of turbulent velocities and scales η_t/ρ ranges from 10^2 to 10^4 cm^2/s, to be compared with the laminar viscosity coefficient of air $\eta \approx 0.1$ cm^2/s. Thus the transfer of momentum is dominated by turbulence.

Besides eddies, the tropospheric turbulence produces also inhomogeneities in temperature and density which are convected along by the fluctuating flow and move vertically under their buoyancy forces. If n_r is the refractive index, $(n_r - 1)$, proportional to the density, will suffer fluctuations in space and time and affect stochastically the image of a distant source (see Sec. 19.6); in other words, a plane surface of constant phase is warped in crossing the troposphere. The image of a point source gathered by an objective larger than the size of the warps is diffuse and, moreover, moves around in time ("dancing"). This phenomenon is called *scintillation*, or *seeing*, and is of crucial importance in limiting the optical astronomical observations from the ground. In practice, the apparent diameter of a celestial point-like source is usually a few seconds of arc and reaches 1" only in the best seeing conditions.

PROBLEMS

7.1. Evaluate the temperature excursion at a fixed latitude on the earth during the day and the year, assuming a constant gravity profile.

7.2. What is the fractional change in the temperature of the earth due to the eccentricity of its orbit?

7.3. Solve the equilibrium condition for a flat "atmosphere" formed by electrons and protons at uniform temperature.

7.4. Prove eq. (7.16).

7.5. Compute the solar constant for all the planets and evaluate their equilibrium temperature.

7.6. Calculate, as a function of latitude, the displacement of the vertical from the radial direction due to the centrifugal acceleration.

7.7. Solve the momentum equation (7.37) for a uniform, circular flow around an axially symmetric pressure gradient.

*7.8. Using the model of a parallelly stratified, isothermal fluid, derive the dispersion relation for gravity waves.

7.9. Derive Kolmogorov's spectrum (7.51) and (7.52).

*7.10. Compute in the order of magnitude the total angular momentum of the earth's atmosphere, and compare it with the spin angular momentum of the solid earth. Which fraction of the atmospheric angular momentum is variable, assuming "global" east-west (or west-east) winds with a typical speed of 10 m/s ? And how much can this variability change the length of the day ?

7.11. A 100 km/h wind sometimes blows from the eastern United States against the Rocky Mountains. Does it appreciably affect the rotation of the earth?

*7.12. Find the dispersion relation of surface water waves, whose restoring force is gravity.

7.13. Find the average distance from the origin reached by summing N vectors **k** of equal size, but random orientation (*random walk).*

*7.14. As a model of seasons which neglects the daily temperature excursion, one can take a fast rotating planet, in which the parallels are isothermal, but have no temperature exchange between each other. Find the latitudinal temperature profile as a function of the position of the sun.

FURTHER READING

J. T. Houghton, *The Physics of Atmospheres*, Cambridge University Press (1977), is an excellent review, stressing the physical processes at work. The classical textbook on radiative transfer is S. Chandrasekhar, *Radiative Transfer*, Dover, several editions. For more recent developments, see D. Mihalas, *Stellar Atmospheres*, Freeman (1970). W. W. Kellogg and R. Schwarz, *Climate Change and Society. Consequences of Increasing Atmospheric Carbon Dioxide*, Westview Press, Boulder (1981) is a review of the climatic effects of the increase of CO_2 in the atmosphere. More generally, see A. Berger (ed.), *Climatic Variations and Variability: Facts and Theories*, Reidel (1981). About the connection between climate and celestial mechanics, see A. Berger, J. Imbrie, I. Hays, G. Kukla and B. Saltzmann, *Milankovitch and Climate. Understanding the Response to Astronomical Forcing*, Reidel (1983).

8. THE UPPER ATMOSPHERE

Under the powerful ultraviolet radiation from the sun the higher and thinner layers of the atmosphere become ionized and produce important consequences for the propagation of radio waves. More generally, because of the large mean free paths the medium is not in thermal equilibrium and many complex atomic and molecular phenomena take place. This brings about many new effects, in particular electromagnetic, unknown in the lower atmosphere. We have selected three topics of particular relevance: the structure of the ionized layer; the propagation of high frequency electromagnetic waves in an electron plasma; and the motion of neutral species in a very rarefied gas under the influence of gravity.

8.1. IONIZING EFFECTS OF THE SOLAR RADIATION

A flux of solar photons with energy larger than the ionization energy is not always sufficient to produce substantial ionization in the atmosphere. There are two main competing processes for a molecule M, ionization

$$M + h\nu \rightarrow M^+ + c$$

and radiative recombination

$$M^+ + e \rightarrow M + h\nu .$$

The photo-ionization rate per unit volume is $N\Phi\sigma_i$, where Φ is the solar flux relevant to this process and N the number density of the neutral molecules M. For simplicity, we assume the ionization cross-section σ_i to be independent of the energy of the photon; a more realistic treatment requires an integration with respect to such energy.

Recombination contributes with the rate $-\alpha n^2$, where n is the electron density and α the recombination coefficient. Since the gas is electrically neutral and we neglect multiple ionization, n is also the number density of the ionized molecules M^+. The radiative recombination requires that the incoming electrons lose energy by *Brehmsstrahlung* in order to become bound. At higher electron densities and lower temperatures another kind of recombination occurs, where the electron-ion collision takes place near a third charged body – another ion, which absorbs part of the electron energy. One can take into account this three-body recombination by an effective recombination coefficient α_e which depends linearly on n. Many other complex collisions take place in the ionosphere, all under the powerful energy source of the solar ultraviolet radiation: they include dissociation, excitation and their inverses. Their discussion, as well as the chemical structure of the ionosphere, is beyond our scope.

The simplified rate equation (where diffusion is neglected)

$$dn/dt = N\Phi\sigma - \alpha n^2 \tag{8.1}$$

shows that in a stationary regime the fractional ionization

$$\frac{n}{N} = \left(\frac{\Phi\sigma}{\alpha N}\right)^{1/2} \tag{8.2}$$

changes with the height z because of two reasons. N decreases with a scale height

$$H = (d(\ln N)/dz)^{-1} \tag{8.3}$$

(which is itself a function of z). The photon flux Φ, neglecting other losses, is determined by a conservation law (see Fig. 8.1).

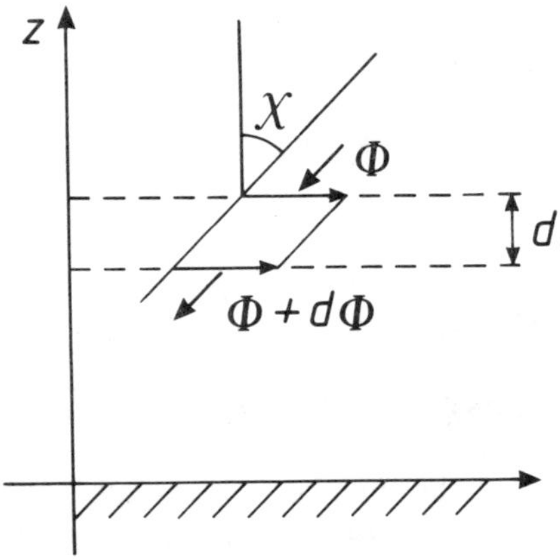

Fig. 8.1. Consider a photon flux $\Phi dS \cos\chi$ impinging with a zenithal angle χ upon a horizontal surface dS. The total cross-section with respect to ionization in a horizontal layer of thickness dz is $\sigma N dS dz$; hence the photon flux changes by $d\Phi = \sigma_i N \Phi dz/\cos\chi$.

For a small fractional ionization N is not affected by this process and we have

$$\Phi(z) = \Phi(\infty)\ \exp\left[-\ \sigma_i \sec\chi \int_z^{\infty} dz N(z)\right] \ . \qquad (8.4)$$

For example, when the scale height H is constant,

$$\Phi(z) = \Phi(\infty)\ \exp[-\ \sigma H N_0 \cdot \sec\chi \cdot \exp(-\ z/H)] \ . \qquad (8.4')$$

We see that, because of the competing effects in eq. (8.2) of the two functions Φ(z) and N(z), the ionization is confined to a layer. The discussion that follows is due to S. Chapman. The maximum electron density occurs at $z = z_m$, say, where ΦN is largest; that is to say, where

$$\sigma H N \sec\chi = 1 \ . \qquad (8.5)$$

The width δ of the layer can be estimated with the relation

$$\Phi_m N_m / 2\delta^2 = [d^2(\Phi N)/dz^2]_m .$$

The calculation, using eq. (8.5) and the assumption H << 1, gives

$$\delta = H/\sqrt{2} \ . \qquad (8.6)$$

At z_m the electron density is proportional (eq. (8.2), (8.4) and (8.5); assume H to be constant) to the square root of:

$$\Phi_m N_m \propto \exp[-\ \sigma_i H N_m \sec\chi \ -\ z_m/H] \propto \exp(-\ z_m/H)\alpha\cos\chi . \quad (8.7)$$

χ changes daily and seasonally (see Problem 8.1); dχ/dt is about the angular velocity of the earth. The change in altitude of the maximum z_m is given by (eq. (8.5))

$$\cos\chi \propto \exp(-\ z_m/H) . \qquad (8.8)$$

To get a feeling of the numbers involved, consider at z ≈ 200 km the solar flux in the region from 250 to 750 Armstrong, with a value of about 10^{10} photons/(cm^2 s). Atomic oxygen is the main constituent there, with a number density 10^9 – 10^{10} cm^{-3} and a scale height H of 50 km; to be specific, let us take

$$N(z) = N_0 \cdot \exp[-\ (z\ -\ z_0)/H] , \qquad (8.9)$$

with $N_0 = 10^{10}$ cm^{-3} and $z_0 = 200$ km. The ionization cross-section σ_i is taken to be $7 \cdot 10^{-18}$ cm^2 and the recombination coefficient α is 10^{-12} cm^3/s. Eq. (8.5) shows that the height z_m of largest ionization changes with the zenithal distance χ of the sun according to

$$\cos\chi \exp[(z_m - z_0)/H] = \sigma H N_0 \approx 0.35 \ . \qquad (8.10)$$

We see that in a few hours z_m changes by an amount of order H; in this stationary model there is no ionization during the night. The

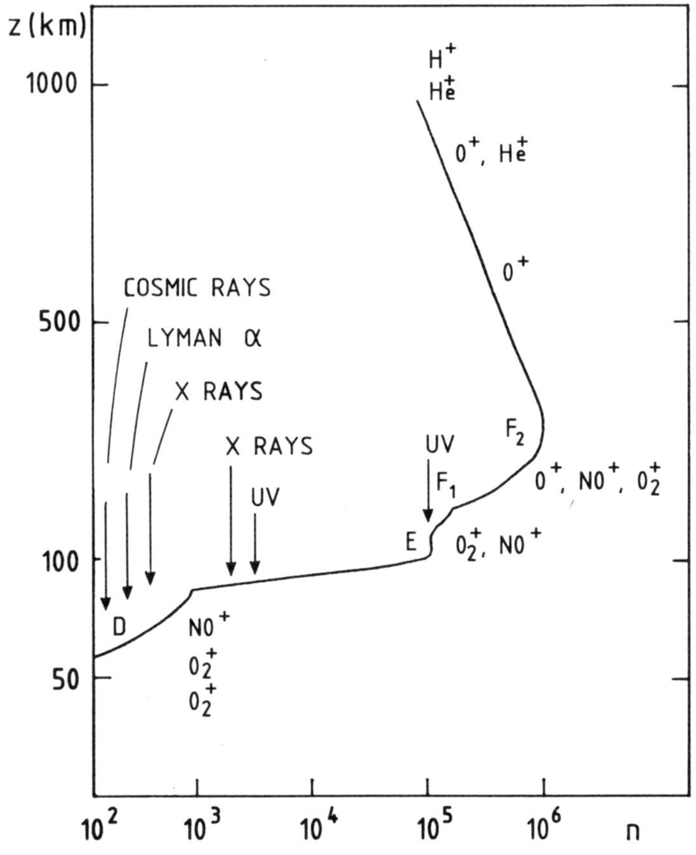

Fig. 8.2. A typical electron concentration profile of the ionosphere. The ionic composition and the main ionization agents at different heights are indicated. (From the review of M. Nicolet.)

maximum electron density, from eq. (8.2), is

$$n_m = (\Phi_m N_m \sigma_i/\alpha)^{1/2} \approx 2.5 \cdot 10^7 \ \mathrm{cm}^{-3},$$

with a fractional ionization of a few thousandths.

The actual electron density profile is much more complicated, mainly because of the many molecular species present, whose concentration varies with height. As a consequence, a given ionization process predominates at different heights, giving rise to ionization layers produced in different ways. They are named with the letters D, E and F (Fig. 8.2). There is also, of course, a strong dependence on latitude due to the variation of the angle χ; moreover, near the poles the precipitation of energetic particles from the solar wind produces the striking auroral effects (see the next chapter).

8.2. ELECTROMAGNETIC WAVES IN AN ELECTRON PLASMA

The ionospheric mantle encloses all the earth and produces striking effects on the propagation of radio waves. A plasma is characterized by a characteristic *plasma frequency*

$$\omega_p = \left(\frac{4\pi e^2 n}{m_e}\right)^{1/2} = 56400 \sqrt{(n \ \mathrm{cm}^3)} \ \mathrm{s}^{-1}; \qquad (8.11)$$

here m_e is the electron mass. With a magnetic field B there is also the *electron cyclotron frequency*

$$\Omega_e = \frac{eB}{m_e c} = 5.8 \cdot 10^6 \ \frac{B}{B_e} \ \mathrm{s}^{-1} \qquad (8.12)$$

and the *ion cyclotron frequency*

$$\Omega_i = \frac{eB}{m_i c} = 2900 \ \frac{ZB}{AB_e} \ \mathrm{s}^{-1}. \qquad (8.13)$$

$m_i = Am_p$ is the mean ionic mass, expressed in terms of the proton mass m_p; Z is the charge number. The numerical value is given in terms of the magnetic field on the equator of the earth, $B_e \simeq 0.3 \ \Gamma$.

A simple and not unreasonable model for electromagnetic waves in the ionosphere is obtained by neglecting the motion of the ions; because of their large mass this is certainly correct at high frequency (i. e., $\gg \Omega_i$). We describe the electron dynamics by the conservation of their charge and their momentum, neglecting also their pressure. It can be shown that the latter assumption is valid provided the phase velocity is much larger than the thermal speed. We also neglect collisions (the collision frequency is much smaller than ω_p). In this approximation the electrons have no velocity dispersion; their fluid velocity **v** fulfils

$$m_e(\partial_t + \mathbf{v} \cdot \nabla)\mathbf{v} = - e(\mathbf{E} + \mathbf{v} \times \mathbf{B}/c) . \tag{8.14}$$

Their charge density $\epsilon = -$ en and their current density $\mathbf{J} = -$ e$n\mathbf{v}$ are conserved (see eq. (1.21')):

$$\partial_t \epsilon + \nabla \cdot \mathbf{J} = 0 . \tag{8.15}$$

In addition, we have Maxwell's equations (1.52).

We study now small perturbations of an equilibrium situation in which the electron gas is at rest, with the same density n_0 of the ions (which we assume, for simplicity, to be singly charged). The magnetic field B_0 is constant and uniform. In the *normal mode approach* the perturbed quantities, like

$$n_1 = n - n_0 , \tag{8.16}$$

are proportional to $\exp[i(\omega t - \mathbf{k} \cdot \mathbf{r})]$ and correspond to a frequency ω and a wave number **k**. Of course, the physical quantities are the real parts of these complex expressions. The equations then become linear:

$$i m_e \omega \mathbf{J} = e^2 n_0 \mathbf{E} - e\mathbf{J} \times \mathbf{B}_0/c , \tag{8.17$_1$}$$

$$-i\mathbf{k} \times \mathbf{B}_1 = i\omega \mathbf{E}/c + 4\pi \mathbf{J}/c , \quad \mathbf{k} \times \mathbf{E} = \omega \mathbf{B}_1/c . \tag{8.17$_2$}$$

The index $_1$ has been safely dropped from **E**, σ and **J**. Note also that the two divergence equations for **B** and **E** now follow from (8.17$_2$) and the conservation equation (8.15)

$$\omega\epsilon = \mathbf{k} \cdot \mathbf{J} \tag{8.15'}$$

(which is not needed any more).

The momentum law (8.17$_1$) gives the current density as a function of the electric field – the *generalized Ohm's law* –

$$J_i = \sigma_{ij} E_j . \tag{8.18}$$

When the magnetic field is along z, the *conductivity tensor* (with dimension of a frequency) reads:

$$\sigma_{ij} = -\, i\frac{n_0 e^2}{m_e\omega}\,\frac{1}{1-\Omega^2/\omega^2}\left|\begin{matrix} 1 & i\Omega/\omega & 0 \\ -\,i\Omega/\omega & 1 & 0 \\ 0 & 0 & 1-\Omega^2/\omega^2 \end{matrix}\right| . \qquad (8.19)$$

This matrix is antihermitean, to wit,

$$\sigma_{ij} = -\,\sigma_{ji}^{*} . \qquad (8.19')$$

As a consequence, the quantity $Q = \mathrm{Re}(\mathbf{E})\cdot\mathrm{Re}(\mathbf{J})$, which measures the density of the Joule heat developed, vanishes when averaged over a period. (This of course is not surprising, since dissipation processes have been neglected from the beginning.) Indeed, Q is the sum of two parts, corresponding to the splitting

$$\mathrm{Re}(J_i) = \mathrm{Re}(\sigma_{ij})\mathrm{Re}(E_j) - \mathrm{Im}(\sigma_{ij})\mathrm{Im}(E_j);$$

the first part gives no contribution because the real part of the conductivity is skew; in the second part the product $\mathrm{Re}(E_j)\cdot\mathrm{Im}(E_i)$ has a vanishing average.

It is convenient to introduce the *refraction index*

$$n_r = ck/\omega , \qquad (8.20)$$

the ratio of the light speed to the *phase velocity* ω/k. Maxwell's equations (8.17_2) give the current as a function of the electric field:

$$(1 - n_r^2)E_i + n_r^2 k_i k_j E_j/k^2 = 4\pi J_i/\omega . \qquad (8.21)$$

When there is no external magnetic field the conductivity (8.19) is a scalar:

$$4\pi\sigma = -\, i\omega_p^2/\omega . \qquad (8.22)$$

There are two modes of propagation. If the electric field is parallel to **k** (*longitudinal plasma waves*) there is no magnetic field and the electric field is given by

$$i\omega E + 4\pi J = 0 .$$

Together with the previous equation this yields a mode of oscillation with a fixed (i. e., independent of **k**) frequency (8.11); this is a peculiar behaviour which makes the plasma differ from most material systems whose resonance frequency depends upon the size. When thermal effects are taken into account this degeneracy is removed and we have, at low frequency, an acoustic mode.

If the electric field is orthogonal to **k** (*transversal plasma waves*) we have an electromagnetic mode of propagation. Maxwell's equations (8.21) give

$$(1 - n_r^2)\mathbf{E} = 4\pi i \mathbf{J}/\omega \ .$$

Together with the conductivity equation we then obtain the dispersion relation in the two forms

$$n_r^2 = 1 - \omega_p^2/\omega^2 \ , \qquad (8.23)$$

$$\omega^2 = \omega_p^2 + k^2c^2 \ . \qquad (8.23')$$

It is easily verified that the geometric mean between the phase velocity ω/k and the *group velocity* $v_g = d\omega/dk$ is the velocity of light:

$$v_g\omega/k = c^2 \ ;$$

this confirms that, although the phase velocity is larger than c, signals are slower than light, in agreement with relativity. It is interesting to note that the dispersion relation (8.32') has the same form as in De Broglie's "matter waves" for a relativistic particle with a rest energy equal to $\hbar\omega_p$.

At very large frequencies we recover the electromagnetic waves. At frequencies of the order of the plasma frequency the plasma deeply affects the propagation: as $\omega \to \omega_p$ from above, the wavelength becomes unbounded; if $\omega < \omega_p$, it is imaginary. Therefore in this case an electromagnetic wave impinging normally upon a plasma confined at $z > 0$, say, decays over the distance $c/\surd(\omega_p^2 - \omega^2)$. In absence of collisions the wave is reflected back into the vacuum. When $\omega > \omega_p$ and hence $0 < n_r^2 < 1$, we have total reflection for incidence angles ι greater than

$$\iota_c = \arcsin(n_r) = \arcsin\surd(1 - \omega_p^2/\omega^2) . \qquad (8.24)$$

In general the dispersion relation is obtained by combining eqs. (8.18, 21); we obtain the linear, homogeneous equation:

$$[\delta_{ij} - i4\pi\sigma_{ij}/\omega - n_r^2(\delta_{ij} - k_ik_j/k^2)]E_j = 0.$$

The vanishing of its determinant is the required *plasma dispersion relation* and determines the refractive index as a function of frequency.

Its detailed discussion is beyond our scope (for a full treatment see, e. g., the book by Stix). Calculation shows that it is of the form

$$An_r^4 - Bn_r^2 + C = 0 \ , \tag{8.25}$$

with the following abbreviations:

$$A = 0.5(R + L)\sin^2\theta + P\cos^2\theta \ , \tag{8.26$_1$}$$

$$B = RL\sin^2\theta + 0.5P(R + L)(1 + \cos^2\theta) \ , \tag{8.26$_2$}$$

$$C = PRL \ ; \tag{8.26$_3$}$$

$$P = 1 - \omega_p^2/\omega^2 \ , \tag{8.27$_1$}$$

$$R = 1 - \omega_p^2/\omega(\omega - \Omega) \ , \tag{8.27$_2$}$$

$$L = 1 - \omega_p^2/\omega(\omega + \Omega) \ . \tag{8.27$_3$}$$

θ is the angle between $\mathbf{B}_0$ and $\mathbf{k}$.

When $\theta = 0$, eq. (8.25) factorizes into

$$P = 0, \qquad n_r^2 = R, \qquad n_r^2 = L \ . \tag{8.28}$$

The first is the longitudinal plasma wave, unaffected by the magnetic field. The transversal modes are modified by $\mathbf{B}_0$ and correspond to different states of polarization: the plasma is birefringent. The second mode shows a divergence (*resonance*) in the refractive index at $\omega = \Omega$ (the *electron cyclotron resonance*); for both of them η_r vanishes at a critical frequency (*cutoff*). When η_r is imaginary the propagation is forbidden, as for the transversal plasma waves; but here, in addition, η_r^2 changes sign also at the resonance, so that the forbidden bands are more complex.

For a generic direction of propagation, from eq. (8.25), the cutoffs appear where $C = 0$, to wit,

$$P = 0 \qquad \text{or} \qquad R = 0 \qquad \text{or} \qquad L = 0 \ . \tag{8.29}$$

They correspond to a reflection of the wave. Resonances occur where $A = 0$, to wit,

$$\tan^2\theta = -\,2P/(R + L) \ . \tag{8.30}$$

Because of dissipation, near a resonance both reflection and absorption occur. When $\Omega = 0$ we recover the two modes previously described.

The two modes of propagation corresponding to the two roots for n_r^2 have a complex classification determined by their resonances and cutoffs as a function of frequency and angle of propagation. The *Clemmow and Mullaly plot* (see Stix) of phase velocity in the parameter space of the plasma is of great help in understanding this complexity.

$n_r \to 1$ from below when $\omega \to \infty$, recovering vacuum propagation. It is interesting to obtain the asymptotic expression of the refractive index. The main correction is electrical (eq. (8.23)); let us see what is the effect of the magnetic field. The expansion

$$n_r^2 = 1 - \frac{\omega_p^2}{\omega^2} + a\,\frac{\Omega\omega_p^2}{\omega^3} + O(\omega^{-4}) \qquad (8.31)$$

is appropriate; we have to find the value of the dimensionless coefficient a. From eq. (8.25) we get $a = \pm \cos\theta$. It can be shown that these modes have opposite circular polarizations.

8.3. ELECTROMAGNETIC PROPAGATION IN THE IONOSPHERE

The model described in the previous Section, although it requires caution and does not cover all aspects of the problem, provides a good understanding of the propagation of radio waves in the ionosphere.

The unphysical singularities of the refractive index are remedied when a collision frequency ν_c is introduced. When $\nu_c \to 0$ this affects the dispersion curve in a neighbourhood of order ν_c around the singularities. Note also that near the origin the effect of the ion cyclotron frequency becomes important and introduces a new kind of waves, named after *Alfvèn*. They will be discussed in the next chapter in the magnetohydrodynamical approximation.

In the low frequency approximation ($\omega \ll \Omega, \omega_p$), for propagation along the magnetic field, eq. (8.28) gives a single mode

$$n_r^2 = \omega_p^2/\omega\Omega \qquad (8.32)$$

in which the frequency is proportional to k^2:

$$\omega = c^2\Omega k^2/\omega_p^2 \ . \qquad (8.32')$$

This is the *whistler mode*, in which the group velocity is proportional to the square root of the frequency. In this case the travel time of a Fourier component of a pulse is inversely proportional to the square root of its frequency; hence the higher frequencies arrive first, giving rise to the typical "whistling sound". These modes are very little damped

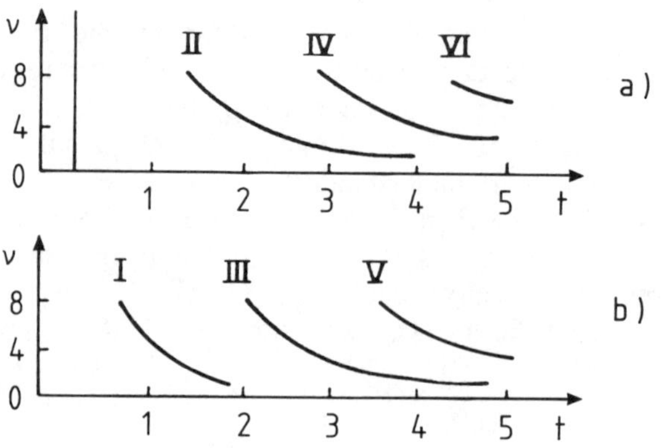

Fig. 8.3. The whistler effect. In a) the reception occurs near the source; in b) near the conjugate point. The numbers indicate the one-way trips between them. The signal profile, initially very short, is drawn out by the propagation in such a way that the higher frequencies arrive first. The frequency ν is given in MHz; the time t is in sec.

and can propagate to long distances along the lines of force. If they are emitted from a point in the northern hemisphere, say, they can be detected at its conjugate end in the southern hemisphere. The whistler signal can also be reflected back and be received several times in a multiple echo. The path of propagation extends very far out into the magnetosphere and provides an easy way to measure its electron density by the analysis of the distortion of the signal.

The presence of forbidden bands and of total reflections in the ionosphere drastically affect the propagation of radio waves. At sufficiently high frequencies – say, above 10 MHz, the radio beam travels more or less freely and, if directed above, escapes into outer space. At lower frequencies the reflection causes the beam to remain near the ground, thereby allowing the reception at large distances, even all around the world. Of course this effect changes from day to night with latitude and as a consequence of the solar activity. During a solar flare the X-ray flux enhances the ionization of the D layer, making it more reflecting and absorbing for radio waves.

If a pulse of a given frequency ω is emitted vertically, it is returned to the ground from the altitude z where $\omega = \omega_c(z)$. By measuring the return times of such pulses as functions of frequency and using the dispersion relation, it is possible to get the electron density profile $n(z)$ for the altitudes below the ionospheric maximum z_m. No information can be gained from the ground about higher altitudes; this region is also not accessible with satellites and only inadequate information is available about it.

To study this kind of experiment we must get the value ω_c of the critical frequency where $n_r = 0$ (eq. (8.21).) It turns out that ω_c does not depend on the angle θ; hence it is given by the expression (8.23) computed for $\theta = 0$. From the known expression of the magnetic field as a function of height, one can then find out the electron density.

More generally, the propagation of an arbitrary electromagnetic wave in a parallely stratified medium, in which the index of refraction $n_r(\omega, z)$ is given as a function of frequency and altitude, can easily be solved in the limit of the geometrical optics. In this case the propagation is entirely described by the angle $i(z)$ that the ray makes with the vertical direction. By applying the elementary Snell's law to an infinitesimal jump from n_r to $(n_r + dn_r)$ we get $n_r \cdot \mathrm{ctg}(i)di + dn_r = 0$. This integrates directly to

$$n_r \sin(i) = \mathrm{const.} \tag{8.33}$$

In the absence of a magnetic field the refractive index decreases with height from its ground value (almost unity); hence the ray gets further from the zenith, until it overcomes the ionization peak at z_m or gets completely reflected back to the ground. The frequency dependence of the refractive index produces a dispersion of the different Fourier components in direction and in travel time.

At high altitude the radius of the earth must be taken into account. As explained in Ch. 5, (see eq. (5.3)) from geometrical optics the constancy of

$$p = rn_r \cdot \sin(i) \tag{8.34}$$

follows. Eq. (8.33) is a particular case of this when r does not change much. The trajectory of the ray is then described by eq. (5.5); the constant p is determined by the zenithal distance at the initial radius (say, on the ground). Total reflection occurs when $n_r \cdot r = p$.

This formalism allows one to evaluate the change in the elevation of an extraterrestrial object seen through the troposphere and the upper atmosphere. This change is, of course, of great importance for astronomical observations in the optical and the radio bands (see Problems 8.7 and 8.8).

When the earth's magnetic field is taken into account, more complicated situations arise; in particular, the medium becomes birefringent, so that different polarizations propagate with different speeds. At high frequencies the effect is described by eqs. (8.28, 29); they show that the phase velocities of two beams with opposite circular polarizations differ by

$$\Delta\left(\frac{\omega}{k}\right) = c\,\frac{\Omega\omega_p{}^2}{\omega^3}\cos\theta\ . \tag{8.35}$$

Therefore a linearly polarized signal, arising from the superposition of two such beams of opposite sense and equal amplitude, changes the direction of its polarization as it crosses a magnetic field. This is the *Faraday effect*; it is used to measure the total electron content of the ionosphere in the following way.

A coherent, linearly polarized signal is emitted by a spacecraft to the earth, where the rotation $\Delta\varphi$ of its plane of polarization is measured. When the properties of the medium change very little in a wavelength the phase of the signal is not given by eq. (8.17), but by

$$\varphi = \omega t - \int d\mathbf{r}\cdot\mathbf{k}(\mathbf{r})\ , \tag{8.36}$$

where the line integral is performed along the path. This says that the phase at a given point and a given time is obtained by adding up all the small changes produced in each part of the ray where the medium is uniform. $\mathbf{k}(\mathbf{r})$ is the local wave vector, determined in direction by the application of Snell's law (or Fermat's variational principle) and in modulus by the index of refraction:

$$k(\mathbf{r}) = \omega n_r(\mathbf{r})/c\ . \tag{8.37}$$

φ is also the angle that the electric field in a circularly polarized beam makes with a fixed direction orthogonal to the propagation; hence the change in the plane of polarization of a linearly polarized beam induced by the birefringence is

$$\Delta\varphi = k\int ds\Delta n_r(\mathbf{r}(s)).$$

The integration is performed along the path $\mathbf{r}(s)$, with parameter s. We have neglected the the difference in the paths of the two component beams. At high frequency, using eqs. (8.28) and (8.29), we get

$$\Delta\varphi = k\int ds\Omega\omega_p{}^2\cos\theta/\omega^3\ . \tag{8.38}$$

We see that the rotation angle decreases with frequency as ω^{-2} and is proportional to the magnetic field and to $\int n ds$, the total electron content along the path (*columnar density*).

8.4. STRUCTURE AND MASS MOTION

A rough picture of the energy balance in the upper atmosphere emerges from the temperature profile as a function of height (Fig. 8.4). At low altitudes, in the troposphere, as explained in the previous chapter, most of the thermal energy comes from the ground in the infrared; the temperature gradient dT/dz is negative because more heat is radiatively lost to outer space at higher altitudes. Above 10 km, in the stratosphere, solar heating prevails and the gradient is reversed; in this region the thermal inertia and the currents do not allow a great variation in temperature between day and night. As the gas becomes thinner, the absorbed radiation from the sun becomes inadequate to balance the cooling into space and in the mesosphere we get again a negative temperature gradient. Finally, for very low densities above it even a small fraction of the solar radiation in the far ultraviolet is sufficient to heat again the gas at high temperatures; because it is a direct effect, the temperature changes very much from day to night (*thermosphere*).

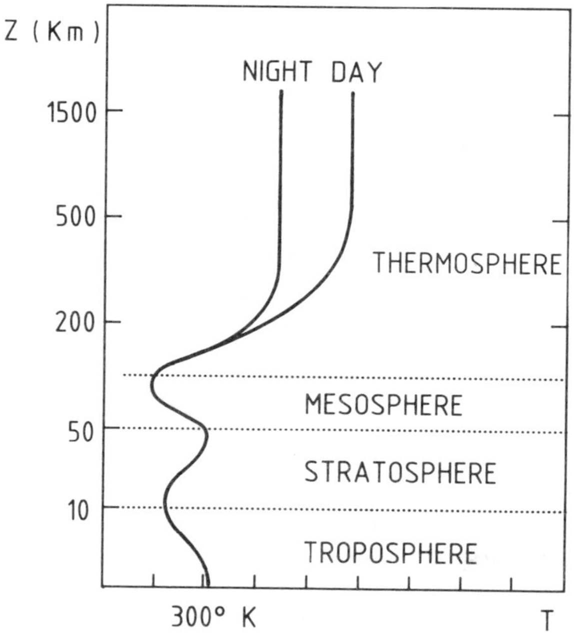

Fig. 8.4. The temperature profile of the atmosphere, which defines four different regions according to the different kind of radiation balance. The plot is indicative only. From a ground value of, say, 300 K, we have two minima of about 200 K in the tropopause between the troposphere and the stratosphere and in the stratopause between the stratosphere and the mesosphere. In the thermosphere the temperature ranges from 600 K to 1500 K during the night and from 1000 K to 2000 K in the day. The transition between homosphere and heterosphere occurs at about 100 km.

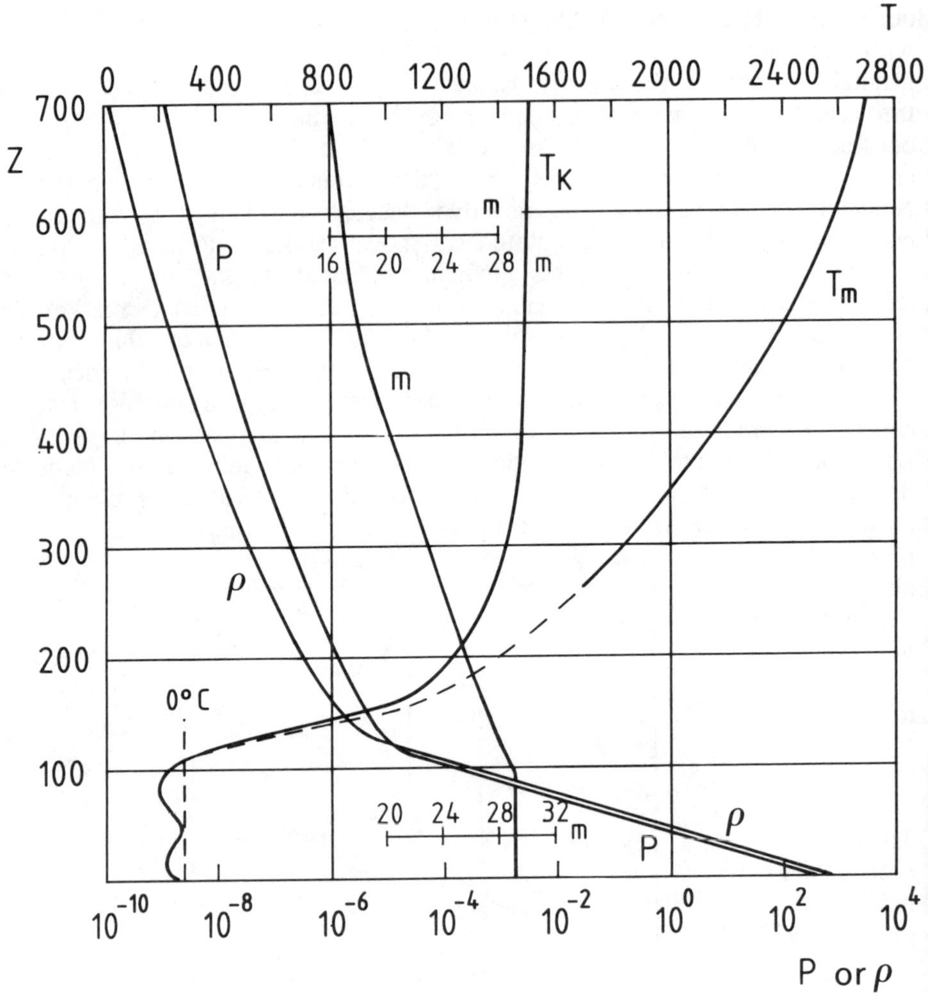

Fig. 8.5. The main physical quantities in the upper atmosphere as a function of height z (in km): the temperature T (in K); the mean molecular weight m; the density ρ in 10^{-6} g cm^{-3}; the pressure in 10^3 dyne cm^{-2}. We have also distinguished between the kinetic temperature T_k and the molecular temperature T_m.

To complete this descriptive information we refer to Fig. 8.5, with plots of the five main physical quantities. In the homosphere we have roughly the same molecular composition, with mean molecular weight 29; outside, in the heterosphere the composition changes and the mean molecular weight m decreases.

As explained in Sec. 1.6, fluid dynamics becomes particularly simple in local thermodynamical equilibrium, that is to say, when the mean free path is smaller than the characteristic scale of variation, in our case the scale height H. This condition, however, must be completed with the specification of the particular kind of process one considers: in general, different processes have different mean free paths and lead to the establishment of local thermodynamical equilibrium for each degree of freedom below different altitudes. When a degree of freedom is in equilibrium, the corresponding particle distribution is maxwellian. Neglecting these subtleties, we define a critical altitude z_{eq} by the condition

$$\ell_c = H(z_{eq}) = H_{eq}. \tag{8.39}$$

Above z_{eq} (typically, a few hundred km) we have the *exosphere.*

Below this height the atmosphere behaves as an ordinary, multiple component fluid (Free electrons and ions, of course, are not in thermal equilibrium with the neutral gas, but do not affect its dynamics.) The density profile of each component results from a balance between the outward diffusion and the inward settling down due to gravity. It is easy to see that this results in a change of the mean molecular weight with altitude.

Let us first derive the diffusion equilibrium condition of a species with mass m_1 and density N_1, embedded within the main species of mass m and density N. The argument of Fig. 1.3 can easily be adapted to get the diffusion flux

$$F_{1d} = \ell_c v_{T1} dN_1/dz . \tag{8.40}$$

To this we must add the downward net flux due to the fact that a molecule starting at $(z - \ell_c/2)$ with the thermal velocity v_{T1} crosses the height z with the velocity $(v_{T1} - g\ell_c/2v_{T1})$. This gives a net gravitational flux

$$F_{1g} = N_1 g\ell_c/v_{T1} . \tag{8.41}$$

Equating to zero the total flux and setting $v_{T1}^2 = kT/m_1$, we get, in isothermal conditions, the usual exponential density profile with the scale height

$$H_1 = v_{T1}^2/g = kT/m_1 g = mH/m_1 . \tag{8.42}$$

Therefore heavier species settle at lower altitudes, unless there is a substantial mixing due, for instance, to turbulent motion.

When $\ell_c \succ H$, a molecule which had its last collision at an altitude with background density n and travels upwards, cannot undergo the next collision after a distance $\ell_c = 1/n\sigma$, computed with the original density;

after having travelled ℓ_c it just does not find enough molecules to collide with and continues its path undisturbed. Above the critical altitude z_{eq} we have therefore, in general, a net outward flow. Note also that even below this altitude, a minor species may be produced or destroyed by external factors and have a net flux.

As a consequence, the atmosphere loses all the time molecules to outer space. To evaluate this loss, assume that a molecule does not undergo any collision above the critical height z_{eq}. One can say, at this height molecules are in dynamical contact with outer space and move in free keplerian orbits. They are lost if their total energy is positive, that is to say, if

$$\tfrac{1}{2}mv^2 > GM/(R + z_{eq}). \qquad (8.43)$$

In the velocity interval (v, v+dv) the number of lost molecules is

$$n_{eq}\left(\frac{m}{2\pi kT_{eq}}\right)^{3/2} \exp(-\,mv^2/2kT_{eq})\cdot 2\pi v^2 dv \;\; ;$$

integration up to the limit (8.43) yields:

$$F_{esc} = n_{eq}(GM/H_{eq})^{\frac{1}{2}}\; [1+H_{eq}/(R+z_{eq})]\cdot\exp[-\,(R+z_{eq})/H_{eq}]. \qquad (8.44)$$

This formula, of course, does not hold only for the main component; for another, minor component one must use the appropriate scale height (8.42). The lost flux is strongly dependent on the mass, through the scale height which appears in the exponential factor. Another way to express this result is to define the mean escape velocity

$$v_{esc} = F_{esc}/n_{eq}\;. \qquad (8.45)$$

A thermal velocity of this magnitude would give, in absence of gravity, the same escape flux. One could say that, a molecule at the critical altitude z_{eq} has the mean lifetime

$$\tau_{esc} = H_{eq}/v_{esc}\;. \qquad (8.46)$$

We give in the Table 8.1 some numerical values. They forcefully point out an important problem in the mass balance of the atmosphere, how these lost components are continuously replenished.

Species	H (km)	τ_{esc}
H_e^4	422	11 years
H_e^3	563	100 days
H^2	844	1 day
H^1	1688	4 hours

Table 8.1. Mean lifetimes of different molecules at z_{eq} = 500 km (from M. Nicolet).

PROBLEMS

*8.1. Express the zenithal distance of the sun as a function of the hour of the day and the day of the year; using eq. (8.8), find out the daily and the yearly amplitude of the oscillation of the height of the ionosphere. The first part of this problem is the basis for the construction of a sundial.

*8.2. Study the biquadratic equation (8.21) (Booker's equation) in various limits: $\omega \to \infty$; $\omega_p >> \Omega$; $\omega \to 0$; near resonance; etc.

8.3. Plot the electron density as a function of height using Chapman's approximation (8.4') and (8.9).

8.4. Plot ω_p as a function of z (eq. (8.11)) assuming a simple shape for the ionosphere layer (Problem 8.3).

*8.5. An easy and simplified way to describe the effect of collisions on the electron gas is to introduce in the conservation of momentum (8.14) a phenomenological friction term, replacing the time derivative ∂_t by $(\partial_t + \nu_c)$. ν_c is the collision frequency with the ions and the neutrals, assumed at rest. Study the dispersion relation (8.21) when ν_c is much smaller than ω, ω_p and Ω; the interesting regions are the neighbourhoods of the resonances.

8.6. Derive the table at the end of the chapter.

8.7. The troposphere produces a small deviation in the apparent position of an extraterrestrial source. Show that this deviation is proportional to the cotangent of the elevation over the horizon. Evaluate the corresponding delay in sunset (Hint: the curvature of the earth must be taken into account !.)

*8.8. Calculate the change in elevation of an extraterrestrial source in a spherically stratified atmosphere. Note that a straight line which meets the polar axis at a distance r_0 at an angle θ_0 has the equation

$$r = r_0 \sin\theta_0 / \sin(\theta - \theta_0) \ .$$

When r is large and η_r is unity, eq. (8.36) is indeed solved by this function. The angle θ_0 so determined measures the real zenithal distance of the source, to be compared with the apparent value θ_a, determined by $R = r(\theta_a)$.

FURTHER READING

A comprehensive text on plasma wave propagation is T.H. Stix, *The Theory of Plasma Waves*, McGraw-Hill (1962). About wave propagation in the ionosphere, see, e. g., the lectures on Magnetoionic Theory by K. G. Budden in C. DeWitt, J. Hieblot and A. Lebeau (eds.), *Geophysique exterieure*, Gordon and Breach (1962). In the same volume there is also a paper by M. Nicolet on *Composition and Constitution of the Earth's Environment.* Other references: S. Kato, *Dynamics of the Upper Atmosphere*, Reidel (1980); J.K. Hargreaves, *The Upper Atmosphere and Solar Terrestrial Relations. An Introduction to the Aerospace Environment*, Van Nostrand Reinhold (1979); S.J. Bauer, *Physics of Planetary Ionospheres*, Springer-Verlag (1973).

9. *MAGNETOSPHERE*

In the exosphere, where the very low collision frequency prevents an effective thermal contact with the atmosphere, the gas has a very high degree of ionization; its behaviour is determined mainly by the earth's magnetic field and the supersonic flow of charged particles emitted by the sun (solar wind). Their complex interaction takes place in the magnetosphere and can be understood on the basis of two physical processes, to be discussed in this chapter. The low energy charged particles are trapped in the equatorial belt of the magnetic field of the earth, but can precipitate along the lines of force on the polar regions of the ionosphere. Their behaviour can be described with the method of the centres of gyration. The overall magnetic field configuration results from the interaction between the bulk kinetic energy of the solar wind and the magnetic pressure of the dipole field, which acts as an obstacle to the supersonic flow. As a consequence, on the bow a shock wave develops and on the stern a wake arises, extending to very large distances downstream. This magnetospheric configuration is changed seasonally because of the different angle between the magnetic axis and the solar wind; it also reacts with complex wave processes to the solar activity. Charged, high energy cosmic particles are affected by the dipole field in a more complicated way; the regions forbidden to their penetration can be understood on the basis of the theory due to Størmer.

9.1. THE GUIDING CENTRE APPROXIMATION

The bulk of the population of charged particles in the magnetosphere can be understood on the basis of the guiding centre approximation. When the magnetic field is large enough the motion is the superposition of two different components: a fast circular motion in the plane orthogonal to the magnetic field and a slow motion of its centre, determined by the electric field and the gradient of the magnetic field. By considering a distribution of guiding centres one can achieve a satisfactory statistical description of the plasma.

The basic configuration we consider is a uniform and constant electromagnetic field, which determines the motion of a particle with charge q and mass m by means of the equation

$$m \frac{d\mathbf{v}}{dt} = q(\mathbf{E} + \frac{1}{c} \mathbf{v} \times \mathbf{B}) \ . \qquad (9.1)$$

The motion along the lines of force is determined only by the electric field $E_{\parallel} = \mathbf{E} \cdot \mathbf{B}/B$ along them:

$$m \, dv_{\parallel}/dt = qE_{\parallel} . \qquad (9.2)$$

To describe the perpendicular motion, consider a frame of reference moving with the *drift velocity*

$$\mathbf{v}_D = c\mathbf{E} \times \mathbf{B}/B^2 , \qquad (9.3)$$

which is orthogonal to the magnetic field and does not depend on the parallel electric field. In this frame the particle has a velocity $\mathbf{u} = (\mathbf{v} - \mathbf{v}_D)$ and obeys the simpler equation

$$\frac{d\mathbf{u}}{dt} = \frac{q}{mc} \mathbf{u} \times \mathbf{B} \ . \qquad (9.4)$$

This is a circular, uniform motion about the vector **B** with the angular velocity

$$\Omega_c = |q|B/mc \qquad (9.5)$$

(the *cyclotron frequency*). Its radius is the *Larmor radius*

$$r_L = u/\Omega_c \ . \qquad (9.6)$$

The centre of the circle is the *guiding centre*. Indeed, from eq. (9.4) the acceleration is normal to the orbit; this requires a constant speed u. Introducing the tangent unit vector $\mathbf{t} = \mathbf{u}/u$ and the arc length s we have

$$u|d\mathbf{t}/ds| = |q\mathbf{u} \times \mathbf{B}/mc| = |quB/mc| .$$

But this means that the radius of curvature of the orbit is constant and is given by eq. (9.6). When $q > 0$ the sense of circulation is clockwise for an observer who stands along the line of force. The overall motion is helical, with a pitch angle given by:

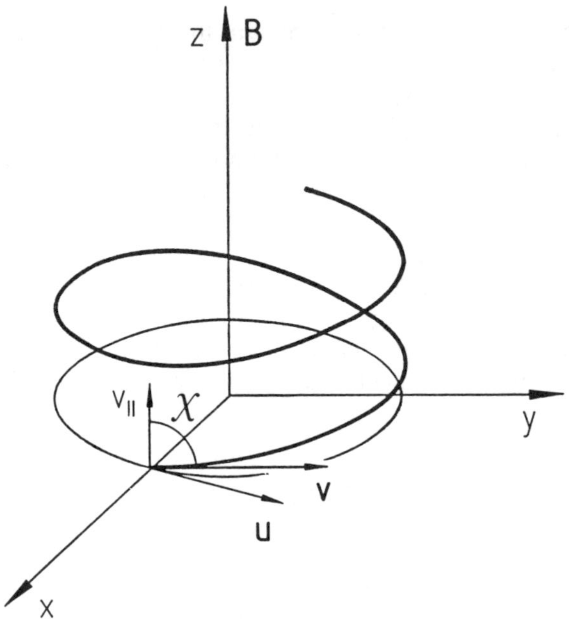

Fig. 9.1. The motion of a charged particle in a uniform magnetic field. χ is the pitch angle. The velocity **v** is decomposed in its longitudinal component $v_{\|}$ and its perpendicular component **u**.

$$\mathrm{ctg}\chi = v_{\|}/u = v_{\|}/r_L\Omega_c \; ; \tag{9.7}$$

χ measures the (constant) angle between the velocity vector and the magnetic field (Fig. 9.1).

This fundamental solution is a good description of the motion when the electromagnetic field does not vary appreciably in a Larmor radius and in a cyclotron period:

$$\left|\frac{1}{\Omega B}\,\frac{\partial B}{\partial t}\right| \ll 1 \; , \qquad \left|r_L\,\frac{\nabla B}{B}\right| \ll 1 \; ; \tag{9.8}$$

in a gyration the particle feels a field which is essentially constant. We are particularly concerned with the second condition; for the geomagnetic dipole, for example, the field is inversely proportional to the cube of the distance r from the centre of the earth and the condition reads $3r_L \ll r$. Of course, for a given velocity, r_L itself varies with the distance as r^3. For protons with a velocity equal to the solar wind speed (about 400 km/s) this is fulfilled below 100 earth radii; this comforts us that we have a good model. In this region the particles are bound to a line of force by their helical motion; the parallel motion is free, as if the line of force were a track.

We must now take into account the effect of small spatial changes of the magnetic field upon the motion of the guiding centre. These changes are of three kinds: variations in size along the line of force; perpendicular variations and variations in direction. To understand the effect of changes of the first kind it is useful to study first the motion of a charged particle in a magnetic field uniform in space, along a unit vector **n**, but slowly changing with time; with a change of frame the result will be applied to the case of our concern.

In this auxiliary problem Maxwell's equation

$$\nabla \times \mathbf{E} = -\frac{1}{c}\frac{dB}{dt}\mathbf{n} \tag{9.9}$$

requires an electric field orthogonal to the magnetic field. This electric field has a component along the perpendicular velocity **u** which alters the perpendicular kinetic energy. This change during a cyclotron period is easily evaluated by means of Stokes' formula and eq. (9.9):

$$mu\delta u = -q\int d\mathbf{s} \cdot \mathbf{E} = -q\int ds\mathbf{n} \cdot \nabla \times \mathbf{E} = \frac{q}{c}\pi r_L^2 \frac{dB}{dt} .$$

Here the line integral is taken around the gyration circle in the appropriate, counterclockwise sense; the minus sign is required because when the field is directed upwards (which we assume for simplicity and with no lack of generality) the actual motion is clockwise. Introducing the change δB of the magnetic field in a cyclotron period, we get

$$0 = \frac{\delta u}{u} - \frac{1}{2}\frac{\delta\Omega_c}{\Omega_c} = \delta\left(\frac{u^2}{\Omega_c}\right) = \delta(r_L^2 B) .$$

In other words, while the magnetic field changes, the orbit adjusts itself in such a way as to keep constant the magnetic flux embraced in the gyration. This flux is proportional to the *magnetic moment*

$$\mu_m = \frac{\text{perpendicular kinetic energy}}{\text{cyclotron frequency}} = \frac{mu^2}{2\Omega_c} . \tag{9.10}$$

A more accurate analysis is required to assess the limits of this approximation. This entails the consideration of the smallness parameter

$$\epsilon = \left| \frac{1}{\Omega_c^2} \frac{d\Omega_c}{dt} \right| ;$$

it turns out that in a period $\delta\mu_m/\mu_m = O(\epsilon)$. The proof requires the methods of *singular perturbations theory*: we have a system of differential equations whose solutions are not analytic functions of the small parameter ϵ as it tends to zero. This is equivalent to the limit $\Omega_c \rightarrow \infty$, keeping all other quantities fixed; in this limit the radius of the Larmor circle tends to zero. It is interesting to note that the expression of the magnetic moment can be corrected to yield quantities which are constant to an arbitrary order in ϵ.

The magnetic moment is an *adiabatic invariant*, referring to the slow change of the external magnetic field. This very important result has a close analogy in the theory of a harmonic oscillator with a slowly varying frequency

$$\ddot{x} + \Omega^2(t)x = 0 . \tag{9.11}$$

It can be shown that, if the corresponding ϵ (the fractional change of the frequency in a period) is small, the quantity

$$\mu = \frac{\text{average kinetic energy}}{\text{frequency}} = \frac{\langle \dot{x}^2 \rangle}{2\Omega} \tag{9.12}$$

is indeed an adiabatic invariant (see Problem 9.5). Therefore the amplitude of a harmonic oscillator is proportional to the square root of the varying frequency.

We should also remark that the appropriate mathematical level to deal with adiabatic invariants is the Hamiltonian theory of quasi-periodic systems. Consider, for example, a system with one degree of freedom and periodic solutions; the relevant quantity is the action

$$S = \int p dq , \tag{9.13}$$

i. e., the area in the phase plane (p, q) enclosed in a period. It can be shown that when the Hamiltonian function varies slowly with time, S is adiabatically invariant. This provides, by the way, the necessary justification for the Bohr-Sommerfeld quantization rule, which requires the action to be a multiple of Planck's constant. This rule would be unacceptable if small changes of the external conditions would change it continuously: the only changes allowed to S by quantum mechanics are discrete.

The case in which the magnetic intensity varies along the (straight) line of force is straightforward: just go over to a frame of reference which moves with the particle's parallel velocity. In this case the motion is entirely perpendicular, but the magnetic intensity is variable in time. The transformation does not alter the perpendicular velocity, so that the constancy of the magnetic moment (9.10) is still ensured.

Let us now follow the motion of a particle along a line of force with arc length s. Since no work is performed on the particle, to compensate the change in the transversal kinetic energy the longitudinal kinetic energy must change as well:

$$mv_{\parallel}\frac{\partial v_{\parallel}}{\partial s} = m\frac{dv_{\parallel}}{dt} = -\,mu\frac{du}{ds} = -\,\mu_m\frac{d\Omega_c}{ds}\ . \tag{9.14}$$

We have an effective longitudinal force (independent of the charge)

$$F_{\parallel} = -\frac{mu^2}{2B}\frac{dB}{ds} \tag{9.14'}$$

and an effective potential energy

$$U_{eff} = \mu_m\,\Omega_c(s). \tag{9.15}$$

The quantity

$$E_{eff} = \tfrac{1}{2}mv_{\parallel}^2 + \mu_m\Omega_c(s) \tag{9.16}$$

is adiabatically conserved during the motion.

The general theory of the motion of the guiding centres can be constructed by writing the position of the particle

$$\mathbf{r}(t) = \mathbf{r}_o(t) + \mathbf{r}_L(t) \tag{9.17}$$

as the sum of the position of the centre $\mathbf{r}_o(t)$ and the Larmor gyration $\mathbf{r}_L(t)$, which describes a small circular motion of radius r_L around $\mathbf{r}_o$. The magnetic field is expanded around $\mathbf{r}_o$ in powers of r_L and the equation of motion is averaged over a gyration period.

Let us discuss the result of this analysis first when the parallel velocity vanishes; we have an effective new force

$$\mathbf{F} = -\frac{m\mu^2}{2B}\nabla B\ , \qquad (9.18)$$

which reduces to the previous expression (9.14') when the magnetic intensity changes only along the lines of force. The perpendicular part has an effect similar to the electric force: it produces a drift $c\mathbf{F} \times \mathbf{B}/qB^2$, to be added to (9.3):

$$\mathbf{v}_D = c\,\frac{\mathbf{E} \times \mathbf{B}}{B^2} - \frac{mu^2 c}{2qB^3}\nabla B \times \mathbf{B}\ . \qquad (9.19)$$

The new drift depends on the charge and therefore generates an electric current. In the dipole field of the earth this current density is of order $J \approx nmu^2 c/Br$ and produces an additional magnetic field

$$\delta B \approx BnK/B^2\ ,$$

where n is the electron density and K is the kinetic energy per particle (*Dessler-Parker relation*). The intensity of the field is thereby diminished. This current flows along the parallels (*ring current*).

The case in which the longitudinal velocity does not vanish can be dealt with by going to a frame of reference in which the guiding centre is at rest. If the longitudinal velocity is constant we need just a Galilean transformation, already discussed. In general, however, the new frame is not inertial and apparent forces arise. The averaging procedure produces a contribution from the centrifugal acceleration

$$\mathbf{F}_{cent} = -mv_{\parallel}^2\ [\mathbf{B}\cdot\nabla\mathbf{B}]_{\perp}/B^2 \qquad (9.20)$$

and a corresponding drift $c\mathbf{F}_{cent} \times \mathbf{B}/qB^2$. One can show that, in the interesting situation of no electric currents, this drift is proportional to ∇B. The corresponding drift adds to (9.19) to yield the final expression

$$\mathbf{v}_D = c\,\frac{\mathbf{E} \times \mathbf{B}}{B^2} - \frac{mc}{qB^3}\left(\tfrac{1}{2}u^2 + v_{\parallel}^2\right)\nabla B \times \mathbf{B}\ . \qquad (9.21)$$

9.2. TRAPPED PARTICLES

The constancy of the magnetic moment (9.10) slows down the longitudinal motion of a charged particle in the dipole field as it moves to higher latitudes until $U_{eff} = E_{eff}$ and the longitudinal velocity vanishes: we have there a turning point. This concentrates a population of charged particles at lower latitudes (the *van Allen belt*). Since the velocity is constant in size, it is convenient to describe the effect by means of the pitch angle (9.7). From eq. (9.10) we have

$$\sin^2\chi = \sin^2\chi_o \cdot (\Omega_c/\Omega_{co}), \qquad (9.22)$$

where the index o refers to the magnetic equator. The turning point occurs where

$$\Omega_c(s) \cdot \sin^2\chi_o = \Omega_{co} \, . \qquad (9.23)$$

The largest magnetic field on a given line of force may occur at the top of the ionosphere; then the motion is hindered by collisions and the theory developed in this Section does not apply. If the index 1 refers to this maximum, the population of trapped particles is determined by the relationship

$$\sin\chi_o > \sqrt{(B_o/B_1)} = \sqrt{(\Omega_{co}/\Omega_{c1})} = \sin\chi_1 \, . \qquad (9.24)$$

In this model the interior of the cone $\chi_o < \chi_1$ (the *loss cone*) is empty. In practice collisions and other perturbations throw some particles inside the loss cone, but this picture is generally valid. It does not hold for energetic particles which violate the adiabatic condition (9.8). Notice also that this effect destroys any isotropy in velocity that the particle's population may have; we are not allowed, therefore, to assume a thermal equilibrium. Particles injected on the equator may escape from inside the loss cone and precipitate onto the polar ionosphere inducing there, by collisional excitation of the neutral molecules, the auroral glow (Problem 9.2.)

In this simple picture, assuming an axially symmetric field, the particle density is independent of longitude. To have a rough idea of the numbers involved, note that – as we shall see later – the kinetic energy of the particles must be appreciably less than that of the trapping field, about $8.4 \cdot 10^{24}$ erg (see Problem 9.1); indeed, this energy is about $6 \cdot 10^{22}$ erg. Now aurorae require a widely varying energy deposition rate, from, say 3 to 3000 erg/(cm$^2\cdot$s), over, say, 0.1 steradians, corresponding to a power of 10^{17} to 10^{20} erg/s. Therefore the life-time of the trapped particles ranges from $6 \cdot 10^5$ s to $6 \cdot 10^2$ s.

Another application of the adiabatic invariants to the ionosphere concerns the case in which the field is not quite axially symmetric or changes slowly with time. A particle is turned around in its longitudinal motion at the point given by eq. (9.23); but as the magnetic field experienced by the particle changes because of time variations or the

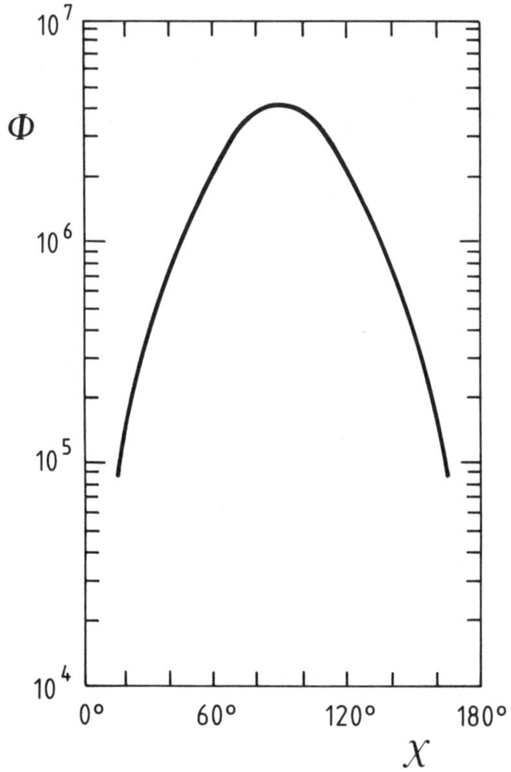

Fig. 9.2. A typical profile of the proton flux Φ as a function of the pitch angle for energies of the order of 1 MeV. The flux is given in protons/(cm^2 $\cdot$s$\cdot$sterad).

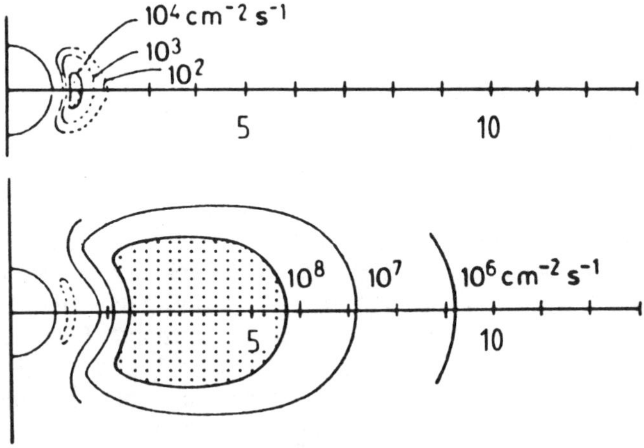

Fig. 9.3. Typical isoflux contours for protons of energy > 30 MeV (upper plot) and between 0.1 and 5 MeV (lower plot). The distance is measured in earth's radii. As expected, the more energetic particles penetrate nearer to the earth. (From J.V. van Allen, "Particle description in the magnetosphere", in *Physics of the Magnetosphere* (R.L. Carovillano, F. McClay and H.R. Radoski, editors), Reidel (1968).)

drift of its gyration centre, the effective energy (9.16) cannot be considered as constant any more. In the lowest approximation we have a periodic motion along the line of force with period

$$P = 2\int_{s_1}^{s_2} ds/v_{\parallel}(s) \ , \qquad (9.25)$$

where s_1 and s_2 are the solutions of eq. (9.23). If $\Omega_c(s)$ changes very little in this period, we can apply the general theory of adiabatic invariants, which states that the action (9.13) or, to within a factor,

$$S = \int_{s_1}^{s_2} ds v_{\parallel}(s) \qquad (9.26)$$

is to a good approximation constant in time. Using the value $B_m = B_0/\sin^2\chi_0$ of the magnetic intensity at the turning points in terms of the pitch angle χ_0 at the equator, from the constancy of the magnetic moment we get:

$$v^2 - v_{\parallel}^2(s) = v^2 B(s)/B_m \ ; \qquad (9.27)$$

so that

$$S = v\int_{s_1}^{s_2} ds\sqrt{(1 - B(s)/B_m)} \ . \qquad (9.28)$$

For example, consider the particles injected in the equatorial region by the solar wind during the day. Their pitch angle χ_0 determines the turning point. The solar wind compresses the equatorial magnetic field, but leaves the field more or less unchanged at high latitudes; moreover, this compression does not change the total velocity v. When the night comes the field lines expand and the equatorial pitch angle must change in such a way as to keep the action S constant. One can see from simple examples (Problem 9.3) that the equatorial pitch angle grows with the field, so that in night time the particles are able to penetrate deeper into the upper atmosphere at high magnetic latitudes.

9.3. ALFVEN WAVES

Wave propagation plays an essential role in magnetospheric dynamics;

besides electromagnetic oscillations of the electrons, discussed in the previous chapter, we have *Alfvên waves*, which can be regarded as the modification of acoustic waves when the magnetic restoring force is important. Other kinds of waves will not be discussed here. Magnetospheric dynamics is described fairly well by *magnetohydrodynamics* (MHD): the relevant velocities are not relativistic; the conductivity is large; the plasma is neutral.

The linearized equations for the MHD flow are the laws for conservation of mass (1.21')

$$\partial\rho/\partial t + \rho_o \nabla \cdot \mathbf{v} = 0$$

and momentum (1.58):

$$\rho_o \partial\mathbf{v}/\partial t + c_s^2 \nabla P = (\nabla \times \mathbf{B}) \times \mathbf{B}_o/4\pi \ .$$

The suffix o denotes a uniform background at rest; quantities with no index are small corrections. We have assumed an adiabatic equation of state with a sound speed $c_s = \sqrt{(dP_0/d\rho)}$ (eq. (1.49)) and have eliminated the current by means of eq. (1.59_2). We have also the remaining Maxwell's equation (1.59_3)

$$\nabla \times (\mathbf{v} \times \mathbf{B}_O) = \partial\mathbf{B}/\partial t \ , \tag{9.29}$$

where Ohm's law (1.65) has been used. We can also eliminate the magnetic field perturbation, to get

$$\partial^2\mathbf{v}/\partial t^2 - c_S^2 \nabla\nabla \cdot \mathbf{v} = V_A^2 [\nabla \times (\nabla \times (\mathbf{v} \times \mathbf{e}_Z))] \times \mathbf{e}_Z =$$

$$= V_A^2 [\nabla \times (\partial_Z \mathbf{v} - \mathbf{e}_Z \nabla \cdot \mathbf{v})] \times \mathbf{e}. \tag{9.30}$$

We have taken here the unperturbed, constant magnetic field B_0 along the z-axis, and introduced a new characteristic velocity

$$V_A = \sqrt{(B_O^2/4\pi\rho_O)} \ , \tag{9.31}$$

Alfvên's speed.

When $B_0 = 0$, **v** is a solenoidal vector and the only surviving (and longitudinal) component obeys d'Alembert equation with the propagation speed c_s. These are the acoustic modes. The magnetic field changes their dispersion relation and introduces two new modes.

With the usual sinusoidal assumption eq. (9.32) becomes algebraic:

$$\omega^2\mathbf{v} - (c_S^2 + V_A^2)\mathbf{k}\ \mathbf{k}\cdot\mathbf{v} = k_Z V_A^2 (k_Z \mathbf{v} - \mathbf{k} v_Z - \mathbf{k}\cdot\mathbf{v}\ \mathbf{e}_Z). \tag{9.32}$$

If the propagation vector is orthogonal to the magnetic field ($k_z = 0$) the right-hand side vanishes and we have a longitudinal magnetosonic wave with dispersion relation

$$\omega^2 = (c_S^2 + V_A^2)k^2; \qquad (9.33)$$

from eq. (9.31) one sees that the perturbed magnetic field is still parallel to the z-axis: we have just a compression and a rarefaction of the field governed by the magnetic pressure. If the propagation is along the magnetic field, we have

$$(\omega^2 - k^2V_A^2)\mathbf{v} + (V_A^2 - c_S^2)k^2v_Z\mathbf{e}_Z = 0 ,$$

with two solutions. When v_z only is different from zero, we recover the ordinary magnetosonic mode $\omega = \pm kc_s$; but when $v_z = 0$, there is transverse wave with phase velocity V_A. In this case the magnetic lines of force oscillate and their tension provides the restoring force.

It can be shown (Problem 9.4) that for a generic propagation vector making an angle θ with the z-axis there are three different modes with dispersion relations

$$\omega^2 = (kV_A\cos\theta)^2 , \qquad (9.34)$$

$$\omega^2 = \tfrac{1}{2}k^2\{c_S^2 + V_A^2 \pm \surd[(c_S^2+V_A^2)^2 - 4c_S^2V_A^2\cos^2\theta]\}. \qquad (9.35)$$

When $V_A \gg c_s$ the upper sign gives a phase velocity equal to V_A.

The ratio of the two characteristic velocities determines the general features of the propagation. More generally, one considers the dimensionless parameter

$$\beta = 8\pi nkT/B^2 , \qquad (9.36)$$

ratio of thermal to magnetic energy. Acoustic and purely alfvénic disturbances are recovered when $\beta \ll 1$ and $\beta \gg 1$, respectively. When $\beta \approx 1$, there is continuous energy exchange between the particles and the field, drastically affected by their motion. In the magnetosphere of the earth, especially at low altitudes, $\beta \ll 1$ indicates that the magnetic field is mainly of external origin. Plasma disturbances in the magnetosphere and the solar wind are very often of this kind.

9.4. THE BOW SHOCK OF THE MAGNETOSPHERE

With these concepts in mind, let us turn to the solar wind, which at 1 AU consists mainly of electrons and protons, with a bulk velocity in the range 200 to 700 km/s and a temperature about 1% of the directed proton energy, say 500 eV. It carries along a residual magnetic field of solar origin of a few 10^{-5} Γ. We can therefore define a sonic and an alfvénic Mach number

$$M_S = V/c_S \ , \qquad M_A = V/V_A \ ; \tag{9.37}$$

they are about 10. From the hydrodynamical point of view, the main feature of the solar wind is that its speed is larger than the relevant phase velocities.

When a point obstacle is placed in an inviscid supersonic flow with Mach number M, its disturbances – which propagate with respect to the medium with a phase velocity V/M, say – cannot propagate upstream. The propagation velocity in the rest frame, being the vectorial sum of **V** and a vector of size V/M with arbitrary direction, always lies within a cone – Mach's cone – with semiaperture arcsin(1/M) (equal to $5^\circ.74$ when M = 10) and containing the wake. One can say, the flow does not "see" the obstacle before actually meeting it. In reality the situation is more complex because the obstacle is finite and may be modified by the wind itself; but this basic feature is evident for all planets and natural satellites in the solar system.

The interaction between the solar wind and the dipole field of the earth is determined by the balance between its kinetic energy and the magnetic pressure $B^2/4\pi$ (Sec. 1.6). In terms of the ground equatorial value B_e (at r = R) this balance occurs at the distance

$$D = R[B_e{}^2/2\pi n m_p V^2]^{1/6} . \tag{9.38}$$

With a typical solar wind density n = 5 protons/cm^3 it is about 10 R.

A finite obstacle generates also a *shock wave* upstream. Consider the very simple example of a piston accelerated in a fluid at rest. Initially, when its motion is subsonic, the disturbance travels ahead of the piston with the sonic speed determined by the fluid state still at rest and unaffected. If the piston moves with a supersonic speed, it eventually overcomes the front of the disturbance; at this point the dissipationless fluid laws become inconsistent. One expects a similar phenomenon when the supersonic solar wind hits the magnetosphere's boundary.

In inviscid fluid dynamics the shock can be described with a discontinuity which propagates ahead of the piston and separates the perturbed and the unperturbed regions. The flow across it conserves the mass, the momentum and the energy. This mathematical solution is interpreted physically in the following way. The assumption of inviscid flow may become invalid in regions with very large gradients; in fact the viscous force, proportional to the second derivatives of the velocity field, may be quite large there, even though the viscosity coefficient is small. A thin region arises where the viscous forces are important. The shock thickness and its structure are indeed determined by viscosity.

In the case of the interaction between the solar wind and the earth viscosity and other conventional dissipation processes do not play a relevant role; indeed, on small scales even the MHD equations are not appropriate and must be replaced with a kinetic description in phase space (see Sec. 1.8). The measurements of particle densities and electric fields by spacecraft *in situ* have shown that across the bow shock of the earth large, high frequency oscillations arise. It is believed that their random, non-linear interactions provide the appropriate dissipation processes and determine the shock thickness. This is an example of *collisionless dissipation* and must be described with the Boltzmann-Vlasov equation in the non-linear regime (see Sec. 1.9).

Upon crossing the bow shock the flow becomes slower, denser and hotter and suffers a deviation in direction. Its kinetic pressure compresses the earth's magnetic field and the deviation increases further until the flow grazes a cavity in which, in the fluid dynamic approximation, there is no gas. The boundary of this cavity is the *magnetopause*. In reality, particles do penetrate in this cavity through the polar region and by other mechanisms. This fluid dynamic problem has been solved and has given reasonable results (Fig. 9.4).

The basic features of the bow shock can be understood from first principles. We neglect the magnetic field outside the magnetopause; this does some violence to the physical model (in which the magnetic and thermal energies of the solar wind are comparable!), but greatly simplifies the geometry. We also confine ourselves to a limited region of the shock and neglect all tangential changes: this is then the basic problem of an inviscid flow going through an oblique, plane surface of discontinuity at rest. Let Ψ be the angle between its normal and the oncoming flow, with velocity V. Consider now the flow of the relevant conserved quantities, given by the amount contained in an oblique cylinder of length V, with base 1 cm^2 on the plane of discontinuity and volume $V\cos\Psi$. The flow of matter $F = \rho V\cos\Psi$ is of course the same upstream and downstream. Tangential momentum, carried through with a flux $FV\sin\Psi = \rho V^2 \sin\Psi\cos\Psi$, is also unchanged, since there are no tangential forces; hence the tangential velocity $V\sin\Psi$ is constant across the shock. This allows a great simplification: in the frame of reference moving with such a velocity the shock is normal to the flow and the problem becomes one-dimensional.

Denoting with the suffixes 1 and 2, respectively, the upstream and downstream quantities in this frame, we have $v_1 = V\sin\Psi$; hence

$$F = \rho_1 v_1 = \rho_2 v_2 \; . \tag{9.39}$$

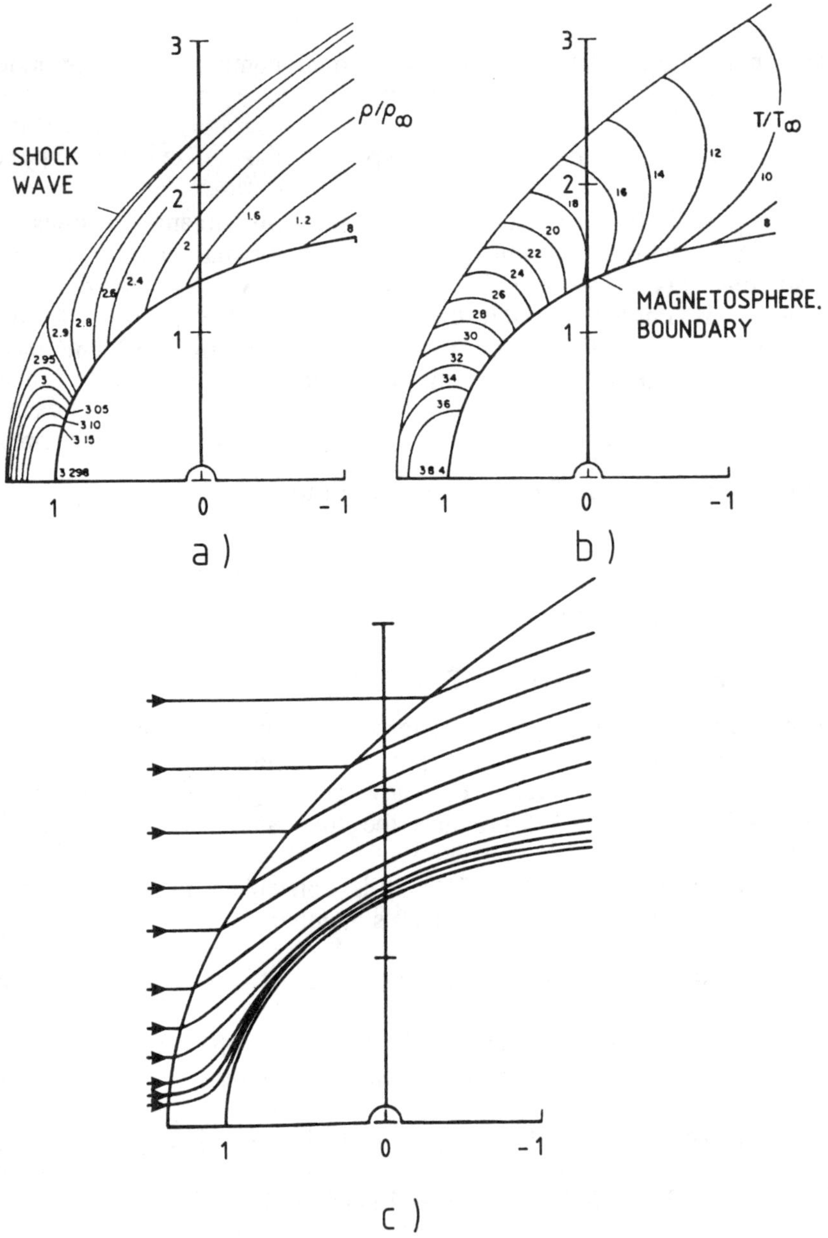

Fig 9.4. The compression a) and the temperature ratio b) in the region behind the bow shock, as computed with the magnetohydrodynamical equations. The upstream Mach's number is 8.71; the magnetic field was neglected outside the magnetopause. The polytropic index γ is 2. Fig. 9.5c gives the flow pattern (from J. R. Spreiter et al., in *Physics of the Magnetosphere* (R. L. Carovillano, ed.), Reidel (1968).) In agreement with eq. (9.37), the unit of length is chosen to be $D = R(B_e^2/5.289\ V^2)^{1/6}$. The bow shock stands at a distance of about 0.3 D from the nose of the magnetosphere.

The conservation of the flux of the normal component of the momentum gives

$$\rho_1 v_1^2 + P_1 = \rho_2 v_2^2 + P_2 . \qquad (9.40)$$

where P_1 and P_2 are the upstream and downstream pressures. To get the energy condition we must introduce the internal energy per unit mass ϵ, so that $F(\frac{1}{2}v_1^2 + \epsilon_1)$ is the energy flow. As the cylinder with height v_1 is pushed through the discontinuity, the pressure P_1 performs the work $P_1 v_1 = FP_1/\rho_1$; on the other side, during the same interval of time, the pressure P_2 performs the work FP_2/ρ_2. Therefore

$$\tfrac{1}{2}v_1^2 + \epsilon_1 + P_1/\rho_1 = \tfrac{1}{2}v_2^2 + \epsilon_2 + P_2/\rho_2 . \qquad (9.41)$$

From eqs. (9.39) and (9.40), we get the flux

$$F^2 = (P_2 - P_1)/(1/\rho_1 - 1/\rho_2). \qquad (9.42)$$

The energy equation (9.41) reads also

$$\epsilon_1 - \epsilon_2 + \tfrac{1}{2}(P_1 + P_2)/(1/\rho_1 - 1/\rho_2) = 0 , \qquad (9.43)$$

which says that the total increase in internal energy equals the compression work done by the pressure averaged between the upstream and the downstream values. For any given gas, with a given internal energy $\epsilon(P, \rho)$, eq. (9.43) provides the final state when the other side is given.

In a perfect gas (eq. (1.51)) it is convenient to use as independent parameter the upstream normal Mach's number

$$M_1 = v_1 \surd(\rho_1/\gamma P_1) , \qquad (9.44)$$

where γ is the adiabatic index. One arrives at the following expressions for the ratios of densities, pressures and temperatures:

$$\rho_2/\rho_1 = (\gamma + 1)M_1^2/[(\gamma - 1)M_1^2 + 2] , \qquad (9.45)$$

$$P_2/P_1 = [2\gamma M_1^2 - \gamma + 1]/(\gamma + 1) , \qquad (9.46)$$

$$T_2/T_1 = [2\gamma M_1^2 - \gamma + 1][(\gamma - 1)M_1^2 + 2]/(\gamma + 1)^2 M_1^2 . \qquad (9.47)$$

In the bow shock the approximation $M_1 \gg 1$ (strong shock) is relevant and we get

$$\rho_2/\rho_1 = (\gamma + 1)/(\gamma - 1) \qquad (9.45')$$

$$P_2/P_1 = 2\gamma M_1^2/(\gamma + 1) \qquad (9.46')$$

$$T_2/T_1 = 2\gamma(\gamma - 1)M_1{}^2/(\gamma + 1)^2 \ . \qquad (9.47')$$

The compression ratio tends to a finite value, equal to 4 for the usual value $\gamma = 5/3$; but the heating becomes very large, with a temperature ratio proportional to the square of Mach's number. Let us also record the normal Mach's number downstream

$$M_2{}^2 = [2 + (\gamma - 1)M_1{}^2]/[2\gamma M_1{}^2 - \gamma + 1] \ , \qquad (9.48)$$

becomimg for a strong shock

$$M_2{}^2 = (\gamma - 1)/2\gamma. \qquad (9.48')$$

This solution is easily generalized to the case in which a magnetic field **B** is present. Since its divergence vanishes, its normal component is continuous across the shock; similarly, since $\nabla \times \mathbf{E} = 0$, the tangential component $\mathbf{E}_t$ of the electric field is continuous also. We can always adopt a frame in which $\mathbf{E}_t = 0$; Ohm's law (1.65) then determines the tangential velocity $\mathbf{v}_t$. An interesting, particular solution is when the tangential magnetic field $\mathbf{B}_t$ vanishes on both sides; then the jump conditions previously derived still hold (Problem 9.6).

9.5. INTERACTION BETWEEN THE SOLAR WIND AND THE MAGNETOSPHERE

The two basic processes described in detail in the previous sections – the trapping of charged particles in the main geomagnetic field and the deflection and compression of the solar wind hitting the magnetosphere – are essential to get a rough picture of this interaction, in reality much more complex.

The lines of force can be topologically classified in four regions. In I an interplanetary field line eventually goes back to interplanetary space; in II it terminates on the earth. There are lines which close up on the earth (III) and, finally, isolated closed loops (IV) (Fig. 9.5). These four regions are highly variable. The solar wind itself is not steady. The angles that the solar wind makes with the interplanetary magnetic field (Sec. 13.3) and the magnetic axis of the planet change seasonally. The rotation of the earth is transmitted to the lines of force of class II and twists them around in the magnetic wake. More important, lines can change their topological class by the *reconnection process*. This is accomplished, for example, when at the boundary between two regions the magnetic field changes its polarity passing through a zero (*neutral sheet*); an instability may arise which interchanges the connection between the lines of force.

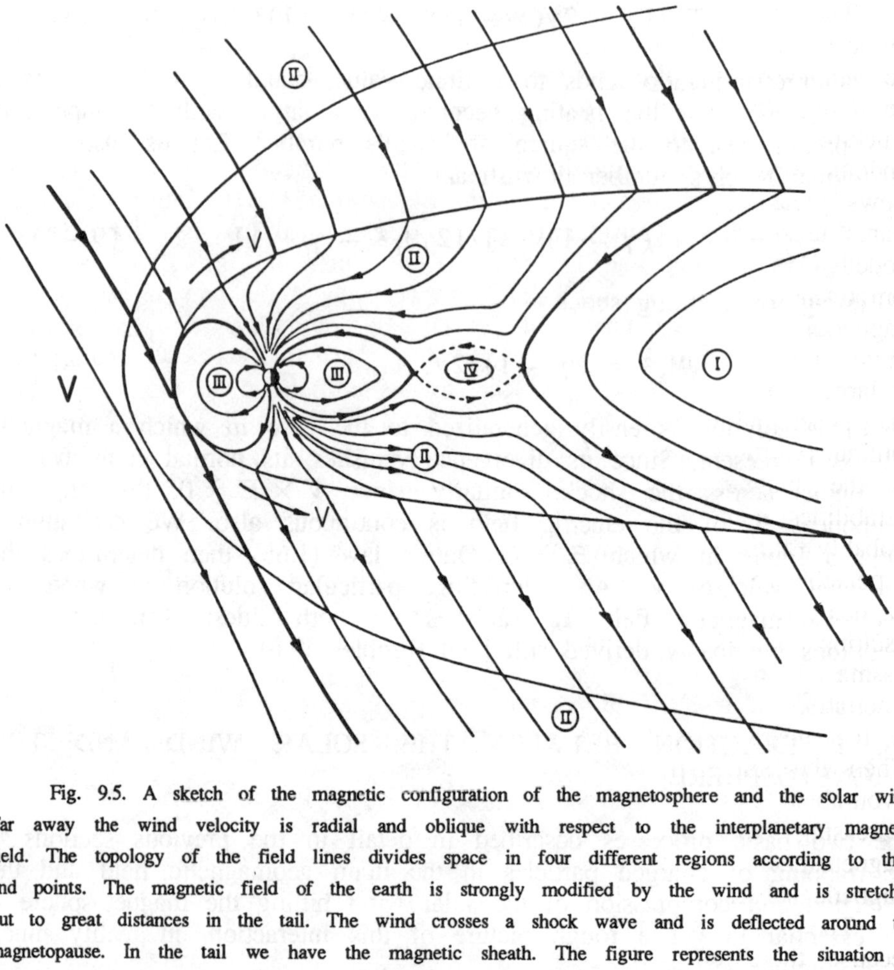

Fig. 9.5. A sketch of the magnetic configuration of the magnetosphere and the solar wind. Far away the wind velocity is radial and oblique with respect to the interplanetary magnetic field. The topology of the field lines divides space in four different regions according to their end points. The magnetic field of the earth is strongly modified by the wind and is stretched out to great distances in the tail. The wind crosses a shock wave and is deflected around the magnetopause. In the tail we have the magnetic sheath. The figure represents the situation at equinox, when the magnetic axis is perpendicular to the solar wind.

Consider the magnetopause when, as in Fig. 9.5, the interplanetary magnetic field is directed downwards; then it is energetically favourable to transform some lines of class III into class II, thereby decreasing the magnetic energy and exposing a little of the magnetosphere to the solar wind. Similarly, closed loops of class IV are subject to the tension of the lines of force and tend to collapse, transforming their energy into kinetic energy and Alfvèn waves. In the magnetic tail the geomagnetic lines of force, stretched along by the solar wind, are generally inward bound in the upper hemisphere and viceversa, generating a plasma sheet. This is a favourable situation for reconnection and generation of bursts of energetic particles, which may impinge on the earth from the night side. Similar events have been observed also in the magnetospheres of Mercury and the outer planets.

The magnetosphere acts on the solar wind as a compressible obstacle, supported by the geomagnetic field; it deviates its flow around the *magnetopause*, after having reduced its velocity to subsonic values in the *bow shock*, which sits at about 15 R on the noon side. The structure of the bow shock, determined by dissipation processes due to random wave-wave interactions, rather than collisions, is turbulent, and shows precursor phenomena in form of emitted waves and reflected ion beams upstream of the main transition region. The magnetopause can be modelled as a tangential discontinuity, in which the flow has no normal component and we have a balance between the total pressures (fluid and magnetic) inside and outside. This discontinuity has a complex microscopic structure due also to the fact that ions, whose Larmor radius is larger, penetrate more into the magnetic field than electrons and a charge separation arises. The region in between – the *magnetosheath* – continues through the *polar cusps* and the *plasma mantle* around the tail. In the magnetopause, where a strong velocity gradient is present, instabilities may develop which release the shear kinetic energy available, probably through the *Kelvin-Helmoltz* mechanism. They are probably responsible for the generation of low frequency oscillations of the magnetopause boundary layer, with periods of the order of minutes, resulting also in pulsations of the magnetic field at the ground. The plasma blobs created by the Kelvin-Helmoltz process allow the penetration of solar wind plasma into the magnetosphere at low latitudes. In the magnetopause also important reconnection processes take place. When the interplanetary field is directed downwards, it is energetically favourable to transform some lines of class III into class II, thereby decreasing the magnetic energy and exposing a little of the magnetosphere to the solar wind. The ensuing instabilities may produce magnetic substorms.

The magnetosheath near the polar cusps has a complex structure because there we have also field lines of class II, which allow the transfer of material into the ionosphere and produce there auroral phenomena. The outer flow in the cusps meets a magnetic "obstacle" whose surface changes abruptly, similarly to the flow around a concave corner; it is possible that a free flow surface develops, which separates an inner region of stagnating and turbulent plasma from the outer, free flow region.

Further away, the tail extends to very large distances in the direction opposite to the sun. The lines of force, stretched along by the solar wind, are generally inward bound in the northern hemisphere and *vice versa*. The tail has been observed up to 240 R, with a thickness about 5 R. Just like in ordinary wakes, vortices develop in the tail; moreover, in the neutral plasma sheet in the middle, where the magnetic polarity reverses, reconnection occurs producing plasma velocities up to 1000 km/s. As a result, plasma is ejected both outwards and inwards, onto the earth. Similar events have also been observed in the magnetic tail of Mercury.

Finally, we mention the modifications of the inner magnetosphere. On the noon side the dipole field is compressed, as expected, but below 5 R the magnetic intensity is *smaller* than the dipole value, by about $\delta B \approx 30 \cdot 10^{-5}\ \Gamma$ or more; this is caused by the *ring current*, a flow of energetic protons at low latitudes. During a *magnetic storm*, when a gust in the solar wind injects more energetic particles into the magnetosphere, the ring current increases and the decrease δB in the magnetic intensity is larger. If the storm is slow enough, the adiabatic invariant (9.10) is constant and the particle energy in the magnetosphere decreases.

Our current understanding of the magnetosphere and its interaction with the solar wind is mainly the result of extensive measurements carried out with spacecraft, in particular with electrostatic detectors to determine the density and the velocity distribution of the different kinds of charged particles and with magnetometers. Electromagnetic emission by the plasma is also detected with radio antennas on board. Active experiments have been performed, e. g., by releasing in space small quantities of easily ionizable material and observing its diffusion. The work done by the *International Sun-Earth Explorer* satellites (ISEE) is particularly remarkable. Aside from the general morphological description, this work requires the identification of the physical nature of the wave phenomena. They appear at the spacecraft as an oscillating quantity with frequency $(\omega - \mathbf{k} \cdot \mathbf{V})$, where ω is the frequency in the rest frame of the plasma and $\mathbf{V}$ is the relative velocity of the spacecraft, usually moving slowly with respect to the plasma flow. In order to get the relevant quantity ω we should measure the wave vector $\mathbf{k}$; more generally, the identification is accomplished when we know the dispersion relation $\omega(\mathbf{k})$. This can be done with two spacecraft, capable of measuring the space dependence of the wave as well. Generally speaking, the study of space plasma places severe requirements on information processing and the transmission bit rate to the ground.

*9.6. ENERGETIC PARTICLES IN THE MAGNETOSPHERE

Besides the solar wind, energetic particles of different nature impinge on the magnetosphere and, if charged, are deflected by the magnetic force. Acceleration processes are at work in the sun, the stars, the galaxy and the universe itself and give rise to a complex spectrum of cosmic rays, whose energy can reach exceedingly large values (10^{11} GeV!). They interact with the radiation and the matter they encounter; in particular, a primary cosmic ray beam generates secondary particles of various kinds in the atmosphere. The primary beam is composed mainly of protons and alpha particles, with other nuclei as well. The origin and the acceleration processes of cosmic rays are still obscure; the table below lists four main trapping regions and the corresponding parameters. It is interesting to look at the way how they can penetrate into the magnetosphere, a topic which has practical importance as well, e. g. for the safety of the astronauts.

Trapping region	Size	Magnetic field (gauss)	Typical energy (GeV/nucleon)	Flux ($cm^{-2} s^{-1}$)	Energy density (eV/cm^3)
Magneto-sphere	10^9 cm	0.1	<0.1	10^3	100
Solar neigh-bourhood	10^3 ly	10^{-5}	10-100	0.1	1
Galactic halo	10^5 ly	10^{-6}	10^3-10^5	10^{-6}	10^{-2}
Cluster of gala-xies	10^7 ly	10^{-8} (?)	$>10^7$-10^8	10^{-10}	$<10^{-5}$

Table 9.1. The four main trapping regions of cosmic rays and their parameters.

Particles of energy much higher than those of the solar wind can penetrate the undisturbed region of the magnetosphere and violate the adiabatic invariance. However, relevant and general conclusions can be obtained for the motion in an unperturbed dipole field. We also neglect collisions, an assumption certainly inadequate in the atmosphere. We follow – and simplify – the treatment by Størmer, a nice application of analytical dynamics to the motion in an axially symmetric field.

The inclusion of relativistic effects is trivial. They show up in the definition of momentum

$$p = mv, \tag{9.49}$$

where in place of the rest mass m_0 we have the total relativistic mass

$$m = m_0/\sqrt{(1 - v^2/c^2)} = m_0\gamma. \tag{9.50}$$

However, since the magnetic force, orthogonal to the velocity vector **v**, does not perform any work, the magnitude of the momentum, and hence of the velocity, remains unchanged. We therefore go from the equations for the classical, non-relativistic motion to the relativistic case simply by replacing m_0 with m.

In cylindrical coordinates (ρ, z, φ) an axially symmetric, poloidal magnetic field is described by a vector potential A along the longitude direction φ function of ρ and z alone. The dipole field (6.3) corresponds to

$$A = A_\varphi = \rho d/r^3 \quad . \qquad (9.51)$$

The Lagrange function of a particle with mass m_0 and charge q reads

$$L = \tfrac{1}{2}\gamma m_0 v^2 + q\mathbf{v} \cdot \mathbf{A}/c = \tfrac{1}{2}\gamma m_0 (\dot{\rho}^2+\dot{z}^2+\rho^2\dot{\varphi}^2) + q\rho\dot{\varphi}A/c. \qquad (9.52)$$

An axially symmetric field has another conserved quantity corresponding to the angular momentum. Formally, this follows from the fact that A does not depend on φ, so that the corresponding canonical momentum

$$p_\varphi = \partial L/\partial\dot{\varphi} = m_0\gamma\rho^2\dot{\varphi} + q\rho A/c \qquad (9.53)$$

is a constant of the motion. The longitude φ is therefore an *ignorable* coordinate and can be eliminated .by expressing the reduced Lagrange function

$$L' = L - \dot{\varphi}p_\varphi \qquad (9.54)$$

in terms of p_φ. This gives the new Lagrange function

$$L' = \gamma m_0(\dot{\rho}^2 + \dot{z}^2)/2 - \frac{1}{2\gamma m_0}\left(\frac{p_\varphi}{\rho} - \frac{qA}{c}\right)^2 =$$
$$= \gamma m_0(\dot{\rho}^2 + \dot{z}^2)/2 - Q(\rho,z). \qquad (9.54')$$

The motion in the meridian plane (ρ, z) is conservative and entirely described by the effective, non-negative potential function Q. The total energy

$$W = \gamma m_0(\dot{\rho}^2 + \dot{z}^2)/2 + Q(\rho, z) \qquad (9.55)$$

is a constant. The constant p_φ is the asymptotic component of the angular momentum along the magnetic axis. It ranges over the real line, but the negative values are obtained from the positive values by inverting the z-axis and we can stick to positive values. For any assignment of W and p_φ, the region where $W < Q$ is forbidden. Similarly to what we shall do with the three-body problem (Ch. 12) from the expression (9.54') of Q one can already have a general idea of this region. Since $A = O(1/r^2)$, Q vanishes at infinity; in the equatorial plane the term p_φ/ρ prevails at large distances, while nearby qA/c is predominant. Therefore Q must have a zero, nay an absolute minimum, in the equatorial plane. Beyond it there is a potential barrier which can be overcome only if there is sufficient energy; around the zero we have a trapping region.

We are interested, in particular, in particles coming from infinity with an impact parameter about the size of the magnetosphere; at large distances their velocity is essentially radial and $v^2 = (\dot\rho^2 + \dot z^2)$. This determines the value of the total energy in eq. (9.55):

$$W = m_0 \gamma v^2/2. \tag{9.56}$$

The formal solutions of the dynamical problem (9.54') are constrained by the requirement $0 \leqslant (\dot\rho^2 + \dot z^2) \leqslant v^2$. The equality signs correspond, respectively, to a velocity vector orthogonal to, or lying in, the meridian plane (ρ, z). From the energy conservation law (9.55) the former condition translates into

$$\left(\frac{p_\varphi}{\gamma m_0 \rho} - \frac{qA}{\gamma m_0 c}\right)^2 \leqslant v^2. \tag{9.57}$$

(The second inequality is always fulfilled.) This defines the allowed region in the meridian plane.

We now specialize to the dipole field (9.51) and to the case of protons and introduce the characteristic cyclotron frequency

$$\Omega_C = ed/R^3 m_0 c \cong 3000 \text{ rad/s} \tag{9.58}$$

in terms of the earth's radius R and dipole moment d (see eq. (9.51)). The problem is entirely determined by a dimensionless constant

$$K = \Omega_C R/\gamma v \tag{9.59}$$

and a length

$$b = p_\varphi/(\gamma m_0 v). \tag{9.60}$$

For an unbound particle this is the impact parameter of the projection on the equatorial plane. Eq. (9.57) reads now

$$\frac{Q}{v^2} = \left[\frac{b}{\rho} - K\frac{R^2\rho}{r^3}\right]^2 \leqslant 1 . \tag{9.57'}$$

The equatorial absolute minimum lies at

$$\rho = \rho_m = KR^2/b = \Omega_c R^3 m_0/p_\varphi . \tag{9.61}$$

Its precise meaning becomes apparent with the energy conservation law (9.55), valid also when the particle is bound and the formal parameter v appearing in eq. (9.56) is not a true velocity: if $W = 0$ the velocity is orthogonal to the meridian plane. We have therefore a circular orbit at the distance ρ_m from the centre. We must now discuss how cosmic rays coming from infinity can overcome the potential barrier.

When $\gamma >> 1$, v is essentially equal to c and (eqs. (9.58, 59))

$$K \cong 64/\gamma . \tag{9.59'}$$

It is interesting to consider the case in which this is of order unity and b is of the order of the earth's radius R. In terms of the colatitude $\theta = \arcsin(\rho/r)$, eq. (9.57') reads:

$$\left[\frac{b}{r\sin\theta} - K\frac{R^2\sin\theta}{r^2}\right]^2 \leqslant 1 . \tag{9.57''}$$

We can restrict ourselves to the range $0 \leqslant \theta \leqslant \pi/2$.

Take note first of the curve $Q = 0$, or

$$r/R = K(R/b)\sin\theta = (\Omega_c R^2 m_0/p_\varphi)\sin\theta . \tag{9.62}$$

On this curve the velocity lies in the meridian plane. The constraint (9.57') can be violated both inside and outside it; this can happen when the round bracketed expression in it is equal to -1 or $+1$. In the former case, corresponding to the velocity vector along the $-\varphi$ coordinate, eq. (9.57'') gives a quadratic equation for r whose discriminant

$$\Delta = -4KR^2\sin\theta + b^2/\sin^2\theta$$

is always positive and there is always one and only one positive (and acceptable) solution. The corresponding curve $r(\theta)$ is an oval, similar to (9.61). In the second case, corresponding to the velocity vector along the φ coordinate, the discriminant Δ may be negative. If it is positive, there are two positive and acceptable roots for v. This happens when $\sin^3\theta < b^2/4KR^2$. If $b^2 > 4KR^2$ (or $p_\varphi^2 > 4\gamma v\Omega_c R^3 m_0^2$, i. e., large values of the angular momentum), Δ is always positive, even when $\theta \doteq \pi/2$. As a consequence, the inner admissible zone is surrounded by an unsurmountable barrier. When $b^2 < 4KR^2$ (or $p_\varphi^2 < 4\gamma v\Omega_c R^3 m_0^2$) there is a critical colatitude θ_c, given by

$$\sin^3\theta_c = \frac{b^2}{4KR^2} = \frac{p_\varphi^2}{4\gamma v m_0^2 \Omega_c R^3}, \tag{9.63}$$

below which there are two real roots; the inner zone is accessible from infinity.

The two types of configuration are sketched in Fig. 9.6. Of course, whether and where a particle actually reaches the earth depends upon whether a point on its surface lies within the allowed region. Consider the critical case

$$b^2 = 4KR^2, \quad \text{or} \quad p_\varphi^2 = 4\gamma v\Omega_c R^3 m_0^2, \tag{9.64}$$

in which the two "jaws" of the forbidden region meet at the point $r = \sqrt{R(K)}$, solution of $Q = v^2$. Such a point is physically relevant only

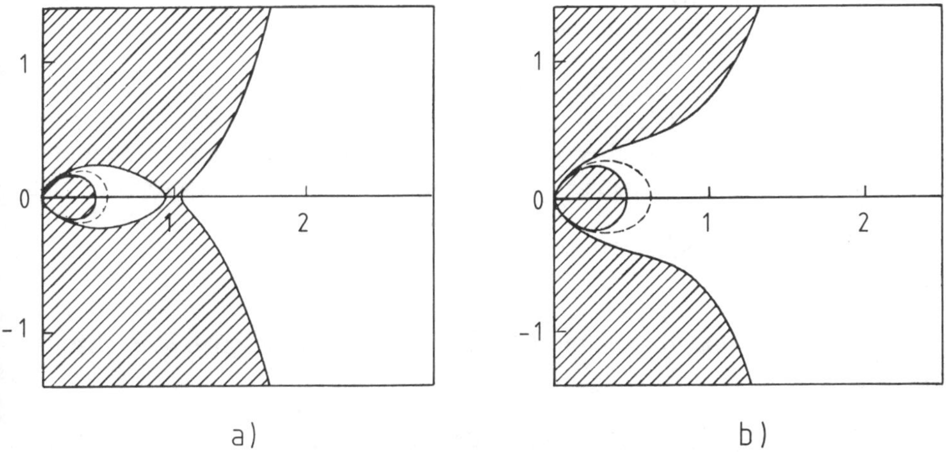

Fig. 9.6. The forbidden regions in the motion of a charged particle in the dipole field. There are two topologically distinct configurations, separated by the critical value (9.64) of the angular momentum component p_φ along the symmetry axis. For larger values - case a) - there is an inner allowed region separated from infinity; for smaller values the two allowed regions are connected. We have indicated also the line (dotted) where the velocity lies in the meridian plane. The units of length are not specified.

if $K < 1$, corresponding to $\gamma > \Omega_c R/v$ for protons. For smaller energies, the saddle point is inside the earth, hence is irrelevant. For higher energies, the condition of closing up of the jaws is determined by the quantity (9.63). For particles other than protons the condition $K > 1$ reads

$$c m_0 \gamma v / q < BR \; . \qquad (9.65)$$

On the left we have the ratio c•momentum/charge, known as *magnetic rigidity*. Its critical value for a proton is 59.6 GeV/c.

With similar considerations one can also determine the domain of forbidden directions at any given point; indeed, if an isotropic flux of cosmic rays bathes the earth, the intensity on the ground is strongly direction dependent as a consequence of the geomagnetic field. Størmer's theory has played an important part in the understanding of the cosmic ray flux and its time variations.

PROBLEMS

9.1. Calculate the energy of the magnetic dipole field outside a sphere and evaluate it for the earth.

9.2. What is the aperture of the equatorial loss cone as a function of r_e, the distance of the equatorial interception of a line of force?

*9.3. A model for the change in the trapped particle distribution when the magnetic field is compressed. We must choose a magnetic profile with a minimum at $s = 0$ and wings which do not change much when the minimum is changed. A simple example (with appropriate units of length) is $B = B_0 + B_1 s^2$. Show that the second invariant S (eq. (9.28)) is proportional to $\surd(B_0/B_1)\cos^2\chi_0/\sin\chi_0$ and plot χ_0 as a function of B_0.

9.4. Derive the dispersion relations (9.36) and describe the character of each wave.

*9.5. Show that (9.12) is adiabatically invariant for the harmonic oscillator. Hint: the limit $\Omega \to \infty$ (or $\Omega^2 \gg |\dot\Omega|$) of eq. (9.11) is a problem of *singular perturbation theory*. It can formally be reduced to solving, in the limit $\epsilon \to 0$, the equation

$$\epsilon^2 \ddot{x} + \Omega^2 x = 0$$

with the Ansatz $x = \exp[i(S_0 + \epsilon S_1 + ...)/\epsilon]$.

*9.6. Study the jump conditions for a MHD shock.

9.7. Two spacecraft measure the variations in the local plasma density due to a linear wave. What information can one derive about the dispersion relation? Hint: study the phase difference in each Fourier component.

9.8. Estimate the total ring current, assuming a thickness of 2 earth's radii for the region of magnetic depression.

FURTHER READING

A good textbook on plasma physics is P.C. Clemmow and J. P. Dougherty, *Electrodynamics of Particles and Plasmas*, Addison-Wesley (1969); we should also quote the short classic by L. Spitzer, Jr, *Physics of Fully Ionized Gases*, Interscience Publishers (1962). On the complex material of Sec. 9.5 one can read the overall synthesis by G. Haerendel and G. Paschmann, *Interaction of the solar wind with the dayside magnetosphere*, in "Magnetospheric Plasma Physics", edited by A. Nishida, Reidel (1984); L.R. Lyons and D.J. Williams, *Quantitative Aspects of Magnetospheric Physics*, Reidel (1984); and J.A. Ratcliffe, *An Introduction to the Ionosphere and Magnetosphere*, Cambridge University Press (1972).

The material of the last Section is related to the cosmic rays problem, which is beyond our scope; for the classical mathematical theory, see, e.g., T.H. Johnson, *Cosmic Ray Intensity and Geomagnetic Effects*, Rev. Mod. Phys., **10**, 193-244 (1938).

10. THE TWO-BODY PROBLEM

For about two thousand years the efforts of western astronomers have been largely devoted to the purpose of understanding the motion in the sky of the sun, the moon and the planets. More and more complex geometrical and kinematical models, both geocentric and heliocentric, were devised in order to reproduce observational data of increasing accuracy. The final outcome of this effort, and at the same time the starting point for an outstanding scientific revolution, can be traced back to Kepler's theory of planetary motions. Giving up the long-standing *a priori* requirement of circular paths covered with unifom velocities (or of finite combinations of them, resulting into epicyclic trajectories), Kepler's three laws elliptic orbits with the sun at one focus; constant areal velocities; 3/2-power dependence of orbital periods on semimajor axes fitted the astronomical observations with unprecedented accuracy, and at the same time summarized into a few simple mathematical relationships most of the available kinematical observations. Within less than a hundred years, the Newtonian synthesis exploited Kepler's results starting from a completely new perspective, that of producing a conceptually simple, but very general *dynamical* theory of which Kepler's laws (as well as many other physical phenomena) could be seen as specific - and approximate consequences. Newton's theory of gravitational forces found its first and simplest challenge in the so-called *two-body problem*: given at some instant of time the positions and velocities of two massive point-like particles moving under the exclusive influence of their mutual gravitational forces, find their positions and velocities at any other time. This is the basis of celestial mechanics and the subject of this chapter.

10.1. REDUCTION TO A CENTRAL FORCE PROBLEM

It is important to note that the two-body problem is highly idealized: the two interacting masses are assumed to be point-like (or spherically symmetric; recall the gravitational Gauss' theorem, Sec. 1.1) and no other body or medium exerts significant forces upon them. It is a lucky coincidence that the planetary motions are well represented, at least at a first approximation, by solutions of the two-body problem. This depends on the fact that the planetary masses are very small (less than 10^{-3}) with respect to the solar mass, and that no close encounter occurs among the major bodies of the solar system. In these conditions, each planet is affected predominantly by the sun's gravitational pull, while the other, much smaller forces that are present cause only minor disturbances of the motion called *perturbations* (see Ch. 11).

Given two masses m_1 and m_2, by Newton's laws we have

$$\mathbf{F}_1 = G\frac{m_1 m_2}{r^2}\frac{\mathbf{r}}{r}, \qquad \mathbf{F}_2 = -\mathbf{F}_1 . \tag{10.1,2}$$

Here $\mathbf{F}_1$ and $\mathbf{F}_2$ are the gravitational forces acting upon the two particles, G is the *gravitational constant* ($\cong 6.67\cdot10^{-8}$ in cgs units) and $\mathbf{r}$ is the vector from m_1 to m_2. If $\mathbf{r}_1$, $\mathbf{r}_2$ are the position vectors of the two particles with respect to the origin of an inertial reference frame and $\mathbf{r} = (\mathbf{r}_2 - \mathbf{r}_1)$ their relative position vector, we have:

$$m_1\ddot{\mathbf{r}}_1 = G\frac{m_1 m_2}{r^2}\frac{\mathbf{r}}{r}, \qquad m_2\ddot{\mathbf{r}}_2 = -G\frac{m_1 m_2}{r^2}\frac{\mathbf{r}}{r}. \tag{10.3,4}$$

Adding these two equations, we get

$$m_1\dot{\mathbf{r}}_1 + m_2\dot{\mathbf{r}}_2 = \text{const}, \tag{10.5}$$

which expresses the conservation of the total momentum of the system, i.e., the linear and uniform motion of the centre of mass. Dividing eqs. (10.3) and (10.4) by m_1 and m_2, respectively and subtracting, we obtain

$$\ddot{\mathbf{r}} = -\frac{Gm}{r^3}\mathbf{r}, \tag{10.6}$$

where $m = m_1 + m_2$ is the total mass. This equation involves only the vector $\mathbf{r}(t)$, the relative position of the two bodies. Thus, once the motion of the centre of mass is eliminated by a suitable choice of the reference system, the problem is completely equivalent to that of a single particle moving under a central force proportional to the power (– 2) of the distance from the attracting centre. For any radial force the angular momentum integral (per unit mass)

$$\mathbf{h} = \mathbf{r} \times \dot{\mathbf{r}} \tag{10.7}$$

is a constant of the motion. Then the motion is planar (it always occurs in the plane perpendicular to **h**) and can be studied using polar coordinates (r, θ) in this plane. The velocity has a radial component $v_r = \dot{r}$ and a transverse component $v_t = r d\theta/dt$. The constancy of

$$h = r v_t = r^2 d\theta/dt \qquad (10.8)$$

is equivalent to *Kepler's second law*, since h can easily be shown to be equal to twice the *areal velocity*, the rate of change of the area described by the radius vector **r**. A second integral of the motion is provided by the conservation of energy. Taking the scalar product of d**r**/dt with eq. (10.6) and integrating with respect to time, we obtain

$$E = v^2/2 - Gm/r = \text{constant}, \qquad (10.9)$$

where E is the energy per unit mass and $v^2 = \dot{\mathbf{r}} \cdot \dot{\mathbf{r}} = v_r^2 + v_t^2$. Since the radial component of the acceleration is $\ddot{r} - r(d\theta/dt)^2$, we can write the radial part of eq. (10.6) in the form

$$\frac{d^2 r}{dt^2} = \frac{h^2}{r^3} - \frac{Gm}{r^2} = -\frac{dV(r)}{dr}. \qquad (10.10)$$

Here dθ/dt has been eliminated using eq. (10.8). Note that the radial motion is determined by the competition of the gravitational (attractive) force with an apparent *centrifugal* (repulsive) force, with the two terms having different power dependences on r. Both these forces can be derived from a potential, so that the problem can be qualitatively studied with the usual techniques applicable to one-dimensional conservative systems, i.e., examining the dependence on r of the *effective potential energy*

$$V(r) = -Gm/r + h^2/2r^2. \qquad (10.11)$$

The centrifugal contribution prevails at small distances, the gravitational one at large distances; hence V(r) has a negative minimum at $r = h^2/Gm$. This corresponds to a circular orbit, whose radius is proportional to the − 2/3 power of the angular velocity (from eq. (10.8)), in agreement with *Kepler's third law*.

10.2. KEPLERIAN ORBITS

The differential equation (10.10) can be solved directly by introducing, with eq. (10.8), the angle θ as the independent variable instead of the time and using the substitution $u = 1/r$. We obtain

$$\frac{d^2u}{d\theta^2} + u = \frac{Gm}{h^2} . \qquad (10.12)$$

This is the equation of the harmonic oscillator with a constant external force; it has the general solution:

$$u = Gm/h^2 + A\cos(\theta - \omega), \qquad (10.13)$$

where A and ω are two constants of integration. Going back to the variable r, eq. (10.13) becomes

$$r = \frac{h^2/Gm}{1 + e\cos(\theta - \omega)} , \qquad (10.14)$$

where we have introduced the new constant e (the orbital *eccentricity*), given by

$$e = Ah^2/Gm. \qquad (10.15)$$

Eq. (10.14) gives the general form of a conic section in polar coordinates. The value of the eccentricity discriminates among ellipses ($e < 1$), parabolae ($e = 1$) and hyperbolae ($e > 1$). When $e = 0$ the ellipses degenerate into circular orbits; if $e < 0$, by replacing ω with $(\omega + \pi)$, we can go back to the previous cases. The minimum value of r (the *pericentre distance* r_p) is reached when $\theta = \omega$, implying $r_p = h^2/Gm(1 + e)$; since at this distance $\dot{r} = 0$, the velocity has a tangential direction and is given by $v_p = h/r_p = Gm(1 + e)/h$. Substituting these values into eq. (10.9) we obtain

$$e^2 = 1 + 2Eh^2/(Gm)^2, \qquad (10.16)$$

showing that the energy is negative, zero and positive for elliptic, parabolic and hyperbolic orbits, respectively.

These results can be obtained in a different way by analysing the properties of the so-called *Lenz vector* **e** defined as

$$\mathbf{e} = \frac{1}{Gm}(\dot{\mathbf{r}} \times \mathbf{h}) - \frac{\mathbf{r}}{r} . \tag{10.17}$$

As a consequence of eqs. (10.6, 7, 9, 17), the following relationships can easily be shown to hold:

$$d\mathbf{e}/dt = 0 \; ; \quad \mathbf{e} \cdot \mathbf{h} = 0 \; ;$$

$$\mathbf{e} \cdot \mathbf{r} = h^2/Gm - r \; ; \tag{10.18}$$

$$\mathbf{e} \cdot \mathbf{e} = 1 + 2Eh^2/G^2m^2 = e^2 . \tag{10.19}$$

Therefore **e** is constant, lies on the orbital plane and its magnitude is just the eccentricity. If we define the *true anomaly* f as the angle between **e** and **r**, from eq. (10.18) we find again the polar equation of the conic sections. By comparison with eq. (10.14) we obtain $f = \theta - \omega$; that this to say, the Lenz vector is directed along the *apsidal line*, namely towards the pericentre of the orbit.

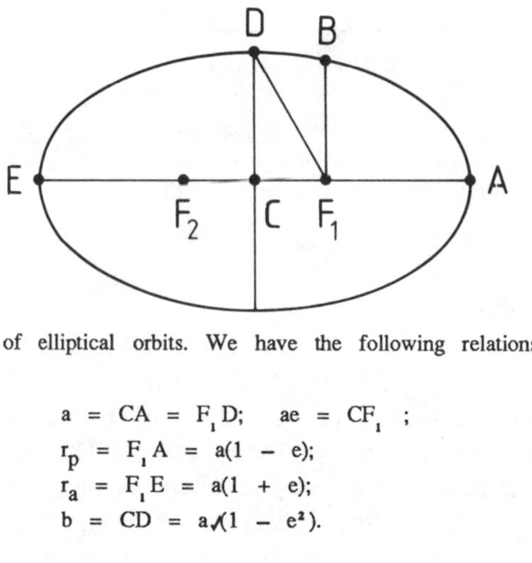

Fig. 10.1. Geometry of elliptical orbits. We have the following relationships:

$$a = CA = F_1D; \quad ae = CF_1 \; ;$$
$$r_p = F_1A = a(1 - e);$$
$$r_a = F_1E = a(1 + e);$$
$$b = CD = a\sqrt{(1 - e^2)}.$$

Let us describe in some more detail the most interesting case, that of elliptic motion. It is easy to show (see Fig. 10.1) that

$$a = \frac{r_p + r_a}{2} = \frac{h^2}{Gm(1 - e^2)} \; ; \quad b = a\sqrt{(1 - e^2)}; \tag{10.20,21}$$

$$r_p = a(1 - e) \; ; \quad r_a = a(1 + e) \tag{10.22,23}$$

are, respectively, the semimajor and semiminor axes, the pericentre ($f = 0$) and apocentre ($f = \pi$) distances. Since the total area of the ellipse πab is swept by the radius vector in one orbital period T, by Kepler's second law we have

$$T = \frac{2\pi a^2 \sqrt{(1 - e^2)}}{h} = 2\pi\sqrt{(a^3/Gm)} \ , \qquad (10.24)$$

which shows that the period does not depend on the eccentricity, but only on the 3/2 power of the semimajor axis and the sum of the masses. In the solar system, this sum is always approximately equal (to within $\approx 0.1\%$) to the mass of the sun, so that *Kepler's third law* represents very closely the observed relationship between semimajor axes and orbital periods. The same holds within the planet-satellite systems. Eq. (10.24) shows also that a very general method to measure the mass of a celestial body is based on the study of the orbit of a satellite. Once its orbital period and semimajor axis are both known, if the satellite's mass is known to be small, we obtain directly the mass of the central body multiplied by the gravitational constant. In order to get the mass one must use the laboratory value of the gravitational constant G, which can be obtained by Cavendish-type experiments, using attracting bodies of known mass. This method has wide applications not only in the solar system, but also in astrophysics. Conversely, if the total mass is known, eq. (10.24) provides a good clock. The earth-moon system, whose perturbations are relatively well known, has been used for a long time as an astronomical clock. Another outstanding example of a gravitational clock based upon eq. (10.24) is provided by the binary pulsar PSR 1913 + 16, a binary system in which one member, being a pulsar, provides a very precise way to measure the orbital period P = 27908 s. It was found that this period decreases at the rate $2.1 \cdot 10^{-12}$ and the corresponding energy loss (see eq. 10.25) was established from the decrease in the semimajor axis. This loss is most likely due to emission of *gravitational waves.*

The semimajor axis of an elliptical orbit is related to the energy integral in a very simple way:

$$a = \frac{h^2}{Gm(1 - e^2)} = \frac{h^2/Gm}{(-2Eh^2/G^2m^2)} = -\frac{Gm}{2E} \ , \qquad (10.25)$$

by eq. (10.16). Thus the semimajor axis depends only on the energy, but not on the angular momentum. From eqs. (10.9,25) we can obtain the velocity as a function of r in an elliptic orbit

$$v^2 = Gm\left(\frac{2}{r} - \frac{1}{a}\right) . \tag{10.26}$$

It is noteworthy that if the particle starts at the distance r, its semimajor axis (and therefore its orbital period) does not depend on the direction, but only on the magnitude of the initial velocity. Of course the eccentricity, and the shape of the orbit, will depend on the initial direction of the motion. The radial and transverse components of the velocity vector are also known as functions of the true anomaly f from eqs. (10.8,13):

$$v_r = \dot{r} = \frac{h^2}{Gm}\,\frac{e\sin f}{(1 + e\cos f)^2}\,\frac{df}{dt} = \frac{Gm}{h}\,e\sin f , \tag{10.27}$$

$$v_t = \frac{h}{r} = \frac{Gm}{h}(1 + e\cos f) . \tag{10.28}$$

Besides the geometrical shape of the orbit, we are interested in the time law of the motion, e. g., the function f(t). To get it, we introduce

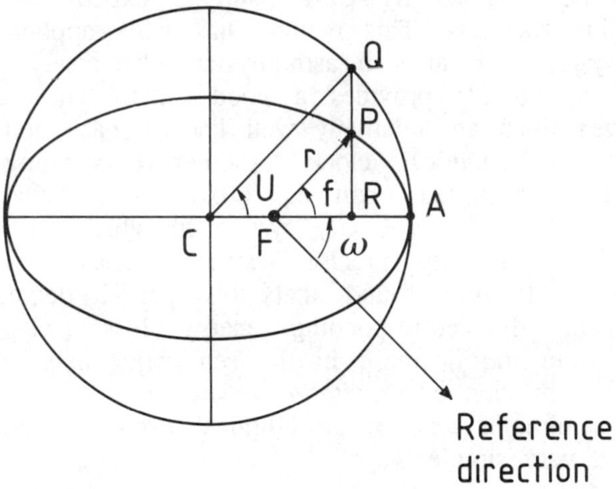

Fig. 10.2 Geometric interpretation of the eccentric anomaly U. From the two relationships:

$$FR = r\cos f = CR - CF = a\cos U - ae$$
$$PR = r\sin f = (b/a)QR = \sqrt{(1 - e^2)}\,a\sin U$$

eq. (10.29) immediately follows. Notice that here we have a parametric representation of the ellipse with the origin at the centre.

an additional angular variable U (the *eccentric anomaly*) such that

$$r = a(1 - e\cos U) . \tag{10.29}$$

U is interpreted geometrically in Fig. 10.2. From the relationships between r and f and between r and U, it also follows that

$$tg(f/2) = \left(\frac{1 + e}{1 - e}\right)^{\frac{1}{2}} tg(U/2). \tag{10.30}$$

From the conservation of angular momentum (eq. (10.8)), we have now

$$h(t - t_0) = \int_0^f df' r^2(f') , \tag{10.31}$$

where t_0 is the time of a passage at pericentre. By eq. (10.30) it can be easily shown that

$$df = \frac{\sqrt{(1 - e^2)}}{1 - e\cos U} dU \tag{10.32}$$

and therefore we have

$$h(t - t_0) = \int_0^U dU' \frac{\sqrt{(1 - e^2)}}{1 - e\cos U'} a^2(1 - e\cos U')^2 . \tag{10.33}$$

Since h is twice the areal velocity, denoting by $n = 2\pi/T$ the *mean motion* (i. e., the mean angular velocity along the orbit), we have

$$M = n(t - t_0) = U - e\sin U. \tag{10.34}$$

Here M is the *mean anomaly*, by definition a linear function of time which increases by 2π (as f and U also do) per revolution. If we know the time elapsed since the pericentre passage, i. e. M/n, by inverting eq. (10.34) (known as *Kepler's equation*) we can obtain U. Then we can derive f from eq. (10.30) and finally the position along the orbit. Kepler's equation can be solved numerically by several methods of successive approximations, which generally converge very fast provided the eccentricity is not too large.

10.3. THREE-DIMENSIONAL ORBITAL ELEMENTS

To entirely specify the orbital motion one can use a set of six quantities, the so-called *Keplerian elements*. Three of them are needed to identify the position on the orbital plane. For example, we can give the semimajor axis a, the eccentricity e and the true anomaly at a specified time $f(t_1)$ (or, equivalently, the *mean anomaly at epoch* $M(t_1)$). Let us consider for instance the orbit-bound reference frame having the origin in the central mass, the y_0-axis along the direction of pericentre (i. e., the vector **e**) and the z_0-axis along the normal to the orbital plane (i. e., along **h**; see Fig. 10.3). In this frame the coordinates of the orbiting body are (rcosf, rsinf, 0), with r given by eq. (10.14).

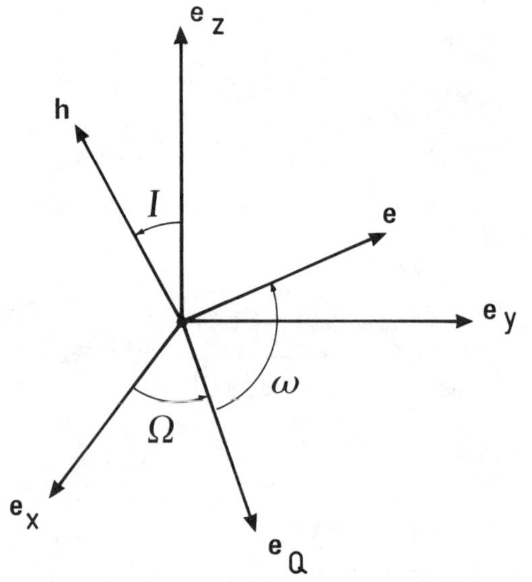

Fig. 10.3 The orbit-bound vectors needed to define the angular orbital elements ω, Ω and I with respect to an arbitrary reference system (x, y, z).

As shown by Fig. (10.3), three additional elements are now required to specify the orientation of the orbit with respect to an arbitrary orthonormal reference frame (x, y, z). Although the choice of the latter frame is in principle arbitrary, when additional forces are present some particular choice becomes frequently suitable; for example, in the solar system one often takes the z-axis normal to the orbital plane of the earth at a given epoch (*ecliptic*) and the x-axis along the intersection of this plane with the equator at the same time (the so-called *vernal equinox* or *γ-point*).

Let us now define the unit vector $\mathbf{e}_\Omega$ as the direction of the ascending node, that is, along the intersection of the orbital plane with the (x, y) plane and oriented towards the point where the orbiting body crosses such plane from below:

$$\mathbf{e}_Q = \mathbf{e}_z \times \mathbf{h}/(|\mathbf{e}_z \times \mathbf{h}|). \tag{10.35}$$

The *longitude of the ascending node* Ω is the angle between the x axis and $\mathbf{e}_Q$:

$$\mathbf{e}_Q = \mathbf{e}_x \cos\Omega + \mathbf{e}_y \sin\Omega. \tag{10.36}$$

The *inclination* I is the angle between $\mathbf{e}_z$ and $\mathbf{h}$:

$$\cos I = \mathbf{e}_z \cdot \mathbf{h}/h . \tag{10.37}$$

Usually I is assumed to range from 0 to π. Finally, the *argument of pericentre* ω is the angle between $\mathbf{e}_Q$ and $\mathbf{e}$:

$$\cos\omega = \mathbf{e}_Q \cdot \mathbf{e}/e. \tag{10.38}$$

Fig. 10.3 summarizes these definitions, and Fig. (10.4) shows an elliptical orbit in space. The five parameters a, e, I, Ω and ω specify the orbit completwly; for the same purpose we could also take the two vectors **e**, **h** (whose six components are not independent because of the condition $\mathbf{e} \cdot \mathbf{h} = 0$). As we already said, the sixth keplerian element is simply t_0 or, equivalently, the mean or true anomaly at some given epoch. The six keplerian elements can be readily obtained from the initial conditions by the integrals of motion; for instance, if one knows the six components of the vectors **r** and $\dot{\mathbf{r}}$ at a given time, E, **h** and **e** can be computed and the elements can be derived from them.

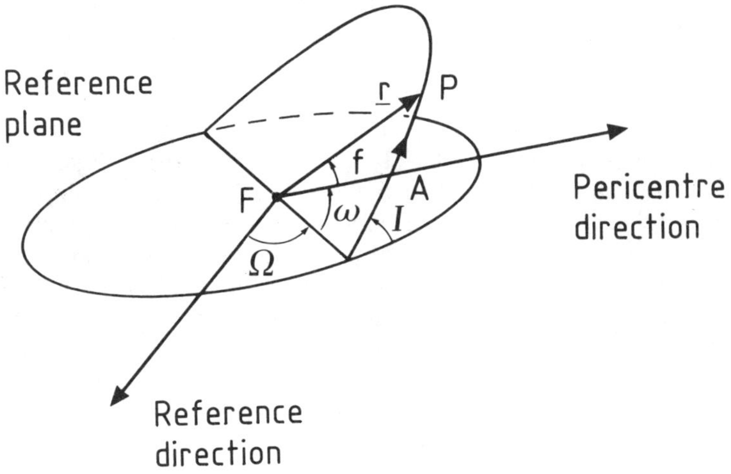

Fig. 10.4 Geometry of an elliptical orbit in space.

10.4. UNBOUND ORBITS

There are some instances in solar system and spacecraft dynamics when unbound keplerian orbits, either parabolic ($e = 1$) or hyperbolic ($e > 1$), have a practical interest. Such are for example the cases or comets or of planetary fly-by's by space probes.

For a parabolic orbit, the equation in polar coordinates corresponding to eq. (10.14) is

$$r = \frac{2r_p}{1 + \cos f} = \frac{r_p}{\cos^2(f/2)} , \qquad (10.39)$$

where r_p is the distance from the origin (the focus) to the pericentre, the vertex of the parabola (see Fig. 10.5) and the true anomaly f is measured from the pericentre. At $r = r_p$ we find by eqs. (10.8,9), with $E = 0$, that $h = \sqrt{(2Gmr_p)}$ (since the velocity vector is perpendicular to the radius vector). From eqs. (10.39) and (10.8) we get

$$\frac{df}{dt} \frac{r_p^2}{\cos^4(f/2)} = h \qquad (10.40)$$

and integrating by separating the variables

$$\mathrm{tg}\left(\frac{f}{2}\right) + \frac{1}{3}\,\mathrm{tg}^3\left(\frac{f}{2}\right) = \left(\frac{Gm}{2r_p^3}\right)^{\frac{1}{2}} (t - t_0), \qquad (10.41)$$

where t_0 is the time of pericentre passage. This cubic equation can be solved numerically (by some iterative method, like for Kepler's equation) to obtain f as a function of t, namely the law of motion.

For hyperbolic orbits, $E > 0$ and as a consequence we obtain that $a < 0$ and $e > 1$ (see eqs. (10.25) and (10.16)). Eqs. (10.21,22,25, 26,27,28) are also still valid. Some useful relationships hold for hyperbolae, relating the range of variation of the true anomaly δf (which is just 2π minus the angle between the two asymptotes, i. e., the deviation angle between the two directions of motion at infinity α plus π), the pericentre distance r_p, the velocity at infinity v_∞ and the perpendicular distance p from the focus to the asymptote, or *impact parameter* (see Fig. 10.5). Since for $r \to \infty$, $f \to \arccos(-1/e)$, we have

$$\begin{aligned} \delta f &= 2\pi - 2\arccos(1/e) \\ &= 2\pi - 2\arccos[Gm/(Gm + r_p v_\infty^2)] , \end{aligned} \qquad (10.42)$$

$$r_p = \left(p^2 + \frac{G^2m^2}{v_\infty^4}\right)^{\frac{1}{2}} - \frac{Gm}{v_\infty^2}\,, \tag{10.43}$$

$$\cos^2\left(\frac{\delta f}{2}\right) = \sin^2\left(\frac{\alpha}{2}\right) = \left(1 + \frac{p^2 v_\infty^4}{G^2m^2}\right)^{-1}\,, \tag{10.44}$$

These relationships can easily be proven by using the angular momentum ($h = pv_\infty$) and energy ($E = v_\infty^2/2$) integrals.

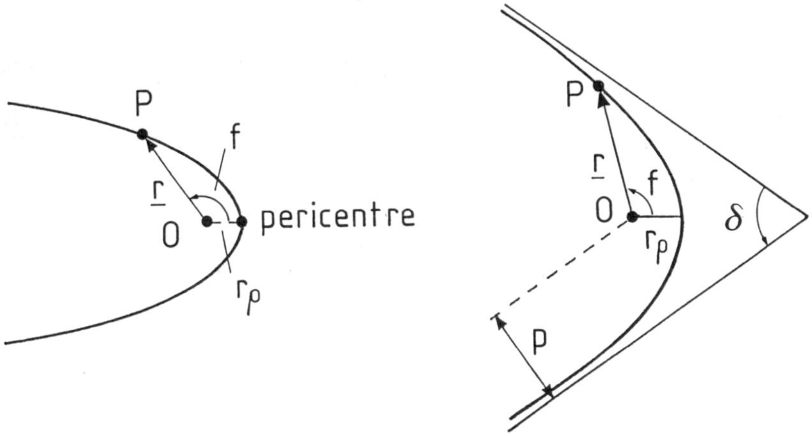

Fig. 10.5 Geometry of parabolic and hyperbolic orbits.

From the relationship (10.8), or $dt = r^2 df/h$, we have also

$$\left(\frac{Gm}{-a^3(e^2-1)^3}\right)^{\frac{1}{2}} dt = \frac{df}{(1 + e\cos f)^2}\;; \tag{10.45}$$

integrating from t_0 to t with the substitution

$$\mathrm{tg}\left(\frac{f}{2}\right) = \left(\frac{e+1}{e-1}\right)^{\frac{1}{2}} \mathrm{tgh}\left(\frac{F}{2}\right), \tag{10.46}$$

we find that

$$\sqrt{[Gm/(-a^3)]}(t - t_0) = e\sinh F - F\,, \tag{10.47}$$

analogous to Kepler's equation (10.34) for the elliptical orbits. From F one can find r via f or directly through the equation

$$r = -a(e\cosh F - 1) \ , \qquad (10.48)$$

which is equivalent to (10.47) and corresponds to eq. (10.29).

PROBLEMS

10.1. How great should two equal charges placed on the earth and on the sun be to balance their gravitational force?

10.2. The properties of the two-body circular orbits can easily be derived from the balance between gravitational and centrifugal forces in a rotating reference frame.

10.3. Calculate the semimajor axes and the heights of circular earth orbits with a period equal to 1/j of a day, from j = 1 to the highest integer value compatible with atmospheric friction, which becomes excessive at about 150 km from the earth's surface.

10.4. Prove the properties of the Lenz vector listed in the text using the equations of motion and the energy and angular momentum integrals.

10.5. Write a simple computer program to solve iteratively Kepler's equation by the following procedure: $U_0 = M$, $U_1 = M + e\sin U_0$, etc. Calculate within 10 arcsec the eccentric anomaly of Jupiter 3 years after its perihelion passage, knowing that T = 11.8622 years and e = 0.048.

*10.6. The energy flux of the solar radiation at 1 AU from the sun is $1.38 \cdot 10^6$ erg/(cm^2 s) and, of course, is inversely proportional to the square of the distance r from the sun. Consider a perfectly absorbing body of mass m and cross-section S; if its velocity is much less than the speed of light, the force exerted by the solar radiation is radial and proportional to $1/r^2$. Show that its effect is equivalent to a change δG in the gravitational constant and calculate this change.

10.7. Solve the two-body problem when the force is proportional to $1/r^2$ but is *repulsive*. Find the angle between the two asymptotic directions of the vector **r**, assuming that initially r is very large, the relative velocity is v_∞ and the impact parameter (i.e., the minimum distance between the two particles were their trajectories unaffected by the interaction) is b.

*10.8. Two point-like bodies collide head-on under their reciprocal gravitational attraction. Solve for their motion (this the *one-dimensional two-body problem*). Show that close to the collision (occurring for $t = t_0$), their distance is proportional to $(t_0 - t)^{2/3}$. If the earth were stopped in its orbit at 1 AU from the sun, how long would it take to fall onto it? (This problem is equivalent to the dynamics of an isotropic and homogeneous universe under its own gravitational force, when its mass-energy content is entirely the rest mass.)

10.9. Assume that an artificial satellite is placed at distance r_0 from the earth's centre (larger than the earth's radius R), with a purely tangential velocity of magnitude v_0. In what conditions is the satellite bound to the earth, i.e., is its orbit elliptical ? What are the semimajor axis and eccentricity of the orbit ? And what is the minimum value of v_0 for which no collision with the earth's surface occurs ?

*10.10. Determine the trajectory of the warhead of an intercontinental ballistic missile, released by its booster at a height h above the earth's surface, with a speed v_0 directed at an angle γ from the radial direction. Which is the value of γ which yields the maximum range of the warhead ? Show that the same initial conditions correspond to the trajectory of minimum energy for a given range.

10.11. A test body moves around the sun in an almost circular orbit. Find its motion by linearizing the equations of motion about the circular solution: one obtains three linear equations which are easy to solve. (Hint: work in the appropriate rotating reference frame.) What happens if the force depends upon the distance according to a power law with exponent different from (-2) ?

10.12. Find out the energy loss of the binary pulsar (Sec. 10.2).

10.13. Find the cartesian coordinates of a satellite on a circular orbit of radius r, inclination I and nodal longitude Ω, when its angular distance from the ascending node is λ. Show that, for every value of λ, $\mathbf{r}$ is perpendicular to the unit vector $(\sin I \sin\Omega, -\sin I \cos\Omega, \cos I)$, which is parallel to the angular momentum vector.

10.14. Study the effective potential (eq. (10.11)) for an arbitrary power law force.

*10.15. Derive the function $\mathbf{r}(\tau)$ obtained by substituting a *regularized* time τ, defined by $dt = r d\tau$, into the equations of motion of the two-body problem. Show that the integration of this relationship with respect to the variable τ provides an alternative derivation of Kepler's equation.

10.16. Evaluate the orbital average of a second order polynomial in cosU and sinU.

11. PERTURBATION THEORY

In this section we will discuss the changes in the orbital elements produced by the action of a perturbing acceleration **F**, corresponding to the equations of motion in cartesian coordinates:

$$\frac{d^2\mathbf{r}}{dt^2} = -\frac{Gm\mathbf{r}}{r^3} + \mathbf{F} \,. \qquad (11.1)$$

Here the *perturbing acceleration* **F** is assumed to be a known function of **r**, $\dot{\mathbf{r}}$ and t. The main idea underlying the perturbative equations is the *osculating orbital elements*: given an orbit **r**(t) solution of eq. (11.1), for every time t' we consider the keplerian elements a, e, I,..., of a keplerian orbit with initial conditions **r**(t'), $\dot{\mathbf{r}}$(t'). These osculating elements give the orbit that the body would follow if at time t' the perturbation **F** were suddenly turned off; of course, they are functions of the time t'. As a consequence, the osculating orbit changes with time, but does so slowly, provided the magnitude of **F** is small with respect to the *monopole* gravitational acceleration Gm/r^2 and its time scale is longer. We want the differential equations which describe the time dependence of the osculating elements. In this chapter we shall derive them in two different forms (the so called *Gauss'* and *Lagrange's equations*), and will show how they can be applied to solve two classical problems of artificial satellite dynamics, i.e. the orbital effects of drag and of the oblateness of the planet. The approximation methods commonly used to obtain the solutions are also discussed, as well as the problems and limitations of these methods.

11.1. GAUSS' PERTURBATION EQUATIONS

The effects of **F** can be very different according to its direction and its relationship with the orbital geometry. Hence, it is useful to decompose it in three orthogonal components: R, acting in the radial direction (along **r**); W, in the out-of-plane direction (along the orbital angular momentum per unit mass **h**); and T, transversally in the orbital plane (along **h** × **r**):

$$R = \mathbf{F} \cdot \mathbf{r}/r \ , \quad W = \mathbf{F} \cdot \mathbf{h}/h \ , \quad T = \mathbf{F} \cdot (\mathbf{h} \times \mathbf{r})/hr. \quad (11.2)$$

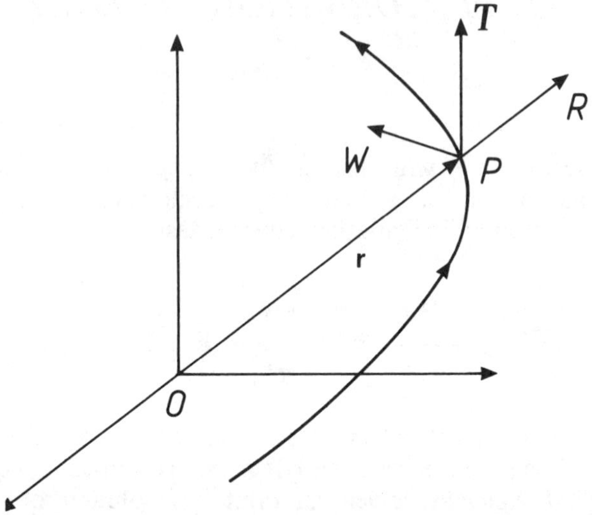

Fig. 11.1. The standard decomposition of the perturbing force.

The osculating orbital energy $E = v^2/2 - Gm/r$ (which is not the true total energy, since it neglects the perturbing force) changes at the rate $dE/dt = \mathbf{F} \cdot (d\mathbf{r}/dt)$. By eqs. (10.27, 28) we obtain

$$dE/dt = Rv_r + Tv_t = [R(e\sin f) + T(1 + e\cos f)]Gm/h. \quad (11.3)$$

Introducing the osculating semimajor axis $a = -Gm/2E$ (eq. (10.25)), we get

$$\frac{da}{dt} = \frac{2a^2}{Gm}\frac{dE}{dt} = \frac{2}{n(1 - e^2)^{\frac{1}{2}}} [T + e(T\cos f + R\sin f)], \quad (11.4)$$

where we have used also eq. (10.20), namely $h = [Gma(1 - e^2)]^{\frac{1}{2}}$. Whenever the eccentricity is small, the transverse component T is the most effective one in changing the size and (by Kepler's third law) the period of the orbit. This fact is important, because it implies that in presence of a perturbation having a nonzero T component the position of the body along the orbit can be affected in a very significant way. For instance, in the case of a circular orbit with a constant T, the orbital radius and the orbital period change linearly with time; as a consequence, the true anomaly is displaced with respect to the unperturbed case by an amount which accumulates quadratically with time, and thus can become quite important. It is intriguing to notice that a drag force, which has a negative T component and causes an overall energy loss, by eq. (11.4) produces a decrease in the semimajor axis; by Kepler's third law this implies an increase of the orbital velocity and of the kinetic energy. Thus orbiting bodies, contrary to common sense intuition, *increase their velocity when subjected to drag* (see Problem 11.1).

To derive the equation for the change in the eccentricity, we can use the fact that the magnitude h of the angular momentum can be changed only by a torque having the same direction of **h**, so that dh/dt = rT. On the other hand, differentiating eq. (10.20) we get

$$rT = dh/dt = [Gm\,(1 - e^2)da/dt - 2eGmade/dt]/2h. \quad (11.5)$$

We can solve eq. (11.5) for de/dt, substituting r and $\dot{a}$ from eqs. (10.29) and (11.4) respectively, and obtain

$$\frac{de}{dt} = \frac{\sqrt{(1 - e^2)}}{na}\,[R\sin f + T(\cos f + \cos U)], \quad (11.6)$$

where cosU = (e + cosf)/(1 + e cosf) (see Fig. 10.2 and eq. (10.29)). Note that only forces acting in the orbital plane can change both a and e, that is the size and the shape of the orbit.

On the other hand, the orientation of the orbital plane, i. e. of the vector **h**, is changed only if an out-of-plane component W is present; so do the osculating inclination I and osculating nodal longitude Ω. The torque rW associated with W lies on the orbital plane and is perpendicular to **h**. It changes **h** perpendicularly to itself, keeping its magnitude h constant; that is, we can write d**h**/dt = $\mathbf{\Omega}_{pr} \times \mathbf{h}$. This means that **h** precesses around the $\mathbf{\Omega}_{pr}$ vector with angular velocity $\Omega_{pr} = |d\mathbf{h}/dt|/h\,\sin(\mathbf{\Omega}_{pr}, \mathbf{h})$, where $|d\mathbf{h}/dt| = rW$. Now let us split the torque into the component along the nodal line, rW sin(ω + f), and the component perpendicular to it in the orbital plane, rW cos(ω + f) (recall that the torque is always perpendicular to **r**). The former is perpendicular to **h** and to $\mathbf{e}_z$ (the z-axis is the reference axis for the inclination), hence it causes **h** to precess about $\mathbf{e}_z$ with angular velocity

$$\frac{d\Omega}{dt} = \frac{rW\,\sin(\omega + f)}{h\,\sin I}, \quad (11.7)$$

since I is the angle between $\mathbf{e}_z$ and $\mathbf{h}$. The latter component turns $\mathbf{h}$ around the nodal line, namely it just changes the inclination at the rate

$$\frac{dI}{dt} = \frac{rW\cos(\omega + f)}{h} , \qquad (11.8)$$

since the angle between $\mathbf{h}$ and the nodal line is, by definition, $\pi/2$. Eq. (11.7) should be compared with eq. (4.7), yielding the lunisolar precession rate. More in general, eqs. (11.7, 8) can be interpreted as the effect of an external torque upon a gyroscope with angular momentum $\mathbf{h}$ aligned along the normal to the orbital plane.

The in-plane orientation of the orbit cannot be changed by W. However, if a W component is present and displaces the nodal line, it affects ω, the angle between nodal line and pericentre. So, let us assume for the moment that W = 0. If we differentiate with respect to time the polar equation of the ellipse (10.14), we obtain:

$$de/dt \cdot \cos f + e \cdot \sin f(df/dt) = 2hT/Gm - he(\sin f)/r^2 , \qquad (11.9)$$

where we have used $dh/dt = rT$ and $dr/dt = (Gme/h)\cdot\sin f$ (eq. (10.27)). Now, df/dt is not given by $v_t/r = h/r^2$, like in the unperturbed problem, because the time derivative of the osculating true anomaly is not the same as the time derivative of the true anomaly along the unperturbed orbit. When the orbit is perturbed, ω also is changing, and the transverse velocity is $v_t = r(df/dt) + rd\omega/dt$. As a consequence, $df/dt = (h/r^2) - d\omega/dt$. Substituting this relationship in eq. (11.9) and using eq. (11.6) for de/dt, we can obtain $d\omega/dt$. Now, if Ω changes by $\delta\Omega$ because of the presence of a W component, ω will change by $\delta\omega = -\delta\Omega\cdot\cos I$, and therefore we finally get:

$$\frac{d\omega}{dt} = \frac{(1-e^2)^{\frac{1}{2}}}{nae}\left[-R\cos f + T\sin f\left(\frac{2+e\cos f}{1+e\cos f}\right)\right] - \frac{d\Omega}{dt}\cos I . \qquad (11.10)$$

The last Gauss' equation must yield the change in an element which specifies the position along the orbit at a given time. As for the other five elements, it is suitable to define a quantity which is constant in absence of perturbations, unlike f or M. We can use, for instance,

$$\epsilon = \omega + \Omega + M - \int_{t_0}^{t} n(t')dt' ; \qquad (11.11)$$

t_0 is a time of passage through the pericentre and the integral is just M in the unperturbed problem. We obtain

$$d\epsilon/dt = d\omega/dt + d\Omega/dt + dM/dt - n =$$

$$= d\omega/dt + d\Omega/dt + (1 - e\cos U)dU/dt - \sin U de/dt - n, \quad (11.12)$$

where we have substituted for dM/dt the value obtained by differentiating Kepler's equation (10.34) with respect to time. In this equation we now substitute (11.6) for de/dt, while dU/dt can be obtained by differentiating eq. (10.29) and using eqs. (10.27) for dr/dt, (11.4) for da/dt and again (11.6) for de/dt. In terms of the derivatives of the other elements, the result can also be expressed as

$$\frac{d\epsilon}{dt} = \frac{d\omega}{dt} + \frac{d\Omega}{dt} - \frac{(1 - e^2)^{3/2}}{ae\cdot\sin f(1 + e\cos f)}\frac{da}{dt} +$$

$$+ \cot g f \frac{\sqrt{(1 - e^2)}}{e}\frac{de}{dt}, \quad (11.13)$$

where we have used the relationship between U and f (Sec. 10.2.)

11.2. QUALITATIVE DISCUSSION OF SOME PERTURBATIONS

We shall now apply the equations derived above to qualitatively understand some typical orbital changes caused by perturbations. For most of these items we shall come back in following Sections for a more detailed discussion.

Let us consider first the motion of a satellite in the gravitational field of an *oblate* planet. In Sec. 4.1 we computed the torque exerted on the (flattened) earth by an external orbiting body, and found that after averaging over one orbital period its only non-zero component is perpendicular both to the earth's spin axis and to the orbital angular momentum. By Newton's third law, an equal and opposite torque is exerted by the earth on a satellite. As a consequence, apart from effects with periods comparable with the orbital period, the angular momentum **h** will have a constant magnitude, and will rotate about the earth's spin axis with a rate dΩ/dt such that

$$\frac{d\Omega}{dt} h\sin I = -\frac{3}{2}\frac{Gm}{r^3}(C - A)\sin I\cos I \quad (11.14)$$

(compare with eq. (4.5)). Thus over a long period we have

$$d\Omega/dt = -\ 3J_2 nR_\oplus^2 \cdot \cos I/2a^2(1-e^2)^2 \ , \qquad (11.15)$$

where we have introduced the quadrupole coefficient J_2 (and the equatorial radius $R_\oplus$) by eq. (2.28). This nodal rate is several degrees per day for a satellite close to the earth (unless the inclination is near 90^0). When the gravitational field is axially symmetric, the component h_z of **h** along the earth's polar axis is rigorously conserved, since the torque has no component in this direction. (The same result can be derived with lagrangian dynamics by noting that the longitude angle is an *ignorable* coordinate – see Sec. 9.6.) Thus, since $\cos I = h_z/h$, the inclination is constant, apart (again) from short-period terms.

What about a and e ? Since the force can be derived from a time independent potential, the total energy is conserved. The change of the osculating energy, equal to the work done by the perturbing force, vanishes in an orbital period; hence, apart from short-period terms, by eqs.(11.4, 5) the semimajor axis and the eccentricity are constant. Finally, since the perturbing force is not newtonian, the orbit is no longer a close (elliptical) path; this is reflected in the fact that $d\omega/dt$ does not vanish on the long term. Instead, the orbit undergoes a rotation in its own plane whose rate has the same order of magnitude as $d\Omega/dt$.

Similar arguments can be applied to orbits perturbed by the presence of a third body: if we are interested in the orbital evolution over times much longer than the orbital period of the third body, this latter can be modelled as a static ring of material lying along its orbit (with an axial symmetry only if the orbit is circular). The effect of this ring is equivalent to that of an oblate planet, with a symmetry around an axis normal to the orbital plane, and can be estimated by choosing a suitable value of an effective J_2.

Another interesting case is that of the *drag force* produced by a gaseous medium surrounding the primary (see Sec. 11.3). Dissipative effects decrease the orbital energy; hence, by eq. (11.4), the semimajor axis diminishes and the orbit shrinks. To understand the evolution of the eccentricity, it is useful to roughly model the drag with two impulsive kicks at pericentre and apocentre. If the former is larger (as it happens when the gas density decreases outward), from eq. (11.6) it follows that de/dt is, on the average, negative; the orbit is circularized. In first approximation, the W component of the drag force vanishes and both the inclination and the nodal longitude are constant.

A third example of important perturbative effects is that of *tidal evolution* in planet-satellite systems (see Sec. 15.3). A massive satellite distorts the shape of its primary, due to the gravity gradient acting along the direction connecting the centres of the two bodies. However, the corresponding tidal bulge of a spinning primary is not perfectly aligned with this direction, since dissipative processes brake the displacement of the bulge in the orbital plane. If the satellite's mean motion is smaller than the planet's rotation rate (as it is the case for most natural satellites), the tidal bulge leads the satellite by a small phase shift; as a consequence, the satellite feels a positive T perturbation and its semimajor axis grows. The eccentricity must also increase, because the perturbation is smaller at apocentre than at pericentre, and the argument we used for drag can be applied again (but here the sign of T is reversed!) As for I and Ω , they change because the misaligned bulge is driven out of the satellite's orbital plane, and a W component arises. In a few cases (the Martian moon Phobos, with a period smaller than one Martian "day", and Neptune's retrograde satellite Triton) the tidal bulge trails the satellite and as a consequence the orbit shrinks and is circularized. Other tidal effects are due to the deformation of the satellites by their planets: owing to their smaller mass, the rotation of satellites is usually synchronized with the orbital period by dissipative forces in a comparatively short time; however, if the orbit is not circular, the tidal bulge of the satellite changes its amplitude when the body moves from apocentre to pericentre and *vice versa.* This again dissipates energy and damps the eccentricity. This effect is usually larger than that of tides in the planet, thus explaining why all close satellites in the solar system have nearly circular orbits.

A final problem which can be briefly addressed in the perturbative formalism is that of *orbital commensurabilities* and *resonances* (see also Sec. 15.2.) When the perturbing acceleration is periodic in time (a typical example being that of the gravitational perturbations caused by a third body), it is possible to expand it (and its R, T and W components) in a trigonometric series of the form

$$\Sigma_{i,j}[C_{ij}\cos(int + jn't) + S_{ij}\sin(int + jn't)], \qquad (11.16)$$

where i and j are integer numbers, n is the mean motion of the orbiting body and n' is the frequency of the perturbation. In order to obtain the evolution of the elements, we can integrate the perturbation equations with respect to time. In each term of the series (11.16) we get (in + jn') as divisors of the coefficients C_{ij}, S_{ij}. The order of magnitude of the new coefficients can be much greater than their generic value when n and n' are nearly in the ratio of two integer numbers, that is, when the perturbation is nearly in resonance with the orbital period of the body. This happens when i and j are opposite in sign and $|in + jn'| << n, n'$; in this case we have long period changes in the elements with quite large amplitudes. (In fact, in the limit case of a "perfect" resonance in + jn' = 0, the perturbations would grow linearly with time.) Although formally the ratio n/n' can always be approximated by a rational number, usually the coefficients C_{ij} and S_{ij} decrease rapidly when i and j grow; only when i and j are small integers the resonance can produce significant effects. When this occurs, the character of the orbital motion is deeply affected and new stability (or instability) properties arise. In Sec. 15.2 we shall discuss several examples of this phenomenon, which has strongly affected the dynamical structure of the solar system.

11.3. EFFECTS OF ATMOSPHERIC DRAG

When an artificial satellite orbits not too far from the earth's surface, it is subjected to the drag due to interaction with the atmospheric gas particles. A knowledge of the orbital effects of drag is important both to avoid degradation of the orbital predictions and determinations (see Sec. 18.1) and to obtain some information on the properties (density, temperature, motions) of the upper atmosphere itself. If the gas (of density ρ) is at rest in an inertial reference frame and the satellite is an object of mass m and cross section S (normal to its velocity vector $\mathbf{v}$), which absorbs all the impinging gas molecules, the drag force is given by the amount of momentum absorbed per unit time, that is $-(Sv\rho)\mathbf{v}$. Therefore, the drag acceleration undergone by a satellite is usually written as

$$\mathbf{F}_D = -\tfrac{1}{2}C_D S\rho v\mathbf{v}/m. \qquad (11.17)$$

Here the numerical coefficient (of order unity) $\frac{1}{2}C_D$ accounts for the real shape of the body and for the way it interacts with the molecules, since of course perfect absorpbtion is not realistic. In fact, every surface element of the spacecraft can absorb, reflect or diffuse the incident molecules in a way depending on its composition and microscopic structure; therefore, a complex integration is needed over the surface of the body exposed to the molecular flux. Moreover, whenever the thermal motion of the molecules is not negligible, C_D is also a function of the ratio between v and the thermal velocity. Finally, note that if the atmosphere is not at rest, but rotates more or less rigidly with the earth and has therefore a velocity $\mathbf{V}_A = \boldsymbol{\omega}_A \times \mathbf{r}$, with $\boldsymbol{\omega}_A$ close to the earth's angular velocity $\boldsymbol{\omega}$, $(\mathbf{v} - \mathbf{V}_A)$ should take the place of $\mathbf{v}$ in eq.(11.17). As a consequence, in this case the magnitude and the direction of the drag force depend in a complex way on the geometry and orientation of the orbit. For the moment, we shall neglect these complications, and just use eq. (11.17) as such, with $D \equiv C_D S/m$ assumed to be constant.

In order to evaluate with Gauss' equations the rate of change of the orbital elements, we note that, since $\mathbf{F}_D$ is directed along $-\mathbf{v}$, its T and R components read $T = -F_D v_t/v$ and $R = -F_D v_r/v$, respectively, where v_r and v_t are the radial and transverse components of the orbital velocity. Using eqs. (10.27, 28) for v_r and v_t, from eq. (11.4) we get

$$\frac{da}{dt} = -\frac{2a^2}{Gm} F_D v = -\frac{D\rho v^3}{n^2 a} . \qquad (11.18)$$

For the eccentricity, from eqs. (10.27, 28) and (11.6), we get

$$de/dt = -(e + \cos f)\rho D v . \qquad (11.19)$$

To compute the long-term perturbations we average over one revolution (see Sec. 11.5); it is simpler to express the resulting integrals in terms of the eccentric anomaly U. The energy integral yields:

$$v^2 = Gm\left(\frac{2}{r} - \frac{1}{a}\right) = n^2 a^2 \frac{1 + e\cos U}{1 - e\sin U} , \qquad (11.20)$$

where n is the mean motion. To change the variable in the integrals we use

$$\frac{da}{dU} = \frac{da/dt}{dU/dt} = \frac{da/dt}{n}(1 - e\cos U) . \qquad (11.21)$$

By substituting into eq. (11.18) and integrating over one orbital period $2\pi/n$ we obtain the change of the semimajor axis during one orbit

$$\Delta a = -Da^2 \int_0^{2\pi} \frac{(1 + e\cos U)^{3/2}}{(1 - e\cos U)^{1/2}} \rho dU . \qquad (11.22)$$

A similar computation for e, using eq. (11.19), gives

$$\Delta e = -Da(1 - e^2)\int_0^{2\pi} \frac{\surd(1 + e\cos U)}{\surd(1 - e\cos U)} \rho \cos U dU . \qquad (11.23)$$

From these equations we can already draw some conclusions about the qualitative evolution of the orbit under the infuence of drag. First, $\Delta a < 0$, that is the semimajor axis decreases and so does the orbital period; this could have been easily anticipated, considering that a dissipative force must decrease the total energy, proportional to $(-1/a)$. Second, for every reasonable density distribution (provided only it is a decreasing function of the radius r), $\Delta e < 0$, unless e is 0 from the beginning. This can be seen from the integral in eq. (11.23), since the integrand is positive for $-\pi/2 < u < \pi/2$ and negative for $\pi/2 < U < 3\pi/2$, while both the fraction in it and ρ are larger in the section of the orbit closer to perigee. This corresponds to the fact that if the greatest drag is experienced near perigee, the satellite does not swing out so far on the opposite side of the orbit, and as a result the maximum altitude is reduced, without affecting too much the perigee distance itself.

To further discuss the effects of drag we introduce a *locally exponential atmosphere*, in which the *scale height* (see Sec. 7.1)

$$H(r) = 1/[-d(\ln\rho)/dr] \tag{11.24}$$

changes over a scale ℓ greater than H itself:

$$dH/dr \approx H/\ell \ll 1 . \tag{11.25}$$

As explained in Secs. 7.1 and 8.4, this is a good approximation. Moreover, for a thin atmosphere, in eq. (11.18) we can evaluate $(v^3/n^2 a)$ at the earth radius R and use the simpler form

$$da/dt = -\rho(a)D\sqrt{(GmR)} . \tag{11.18'}$$

We see that the re-entry time scale

$$\tau = 1/[-da/adt] = \sqrt{(R/Gm)}/D\rho(a) \tag{11.26}$$

decreases very fast with height. The relevant integral of the reciprocal density in eq. (11.18') can be approximated by using, in analogy with the optical depth (7.12), the dimensionless variable

$$\xi = \int_R^r dr'/H(r'), \qquad d\xi = dr/H,$$

in terms of which the density is $\rho = \rho_0 \exp(-\xi)$. Integration of eq. (11.18') yields then

$$-Dt\sqrt{(GmR)} = \int da/\rho(a) = 1/\rho_0 \int_0^\xi d\xi' H(\xi')\exp(\xi') .$$

$t = 0$ is the time of impact on the ground and the index $_0$ refers to r = R. The primitive of $H(\xi)\exp(\xi)$ can be expressed as a series in the derivatives of H, in which each term is smaller by the factor H/ℓ:

$$d[e^{\xi}(H - H' + H'' - \ldots)]/d\xi = He^{\xi}.$$

For an exactly exponential atmosphere ($H = H_0$ = const), we have a "logarithmic" re-entry

$$\xi = (a - R)/H_0 = \ln[1 - (\rho_0 Dt/H_0)\sqrt{(GmR)}] \; ; \qquad (11.27)$$

in general, for a given atmospheric model, here described by its scale height H(r), one should solve the equation

$$e^{\xi}H(\xi) - H_0 = - \rho_0 Dt\sqrt{(GmR)} \; , \qquad (11.28)$$

where terms of order H/ℓ have been neglected. As shown in Fig. 11.2, the orbital decay is indeed "catastrophic".

Our poor knowledge of the atmospheric density profile is due also to its time variation, even by more than one order of magnitude, correlated with solar and geomagnetic activity; significant daily, seasonal and latitudinal variations are present as well below 1000 km. As a result, at heights between 100 and 200 km, where the catastrophic decay starts, reliable and accurate predictions are impossible for the atmospheric density; hence, the exact re-entry time and its location for an uncontrolled satellite in general cannot be predicted well in advance. Such have been the cases of *Skylab* and of *Cosmos 954, 1402* and *1900*; particular concern arose in the cases of spacecraft carrying radioactive material for radioisotopic electricity generators or nuclear reactors.

Contrasting with semimajor axis and eccentricity, the other orbital elements do not undergo long-term perturbations due to drag: as an example, with a drag-like force of the type given by eq. (11.17) and with ρ an arbitrary function of r, there is no secular perturbation in ω. This can be proved by substituting in eq. (11.10) $d\Omega/dt = 0$ (because $W = 0$) and the expressions for T and R computed as above:

$$\frac{d\omega}{dt} = - \frac{D\rho v}{2e}\left[\sin f + \frac{e}{\sqrt{(1 - e^2)}}\cos f \,\sin u\right] . \qquad (11.30)$$

Since $d\omega/dt$ is an odd function of the anomalies, its average vanishes.

Even simpler is the case of the inclination I and the nodal argument Ω: with a drag-like force acting only in the orbital plane, the out-of-plane component W is zero and the orbital plane does not change. This mathematical result unfortunately relies on physically unrealistic

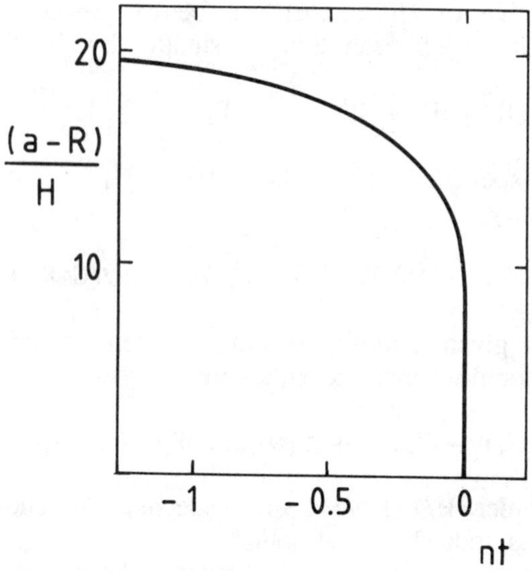

Fig. 11.2 Orbital decay due to drag in an exponential atmosphere. The figure shows a plot of $(a - R)/H_0$ in terms of the dimensionless variable nt (eq. (11.27)). Here H_0 and D are assumed to be equal to 10 km and 0.05 cm^2/g, respectively.

assumptions: even for a spherical satellite experiencing no aerodynamic lift forces, the rotation of the atmosphere induces a component of the relative velocity of the satellite with respect to it normal to the orbital plane. Taking as an example a circular, inclined orbit, the out-of-plane component of the drag is approximately

$$W = - F_D \frac{\omega_A}{n} \sin I \cos(\omega + f), \qquad (11.31)$$

where we have assumed the ratio between the atmospheric velocity and the orbital speed (of order ω_A/n) to be small. We recall that $(\omega + f)$ is the angle between the current position and the ascending node. Using eqs. (11.7, 8) we obtain:

$$\frac{d\Omega}{dt} = - \frac{F_D}{na} \frac{\omega_A}{n} \cos(\omega + f) \sin(\omega + f) , \qquad (11.32)$$

$$\frac{dI}{dt} = - \frac{F_D}{na} \frac{\omega_A}{n} \sin I \cos^2(\omega + f) . \qquad (11.33)$$

From the former equation it is easy to appreciate that the average of $d\Omega/dt$ over one orbit vanishes and so does any long-term effect on the node (this is simply due to the fact that the torque caused by the spinning atmosphere is directed along the earth's polar axis). On the contrary, the change of I in one orbit does not vanish:

$$\Delta I = -\pi \left(\frac{F_D}{n^2 a} \right) \frac{\omega_A}{n} \sin I . \tag{11.34}$$

As a result, the orbital plane approaches the equatorial plane; the factor in brackets is again the ratio of the drag force to the earth's attraction. Starting from *Sputnik* 2, measurements of the rate of decrease of the inclination have yielded information on the atmospheric rotation rate, which has been shown to be significantly greater than that of the solid earth, due to the presence at high altitudes of strong west-to-east winds.

11.4. THE PERTURBING FUNCTION AND LAGRANGE'S PERTURBATION EQUATIONS

In the case of two planets whose heliocentric motions are perturbed by their mutual gravitational interaction it is not easy to directly split the perturbing force into its R, T and W components and a more general method proves useful. Moreover, for practical applications it is often better to take the origin at the sun (and not at the centre of mass) and to use relative coordinates, like we have already done in the two-body problem. As in Sec. 10.1, we can subtract the equation of motion of the sun from that of the first planet and obtain a differential equation for the heliocentric position $\mathbf{r}_1$ of the latter:

$$\frac{d^2 \mathbf{r}_1}{dt^2} + G(M_\odot + m_1)\frac{\mathbf{r}_1}{r_1^{\,3}} = Gm_2 \left[\frac{\mathbf{r}_2 - \mathbf{r}_1}{r_{12}^{\,3}} - \frac{\mathbf{r}_2}{r_2^{\,3}} \right] =$$

$$= -\frac{\partial}{\partial \mathbf{r}} \left[-\frac{G(M_\odot + m_1)}{r_1} + \mathbf{R}_{12} \right] , \tag{11.35}$$

where $r_{12} = |\mathbf{r}_1 - \mathbf{r}_2|$. Of course, interchanging 1 and 2 yields the equation for the second planet. The scalar function

$$\mathbf{R}_{12} = -Gm_2 \left[\frac{1}{r_{12}} - \frac{\mathbf{r}_1 \cdot \mathbf{r}_2}{r_2^{\,3}} \right] \tag{11.36}$$

is the so-called *perturbing function* for the first planet. Its first term (times m_1) is the interaction potential energy; the second takes into account the motion of the sun. An important feature of this way of writing the equations of motion, to be used in Ch. 15, is that if several perturbing bodies are present, it is sufficient to generalize the definition of **R** by summing up terms like (11.36). **R** is usually a small correction to the total potential, of the order of the ratio of the masses of the perturbing planet(s) and the sun.

More in general, whenever the perturbative acceleration is conservative and can be expressed as the gradient of a scalar perturbing function ($-\mathbf{R}$), one can exploit the *hamiltonian* character of the dynamical system and find the time derivatives of the osculating orbital elements c_k in term of the partial derivatives $\partial\mathbf{R}/\partial c_h$ (h, k = 1,..., 6); these are the so-called *Lagrange's perturbation equations*, to be derived in this section.

Consider the general problem of the perturbed orbital motion of a body with cartesian coordinates $q_i = r_i$ (i = 1,2,3). In the hamiltonian formulation, instead of the three, second order differential equations (11.1), one has six first order equations. The *Hamiltonian function* reads in our case

$$H = -\,Gm/r + \mathbf{R} + \tfrac{1}{2}p_i p_i \ , \qquad (11.37)$$

while Hamilton's equations

$$\frac{dp_i}{dt} = -\frac{\partial H}{\partial q_i} \ , \qquad \frac{dq_i}{dt} = \frac{\partial H}{\partial p_i} \qquad (11.38)$$

are easily seen to yield eq. (11.1), with $p_i = dr_i dt$, the momentum per unit mass. The unperturbed Kepler's problem provides an implicit mapping of the phase space (q_i, p_i) onto the six-dimensional space of the osculating elements c_k; using them as independent variables eqs. (11.38) become

$$\frac{\partial p_i}{\partial c_k}\frac{dc_k}{dt} = -\frac{\partial H}{\partial q_i} \ , \qquad \frac{\partial q_i}{\partial c_k}\frac{dc_k}{dt} = \frac{\partial H}{\partial p_i} \ , \qquad (11.39)$$

where we have adopted Einstein's summation convention (implicit summation over repeated indexes). We now multiply the first equation by $(-\,\partial q_i/\partial c_h)$, the second by $\partial p_i/\partial c_h$ and sum them up, obtaining

$$\left[\frac{\partial q_i}{\partial c_k}\frac{\partial p_i}{\partial c_h} - \frac{\partial p_i}{\partial c_k}\frac{\partial q_i}{\partial c_h}\right]\frac{dc_k}{dt} = \frac{\partial p_i}{\partial c_h}\frac{\partial H}{\partial p_i} + \frac{\partial q_i}{\partial c_h}\frac{\partial H}{\partial q_i} = \frac{\partial H}{\partial c_h}. \qquad (11.40)$$

These equations can be rewritten in a more synthetic form as

$$\{c_h, c_k\}\dot{c}_k = -\,\partial H/\partial c_h \,, \tag{11.41}$$

by defining the *Lagrange brackets* $\{c_k, c_h\}$ as the factor in parentheses in the left-hand side of eq. (11.40). The Lagrange brackets are obviously antisymmetric in their indexes, namely

$$\{c_h, c_k\} = -\,\{c_k, c_h\} \,, \quad \{c_k, c_k\} = 0. \tag{11.42}$$

Moreover, it can be shown that they are time independent, provided the dynamical equations (11.38) are fulfilled (see Problem 11.3). When we express the Hamiltonian (11.37) in terms of the orbital elements, we must identify $(\frac{1}{2}p_ip_i - Gm/r)$ with the unperturbed energy:

$$H = -\,Gm/2a + \mathbf{R}. \tag{11.43}$$

The perturbing function $\mathbf{R}$ depends on the orbital elements only through the coordinates $q_i = r_i$ (and not the momenta). Using the cartesian coordinates of a keplerian orbit in terms of the eccentric anomaly, in a frame where the orbit lies in the (x, y) plane with the pericentre along the x-axis, we can write

$$\begin{pmatrix} q_1 \\ q_2 \\ q_3 \end{pmatrix} = \mathbf{T}\cdot\begin{bmatrix} a(\cos U - e) \\ a(1 - e^2)^{\frac{1}{2}}\sin U \\ 0 \end{bmatrix}, \quad \begin{pmatrix} p_1 \\ p_2 \\ p_3 \end{pmatrix} = \mathbf{T}\cdot\begin{bmatrix} -na^2\sin U/r \\ -na^2(1 - e^2)^{\frac{1}{2}}\cos U/r \\ 0 \end{bmatrix}. \tag{11.44}$$

Here the transformation matrix $\mathbf{T}$ is the product of three planar rotations around the normal to the orbit, the nodal line and the reference axis for the inclination:

$$\mathbf{T} = \begin{pmatrix} \cos\Omega & -\sin\Omega & 0 \\ \sin\Omega & \cos\Omega & 0 \\ 0 & 0 & 1 \end{pmatrix}\begin{pmatrix} 1 & 0 & 0 \\ 0 & \cos I & -\sin I \\ 0 & \sin I & \cos I \end{pmatrix}\begin{pmatrix} \cos\omega & -\sin\omega & 0 \\ \sin\omega & \cos\omega & 0 \\ 0 & 0 & 1 \end{pmatrix} \tag{11.44'}$$

(see Fig. 10.3). Since Lagrange's brackets are constant, we can compute them at the pericentre (f = U = M = 0), and after many simplifications we find that only the following six brackets do not vanish:

$$\{a, \Omega\} = -\,\tfrac{1}{2}na(1 - e^2)^{\frac{1}{2}}\cos I \,, \tag{11.45$_1$}$$

$$\{e, \Omega\} = na^2e(1 - e^2)^{-\frac{1}{2}}\cos I \,, \tag{11.45$_2$}$$

$$\{I, \Omega\} = na^2(1 - e^2)^{\frac{1}{2}}\sin I \,, \tag{11.45$_3$}$$

$$\{a,\ \omega\} = -\tfrac{1}{2}na(1-e^2)^{\frac{1}{2}}\ , \qquad (11.45_4)$$

$$\{e,\ \omega) = na^2e(1-e^2)^{-\frac{1}{2}}\ , \qquad (11.45_5)$$

$$\{a,\ M\} = -\ na/2\ . \qquad (11.45_6)$$

Now we can substitute these relationships into the system (11.41), where, as we already noticed, $\partial H/\partial c_k = \partial \mathbf{R}/\partial c_k$ for all the elements, but the semimajor axis, for which $\partial H/\partial a = \partial \mathbf{R}/\partial a + Gm/2a^2$. Solving the system in terms of the derivatives of the elements, we finally obtain a set of equations relating the vector of the time derivatives of the elements (da/adt, de/dt, dI/dt, $d\Omega/dt$, $d\omega/dt$, $dM/dt - n$) to the vector of the derivatives of $\mathbf{R}$ with respect to the elements ($\partial\mathbf{R}/\partial a$, $\partial\mathbf{R}/\partial e$, $\partial\mathbf{R}/\partial I$, $\partial\mathbf{R}/\partial\Omega$, $\partial\mathbf{R}/\partial\omega$, $\partial\mathbf{R}/\partial M$). These six equations read:

$$\frac{1}{a}\frac{da}{dt} = \frac{2}{na^2}\frac{\partial \mathbf{R}}{\partial M}\ , \qquad (11.46_1)$$

$$\frac{de}{dt} = -\frac{\sqrt{(1-e^2)}}{na^2e}\frac{\partial \mathbf{R}}{\partial \omega} + \frac{(1-e^2)}{na^2e}\frac{\partial \mathbf{R}}{\partial M}\ , \qquad (11.46_2)$$

$$\frac{dI}{dt} = -\frac{1}{na^2\sqrt{(1-e^2)}\sin I}\frac{\partial \mathbf{R}}{\partial \Omega} + \frac{\cos I}{na^2\sqrt{(1-e^2)}\sin I}\frac{\partial \mathbf{R}}{\partial \omega}\ , \qquad (11.46_3)$$

$$\frac{d\Omega}{dt} = \frac{1}{na^2\sqrt{(1-e^2)}\sin I}\frac{\partial \mathbf{R}}{\partial I}\ , \qquad (11.46_4)$$

$$\frac{d\omega}{dt} = \frac{\sqrt{(1-e^2)}}{nea^2}\frac{\partial \mathbf{R}}{\partial e} - \frac{\cos I}{na^2\sqrt{(1-e^2)}\sin I}\frac{\partial \mathbf{R}}{\partial I} \qquad (11.46_5)$$

$$\frac{dM}{dt} - n = -\frac{2}{na}\frac{\partial \mathbf{R}}{\partial a} - \frac{(1-e^2)}{na^2e}\frac{\partial \mathbf{R}}{\partial e}\ . \qquad (11.46_6)$$

It is noteworthy that the time derivatives of a, e and I contain only the partial derivatives of $\mathbf{R}$ with respect to Ω, ω and M and *vice versa.* The fact that several entries in the matrix relating the two above mentioned vectors diverge for $e \to 0$ or $\sin I \to 0$ is not due to a physical singularity, but only to the geometrical definition of the two angles ω and Ω, which lose their significance for $e = 0$ and $I = 0$. In fact, when $e \ll 1$ or $I \ll 1$, these two elements can change by large amounts even for small changes of the position or the velocity of the orbiting body. This singularity can be eliminated by a suitable change of variables. If we define the new, *non-singular elements*

$$h = e\,\sin(\omega + \Omega), \qquad k = e\,\cos(\omega + \Omega)\;, \tag{11.47$_1$}$$

$$P = \tan I\,\sin\Omega, \qquad Q = \tan I\,\cos\Omega\;, \tag{11.47$_2$}$$

they may be used to form equations for dh/dt, dk/dt, dP/dt and dQ/dt, replacing those for de/dt, dω/dt, dI/dt and dΩ/dt, respectively. If we neglect terms of order $\geqslant 2$ in e and I, it is easy to show that these equations have the following, nicely symmetrical form

$$dh/dt = (\partial\mathbf{R}/\partial k)/na^2, \quad dk/dt = -\,(\partial\mathbf{R}/\partial h)/na^2, \tag{11.48$_1$}$$

$$dP/dt = (\partial\mathbf{R}/\partial Q)/na^2, \quad dQ/dt = -\,(\partial\mathbf{R}/\partial P)/na^2; \tag{11.48$_2$}$$

i.e., they are again hamiltonian equations, with $H = \mathbf{R}/na^2$. Moreover, instead of the mean anomaly M, we can use the mean longitude $\lambda = M + \omega + \Omega \equiv \rho + \epsilon$ (see eq. (11.11)), where

$$\rho = \int_{t_0}^{t} n(t')dt', \tag{11.49}$$

and therefore ϵ is the mean longitude at the instant t_0. The definition of ρ takes into account the fact that n is related to the semimajor axis by Kepler's third law and therefore is variable. It is easy to see that $\ddot{\rho} = -\,3n\dot{a}/2a$; as we anticipated in Sec. 11.1, a constant drift of the semimajor axis (as caused, for instance, by a constant T component of the perturbation) results into a diplacement in longitude growing proportionally to the square of the elapsed time. The Lagrange equation for dϵ/dt can be derived in a straightforward way from those for dM/dt, dω/dt and dΩ/dt.

11.5. APPROXIMATE SOLUTION METHODS

The equations for the variations of the elements, such as those derived in Secs. 11.1 and 11.4, are valid for every perturbing acceleration **F** (or every perturbing function **R**), no matter what its physical cause or its size are. However, very frequently in practice the perturbation is "small" with respect to the main, newtonian force due to the central mass (generally the sun or the earth). We can therefore define a smallness parameter μ ($\ll 1$) by

$$\mu \equiv \mathbf{R}/(Gm/a) = \mathbf{R}/(n^2 a^2). \qquad (11.50)$$

From Lagrange's equations (11.46) we can see that the rates of change of the elements (taking, instead of da/dt, da/adt) are of order $\mathbf{R}/na^2 = \mu n$. If $\mathbf{R}$ and its derivatives (or equivalently the R, T and W components of the perturbing acceleration) change with time with frequencies of the order of the mean motion n, the elements, too, will vary with the same frequencies and with amplitudes $O(\mu)$. However, if the perturbation does not vanish when averaged over one orbital period, the changes in the keplerian elements after times of the order of $1/\mu$ orbital periods are $O(\mu^0)$. These long-term changes do not occur for the semimajor axis, whose rate of change is proportional to $\partial\mathbf{R}/\partial M$.

These order of magnitude estimates lead to the idea of solving the equations for the variations of the elements by a *perturbative* method. If μ were nil, we would have the zero-order keplerian solution with all of the elements constant (taking ϵ instead of M) and equal to their initial values. To compute a first order solution, we can substitute in the right-hand side of the perturbation equations a set of constant values for the elements (i. e., the osculating elements corresponding to the initial conditions), and then solve the equations by simple integrations with respect to time. For a finite time span this solution will differ from the exact solution by terms $O(\mu^2)$. These corrections can be evaluated, at least in principle, by substituting again the first order solution in the right-hand side of the equations, and so on by iterating the same procedure up to higher and higher orders of μ.

In practice these computations are very cumbersome, because the integrals with respect to time are usually performed by a mixture of Fourier and Taylor series, and the number of terms in these series becomes large very soon. However, an important simplification is possible if we are not interested in short-periodic effects, having time scales of the order of one orbital period, but only in the long-term evolution of the orbit, namely in the so-called *secular* effects. In this case we can average the right-hand sides of the perturbation equations with respect to time, by integrating them from 0 to 2π over all the *fast* angular variables that can be expressed as products of time and a frequency of the order of n (e. g., the mean anomalies of both the perturbed body and the perturbing masses). In other words, we expand the right-hand side into a multiple Fourier series with respect to the fast angular variables, and take only the constant term. This method is very useful because the *averaged equations* are frequently much simpler and easier than the original ones; for instance, in many cases of interest the averages over the fast variables vanish, showing immediately that there is no secular perturbation of the corresponding elements. This is indeed the case for the semimajor axis whenever we have a conservative perturbation, since da/dt is proportional to $\partial R/\partial M$ (as noted earlier).

When there is more than one fast variable over which the averaging procedure is carried out, it is not easy to assess whether the solution of the averaged equations is really a good approximation of the exact solution over a long time span. In other words, in some cases the averaging method may fail; the reason is related to the problem of commensurabilities we have briefly addressed in Sec. 11.2 (see also Sec. 15.2.) Let us assume that there are two fast angular variables M_1 and M_2 with unperturbed frequencies n_1 and n_2. The ratio n_2/n_1 can always be approximated by a rational number p/q, and thus the real orbit will not spend the same amount of time in every surface element $dM_1\,dM_2$ of the (M_1, M_2) surface, but will remain for a very long time close to a line on this surface specified by the equations

$$\begin{aligned} M_1 &= M_1(t_0) + n_1(t - t_0), \\ M_2 &= M_2(t_0) + n_1(p/q)(t - t_0). \end{aligned} \tag{11.51}$$

Therefore the time average along the true solution and the "surface" average over M_1 and M_2 ranging from 0 to 2π may be very different. In other words, since a linear combination $(pM_1 - qM_2)$ of the two fast variables is not fast at all (because the corresponding frequency is very small), then terms do appear in the Fourier expansion of the right–hand side of the perturbation equations which cause perturbations building up over very long times. If the amplitudes of these *resonant* effects are large, they can no longer be treated as small perturbations, and all the approximation methods discussed in this Section cannot be applied (see Sec. 15.2). Luckily, this is not always the case, because it can often be shown that whenever p and or q are not small integer numbers, the corresponding Fourier terms have very small amplitudes. Thus, the failure of the averaging method with commensurable frequencies is in practice really important only for *low-order* commensurabilities, when the ratio of the frequencies is very close to a "simple" fraction of two small integers (1/2, 1/3, 2/3, etc.).

11.6. SECULAR EFFECTS OF THE OBLATENESS OF THE PRIMARY

As regards the gravitational perturbations of a satellite, the dominant part of the perturbing function is usually associated with the J_2 term of the geopotential, which is caused by the polar oblateness of the planet (see Ch. 2). In fact, J_2 is much larger than all the other $J_{\ell m}$ coefficients in the spherical harmonics expansion of the geopotential (see eq. (2.22)). In order to compute the perturbations due to this effect, we have to express in terms of the osculating orbital elements the perturbing function

$$\mathbf{R}_2 = -\frac{GM_\oplus}{r}\left(\frac{R}{r}\right)^2 J_2\left(\frac{3}{2}\cos^2\theta - \frac{1}{2}\right), \tag{11.52}$$

where $M_\oplus$ and R are the earth's mass and equatorial radius, while r and θ are the radial distance and the colatitude of the satellite. By spherical trigonometry (see Fig. 11.3), it can easily be shown that

$$\cos\theta = \sin I \ \sin(\omega + f). \tag{11.53}$$

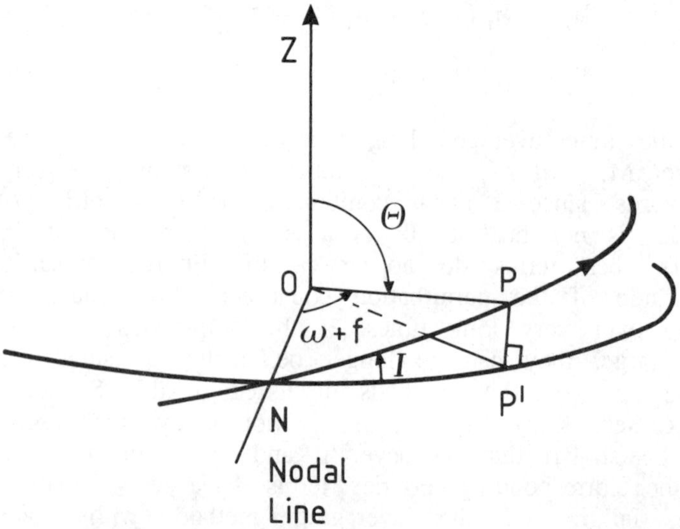

Fig. 11.3 By applying the sine theorem to the spherical triangle N P' P, where P is the orbiting body and P' is its projection on the equatorial plane, eq. (11.53) is immediately derived.

Therefore we get

$$\mathbf{R}_2 = \frac{GM_\oplus}{a}\left(\frac{R}{a}\right)^2 J_2\left[\frac{a^3}{r^3}\left(\frac{1}{2} - \frac{3}{4}\sin^2 I\right) + \frac{3}{4}\sin^2 I\,\frac{a^3}{r^3}\cos(2\omega+2f)\right]. \tag{11.54}$$

The functions a^3/r^3 and $a^3\cos(2\omega + 2f)/r^3$ are periodic in the mean anomaly M. If we are interested only in the secular part of the effect, we can keep in the perturbing function only the averages:

$$\frac{1}{2\pi}\int_0^{2\pi} dM \frac{a^3}{r^3} = (1 - e^2)^{-3/2}, \quad \int_0^{2\pi} dM \frac{a^3}{r^3} \cos(2\omega + 2f) = 0, \qquad (11.55)$$

where we have used the relationships $r = a(1 - e^2)/(1 + e\cos f)$ and $r^2 df/dt = na^2 \sqrt{(1 - e^2)}$ (see Ch. 10). Now we have

$$\langle \mathbf{R}_2 \rangle = \frac{GM_\oplus}{a} \left(\frac{R}{a}\right)^2 J_2 \left[\frac{1}{2} - \frac{3}{4}\sin^2 I\right] (1 - e^2)^{-3/2} \qquad (11.56)$$

and using Lagrange's equations we obtain

$$\langle da/dt \rangle = \langle de/dt \rangle = \langle dI/dt \rangle = 0 \qquad (11.57_1)$$

$$\langle d\Omega/dt \rangle = -\ 3nJ_2R^2\cos I/2a^2(1 - e^2)^2 \qquad (11.57_2)$$

$$\langle d\omega/dt \rangle = -\ 3nJ_2R^2(1 - 5\cos^2 I)/4a^2(1 - e^2)^2 \qquad (11.57_3)$$

$$\langle dM/dt \rangle = n + 3nJ_2R^2(3\cos^2 I - 1)/4a^2(1 - e^2)^{3/2}\ . \qquad (11.57_4)$$

As we discussed in the previous Section, since $J_2 \ll 1$, each right-hand side of eqs. (11.57) can be assumed to be constant in the first approximation. Thus, as anticipated in Sec. 11.2, a, e and I are on the average constant, while Ω and ω grow linearly with time. The derivative of M is not just the two-body mean motion: on the equatorial plane the equatorial bulge causes a force somewhat more intense than for a spherical planet and a higher orbital speed.

The equation for $\langle d\omega/dt \rangle$ shows that the orbit turns in its own plane with a rate which becomes negative for high inclinations. For the *critical inclination* given by $\arccos(1/\sqrt{5}) = 63^\circ\ 26'$, $\langle d\omega/dt \rangle$ vanishes. This fact is exploited by the *Molnyia-type satellites*, which, in addition to the critical inclination, have a large eccentricity (≈ 0.7) and a mean motion equal to 1/2 or 1/3 of the earth's rotation rate.

PROBLEMS

*11.1. The perversity of gravitation shows up in the positive along-track acceleration produced by drag (sec. 11.3). Find out when an arbitrary conservative and radial force is "perverse". (Hint: consider a circular orbit with velocity $v^2 = rdU/dr$, $U(r)$ being the potential, and compute its change when the energy E is slowly changed. Show that the sign of $(dv/dt)/(dE/dt)$ is determined by the dimensionless parameter $r(d^2U/dr^2)/(dU/dr)$.)

11.2. The perturbative force **F** can be split into components related to the orbit in another way, taking T' tangential to the orbit and R' perpendicular to it in the orbital plane (and retaining W). The two sets of components are equal only for $e = 0$. Express R' and T' as a function of R, T, the eccentricity and the true anomaly and write down the modified Gauss' equations containing R' and T' instead of R and T.

*11.3. Compute the time derivative of the Lagrange's brackets and show that they vanish. Use equations (11.38) and the relationship $p_i(\partial p_i/\partial c_k) = \partial(p_i p_i)/2\partial c_k$.

11.4. What is the nodal period for a satellite in a low circular orbit as a function of the inclination?

*11.5. Calculate the secular precession of the node in an elementary way, using the cartesian quadrupole tensor (2.27) and starting from the change it induces in the orbital angular momentum.

11.6. What is the order of magnitude of the perturbative effects in the orbital elements of a two-body system subject to a force linearly dependent on the coordinates? Apply the result to evaluate the effect of the galaxy on the solar system.

11.7. Evaluate, in order of magnitude (for a planet and for the moon) and analytically, the secular effects of the perturbing function

$$(Gm/r)(Gm/rc^2),$$

where M is the central mass and c is the speed of light. This perturbation is typical of relativistic corrections (see Ch. 17).

*11.8. In order to explain, before general relativity, the anomalous advance of the perihelion of Mercury, a gravitational force proportional to $1/r^{2+\epsilon}$ was assumed. Study the secular perturbations of a keplerian orbit when $|\epsilon| \ll 1$.

11.9. Study the secular effects on a circular orbit of a perturbation with a quadrupole dependence from direction (like in tides).

FURTHER READINGS

Useful general textbooks of celestial mechanics, where the various perturbation techniques are treated at increasing levels of depth and mathematical rigour, are: E. Finlay-Freundlich, *Celestial Mechanics*, Pergamon, London (1958); F.R. Moulton, *An Introduction to Celestial Mechanics*, Dover, New York (1970); J. Kovalevsky, *Introduction to Celestial Mechanics*, Reidel, Dordrecht (1967); A.E. Roy, *Orbital Motion*, Hilger, Bristol (1978); and D. Brouwer and G.M. Clemence, *Methods of Celestial Mechanics*, Academic, New York (1961). For many applications to artificial satellites, see *Le Mouvement du Véhicule Spatial en Orbite*, CNES, Toulouse (1980), and A. Milani, A.M. Nobili and P. Farinella, *Non-Gravitational Perturbations and Satellite Geodesy*, Hilger, Bristol (1987). Our derivation of Gauss' equations has been based on that presented in the latter book and on J.A. Burns' paper *Elementary derivation of the perturbation equations of celestial mechanics*, American Journal of Physics, **44**, 944, (1976).

12. THE RESTRICTED THREE-BODY PROBLEM

The determination of the motion of N point-like masses under their mutual gravitational forces is the basic problem in celestial mechanics, with important applications to fundamental astrophysical problems, like the dynamical structure of planetary and satellite systems and the evolution of multiple stellar systems, ranging from multiple stars to stellar clusters and galaxies. However, even for N as small as 3 it is impossible to find general solutions of this problem in terms of simple analytical functions (like in the case of the two-body problem). A great variety of orbits is possible and one can show that only very particular choices of the initial conditions give rise to a periodic behaviour. Even the perturbative techniques worked out in Ch. 11 can be applied only in the particular case of *hierarchical* systems (see Sec. 15.1). As a consequence, the gravitational N-body problem has been studied by two other methods which can yield only partial results, but are in some way complementary. The first method is the numerical integration of the orbits, starting from a given set of initial conditions. In this way the motion can be determined in detail, but only for a limited span of time (due to limitations in computer time and accumulating numerical errors; see Sec. 15.1); and often it is not possible to generalize some property of the chosen orbits to significant regions of the phase space. The second method is the search for general constraints or criteria regarding the qualitative features of the motion, such as its periodic character, its stability, its geometrical and topological properties.

In order to illustrate the typical results that can be achieved by these latter techniques, we discuss in detail the simplest N-body problem (for $N > 2$), namely the so-called planar, *restricted three-body problem*, studied for the first time by Lagrange in 1772. A test body with a negligible mass moves under the gravitational attraction of two finite masses M and m (with $M \geqslant m$) in a circular, keplerian orbit around their centre of mass; the test body is assumed to have initial position and velocity, and therefore to move at all times, in the same plane as the two major masses. The three-body problem is thereby highly simplified and idealized; in the reference system of the centre of mass we have only two degrees of freedom, to be compared with $3(N - 1)$ for the general N-body problem. Nevertheless, the restricted problem has been extensively studied not only because of its simplicity: many of the important results we shall obtain in the following can be generalized or adapted to more complex cases (e. g., to the general 3-body problem with three finite masses attracting each other). Moreover, several interesting astronomical systems approximately fulfil the assumptions of the restricted problem. Three examples are an artificial probe in the earth-moon system, a minor planet in the sun-Jupiter system and a hypothetical planet about binary stars with small eccentricity. In the former two cases, the eccentricity of the relative orbit of the two major masses is about 0.05: thus the restricted problem is a rough, but physically illuminating approximation.

12.1. EQUATIONS OF MOTION AND THE JACOBI CONSTANT

Let us choose the units of mass, distance and time in such a way that G(m + M), the constant distance between the two massive bodies and the gravitational constant G are set to one. Then, by Kepler's third law (10.24) the constant angular velocity (or mean motion) n of the two massive bodies in an inertial reference frame is also 1. In these units m is the ratio of the smaller mass to the sum of the two masses. The distances of M and m from the centre of mass are, respectively, m and (1 − m). We now write the equations of motion in a reference system rotating with angular velocity n = 1 about the normal to the orbital plane through the centre of mass; m and M are at fixed positions on the x-axis (see Fig. 12.1). If (x, y) are the coordinates of the test mass, we have

$$\frac{d^2x}{dt^2} - 2\frac{dy}{dt} = x - \frac{(1-m)(x+m)}{r_1^3} - \frac{m(x-1+m)}{r_2^3}, \qquad (12.1_1)$$

$$\frac{d^2y}{dt^2} + 2\frac{dx}{dt} = y - \frac{(1-m)y}{r_1^3} - \frac{my}{r_2^3}, \qquad (12.1_2)$$

where $r_1 = \sqrt{[(x+m)^2 + y^2]}$ and $r_2 = \sqrt{[(x-1+m)^2 + y^2]}$ are the distances of the test mass from M and m, respectively (Fig. 12.1). In the left-hand sides we have added Coriolis' acceleration, leaving in the right-hand sides the centrifugal acceleration and the gravitational pull of the two massive bodies. All the latter terms are functions of (x, y) only, and can be derived from the potential function

$$W(x,\ y) = -\tfrac{1}{2}(x^2 + y^2) - \frac{(1-m)}{r_1} - \frac{m}{r_2}\ , \tag{12.2}$$

so that eqs. (12.1) become

$$\ddot{x} - 2\dot{y} = -\ \partial W/\partial x\ , \tag{12.3$_1$}$$

$$\ddot{y} + 2\dot{x} = -\ \partial W/\partial y\ . \tag{12.3$_2$}$$

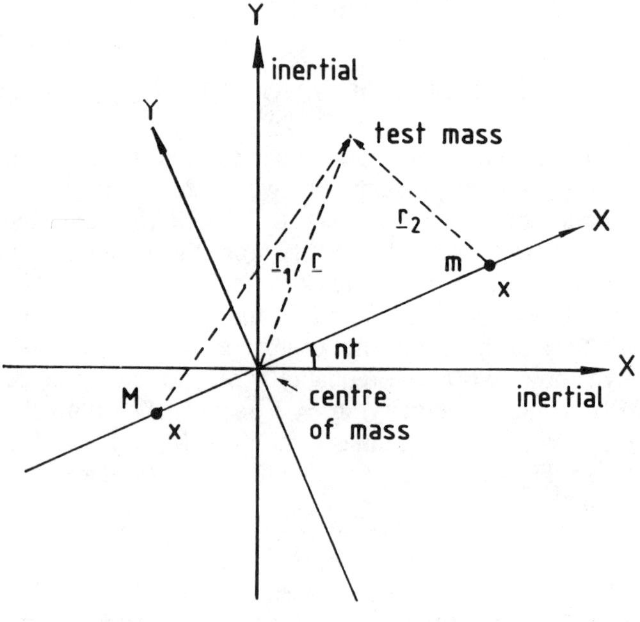

Fig. 12.1. The rotating reference system in which the equations of motion of the restricted three-body problem are written down.

If we multiply eqs. (12.3) by $2\dot{x}$ and $2\dot{y}$, respectively, and integrate with respect to time, we obtain

$$\tfrac{1}{2}(\dot{x}^2 + \dot{y}^2) = -W(x,\ y) + C\ , \tag{12.4}$$

where C is a constant of integration, determined by the initial conditions. The derivation of eq. (12.4) has an obvious analogy with the derivation of the conservation of mechanical energy for a point mass in a curl-free force field; however, the *Jacobi constant* C is *not* the energy, since we are in a rotating frame. Note also that the Coriolis force does not perform work, nor affects the conservation equation (12.4), while still influencing the motion in a crucial way. Eq. (12.1) is similar to the law of motion of a charged particle in a constant and uniform magnetic field and in a conservative electric field (see Sec. 9.6).

An important consequence of eq. (12.4) is that, since the "kinetic energy" term is never negative, the motion can occur only in the regions of the (x, y) plane where

$$W(x, y) \leqslant C. \tag{12.5}$$

This constraint depends on the value of the Jacobi constant, hence on the initial conditions. For a given value of C the equation $W(x, y) = C$ defines the boundary of the "allowed" region, i. e., the so-called *Hill's zero-velocity curves* (since the test mass can reach these curves only when its velocity vanishes in the rotating frame.)

To understand the topology of the "allowed" regions, note first that $-W$ becomes very large whenever $(x^2 + y^2)$ (i. e., the squared distance of the test mass from the origin) is large or, alternatively, when r_1 or r_2 are small. Therefore if $-C$ is large the test mass must remain either outside a large circle defined by $x^2 + y^2 = -2C$, or inside one of two very small circles surrounding the two massive bodies, defined by $r_1 = (m - 1)/C$ and $r_2 = -m/C$ (see Fig. 12.3, I). In the former case the test mass "feels" the two gravitating centres approximately as a single one, in the latter case only one of the masses is important. Since the three regions where the motion is allowed in this limiting case are disconnected, the test mass will remain for ever in the region where it is initially placed.

The planar, restricted three-body problem can easily be generalized to three dimensions by noting that both the centrifugal and the Coriolis accelerations have no z-component (normal to the orbital plane of the two massive bodies). A conservation equation like (12.4) still holds, provided we add $\dot{z}^2/2$ to the right-hand side and consider W as a function of z as well, through r_1 and r_2. The study of the topology of the allowed regions in three dimensions is straightforward, with results very similar to the planar case (see Sec. 12.4 and Problem 12.6).

12.2. LAGRANGIAN POINTS AND ZERO-VELOCITY CURVES

It is useful at this point to determine the equilibrium positions of the problem, namely the points where the test mass feels a zero net force, and thus can stay at rest in the rotating frame. We require that

$$-\nabla W = \mathbf{r} - (1 - m)\frac{\mathbf{r}_1}{r_1^{\,3}} - m\,\frac{\mathbf{r}_2}{r_2^{\,3}} = 0\ , \tag{12.6}$$

where $\mathbf{r} = (x, y)$, $\mathbf{r}_1 = (x + m, y)$, $\mathbf{r}_2 = (x - 1 + m, y)$ (see Fig. 12.1). But since the origin is at the centre of mass, we have

$$\mathbf{r} = (1 - m)\mathbf{r}_1 + m\mathbf{r}_2 \tag{12.7}$$

and, substituting into eq. (12.6), obtain:

$$(1 - m)\left[\frac{1}{r_1^{\,3}} - 1\right]\mathbf{r}_1 + m\left[\frac{1}{r_2^{\,3}} - 1\right]\mathbf{r}_2 = 0 \ . \tag{12.8}$$

If the equilibrium points are not on the x-axis, eq. (12.8) is satisfied only for $r_1 = r_2 = 1$, namely at the two *equilateral triangular points* which are at the same distance from the two massive bodies.

On the other hand, if we look for solutions of eq. (12.8) lying on the x-axis, it is convenient to use the dimensionless variable $\alpha = x + m$, leading to the algebraic equation

$$(1 - m)\frac{\alpha}{|\alpha|^3} + m\,\frac{(\alpha - 1)}{|\alpha - 1|^3} = \alpha - m \ . \tag{12.9}$$

$\alpha = 0$ and $\alpha = 1$ correspond, respectively, to the cases in which the test body is at M or at m. It can be shown that eq. (12.9) always has three real solutions; we shall determine them in the limiting case $m \ll M$, i. e., $m \ll 1$, which is interesting for most solar system applications, such as the sun-Jupiter and the earth-moon pairs.

One obvious solution (L_3) corresponds to the test body almost diametrically opposite to the smaller mass m. We look for this solution by substituting $\alpha = -1 + O(m)$ into eq. (12.9) and obtain:

$$\alpha = -1 + 7m/12 + O(m^3) . \tag{12.10}$$

With one attracting body only, the test body is in equilibrium under the attraction from the main body and the centrifugal force; the presence of the second attracting body gives rise to a new, positive gravitational force and to an increase in the attraction of the primary, while the equilibrium point is slightly displaced to the left; the new forces are balanced by the increase in the centrifugal force.

Two more equilibrium positions are close to the smaller mass, whose perturbation changes the radius of the keplerian circular orbit around M. In this case, we have first to find out the order in m of the small quantity $\alpha \ll 1$. We tentatively set

$$\alpha = 1 + \lambda m^q, \tag{12.11}$$

with $q > 0$ and λ of the order of unity. Then eq. (12.9) reads

$$m^{(1-2q)}\lambda/|\lambda|^3 = 3\lambda m^q + O(m^{2q}) \tag{12.12}$$

and the two larger terms are of the same order for $q = 1/3$. In this case we get $|\lambda|^3 = 1/3$, and the two equilibrium points correspond to

$$\alpha = 1 \pm (m/3)^{1/3} + O(m^{2/3}) , \tag{12.13}$$

and their distance from the smaller mass decreases with m only proportionally to its cubic root. In Fig. 12.2 we have plotted the positions of the five equilibrium (or *lagrangian*) points, traditionally called $L_1,\ldots,L_5$, in the rotating reference system.

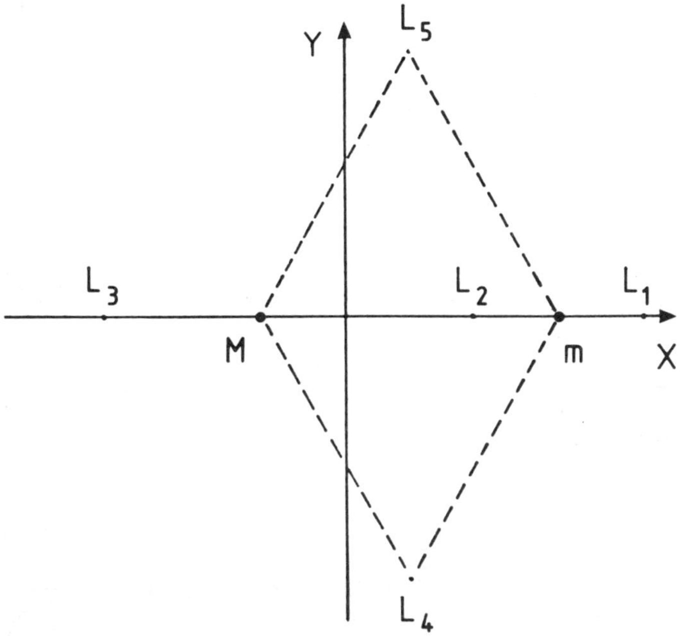

Fig. 12.2. The positions of the five Lagrangian equilibrium points in the rotating reference system. The two massive bodies m and M lie on the x-axis. The coordinates of the five points are:

$L_1 = [1-m+(m/3)^{1/3}, 0]$, $L_2 = [1-m-(m/3)^{1/3}, 0]$, $L_3 = [-1-5m/12, 0]$,

$L_4 = [1/2-m, -\sqrt{3}/2]$, $L_5 = [1/2-m, \sqrt{3}/2]$.

The x-coordinates of L_1 and L_2 neglect terms $O(m^{2/3})$, while for L_3 the neglected terms are $O(m^3)$.

To determine the shape of the zero-velocity curves $W(x, y) =$ constant, it is useful to determine the values of W at the five stationary points $L_1,\ldots,L_5$ and the topological character of these points. This can be done in a straightforward way by computing the second derivatives of W and determining the sign of the Jacobian determinant

$$J = (\partial_{xx}W)(\partial_{yy}W) - (\partial_{xy}W)^2$$

at the lagrangian points $L_1, \ldots, L_5$ (for brevity, hereinafter we shall use the notation ∂_{xx} for $\partial^2/\partial x^2$, etc.). These computations show that

$$W(L_5) = W(L_4) > W(L_3) > W(L_1) > W(L_2), \qquad (12.14)$$

and also that L_4, L_5 are two maxima, while L_1, L_2 and L_3 are saddle points. These results allow us to determine unambiguously the behaviour of the zero-velocity curves when the Jacobi constant C increases from very large negative values (i. e., the limit we discussed earlier) through the critical values given by eq. (12.14) (see Fig. 12.3).

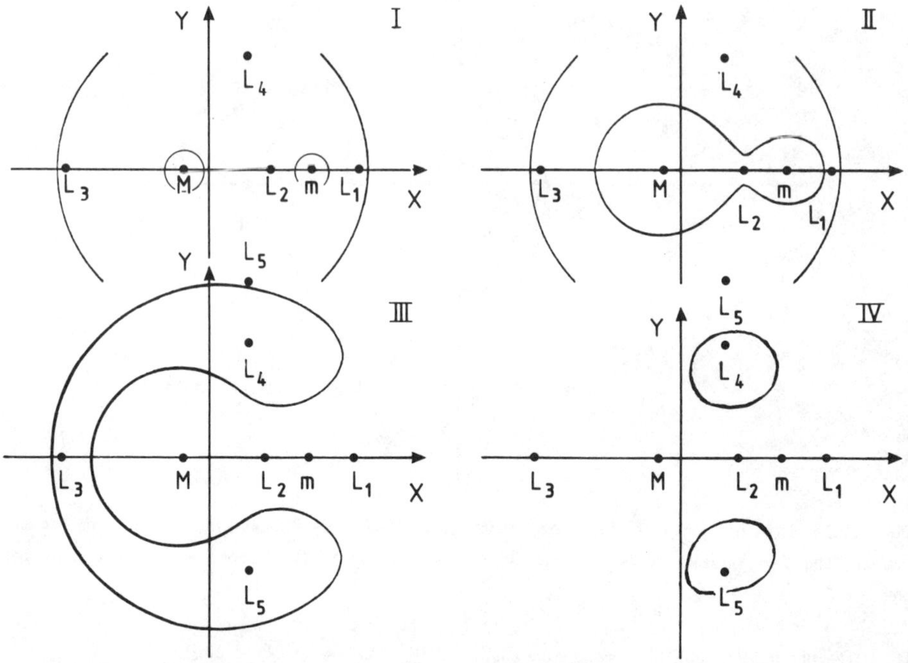

Fig. 12.3. The four different topologies of the zero-velocity curves in the restricted three-body problem for increasing values of the Jacobi constant C. The positions of the two masses M and m and of the five Lagrangian points are indicated. The fifth case, when the motion is allowed in the whole (x, y) plane, is not shown in the figure.

Qualitatively, we have five possibilities:

(I) $C < W(L_2)$: the motion is allowed only within two small regions surrounding the massive bodies or outside a large boundary including both of them (in the former case the test mass can be viewed as a satellite of the pair).

(II) $W(L_1) > C > W(L_2)$: a channel opens up at L_2 between the two allowed regions about the massive bodies, but the test body cannot escape very far from them if it starts in their vicinity. (Satellites can be exchanged, but not ejected from the system; this topology explains also why most earth-to-moon spacecraft trajectories have an eight-like shape, with the crossing point in the neighbourhood of L_2.)

(III) $W(L_3) > C > W(L_1)$: the allowed region opens up behind the smaller mass, and the test body can pass through this hole either to escape from or to penetrate into the neighbourhood of the massive bodies. The forbidden zone is shaped like a horseshoe, with two tops at L_4 and L_5 and a ridge through L_3.

(IV) $W(L_5) = W(L_4) > C > W(L_3)$: a channel opens up also behind M. The zero-velocity curves become again disconnected and surround just the two equilateral points L_4 and L_5, which lie in the only regions of the plane where the motion is now forbidden.

(V) $C > W(L_4) = W(L_5)$: the motion is possible in the whole (x, y) plane.

This classification still holds also when m is not infinitesimal, with the three collinear lagrangian points just displacing along the x-axis as m changes, to satisfy eq. (12.9)). Even when the orbit of the two massive bodies has not too large an eccentricity, a classification in terms of the Jacobi constant is useful to characterize the *hierarchical stability* of the system, namely the possibility that a hierarchical structure like a planet-satellite pair plus a distant disturbing body can be at some time destroyed by perturbations. This is obviously interesting for the cases of natural satellites and asteroids. The value of the Jacobi constant can also discriminate whether the ejection to infinity of the test mass is possible, or not (e.g., whether an interplanetary probe – which generally moves essentially in the field of a planet and the sun – can or cannot leave the solar system).

As we already noted in Sec. 12.1 (see also Sec. 12.4), all the previous results can easily be generalized to the three-dimensional case.

12.3. STABILITY OF THE LAGRANGIAN POINTS

Are the five lagrangian points stable? In other words, were the test mass initially placed at a small distance from one of them, will it remain close to the equilibrium position for ever, or will it get farther and farther from it? Were it not for the presence of the velocity dependent Coriolis' force, the answer would be already provided by the properties of the second derivatives of the potential function W: no lagrangian point would be stable, since none of them corresponds to a minimum of the potential. But we shall show in a moment that the Coriolis force modifies this conclusion in an essential way. Let (x_0, y_0) be one of the five equilibrium points and assume that the test mass starts close to this position. Then we can write

$$x = x_0 + \delta x \ , \qquad y = y_0 + \delta y \ , \tag{12.15}$$

where the δ quantities are assumed to be infinitesimal. Substituting (12.15) into (12.3) and keeping only the first-order terms, we obtain the linearized equations of motion

$$\delta\ddot{x} - 2\delta\dot{y} + \partial_{xx}W\delta x + \partial_{xy}W\delta y = 0 \tag{12.16$_1$}$$

$$\delta\ddot{y} + 2\delta\dot{x} + \partial_{yy}W\delta y + \partial_{xy}W\delta x = 0 \ . \tag{12.16$_2$}$$

The second derivatives of W are computed at the lagrangian point. This linear system can easily be solved by putting

$$\delta x(t) = a\cdot\exp(\sigma t), \qquad \delta y(t) = b\cdot\exp(\sigma t), \tag{12.17}$$

where σ, a and b are complex numbers. The condition for stable motion is obviously that the real part of σ is $\leqslant 0$, corresponding either to asymptotic approach to the equilibrium point or to a two-dimensional harmonic oscillation about it. On the contrary, when the real part of σ is positive, the displacement from (x_0, y_0) increases exponentially with time. We obtain the following algebraic relations for a and b:

$$a\sigma^2 - 2b\sigma + a\partial_{xx}W + b\partial_{xy}W = 0, \tag{12.18$_1$}$$

$$b\sigma^2 + 2a\sigma + b\partial_{yy}W + a\partial_{xy}W = 0. \tag{12.18$_2$}$$

This homogeneous system is solvable only provided the determinant of the coefficients vanishes, namely,

$$\sigma^4 + (\partial_{xx}W + \partial_{yy}W + 4)\sigma^2 + \partial_{xx}W\partial_{yy}W - (\partial_{xy}W)^2 = 0. \tag{12.19}$$

The discriminant of this equation (in the unknown σ^2) is

$$\Delta = (\partial_{xx}W + \partial_{yy}W + 4)^2 - 4J \ , \qquad (12.20)$$

where J is the Jacobian determinant of W computed at the equilibrium point. Since the three collinear points are of the saddle type, for them $J < 0$ and therefore eq. (12.19) has two non-vanishing real solutions for σ^2, with opposite signs. The positive one corresponds to instability. At the two triangular points, on the other hand, eq. (12.19) becomes

$$\sigma^4 + \sigma^2 + 27m(1 - m)/4 = 0 \ , \qquad (12.21)$$

so that

$$\Delta = 1 - 27m(1 - m) . \qquad (12.22)$$

If Δ is negative, we get complex solutions for σ^2 and two solutions for σ having a positive real part, hence giving rise to instability. But if $\Delta \geqslant 0$, the solutions for σ^2 are real and negative, σ is always imaginary and we obtain stable oscillations. It is easy to show that the condition $\Delta \geqslant 0$, with Δ given by (12.22), is equivalent to the condition

$$m \leqslant m_{cr} \cong 0.0385 \qquad (12.23)$$

(since $m \ll 1/2$, eq. (12.22) shows that $m_{cr} \cong 1/27$). Thus L_4 and L_5, though being maxima of the function W, are stable equilibrium points when the mass ratio m is small enough. The condition (12.23) is satisfied both for the sun-Jupiter and for the earth-moon system, for which m is equal to about 1/1047 and 1/81, respectively; however, in practice the stability of L_4 and L_5 is determined also by the extent to which the basic assumptions of the restricted three-body problem are satisfied (e. g., finite eccentricity, other perturbing masses, etc.). In the case of Jupiter's triangular lagrangian points, their stability is empirically proven by the existence of many asteroids (collectively called *Trojans*), which oscillate around two points lying on Jupiter's orbit, leading and trailing the planet by 60°. The amplitude of these oscillations is generally quite sizable. No similar body has been found near the triangular points of the earth-moon system; on the other hand, several saturnian moons have small, Trojan-like companions.

*12.4. TISSERAND'S INVARIANT

The conservation of the Jacobi integral C in the restricted three-body problem is useful also because it provides a constraint on the orbital evolution of the test particle due to the gravitational perturbation from the smaller mass. These orbital changes can be strong when close encounters occur. This is, for instance, the case of comets, for which orbital changes at close encounters with Jupiter can be so drastic that identifying the same comet at different apparitions is sometimes not straightforward. But we can apply the restricted three-body problem to the sun-Jupiter-comet case (neglecting the effects of Jupiter's eccentricity) and compare the pre- and post-encounter orbits of the comet by assuming the constancy of the Jacobi integral. For this purpose, the latter quantity has to be expressed as a suitable function of the comet's heliocentric orbital elements. We have then to switch to a non-rotating reference system. Let (x, y, z) and (x_0, y_0, z_0) be the coordinates of the comet in the frame rotating with Jupiter's mean motion (n) and in the non-rotating frame (with the same origin in the centre of mass), respectively. Then we have

$$\dot{x}^2 + \dot{y}^2 = \dot{x}_0{}^2 + \dot{y}_0{}^2 - 2n(x_0\dot{y}_0 - y_0\dot{x}_0) + n^2(x^2 + y^2) \quad (12.24)$$

and therefore, from eqs. (12.2) and (12.4), we get for the Jacobi integral C (in three dimensions):

$$\tfrac{1}{2}(\dot{x}_0{}^2 + \dot{y}_0{}^2 + \dot{z}_0{}^2) - n(x_0\dot{y}_0 - y_0\dot{x}_0) = \frac{GM}{r_1} + \frac{Gm}{r_2} + C, \quad (12.25)$$

where M, r_1, are now referred to the sun and m, r_2 to Jupiter. Since the sun-Jupiter mass ratio (1047) is large, we can neglect the displacement of the sun from the centre of mass and express the left-hand side of eq. (12.25) as a function of the instantaneous (osculating) heliocentric orbital elements of the comet. From the two-body energy and angular momentum integrals we get

$$\tfrac{1}{2}(\dot{x}_0{}^2 + \dot{y}_0{}^2 + \dot{z}_0{}^2) = GM\left[\frac{1}{r_1} - \frac{1}{2a}\right], \quad (12.26_1)$$

$$x_0\dot{y}_0 - y_0\dot{x}_0 = \sqrt{[GMa(1 - e^2)]}\cos I . \quad (12.26_2)$$

The inclination I is referred to Jupiter's orbital plane. Substitution of these expressions in eq. (12.25), together with Kepler's third law applied to Jupiter's orbit ($n^2 a_J^3$ = GM, again neglecting terms of order m/M), yields

$$T \equiv \frac{a_J}{a} + 2\left[\frac{a(1 - e^2)}{a_J}\right]^{\frac{1}{2}}\cos I = -\frac{2a_J}{GM}\left[\frac{Gm}{r_2} + C\right]. \quad (12.27)$$

If the distance r_2 of the comet from Jupiter is not much smaller than Jupiter's semimajor axis a_J, the term proportional to m/M in the right-hand side of this equation can also be neglected, hence $T \cong -2a_JC/GM$ is approximately constant and called *Tisserand's invariant*. Therefore, if in two different apparitions the comet is far enough from Jupiter, the expression in the left-hand side of eq. (12.27) has to approximately keep the same value, regardless of any close approach to Jupiter occurred in the meantime (which may have substantially modified the individual elements). The constancy of Tisserand's invariant is useful (in the frame of the restricted three-body problem) whenever one needs some constraint on the orbital changes that may be caused by close encounters.

Tisserand's invariant can also be interpreted in another way. Assume that the orbit of the comet crosses the circular orbit of Jupiter and let us compute the relative velocity at intersection. When $r = a_J$, the transverse and radial component of the comet's velocity are given by

$$v_t = \frac{h}{a_J} = \left(\frac{GM}{a_J}\right)^{\frac{1}{2}} \left[\frac{a}{a_J}(1 - e^2)\right]^{\frac{1}{2}}, \qquad (12.28_1)$$

$$v_r = (v^2 - v_t^2)^{\frac{1}{2}} = \left(\frac{GM}{a_J}\right)^{\frac{1}{2}} \left[2 - \frac{a_J}{a} - \frac{a}{a_J}(1 - e^2)\right]^{\frac{1}{2}} \qquad (12.28_2)$$

(see eqs. (10.28, 26).) The transverse component v_t is of course tangent to the planet's circular orbit, and can be decomposed into a part ($v_t \cos I$) in the orbital plane of Jupiter and in a part ($v_t \cdot \sin I$) orthogonal to it. If we subtract from the former component Jupiter's circular velocity $\sqrt{(GM/a_J)}$ (neglecting again terms of the order of m/M), we obtain the magnitude of the relative velocity at the encounter $\mathbf{v}_{rel}$:

$$v_{rel}^2 = \left[v_t \cos I - \left(\frac{GM}{a_J}\right)^{\frac{1}{2}}\right]^2 + (v_t \sin I)^2 + v_r^2 =$$

$$= \frac{GM}{a_J}\left\{3 - \frac{a_J}{a} - 2\left[\frac{a}{a_J}(1 - e^2)\right]^{\frac{1}{2}} \cos I\right\} = \frac{GM}{a_J}(3 - T). \qquad (12.29)$$

Of course, this relative velocity does not take into account the acceleration due to Jupiter's gravitational field, namely it is the velocity "at infinity" with respect to Jupiter, due to the heliocentric motion of the comet. Thus the conservation of Tisserand's invariant simply corresponds to the fact that an encounter can just rotate the direction of the vector $\mathbf{v}_{rel}$, but not change its magnitude since, when Jupiter's gravitational field is predominant, the Jovicentric path of the comet is well approximated by a keplerian hyperbola. Given the initial value of T (or of v_{rel}), a sequence of encounters changes the cometary heliocentric elements because at each intersection the comet's velocity can be expressed as the vector sum between Jupiter's circular velocity and the vector $\mathbf{v}_{rel}$, of variable orientation. Of course, these conclusions are valid only in the (rough) approximation of neglecting the effects of Jupiter's eccentricity.

PROBLEMS

12.1. Show that the equations of motion of the restricted three-body problem are the same (apart from the meaning of the constants) as for a charged particle moving in an electrostatic potential W(x, y) and in a uniform magnetic field directed along the z-axis.

12.2. Compute to first order in m the values of the function W(x, y) at the five lagrangian points.

12.3. What is the position of the five lagrangian points in the case m = 1/2, i.e., when the two massive bodies are equal?

*12.4. Show that the equilateral triangular configuration corresponds to equilibrium also for the general three-body problem, for every choice of the three (finite) masses, provided the system rotates rigidly around its centre of mass with angular velocity $\sqrt{(GM_{tot}/d^3)}$ (M_{tot} is the sum of the masses and d is the distance between two of them). Can this result be generalized to four or more unequal masses lying at the vertexes of regular polygons?

*12.5. Consider the planar, restricted sun-Jupiter-asteroid problem, and assume that an asteroid starts on a keplerian circular heliocentric orbit of radius a_0 on the sun-Jupiter line. Compute the Jacobi constant of the asteroid (neglecting terms proportional to the square of the Jupiter-to-sun mass ratio) and evaluate the maximum value of a_0 for which the asteroid is not allowed to cross Jupiter's orbit. For a generic value of a_0, less than the upper limit quoted above, what is the maximum heliocentric distance the asteroid can ever reach ?

12.6. Discuss the topology of the zero-velocity surfaces for the three-dimensional, restricted three-body problem.

12.7. Assume that Tisserand's invariant of a comet with respect to Jupiter is equl to 2.9. If the effects of Jupiter's eccentricity are neglected, what are the minimum perihelion distance and the maximum aphelion distance and inclination (with respect to Jupiter's orbital plane) that the comet can ever reach? In general, for which values of T can a comet be ejected from the solar system following a close encounter with Jupiter?

12.8 Compute for the earth and for Jupiter the size of the so-called *Hill's lobe*, which extends from the planet to the L_1 (or L_2) lagrangian points of the restricted sun-planet-test particle problem.

*12.9 Consider the three-body problem (*Hill's problem*) in the case of three masses $m_1 \approx m_2 \ll M$, when the separation r between the two small masses is much less than the distance R between each of them and M. Write down the equations of motion in an inertial frame and show that, to first order in (r/R), one can separate the motion of the centre of mass *G* of the pair (m_1, m_2) and the relative motion of m_1 and m_2. If one assumes that *G* moves in a circular keplerian orbit about M, the equations of motion can be written in the rotating frame having the origin in *G*. Show that these equations have a form similar to eqs. (12.1) and that there is an integral of motion analogous to the Jacobi constant.

FURTHER READINGS

The most extensive treaty on the three-body problem is V. Szebehely's *Theory of Orbits*, Academic Press, New York (1967). See also A.E. Roy, *Orbital Motion*, Hilger, Bristol (1978), Ch. 5; F.R. Moulton, *An Introduction to Celestial Mechanics*, Dover, New York (1970), Ch. VIII; and E. Finlay-Freundlich, *Celestial Mechanics*, Pergamon Press, London (1958), Chs. I and II.

13. THE SUN AND THE SOLAR WIND

In this chapter we shall deal with the sun only as far as its effects on the solar system and its evolution are concerned; the sun as a star and its interior structure are more suitably treated in an astrophysical context. The solar radiation had a crucial role in the past in determining the evolution of the primeval nebula and, more recently, the structure of planetary atmospheres, and the origin and evolution of life; its spectrum and its time variation are discussed in Sec. 13.1. An intense flow of charged particles is heated and emitted at supersonic speed by the corona, with a mechanism which is still partly unknown; however, the fluid dynamical model of the solar wind is reasonably satisfactory. This wind pervades the whole solar system and interacts in different ways with the planets and their magnetic fields.

13.1 THE SUN AND ITS RADIATION

The sun is a gaseous mass $M_\odot$ of $2\cdot10^{33}$ g, 743 times heavier than the total planetary mass, with a radius $R_\odot \simeq 7\cdot10^{10}$ cm and a mean density 1.4 g/cm^3. This high density is consistent with the gaseous state because of the very high pressure and temperature in the interior. The structure of the sun is determined by the condition of hydrostatic equilibrium (1.6), which leads to an internal temperature of the order of $3\cdot10^{-9}$ erg $\simeq$ 2 keV (eq. (1.11)). One can arrive at this figure also from the gravitational binding energy $|W_\odot| \simeq 2\cdot10^{48}$ erg (eq. (1.10)), about 10^{-6} times the rest energy $M_\odot c^2 = 2\cdot10^{54}$ erg, assuming an approximate equality between the potential and thermal energies of the constituent particles. At the centre the density $\rho_0 \approx 100$ g/cm^3 is much higher than the mean value; the temperature and pressure there are $T_0 \approx 1.5\cdot10^7$ K and $P_0 \approx 10^{17}$ dyne/cm^2, respectively. The sun is mainly composed of ionized hydrogen and helium (respectively 80% and 18% by weight).

The internal temperature results from a delicate balance between the nuclear fusion reactions in the core, which have a very strong temperature dependence; and the large opacity of the material around, which makes the loss of photons at the surface a very slow process. The large temperature gradient induces convection (Sec. 5.4), which contributes appreciably to the energy transport. This balance regulates the nuclear combustion processes in the sun, which, starting from the centre, slowly turns hydrogen into helium and heavier nuclei. The basic fusion process of four protons into an alpha particle (the "*pp cycle*") releases about 0.3% of the initial rest energy; it is easily seen that the energy available in this way is sufficient to ensure the current luminosity $L_\odot \simeq 4 \cdot 10^{33}$ erg/s for a time much longer than the present age of the sun, about $4.5 \cdot 10^9$ y. The energy generated in the fusion process ultimately determines the radius of the sun through eq. (1.11), an important example of a macroscopic property determined by microphysics.

The escape velocity

$$v_e = \sqrt{(2GM_\odot/R_\odot)} \approx 600 \text{ km/s} \tag{13.1}$$

corresponds to a proton energy of $3 \cdot 10^{-9}$ erg $\simeq$ 2 keV $\simeq 2 \cdot 10^7$ K, much higher than the surface temperature T_s, about 5800 K. This shows the need of an acceleration process to produce the solar wind. The escape velocity is much smaller than c, corresponding to the fact that the radius $R_\odot$ is much larger than the gravitational radius $m_\odot$ (eq. 1.15)). Relativistic effects in the solar system, at a distance r from the sun, are of order $m_\odot/r$ (see Ch. 17); on the surface $m_\odot/R_\odot \simeq 2 \cdot 10^{-6}$.

The surface of the sun rotates with an angular speed ω, which depends on the colatitude θ; one can take the model profile:

$$\omega \simeq 3 \cdot 10^{-6}(1 - 0.2\cos^2\theta)\text{rad/s}. \tag{13.2}$$

The axis makes an angle of 7° 15' with the pole of the ecliptic. The radial dependence of ω is not known; but recent results from the spectral analysis of the solar oscillations suggest a mild increase with depth, up to a factor 2 in the centre. Neglecting differential rotation, the rotational energy is $\simeq 1.3 \cdot 10^{43}$ erg and the angular momentum is $\simeq 9 \cdot 10^{48}$ g cm²/s. The deformation parameter (eq. (3.7)) is

$$\mu = \omega^2 R_\odot^3/GM_\odot = (\omega R_\odot/c)^2(R_\odot/m_\odot) \simeq 2.3 \cdot 10^{-5}. \tag{13.3}$$

The quadrupole moment J_2 is much less, due to the mass concentration near the centre. In the standard, rigidly rotating model of the sun, $J_2 \simeq 1.8 \cdot 10^{-7}$; its direct measurement is an important, still unsolved problem (see Sec. 20.6).

The sun emits $L_\odot = 4 \cdot 10^{33}$ erg/s, with a spectrum (see Fig. 13.1) very near the black body, corresponding to a surface temperature T_s = 5800 ± 100 K (see eq. (7.7)). The radiation flux $\Phi = L_\odot/4\pi r^2$ is, like the gravitational force, inversely proportional to the square of the distance r. In the optical band the radiation is emitted from the uppermost layer of the solar atmosphere, within an optical depth of order unity (the *photosphere*, about 250 km deep). The main microscopic processes responsible for the opacity there are excitations and bound-free transitions, in particular photo-ionization of H and H^-. The solar radiation also shows a marked limb-darkening effect (see Sec. 7.3). The solar optical spectrum shows many small dips corresponding to *absorption lines*, the *Fraunhofer spectrum.* These lines (about 25,000) arise because of resonant photon absorption, corresponding to bound-bound processes in the photosphere. Their formation can be quantitatively described by the photospheric opacity $\kappa(\nu)$, which has peaks near the absorption frequencies. The radiation temperature at a given optical thickness is smaller near a resonance because it corresponds to a smaller depth, producing a weaker intensity (see Sec. 7.2). The sun is also a powerful ultraviolet and X-ray source; the corresponding, effective black body temperature is 10^5 to 10^6 K. These emissions vary very much according to solar activity and the 11 y cycle. In the radio, we have strong bursts and a steady emission as well.

The solar optical spectrum provides essential information about the abundances of the different atomic species and the composition of the primordial solar nebula. The elements heavier than helium are all produced by nucleosynthesis in the burning core of stars, in particular of the sun; the complexity of nuclear reactions and their resonances results in the actual distribution of abundances. Some helium was formed in the cosmological nucleosynthesis, near the big bang. Note also that the sun, a second or third generation star, was made up with material already processed in the interior of other stars. We report below the table of atomic abundances in the solar photosphere of the 13 most abundant (by number) atomic species. If n is the number density and m the atomic

Atomic number	Element	Atomic mass	Numerical abundance	Abundance by mass
1	H	1	1	1
2	He	4.0	$5.88\cdot10^{-2}$	0.23
8	C	12.0	$3.55\cdot10^{-4}$	$4.20\cdot10^{-3}$
6	O	16.0	$6.16\cdot10^{-4}$	$9.86\cdot10^{-3}$
7	N	14.0	$9.55\cdot10^{-5}$	$1.30\cdot10^{-3}$
10	Ne	20.2	$7.41\cdot10^{-5}$	$1.50\cdot10^{-3}$
14	Si	28.1	$3.02\cdot10^{-5}$	$8.48\cdot10^{-4}$
12	Mg	24.3	$2.57\cdot10^{-5}$	$6.25\cdot10^{-4}$
16	S	32.1	$1.86\cdot10^{-5}$	$5.97\cdot10^{-4}$
26	Fe	55.8	$6.60\cdot10^{-6}$	$3.69\cdot10^{-4}$
13	Al	27.0	$2.51\cdot10^{-6}$	$6.77\cdot10^{-5}$
20	Ca	40.0	$2.14\cdot10^{-6}$	$8.57\cdot10^{-5}$
11	Na	23.0	$1.51\cdot10^{-6}$	$3.48\cdot10^{-5}$

Table 13.1. The abundances of the elements in the solar photosphere.

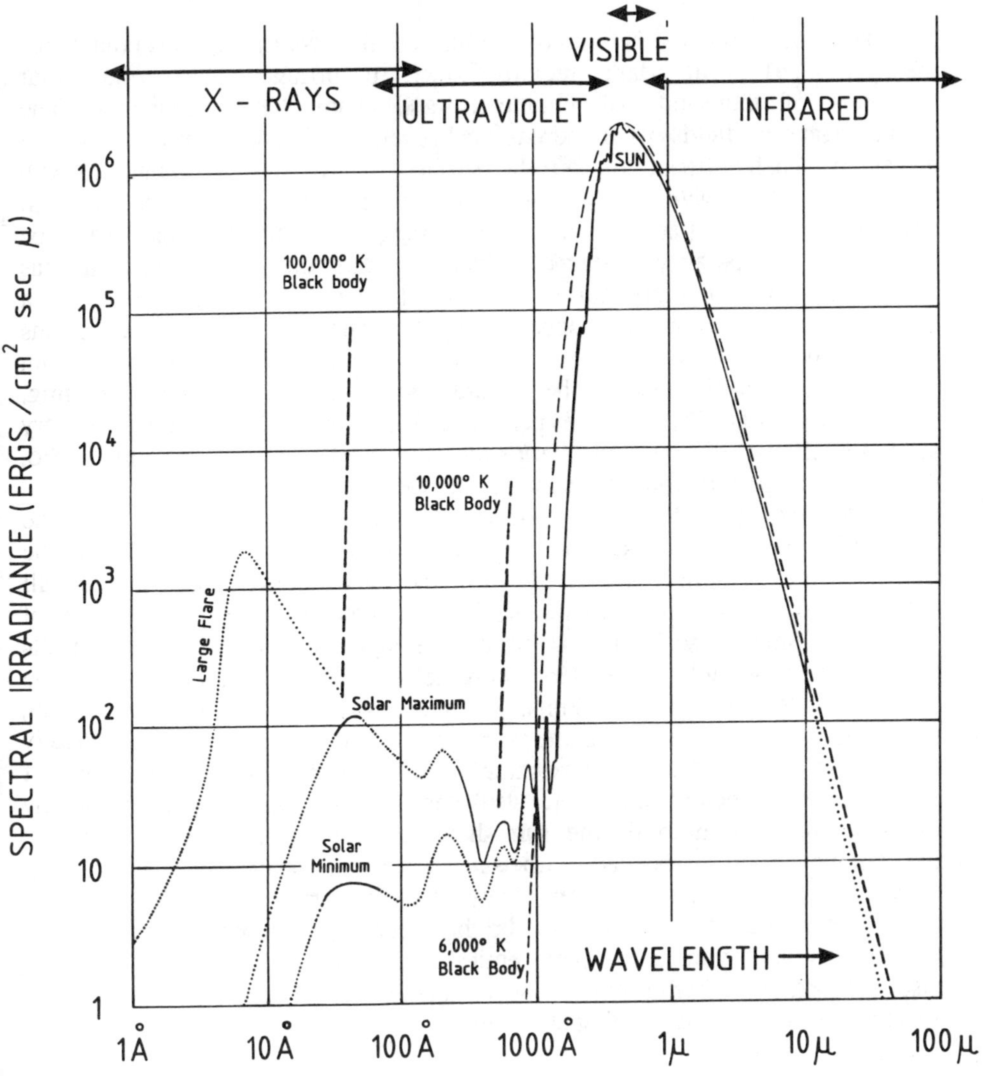

Fig. 13.1. The solar electromagnetic spectrum. For reference the dashed line indicates the black body spectrum corresponding to a surface temperature of 6000 K. In the UV and X bands the radiation is not thermal and very variable. (Adapted from H.H. Malitson, *The Solar Electromagnetic Environment*, Solar Energy, **12**, 197 (1968).)

mass, the last two columns give the abundance by number N/N_H and the abundance by weight mN/m_HN_H relative to hydrogen. This mix of nuclear species is produced by the nuclear reactions in the interior of stars, which at the end of their life throw out some of their content in interstellar space. It is remarkable that these abundances are in good agreement with theoretical predictions on nuclear reactions and stellar evolution.

The *solar activity* is varied and spectacular. *Sunspots*, observed since the year 1611, are dark regions, often occurring in a group with diameters of thousands of kilometers and more. Through the Zeeman effect magnetic fields of hundreds and even thousands gauss have been measured. Their distribution on the surface of the sun is connected with the 11 y *solar cycle*. They usually come in pairs of opposite polarity, aligned along a parallel, and are dragged along by the rotation; moreover, the polarity sequence changes from one hemisphere to the other. The spots initially form at high latitudes and, as the solar cycle advances, move toward the equator. By the end of the cycle, when the spot number is near its minimum, they are at latitudes of about 7°. When a new cycle begins the polarity sequence reverses: the complete cycle really lasts 22y. In the period between about 1645 and 1715 the sunspots almost disappeared (*Maunder mimimum*); it is very remarkable that in this period the temperature of the earth was lower than usual (the *little ice age*). We do not know whether this correlation is due to a a definite physical cause, nor do we know its possible mechanism. Sunspots are certainly connected with the large convective cells in the solar interior, which drive to the surface magnetic tubes.

On a smaller scale and at weaker strength the whole surface of the sun shows a granulation pattern with cells of, say, a thousand km and random velocities of 1 km/s. Sometimes a magnetic field loop protrudes out of this turbulent surface and forms a *bipolar magnetic region*, increasing fast in size until the field wanes out. Regions of magnetic enhancement are seen also near the *faculae*, bright regions in the H_{α} line. The limb of the sun shows dark filaments pushing out into the corona, often associated with sunspots (*prominences*). *Flares* are the major aspect of solar activity. They are observed as rapidly evolving (with time scales of hours), bright regions, extending considerably horizontally and vertically. They emit significant radiation in the X-band also. The large, moving plumes seen against the sky in an eclipse are a spectacular manifestation of solar flares.

13.2. MATHEMATICAL THEORY OF THE SOLAR WIND

The *solar wind* is a supersonic, radial flow of charged particles which carries along the solar magnetic field and exerts a profound influence upon the magnetospheres of the planets and on the comets. At 1 AU it is mainly composed of protons and electrons (together with helium, 23% by weight), with a density of, say, 5 protons/cm^3 and a temperature of 10^5 K. Its velocity V, about 400 km/s, corresponds to a Mach number of about 10. It carries along a magnetic field of about $5 \cdot 10^{-5}$ Γ.

The solar wind is a general phenomenon which appears every time the atmosphere of a gravitating body has sufficient heat conductivity and occurs with stars and galaxies as well. We review the model due to E.

N. Parker. In a thin plane atmosphere, with a given temperature gradient T(z), the pressure profile for a perfect gas in static equilibrium (eq. (7.1)) is proportional to

$$\exp\left[-\int_0^z dz \frac{gm}{kT(z)}\right] .$$

g is the acceleration of gravity and m the mean molecular mass. A finite atmosphere (i. e., when the pressure tends to zero for $z \Rightarrow \infty$) exists only if $T(z) = O(z^n)$, with $n \leq 1$; in particular, an isothermal atmosphere is finite. When the spherical shape of the body is taken into account, however, things are different. The equilibrium condition (1.6) for a perfect gas can be integrated for a given T(r):

$$P(r) = P(R)\exp\left[-\int_R^r dr' \frac{GMm}{kT(r')r'^2}\right] . \qquad (13.4)$$

We have a finite, static atmosphere only if the exponent diverges at large r, that is, if $T(r) = O(r^n)$, with $n \leq -1$; thus any process which makes the temperature gradient less steep favours the onset of a wind.

This is the case of the sun, around which the corona is dynamically heated from below by turbulent processes and the plasma ensures a high thermal conductivity. We must therefore allow a radial flow, whose conservation requires

$$4\pi\rho(r)v(r)r^2 = 4\pi\rho_0 v_0 R^2 = -dM/dt , \qquad (13.5)$$

in terms of the mass loss $-$ dM/dt. With the suffix 0 we indicate the value at the surface r = R. The conservation of momentum requires, in an adiabatic flow with the sound speed $c_s(r)$ (eq. (1.49)),

$$v\frac{dv}{dr} + c_S\frac{d\ln\rho}{dr} + \frac{GM}{r^2} = 0 . \qquad (13.6)$$

Using eq. (13.5) this reads

$$v\frac{dv}{dr}\left(1 - \frac{c_S^2}{v^2}\right) = -\frac{GM}{r^2} + \frac{2c_S^2}{r} \equiv H(r) . \qquad (13.6')$$

When the temperature does not vary much, the function H(r) near the sun is dominated by the first term and is negative there, but becomes positive as r increases. Let r_c be the point where H vanishes. When the sound speed is constant, in terms of the escape velocity (13.1),

$$r_c = 4R(v_e/c_s)^2 . \tag{13.7}$$

We can guess the general features of the solution by looking at the sign of the derivative dv/dr in the (r, v) plane (Fig. 13.2), which changes at $r = r_c$ and at the sonic value $v = v_T$. Starting at the surface with a low, subsonic value v_0, the velocity cannot penetrate in region IV where it would be double-valued. Region II is also excluded because the pressure would diverge at large distances; indeed, integrating eq. (13.5)/P we find that at large distances the velocity must be increasing. Therefore the appropriate solution is the one which crosses from I into III, becoming sonic just at the critical distance r_c; beyond it the velocity increases slowly. The time-reversed solution ($v < 0$) corresponds to an inflow of gas and is important for the accretion problem (see Sec. 16.6).

For a polytropic flow (cq. (1.52)), eq. (13.6) can be integrated and gives *Bernoulli's equation*

$$\frac{v^2}{2} + \frac{\gamma}{\gamma - 1}\frac{P}{\rho} - \frac{GM}{r} = \text{const}. \tag{13.8}$$

This, together with the mass conservation law (13.5), gives the pair (v, $c_s = \surd(\gamma P/\rho)$) at the distance r, with given initial conditions at the surface $r = R$:

$$v = v_0 \left(\frac{R}{r}\right)^2 \left(\frac{c_{s0}}{c_s}\right)^{2/(\gamma-1)}, \tag{13.9$_1$}$$

$$\frac{v^2}{2} + \frac{c_s^2}{\gamma - 1} = \frac{v_0^2}{2} + \frac{c_{s0}^2}{\gamma - 1} - \frac{v_e^2}{2}\left(1 - \frac{R}{r}\right). \tag{13.9$_2$}$$

In the plane (v, c_s) (Fig. 13.3), for a given r, the last equation is an ellipse, while (13.9$_1$) is a power function with negative exponent ($\gamma >$ 1!), whose coefficient is proportional to the mass loss. In both cases v is a monotonic function of c_s.

It can be shown that curves of the two families can have at most two intersections, in which case one is supersonic and the other

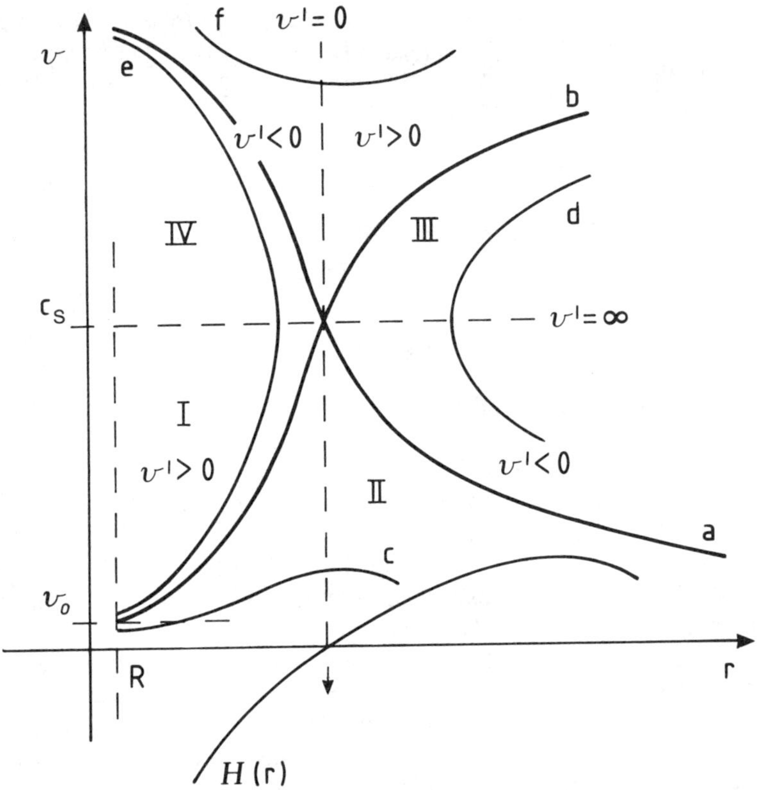

Fig. 13.2. The qualitative features of the solar wind solution (curve a). The curves b or c correspond to the accretion (see Sec. 16.5). The plots correspond to the isothermal case; for a polytropic flow the sound speed is not constant and depends on the solution, but qualitatively we have the same behaviour.

subsonic. If they have just one intersection, it is also a sonic tangent point ($v = v_s = c_s$). Therefore the behaviour of the intersections can be studied by looking at how the sonic point on each curve varies with the distance. On the ellipses the sonic point corresponds to the velocity

$$v_{s,ell} = \left[\frac{2(\gamma - 1)}{\gamma + 1}\left(\frac{c_{s0}^2}{\gamma - 1} + \frac{v_0^2}{2} - \frac{v_e^2}{2}\left(1 - \frac{R}{r}\right)\right)\right]^{\frac{1}{2}}, \quad (13.10_1)$$

and, as r decreases, grows from a finite value to infinity. On the adiabatic lines the sonic point sits at

$$v_{s,ad} = \left[v_0 \left(\frac{R}{r}\right)^2 c_{s0}{}^{2/(\gamma-1)} \right]^{(\gamma-1)/(\gamma+1)} . \qquad (13.10_2)$$

When

$$v_{s,ell} > v_{s,ad}$$

there are two solutions, one subsonic and one supersonic; when

$$v_{s,ell} < v_{s,ad}$$

there is none. From the explicit expression (13.10) of the two functions we see that we have three possibilities, depending on the boundary conditions: A) The two curves (13.10) have no intersections and the former inequality holds; we have a supersonic and a subsonic solution in the whole range (curves c and f of Fig. 13.2). B) The two curves intersect just once at $r = r_c$ and are tangent there, so that the foremr inequality holds everywhere, except at the tangent point; the solutions are the separatrices a and b. C) The two curves intersect twice and we have a pair of solutions below and above a given radius (curves e and d).

From the previous considerations one sees that we must be in case B); the solution is on the subsonic branch for $r < r_c$ and jumps to the supersonic branch for $r > r_c$. This case B) can occur only in the range $1 < \gamma < 5/3$. Equating $v_{s,ell}$ and $v_{s,ad}$ and their derivatives we can compute the critical distance r_c and the critical velocity v_c (Problem 13.2). Here we give their expression when the escape velocity v_e is much larger than v_0 and c_{s0}:

$$r_c = R \frac{5 - 3\gamma}{4(\gamma - 1)} , \qquad (13.11)$$

$$v_c = c_{sc} = \frac{v_e}{2} \left(\frac{R}{r_c}\right)^{\frac{1}{2}} = \frac{v_e}{2} \left(\frac{4(\gamma - 1)}{5 - 3\gamma}\right)^{\frac{1}{2}} . \qquad (13.12)$$

At this point the internal energy of the gas is comparable to its gravitational potential energy (i. e., the mean velocity is of the order of the escape velocity). At great distance ($r \gg R$), from eqs. (13.9), the wind velocity is constant and c_s decreases like $r^{1-\gamma}$, corresponding to the adiabatic law.

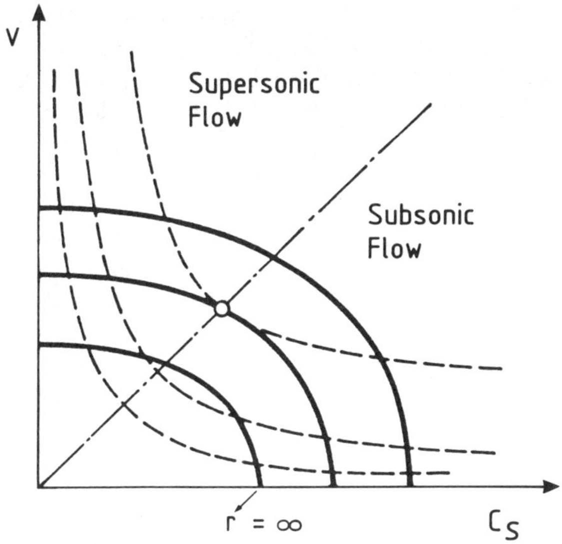

Fig. 13.3. The structure of the solar wind results from the intersections in the (c_s, v) plane between Bernoulli's law (continuous line) and the adiabatic law (dashed line).

The isothermal case corresponds to $\gamma = 1$ and is not included in the previous formulation. Eq. (13.6), however, integrates directly to

$$(v^2 - v_0^2)/2 + c_S^2 \cdot \ln(v_0 R^2/vr^2) - GM/r + GM/R = 0. \quad (13.13)$$

At large distances ($r << R$) the velocity grows very slowly.

The acceleration of the solar wind is similar to the conversion of thermal energy into the kinetic energy of a directed supersonic flow in a *Laval nozzle*. In this device a hot gas in a container leaks out through a pipe whose diameter first decreases and then, after the nozzle, increases. Just like gravity does near the sun, the narrowing section accelerates the flow to the sonic point at the nozzle; then the space available increases and so does the velocity (Problem 13.4).

13.3. STRUCTURE OF THE CORONA AND THE SOLAR WIND

As shown in Fig. 13.4, the temperature in the neighbourhood of the sun reaches a minimum just above the photosphere; beyond, it increases again by more than two orders of magnitude. This hot region – the *corona* – is visible, especially in an eclipse, up to a few solar radii, through the solar radiation scattered by free electrons and emission lines of highly ionized atoms. It is believed to be the cause of the solar wind. The process which is able to deposit far above the surface of the sun such large quantities of energy (along a direction in which the temperature increases !) is uncertain. Of course turbulence and explosive phenomena like flares are evidence of a large reservoir of energy in the photosphere in form of mass motion and magnetic fields far from thermal equilibrium; it is this reservoir, in turn driven by the convective instabilities below the surface, that we must tap to get what we need to heat the corona.

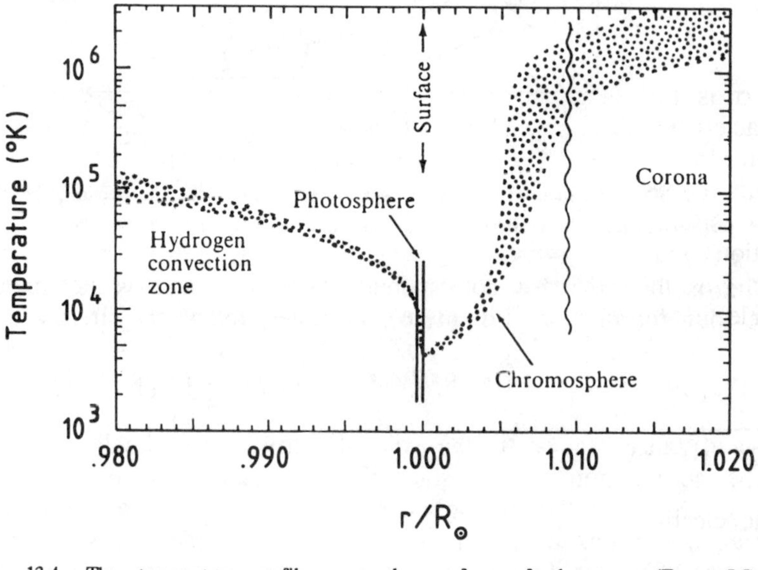

Fig. 13.4. The temperature profile near the surface of the sun. (From J.C. Brandt, *Introduction to the Solar Wind*, Freeman (1970).)

Important evidence about this mechanism recently came from the observation that recurrent, high-speed solar wind streams, often observed in interplanetary space, are correlated with low density, X-band dark regions on the sun (*coronal holes*). They are currently interpreted as regions of open field lines, which end up in interplanetary space; other regions, brighter in X-rays, are supposedly characterized by closed field lines, capable of confining, to some extent, a higher density plasma. This correlation points to the possibility that a high-speed blob of hot plasma is released and accelerated when its magnetic confinement is broken by some instability. A similar process was mentioned in Sec. 9.5 in connection with the regions in the magnetosphere where the field changes sign. In general, a magnetic transition produced by an inhomogeneous magnetic field is capable of transforming magnetic energy in particle and thermal energy in several ways. For example, a magnetic loop protruding out of the sun can confine a hot plasma in its interior; rotation and translation at the surface may produce magnetic transitions (i. e., by "knotting" two loops together) and release large amounts of energy. The study of these processes is largerly facilitated by space observations (in particular, in the X-band).

In the previous Section the theory of the solar wind was developed without reference to the magnetic field. The crucial dichotomy here is, whether resistivity is important in the solar wind flow. As discussed in Sec. 6.4, this is decided by the magnetic Reynolds number

$$R_{\ell m} = 4\pi\ell V\sigma/c^2 \; , \qquad (13.14)$$

where σ is the electrical conductivity, V is the solar wind speed and ℓ a characteristic size. Eq. (1.61) gives its value in terms of the electron collision frequency ν_c. ν_c depends on the cumulative effect of weak Coulomb collisions on an electron; they lead to a random walk in velocity space very similar to what is discussed in Sec. 16.4 in connection with the growth of planetesimals. A precise calculation for a *Lorentz gas*, in which the ions are at rest and electron-electron collisions are neglected, leads to the following value:

$$\nu_c = \frac{\pi^{3/2} n e^4}{2^{5/2}\sqrt{m_e}(kT)^{3/2}} \ln\left(\frac{b_1}{b_2}\right) = 1.5\cdot 10^{-6}(\text{ncm}^3)\left(\frac{\text{eV}}{kT}\right)^{3/2} \Lambda/\text{s}. \qquad (13.15)$$

n is the electron density. The logarithmic factor

$$\Lambda = \ln(b_1/b_2) \qquad (13.15')$$

arises, just like in Sec. 16.4 for the case of the gravitational interaction, because ν_c is an integral with respect to the impact parameter b, with the integrand function proportional to 1/b. b_1 and b_2 are, respectively, the largest and the smallest effective values of b; without them the collision frequency would diverge. Λ does not vary much and is about 20 in the solar corona. ν_c is proportional to the plasma frequency divided by the number of electrons in a cube whose side is the Debye length (but the numerical factor in eq. (13.15) is much less than unity):

$$\frac{\omega_p}{\lambda_D{}^3 n} = \left(\frac{4\pi n e^2}{m_e}\right)^{\frac{1}{2}} \left(\frac{kT}{4\pi n e^2}\right)^{-3/2} \frac{1}{n}$$

In the corona, where $n \approx 10^6/\mathrm{cm}^3$ and $T \approx 100$ eV, the plasma frequency is $6 \cdot 10^7\ \mathrm{s}^{-1}$, the Debye length is ≈ 1 cm and ν_c is about $0.6\ \mathrm{s}^{-1}$. From eq. (1.61) the conductivity is about 10^{15}/s and, with $\ell = 4R_\odot$, we get an enormous magnetic number (about 10^{15}). In these conditions the plasma is a perfect conductor and the magnetic field is frozen in. As shown in Sec. 1.6, this implies that if a small segment lies on a line of force, it continues to do so when dragged along by the flow. We show now that this implies a characteristic spiral shape of the lines of force in interplanetary space.

Consider, for simplicity, a radial flow $v(r)$ in the solar equatorial plane; there the magnetic field has only two components $B_r(r)$ and $B_\varphi(r)$. Note, by the way, that the absence of magnetic poles requires that $r^2 B_r$ be constant. Let

$$q(r) = \mathrm{tg}\psi(r) = r d\varphi/dr = B_\varphi/B_r \qquad (13.16)$$

define the angle $\psi(r)$ between the radial line and the magnetic field (see Fig. 13.5). Two points $P(r, \varphi)$ and $P'(r + r d\varphi/q,\ \varphi + d\varphi)$ lying on a line of force $r(\varphi)$ are displaced by the flow in a time dt into $P_1(r + v dt, \varphi)$ and $P_1'(r + r d\varphi/q + dt(v + r d\varphi v'/q),\ \varphi + d\varphi)$, respectively. At P_1 the line of force must lie along $P_1 P_1'$, so that

$$q(r + v dt) = q + v dt q' = \frac{(r + v dt) d\varphi}{r d\varphi/q + dt\, r d\varphi v'/q} \, .$$

Therefore q fulfils the simple differential equation

$$v q' = q(v/r - v') \, , \qquad (13.17)$$

immediately integrated into

$$q v/r = \mathrm{const.} \qquad (13.17')$$

The value of this constant depends on the conditions at the surface of the sun; its rotation, so to speak, drags along the lines of force. Let us make the simple assumption that at the surface the flow is parallel to the lines of force. Its velocity there, sum of a radial velocity v_0 and a rotational velocity $\Omega R_\odot$, makes with the radial direction an angle

$$\psi_0 = \mathrm{arctg}\ q_0 = \mathrm{arctg}(\Omega R_\odot / v_0);$$

hence

$$q = \mathrm{tg}\psi = \Omega r / v. \qquad (13.17'')$$

When v is a constant this is an Archimedes' spiral.

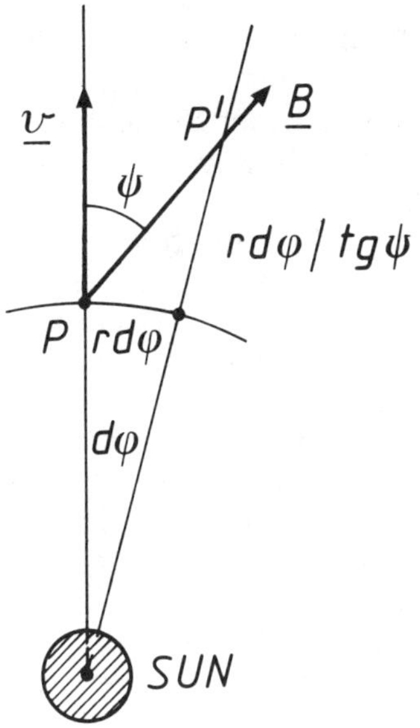

Fig. 13.5. The geometry of the solar wind.

This simplified picture neglects the dynamical effect of the magnetic field on the wind. In the equatorial plane, for example, an azimuthal component v_φ can be expected. The freezing-in condition (13.17) does not specify the sense and the intensity of the magnetic field, which are determined by the conditions at the surface and the way in which the local acceleration processes tear away the lines of force there. Experimentally, in the ecliptic plane we find sectors where the field has the same sign, separated by sharp transitions. The conditions on the surface of the sun also determine local streams and "gusts", with velocity peaks appreciably larger than the average. One can say, the solar wind maps into the heliosphere the state and the dynamics of the photosphere and the solar corona. An important aspect of this problem is the study of the turbulence of the solar wind by statistical methods, i. e., the determination of the relevant spectra.

From eq. (13.15) the collision frequency in the solar wind is about $5 \cdot 10^{-6}$/s, corresponding to a mean free path of, say, 1AU. The flow, therefore, is almost collisionless and the fluid approximation used here can only serve as a guidance and does not provide a rigorous description. Among the effects it fails to describe, the most important are those connected with the definition of temperature. In effect, not only the mean kinetic energy of the electrons and of the ions are different, but also their velocity distribution is not isotropic around their mean velocity and one cannot speak of a scalar pressure. Indeed, the pressure tensor is in general not isotropic, but axially symmetric around the direction of the local line of force, so that one can define a "temperature" $T_{\|}$ parallel to, and a "temperature" $T_{\perp}$ orthogonal to, the magnetic field. The really appropriate tool to describe the dynamics of the solar wind is the set of Vlasov equations (see Sec. 1.8) or, at best, the two-fluid approximation.

The solar system is embedded in an interstellar gas with density of about 1 hydrogen atom per cm^3 (there is also helium, at the cosmic abundance). Its state can be neutral, at a temperature, say, of 100 K; or ionized by the ultraviolet radiation of nearby stars and much hotter. In addition, there are magnetic fields of the order of 10^{-5} Γ and cosmic rays. The interaction of the solar wind with the interstellar gas is determined mainly by the pressure P_i of the latter. We do not have good measurements of the contribution of the four different components quoted above to P_i; moreover, they certainly are dependent on the direction. The magnetic pressure $B^2/8\pi$ for $B = 10^{-5}$ Γ is $4 \cdot 10^{-12}$ dynes/cm^2; a comparable contribution from the cosmic rays (which are confined by the galactic field) is expected; we can therefore assume for P_i something like 10^{-11} dyne/cm^2.

The expansion of the solar wind stops when its kinetic pressure (appreciably larger than its thermal pressure) equates the external pressure, i.e., when

$$nm_pV^2 = 2P_i \quad . \tag{13.18}$$

Assuming a constant $V = 400$ km/s and referring the density n to its value n_0 at 1AU, we obtain the distance of the termination (*heliopause*)

$$r_i = \left(\frac{n_0 m_p V^2}{2P_i}\right)^{\frac{1}{2}} \text{AU} = 25\text{AU}. \tag{13.19}$$

If the cosmic rays and the interstellar magnetic field are neglected we get a larger value. At this distance the supersonic solar wind cannot go further and suffers a shock transition to the much cooler interstellar gas. This shock is likely to be a strong one, with a compression ratio $(\gamma + 1)/(\gamma - 1) = 4$ (see eq. (9.45')).

13.4. PLANETARY MAGNETOSPHERES

From the available estimates of the magnetic dipole field of the planets (Table 6.1) it is possible to draw general conclusions about their magnetospheres. Let us adopt a solar wind model in which the radial flow velocity is constant, so that the plasma density is proportional to

$1/r^2$; with a polytropic constant $\gamma = 5/3$ the temperature decreases like $1/r^{4/3}$ and the solar wind properties are:

electron density $n = 5\ \mathrm{cm}^{-3}/r^2$
(ion) temperature $T = 2\cdot10^4\ \mathrm{K}/r^{4/3}$
wind velocity $v = 4\cdot10^7$ cm/s.

The distance r is measured in AU. The magnetic field is almost orthogonal to the radial direction (eq. (13.17"); since $B_r \propto 1/r^2$, taking into account eq. (13.16), we have a

magnetic intensity $B \simeq B_\varphi \simeq 5\cdot10^{-5}\ \Gamma/r$.

We see that the sonic and the alfvenic Mach's numbers (9.36) increase with the distance, respectively, like $r^{2/3}$ and r.

If a planet – like Venus – does not possess a substantial magnetic field, the solar wind is deflected by its surface, with a bow shock ahead of it, but no magnetosphere. A magnetic dipole $d = B/R^3$ produces a shock at the distance (9.37), where the internal magnetic pressure can withstand the kinetic pressure of the wind:

$$D = [d^2/(2\pi m_p n v^2)]^{1/6}\ . \tag{13.20}$$

We refer this quantity to the earth:

$$D/D_e = (rd/d_e)^{1/3}\ , \tag{13.21}$$

where $D_e \simeq 10R_e = 6.4\cdot10^9$ cm. When D is greater than the radius R of the planet a magnetosphere is present. This corresponds to a minimum surface magnetic field B_{min}, which we express in terms of that of the earth, B_e:

$$B_{min}/B_e \simeq 10^{-3}\ (AU/r). \tag{13.22}$$

This condition is marginally fulfilled for Mercury, which has a weak and thin magnetosphere. The outer planets Jupiter, Saturn, Uranus and Neptune have a much stronger magnetic field and remarkable magnetospheres. Their fast rotation makes the phenomenon of corotation very noticeable, extending the region III (Fig. 9.5) of closed field lines. In Uranus and Neptune the magnetic dipole makes a large angle (60° and 50°, respectively) with the rotation axes, making them *oblique rotators*, in which an electric field of order $Bc\omega/r$ is generated by the corotation; moreover, their inner magnetospheres are exposed twice every rotation period to the solar wind and its properties have a corresponding, periodic modulation. A smaller, but similar effect is present in Jupiter, whose dipole has an obliquity of 9.6°. In all the three giant planets trapped particles belts have been found. Low frequency radio emissions are observed, probably due to high energy magnetospheric particles; in particular, Jupiter has a strong radiation around 1 Mhz.

Io, the innermost galilean satellite of Jupiter, drastically affects its magnetosphere. A strong flux of particles released by the satellite is spread around its orbit, forming *Io's torus*, a cold (about 10 eV) plasma ring mainly made up of oxygen and sulfur. Its density (up to 2000 ions per cm^3 !) is large enough to hamper its dragging around by the rotating magnetic field of the planet; but the residual centrifugal force produces a radial outward drift of particles, which must be continuously replaced by the satellite. This striking feature may be connected with the volcanism of Io (Sec. 15.3); in this case a suitable mechanism (solar radiation, energetic particles, $\mathbf{v} \times \mathbf{B}$ drift) must be invoked to explain the ionization of the material ejected from the volcanoes. Moreover, its temperature must be large enough to escape the satellite's gravity. Another origin of Io's torus has been proposed by T. Gold, exploiting the very large potential drop (about 600 kV) across the satellite, produced by its orbital motion in the magnetic field of the planet. Gold has remarked that in some rocks the resistivity *decreases* with the temperature, facilitating the creation in the body of Io of narrow current ducts through which hot, melted material can flow to the surface and be ejected, just like in an ordinary electric arc. In this view the observed plumes are not of volcanic origin, but are due to these arcs.

PROBLEMS

13.1. Show that the two curves (13.9) have two intersections, one subsonic and one supersonic.

13.2. Draw the two curves (13.10), discuss the alternatives A), B) and C) and prove eqs. (13.11, 12).

13.3. Draw the plot of the solar wind speed in the isothermal regime (eq. (13.13)).

*13.4. Derive the theory of flow in a Laval nozzle. Find the cross-section as a function of distance which simulates the solar wind.

13.5. Evaluate the energy flux from the sun needed to support the coronal loss to the wind.

13.6. Find the cartesian equation of the Archimedes' spiral (13.17").

13.7. Model the ejection of material from the volcanoes of Io. Assuming a free flow, find the flow of escaped material per unit surface at a given temperature.

FURTHER READINGS

J.C. Brandt, *Introduction to the Solar Wind*, Freeman (1970) is an old, but still excellent book. E.N. Parker, *Dynamical Theory of the Solar Wind*, Space Science Rev. **4**, 666-708 (1965) is a very good review on the solar wind; see also M. Kuperus and J.A. Ionsons, *On the Theory of Coronal Heating Mechanisms*, Ann. Rev. Astron. Ap. **19**, 7-40 (1981). The transfer of angular momentum from the sun to the solar wind, not adequately discussed here, is important for the evolution of the solar nebula; the general theory was developed by E.J. Weber and L. Davis, Jr., *The Angular Momentum of the Solar Wind*, Ap. J. **148**, 217 (1967). On the "changing sun" see the papers by J.A. Eddy, in particular J.A. Eddy (editor), *The New Solar Physics*, Westview Press (1978).

Some material of the last Section is taken from C.F. Kennel, *Magnetospheres of the planets*, Space Sci. Rev. **14**, 511 (1973).

14. THE PLANETARY SYSTEM

In this chapter we offer an overview of the properties of our planetary system – both concerning its dynamical structure, namely the orbital and rotational motions of the constituent bodies, and their physical nature and structure, e.g.,size, composition, the presence of an atmosphere and/or of an observable solid surface. These data will mostly be presented in the form of tables and figures with synthetic captions and comments. As discussed in the Introduction, our purpose is not descriptive, but rather to provide an insight on the main physical principles and processes. Therefore, we shall concentrate mainly on the data which are used and/or interpreted in other parts of the book.

14.1. PLANETS

Object	a	P	e	I	P_S	ϵ
Sun	-	-	-	-	25.4	7.25
Mercury	0.387	87.97d	0.206	7.00	58.65	0
Venus	0.723	224.70d	0.007	3.39	243.0	178
Earth	1.000	365.25d	0.017	0.00	1.00	23.4
Mars	1.524	686.98d	0.093	1.85	1.026	25.0
Jupiter	5.203	11.86y	0.048	1.30	0.410	3.08
Saturn	9.539	29.46y	0.056	2.49	0.426	26.7
Uranus	19.182	84.01y	0.047	0.77	0.720	97.9
Neptune	30.058	164.79y	0.009	1.77	0.670	28.8
Pluto	39.44	247.7y	0.250	17.2	6.387	94

Table 14.1. Dynamical properties of the planets and the sun: a is the semimajor axis in AU; P is the orbital period; e the eccentricity; I the inclination in degrees; P_s the spin period in days; ϵ the obliquity to the ecliptic in degrees.

Object	R_e	M	ρ	A	*	**
Sun	695700	$1.99 \cdot 10^{33}$	1.41	-	Ionized H and He	H,He
Mercury	2439	$3.30 \cdot 10^{26}$	5.42	0.12	Igneous rocks	None
Venus	6052	$4.87 \cdot 10^{27}$	5.25	0.59	Basaltic rocks	CO_2
Earth	6378	$5.98 \cdot 10^{27}$	5.52	0.39	Water, basaltic and granitic rocks	N_2, O_2
Mars	3398	$6.42 \cdot 10^{26}$	3.94	0.15	Basaltic rocks, dust	CO_2
Jupiter	71900	$1.90 \cdot 10^{30}$	1.31	0.44	-	H_2, He
Saturn	60330	$5.69 \cdot 10^{29}$	0.69	0.46	-	H_2, He
Uranus	25700	$8.68 \cdot 10^{28}$	1.22	0.56	-	H_2, He, CH_4
Neptune	24750	$1.03 \cdot 10^{29}$	1.66	0.51	-	H_2, He, CH_4
Pluto	1100	$1.2 \cdot 10^{28}$	2.1	0.6	CH_4, H_2O ices	Thin CH_4

Table 14.2. Physical properties of the planets and the sun. The columns show: R_e, the equatorial radius in km; M, the mass in g; ρ, the mean density in g/cm^3; A, the visual albedo; *, the surface materials; **, the main constituents of the atmosphere.

Tables 14.1 and 14.2 present the main dynamical and physical data of the nine planets of the solar system, updated at the beginning of 1990. Pluto is included in this list for traditional reasons; its very small size and its peculiar orbit (locked in a mean motion resonance with Neptune) make it much more alike the minor bodies discussed later in this chapter.

Here are a few comments on these data.

[1] The planetary orbits show in general a fairly regular pattern: eccentricities and inclinations are small; the semimajor axes roughly fit a geometric progression with a ratio close to 2 (when first noticed, in the 18th century, this feature was expressed by the *Titius-Bode law*, see Problem 14.1); the wider gap between Mars and Jupiter corresponds to the location of the asteroid belt. These regularities are in part due to the obvious requirement that the orbits must have been stable, in particular avoiding close encounters, for a time span ranging from 10^7 and 10^{10} orbital periods; however, it is now believed that the process by which the planets have formed (see Ch. 16) had to give rise almost unescapably to such roughly "regular" orbits. It is interesting to notice that the highest eccentricities and inclinations occur at the inner and outer boundaries of the system, i. e., for Mercury and Pluto.

[2] Planetary rotations are much less regular. The three inner planets have been subjected to tidal despinning to a different degree (see Sec. 15.3). Apart from these cases, there is a remarkable clustering of the periods about 1 day, over a range of masses spanning several orders of magnitudes (this applies to asteroids also). Moreover, the polar axes tend to be aligned along the normal to the orbits, an outstanding exception being Uranus. Both these features are probably the outcome of the final phases of the planetary accumulation process, with angular momentum of rotation accumulating at the same time as mass during a sequence of collisional events between planetary "embryos" and smaller bodies orbiting in the same regions.

[3] The cumulative mass of the planets is only 0.15% that of the sun, with most of it concentrated in Jupiter. Jupiter's size is about 10 times smaller than that of the sun and 10 times larger than that of the earth. The bodies forming the solar system are much smaller than their mutual distances; for instance, Jupiter's distance from the sun is about 10^3 times the solar radius. Finally, the rotation of the sun contributes only a small part ($\approx$ 1%) to the total angular momentum of the system; hence it is not surprising that the polar axis of the sun has a significant tilt to the total angular momentum vector.

[4] Inspection of Table 14.2 immediately shows the traditional distinction between the inner, terrestrial type planets and the outer, Jovian type planets. The outer planets are much larger, yet less dense than the inner planets, a consequence of high abundances of the same light gases (hydrogen and helium) which prevail in the sun. This difference is clearly related to the different efficiency with which light gases were accreted by planetary embryos in the primordial solar nebula (see Sec. 16.5). However, within the outer group, Uranus and Neptune are distinctly different from Jupiter and Saturn, with a greater proportion of ices (C, N, O, H compounds). Among inner planets, the observed density gradient is also related to compositional differences, with a larger fraction of iron (the only heavy element abundant enough) in Mercury than in the other planets (especially Mars).

[5] The inner planets, having apparently lost virtually all their primordial light gases, are a kind of "cosmic cinder", with rocky surfaces, moulded by a variable mixture of endogenic (volcanism, tectonics, erosion) and exogenic (impact cratering) processes. Apart from Mercury, they have comparatively thin, *secondary atmospheres* produced by outgassing from rocky materials, which are chemically oxidizing rather than reducing. The opposite is true for the outer planets, whose atmospheres are *primary* (namely, accreted from the nebula itself), extend very deep into the planetary interiors (making any solid layer inaccessible to the observation) and are chemically reducing (since hydrogen has the property of reducing an ore to a metal.)

14.2. SATELLITES

Planet	Satellite	a	P	e	I
Earth	Moon	384.4	27.322	0.0549	5.1*
Jupiter	Amalthea	181.3	0.498	0.003	0.45
Jupiter	Io	421.6	1.769	0.000	0.03
Jupiter	Europa	670.9	3.551	0.000	0.47
Jupiter	Ganymede	1070	7.155	0.001	0.18
Jupiter	Callisto	1880	16.689	0.007	0.25
Saturn	Mimas	185.5	0.942	0.020	1.52
Saturn	Enceladus	238.0	1.370	0.004	0.02
Saturn	Tethys	294.7	1.888	0.000	1.09
Saturn	Dione	377.4	2.737	0.002	0.02
Saturn	Rhea	527.1	4.518	0.001	0.35
Saturn	Titan	1222	15.94	0.029	0.33
Saturn	Hyperion	1481	21.28	0.104	0.4
Saturn	Iapetus	3561	79.33	0.028	14.7*
Saturn	Phoebe	12.95	550.4	0.163	150*
Uranus	Miranda	129.8	1.414	0.017	3.4
Uranus	Ariel	191.2	2.520	0.003	0
Uranus	Umbriel	266.0	4.144	0.003	0
Uranus	Titania	435.8	8.706	0.002	0
Uranus	Oberon	582.6	13.46	0.001	0
Neptune	1989N2	73.6	0.56	0	0
Neptune	1989N1	117.6	1.22	0	0
Neptune	Triton	354.3	5.88	0.000	160
Neptune	Nereid	5510	365.2	0.75	27.7*
Pluto	Charon	19.1	6.387	0	0

Table 14.3. Satellite orbital data. The columns show: a, the semimajor axis in units of 1000 km; P, the period in days; e, the eccentricity; I, the inclination in degrees with respect to the planet's equator, except for some distant satellites (marked by an asterisk), for which it is referred to the planet's orbit.

Tables 14.3 and 14.4 give orbital and physical data on the natural satellites of the solar system, excluding those (frequently of markedly

Planet	Satellite	R	ρ	A	*
Earth	Moon	1738	3.34	0.12	Anorthositic and basaltic rocks, dust
Jupiter	Amalthea	135·85·75	?	0.05	Rocks, sulphurous coating?
Jupiter	Io	1815	3.55	0.63	Sulphur compounds
Jupiter	Europa	1570	3.04	0.64	H_2O ice
Jupiter	Ganymede	2630	1.93	0.43	H_2O ice, dirt
Jupiter	Callisto	2400	1.83	0.17	H_2O ice, dirt
Saturn	Mimas	195	1.2	0.6	H_2O ice
Saturn	Enceladus	250	1.0	1.0	H_2O ice
Saturn	Tethys	530	1.2	0.8	H_2O ice
Saturn	Dione	560	1.4	0.5	Dirty H_2O ice
Saturn	Rhea	765	1.3	0.6	H_2O ice
Saturn	Titan	2575	1.88	0.2	(Liquid?) CH_4, ices; N_2, CH_4 atmosphere
Saturn	Hyperion	185·140·110	?	0.2	Dirty ice?
Saturn	Iapetus	730	1.2	0.04/0.5	H_2O ice, icy dirt?
Saturn	Phoebe	110	?	0.06	Ice, carbonaceous dirt?
Uranus	Miranda	485	1.3	0.34	H_2O ice, dirt
Uranus	Ariel	580	1.6	0.40	H_2O ice, dirt
Uranus	Umbriel	595	1.4	0.19	H_2O ice, dirt
Uranus	Titania	805	1.6	0.28	H_2O ice, dirt
Uranus	Oberon	775	1.5	0.24	H_2O ice, dirt
Neptune	1989N2	100	?	0.06	?
Neptune	1989N1	200	?	0.06	?
Neptune	Triton	1350	2.0	0.7	CH_4, N_2, H_2O ices Thin N_2, CH_4 atmosphere
Neptune	Nereid	170	?	0.14	?
Pluto	Charon	640	?	0.4	H_2O ice?

Table 14.4. Physical properties of satellites. We show: R, the mean radius in km; ρ, the mean density in g/cm^3; A, the visual albedo; * the surface (and atmospheric) composition.

irregular shape, see Sec. 14.7) with mean radii smaller than 100 km. The inventory of such small satellites includes: the two martian moons Phobos and Deimos; 11 Jovian satellites (3 in the inner part of the system and 8 in the outer one); 11 Saturnian satellites (3 at the outer edge of the ring system, 2 coorbitals between the rings and Mimas, 2 Trojans of Tethys and 1 Trojan of Dione); 10 Uranian satellites (all lying between the rings and Miranda); at least 4 Neptunian moons inside the orbit of 1989N2; and, of course, all the planetary ring particles, up to sizes of the order of 10 m (the existence of km-sized moonlets embedded in the rings of the outer planets is also likely.)

Here are a few comments on these data.

[1] The existence of extensive satellite systems around Jupiter, Saturn and Uranus supports the idea that our solar system is not unique: just as the sun formed with planets, so large planets formed with satellites, and we can expect this hierarchy of objects to exist around other stars – in particular single stars – as well. This argument gains further support from the fact that the bulk properties of the large Jovian satellites (the four *Galilean satellites* Io, Europa, Ganymede and Callisto) mimic those of the planets, with the mean density and the abundance of ice decreasing and increasing, respectively, with planetocentric distance. The basic explanation of this pattern is probably similar in the two cases: the region close to the collapsing central body was too hot for the most volatile materials in the proto-solar or proto-planetary nebulae to condense. This is consistent with the absence of a similar pattern in the Saturnian and the Uranian system: the formation of these planets was probably associated with a much weaker energy outflow than for Jupiter. Even from the dynamical point of view, the major satellite systems share the usual features of small eccentricities and inclinations, nearly geometric progression of semimajor axes and more "irregular" orbits at the inner and outer edges.

[2] The Neptunian satellite system has a number of outstanding "irregular" features: (a) only one sizeable satellite, Triton; (b) both Triton and Nereid have large orbital inclinations to the planet's equator; (c) Triton (the seventh largest satellites in the solar system) has a retrograde orbit fairly close to the planet and is subject to strong tidal decay (see Sec. 15.3); (d) the tiny outer moon Nereid has a comet-like eccentricity. The small mass of Pluto and its orbit, which at perihelion is closer to the sun than that of Neptune, have suggested the hypothesis that Pluto be perhaps an escaped Neptunian satellite. The event causing the disruption of a formerly regular Neptunian system might have been the encounter with a massive, stray body or the capture of Triton on an almost parabolic orbit, subsequently decayed to the current one.

[3] The Pluto-Charon and the earth-moon systems are, in fact, "double planets", with large satellite-to-planet mass ratios and strong tidal

interactions. They might have had a peculiar formation mechanism (see Sec. 16.6).

[4] Many small outer satellites of the Jovian planets, which have very irregular orbital parameters, as well as the two Martian moonlets, whose small size and carbonaceous-like composition resemble the properties of many small asteroids, might have been captured from interplanetary orbits. The capture was possibly favoured by non-gravitational forces like gas drag in primordial circumplanetary envelopes or by tidal friction. On the other hand, many small inner satellites close to the rings of Jupiter, Saturn and Uranus are probably fragments generated by collisional disruption of larger precursor bodies.

[5] The natural satellites span a wide range of sizes. We have seven objects (the moon, the four Galilean satellites, Titan and Triton) more than 2500 km across, larger than Pluto and almost reaching the size of Mercury. These bodies are really small "planetary worlds", with complex geological histories and sometimes even an atmosphere (an outstanding such case is that of Titan). All the other $\approx$ 50 known satellites have sizes comparable to those of the observable main belt asteroids, from about 10 to about 1500 km. Thus all of them could be included in a general category of "small solar system bodies", together with asteroids and comets. However, as we shall see in Sec. 14.7, at a size of the order of 100 km an important transition occurs, from irregularly shaped, big "boulders" to small "worlds" dominated by gravity.

14.3. ASTEROIDS

Among the known members of the solar system, namely the objects whose heliocentric or planetocentric orbits have been accurately determined, the asteroids represent by far the most numerous population. The number of catalogued bodies is now approaching 4000, some 95% of which travel around the sun in moderately eccentric ($e \leq 0.15$) and inclined ($I \leq 10^{\circ}$) orbits lying in a large toroidal region between 2.1 and 3.6 AU from the sun, the *main asteroid belt.* In this region another major planet, "predicted" by the empirical Titius-Bode law, is missing. However, some asteroids are known, whose perihelia penetrate inside the orbits of the inner planets (the so-called *Aten-Apollo-Amor objects*), making possible huge sporadic collisions with them. Far beyond the main belt there are the two groups of *Trojan asteroids,* moving near Jupiter's triangular Lagrangian points, and single objects like Hidalgo and Chiron, crossing the orbits of the outer planets.

As shown by Fig. 14.1, the dynamical structure of the asteroid belt is deeply affected by mean motion resonances with Jupiter (see Secs. 11.2 and 15.2), which occur whenever the asteroid mean motion can be expressed as that of Jupiter times the ratio between two small integers. The order of the resonance, i. e. the difference between numerator and denominator of that ratio, is inversely correlated with the magnitude of the corresponding term in the perturbing potential, and hence (although the relationship is not straightforward) to the destabilizing effect of the resonance. Qualitatively, this effect shows up very clearly in Fig. 4.1: in the main belt, low order resonances are isolated, and they correspond to sharp gaps (the so-called *Kirkwood gaps*) in the distribution of mean motions and semimajor axes; on the other hand, between the 2:1 resonance and Jupiter, resonances are clustered together and very few objects are left there (curiously enough, most of these surviving bodies, like the Hilda group, Thule and the Trojans are themselves in resonance).

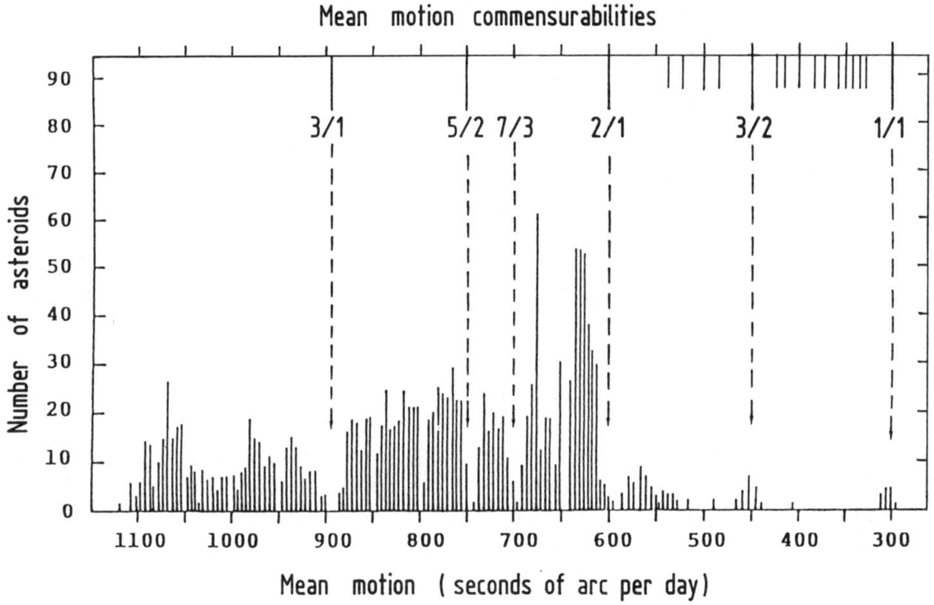

Fig. 14.1. Frequency distribution of the mean motions of asteroids. All the resonances corresponding to mean motion ratios p/q with $p \leq 10$ and order $(p - q) \leq 4$ are marked on the upper horizontal axis. The 3:1, 5:2, 7:3 and 2:1 resonances correspond to the main Kirkwood gaps, while at the 3:2 and 1:1 resonances asteroids of the Hilda and Trojan groups are clustered. Other low-order resonances occur in the depleted region beyond the 2:1 gap.

As discovered in 1918 by K. Hirayama, in the frequency distribution of the orbital parameters of asteroids there are also prominent spikes, especially if the eccentricities and the inclinations are corrected by subtracting the changes due to planetary gravitational perturbations. The outcome of this delicate correction procedure are the so-called *proper elements* (see Sec. 15.2). Hirayama suggested that the origin of these *asteroid families* could be traced back to the explosive breakup of a common parent body, ejecting fragments into independent heliocentric orbits with relative velocities much lower than their orbital speed. In the last few decades, physical studies have shown that the members of the most populous families (associated with the asteroids Themis, Eos and Koronis) have similar surface compositions, supporting the hypothesis of a common origin and opening up the opportunity of probing directly by astronomical observations the interior structure of the parent bodies. Moreover, collisional fragmentation has been shown to represent a plausible formation process for the families, from the point of view of the probability of collisions during the solar system's lifetime, and also of the expected outcomes of such catastrophic events. Asteroidal collisions have been simulated in the laboratory with experiments on high-velocity impacts against solid targets. The most energetic collisions are disruptive, simply because the relative velocities of asteroids, due to their eccentricities and inclinations, may largely exceed their escape velocities; even impacts with projectile to target mass ratios $\approx 0.1\%$ can impart energies exceeding the binding energy of the targets bodies.

Collisional evolution has also substantially affected the mass (and size) distribution of asteroids. The largest asteroid, Ceres, is about 1000 km across. There are about 30 asteroids larger than 200 km, 250 larger than 100 km, 700 larger than 50 km. At smaller diameters D the size distribution is not very well known, but is usually represented by a power law:

$$dN \propto m^{-q}\, dm \propto D^{(2-3q)} dD. \tag{14.1}$$

dN is the number of bodies with mass in the intervals (m, m + dm), (D, D + dD) and q is a characteristic exponent. The observed value of q ranges from 1.6 to 1.9, implying that most of the mass of the asteroids lies in the largest bodies. The total mass is 2 or 3 times the mass of Ceres, about $0.5 \cdot 10^{-3}$ $M_{\oplus}$. With respect to a perfect power law distribution, there is an excess of bodies with $D \approx 100$ km, which may be related to the fact that at this size the self-gravitational binding energy becomes important in determining the outcome of a disruptive impact (by causing reaccumulation of fragments ejected at speeds lower than the escape velocity and formation of "rubble pile" asteroids). The distribution (14.1), with $q \cong 1.8$, has also been observed in fragments produced in laboratory impact experiments. One should not forget, however, that asteroidal collisions involve sizes and energies which are typically 10^6 and 10^{18} times larger than those studied in the laboratory; scaling up the outcomes of collisions by so many orders of magnitude is not straightforward.

Most data on the rotation and shape of the asteroids come from light-curve photometry, which has provided rotational periods for

Number	Name	Semimajor axis (AU)	Eccentricity	Inclination to ecliptic (deg)
1	Ceres	2.768	0.077	10.60
2	Pallas	2.773	0.233	34.80
4	Vesta	2.362	0.090	7.14
31	Euphrosyne	3.148	0.228	26.33
511	Davida	3.181	0.172	15.90
704	Interamnia	3.060	0.153	17.29
87	Sylvia	3.483	0.093	10.88
65	Cybele	3.428	0.110	3.55
15	Eunomia	2.642	0.188	11.76
3	Juno	2.671	0.255	13.00

Table 14.5a. Osculating orbital parameters of the ten largest asteroids.

Number	Name	Diameter (km)	Spin period (hours)	Maximum relative brightness variation (*)
1	Ceres	1025	9.075	1.04
2	Pallas	539	7.811	1.16
4	Vesta	536	5.342	1.12
31	Euphrosyne	333	5.531	1.11
511	Davida	309	5.167	1.26
704	Interamnia	305	8.727	1.11
87	Sylvia	281	5.183	1.45
65	Cybele	280	6.070	1.06
15	Eunomia	262	6.081	1.63
3	Juno	257	7.210	1.15

Table 14.5b. Physical properties of the ten largest asteroids.

(*) For a triaxial ellipsoid of axes $a \geqslant b \geqslant c$, assuming that the brightness is proportional to the cross-section as seen from the earth, the maximum observable relative brightness variation would be just equal to a/b.

about 500 bodies and polar directions for a few tens of asteroids. The average rotation period is about 10 hours, but a large dispersion is present and periods as short as 2 hours or as long as weeks have been observed. The correlations of the spin period with size and taxonomic class (see later) are not yet fully understood, since the asteroid angular momenta have been affected in a complex way by the collisional processes. This latter circumstance is confirmed by the fact that the available data on asteroid poles show no apparent clustering about the pole of the ecliptic, as it would be expected for "primordial" spins.

Finally, in the last decades an intense observational effort has shed light on the problem of asteroid composition, which has been found to be very diverse. The main source of information is spectral analysis of reflected sunlight, but other techniques like radiometry, polarimetry and planetary radar have been applied as well. These data have then been interpreted by comparing them with the properties of minerals found in the different meteorite types.

A clear difference among different types of asteroid surfaces is evidenced by the distribution of albedos. When the observational bias against darker objects is accounted for, some 75% of the asteroid are found to be very dark, with average albedos of $\approx$ 0.04. A distinct group of bodies has moderate albedos of about 0.15, with few asteroids lying in between and a tail of "bright" bodies having albedos up to 0.4 and more. A better discrimination is possible if spectrophotometry data are used, yielding the behaviour of the reflection spectrum over a wide wavelength interval. Some absorption bands are unequivocal evidence of the presence of silicates, water ice and hydrated minerals, but in many cases these prominent features are lacking and any inference on the mineralogical composition must be regarded as conjectural.

Statistical clustering techniques have been applied to sets of observational parameters, potentially relevant for the surface composition of asteroids, in order to define the so-called *taxonomic types*. C-type asteroids have a very low albedo and a flat spectrum throughout the visible and the near infrared; they are very probably similar in composition to carbonaceous chondritic meteorites, which are primitive mineral assemblages subjected to no, or little, metamorphism after their condensation. D-type objects are also dark, but have very red spectra, suggesting the presence of low-temperature organic compounds. These objects are similar to many low-albedo, reddish small bodies found in the outer solar system, including some comets observed at low activity and a few small satellites (e.g., Phoebe). S-types have relatively high albedos, and their spectra show absorption bands due to silicates like pyroxene and olivine. It is debated whether their likely meteorite analogies are the stony-iron meteorites (probably derived from the core-mantle interfaces of differentiated parent bodies), or the ordinary chondrites, interpreted as assemblages of primitive nebular grains of different compositions, subsequently moderately heated and metamorphosed. The M-type asteroids have albedos of about 0.10, with slightly reddish, straight spectra, suggesting a significant content of nickel-iron metal (This interpretation has been confirmed by radar observations.) It is likely that they are akin to iron meteorites, and hence represent pieces of the cores of differentiated precursors.

Of great interest is the fact that different taxonomic types are preferentially located at different heliocentric distances. This orderly progression of types is usually interpreted as reflecting both variations in the composition of the material which condensed in the solar nebula, due to the decrease in temperature with solar distance, and different relevance of subsequent melting events and metamorphism. Indeed, the most primitive types (corresponding to least metamorphosed material) tend to lie in the outer belt, where most asteroids significantly resemble the cometary nuclei.

14.4. COMETS

Bright comets, possibly the most spectacular celestial objects, appear as diffuse sources slowly moving in the sky, visible to the naked eye for times of the order of weeks. Although apparitions of comets had been reported as early as several centuries b.C. by Chinese astronomers, Tycho Brahe first found observational evidence that comets are celestial bodies farther away from the earth than the moon, and E. Halley discovered that they move on highly eccentric keplerian orbits around the sun.

Up to about forty years ago it was not yet clear whether comets belong to the solar system or whether they – at least in part – come from interstellar space. This question was solved in 1950 by J. Oort, who studied a sample of comets with almost parabolic orbits, and pointed out a remarkable clustering of semimajor axes at a solar distance of the order of 50,000 AU (i.e., almost one light-year). This led to the hypothesis of a large reservoir of long-period comets in a vast swarm – the *Oort cloud* – surrounding the solar system at an average distance not much smaller than typical interstellar distances. The few slightly hyperbolic orbits can be explained with the effect of planetary perturbations and of non-gravitational forces (due to anisotropic mass loss of cometary nuclei near the sun). When these effects are taken into account, the cometary binding energies to the sun become slightly negative even when the osculating values are positive. Therefore, although an originally very eccentric, elliptical orbit may sometimes appear hyperbolic when the comet is observed close to perihelion passage, comets do indeed belong to the solar system.

The inclinations of long-period comets have an approximately isotropical distribution with respect to the ecliptic plane; this indicates that the Oort cloud is almost spherically symmetric. This may not always have been the case, however, because perturbations by passing stars are frequent enough to randomize the heliocentric velocities of the comets in the cloud, hence their distribution in space, within a time span far shorter than the age of the solar system. This is important for the origin of comets, because it is now widely believed (see Sec.16.5) that comets formed at the same time as the planets in a thin equatorial layer in the outer zone of the primordial solar nebula and were later ejected into the Oort cloud by close encounters with the outer planets.

The outer boundary of the Oort cloud is set by the increasing probability that comets are expelled into interstellar space by tiny perturbing forces, owing to their low binding energies. Therefore, a significant fraction of the original comets must have been lost from the cloud since its formation. Among the remaining members, only those can reach the inner solar system whose transverse velocities become small enough to yield orbits with perihelion distances comparable with the planetary semimajor axes. From angular momentum conservation, it can easily be seen that this transverse velocity must be at most of the order of 0.02 (the square root of the ratio of perihelion to aphelion distance) times the circular velocity; therefore, for an isotropic distribution of the directions of orbital velocities, at any given time only one comet in 10^4 will have perihelion in the planetary region. This population of 'visible' comets, however, would be depleted in a time not much longer than 10^7 y – the typical orbital period of comets in the cloud – due to planetary perturbations, non-gravitational forces and decay in the vicinity of the sun, were it not continuously replenished from the reservoir in the cloud. This replenishment can be achieved by momentum exchanges with passing stars and with dense interstellar clouds, provided the cloud's population is abundant enough. From the present star density in the solar neighbourhood ($\approx 0.1/pc^3$) and the typical stellar velocity relative to the sun ($\approx$ 20 km/s), it can easily be estimated that about one stellar passage per million years occurs within 10^5 AU from the sun, strongly perturbing the Oort cloud. This must result into the observed flux of about 3 "new comets" per year, which however are estimated to be just a sample of $\approx 10^2$ "new comets" per year that get perihelia inside Saturn's orbit (there is of course a strong observational bias against perihelia larger than $\approx$ 1 AU). Thus every stellar passage must result into $\approx 10^8$ comets getting perihelia in the planetary region and, as we have seen above, this requires a total population of comets in the Oort cloud 10^4 times larger, that is of the order of 10^{12}. Assuming typical masses of cometary nuclei between 10^{15} and 10^{16} g, corresponding to typical sizes of $\approx$ 1 km and densities of the order of 1 g/cm^3, this gives a total mass of the cloud in the order of one earth mass.

Any comet source model must be able to explain not only the flux of "new comets", but also the existence of about 120 (some 20% of the total sample of known cometary orbits) *short-period comets*, with orbital periods less than $\approx$ 200 y and aphelia in the planetary region. The comparatively frequent passages near the sun make their lifetimes with respect to total outgassing and/or disintegration not larger than some 500 revolutions, namely 10^3 to 10^4 y. Therefore their population must also be replenished, presumably from the Oort cloud. The influence of passing stars cannot move distant aphelia into the inner planetary system, because this would require an excessive energy transfer. However, since most short-period comets have periods close to those of the giant planets, aphelia between the orbits of Jupiter and Saturn and fairly small inclinations to the ecliptic plane (the majority of them being in prograde orbits), close gravitational interactions with the giant planets appear as a plausible candidate mechanism to capture new comets in short-period orbits. Numerical integrations of fictitious cometary orbits have shown, however, that close encounters of Oort's cloud comets with Jupiter (or Saturn) in most cases fail to yield a good fit to the observed distribution of orbital parameters. Either a long and complex sequence of encounters with the outer planets is needed, or the existence of a (still unobserved) vast population of "intermediate" comets, orbiting in the zone just beyond Neptune with low inclinations. This population would be the plausible left-over of an aborted planetary accumulation at the outskirts of the solar system, and objects like the "peculiar asteroids" Hidalgo and Chiron might just be its most easily discovered representatives.

Another open problem is the relationship between comets and asteroids. As shown in Fig. 14.2, from the dynamical point of view the two populations of small bodies are generally well separated, with most objects in the main asteroid belt lying in a region of the phase space where close planetary encounters are impossible and the orbits can remain stable for times comparable with the age of the solar system. On the contrary, the high eccentricity of comets results into comparatively frequent planetary approaches, causing drastic orbital changes and a chaotic dynamical evolution. However, some asteroid groups, like the outer belt objects and those which can cross the orbits of the inner planets, are not so easily separated from the comet sample. Taking into account the secular changes of cometary orbits produced by non-gravitational forces, for some of these objects a cometary origin has been suggested. In other words, these asteroids would be inactive comets, either due to their permanently large perihelion distances, or because at the end of the outgassing lifetime the nucleus would have been converted into a "dead" object, with the icy component possibly covered by an insulating dust mantle.

It is impossible to resolve a cometary nucleus inside the gaseous *coma* by earth-bound observations. However, a model of cometary nuclei capable of explaining a large body of observational evidence about the behaviour of comets as they approach the sun, as well as the structure and chemical composition of comae and tails, has been proposed in 1950 by F. Whipple. In this model, frequently called the "dirty snow ball", the nucleus is a small solid body, generally 1 to 10 km across, consisting of some sort of conglomerate of ices and silicate particles. As the comet approaches the sun, the ices are sublimated, and the resultant gas and released meteoric dust are available to form the

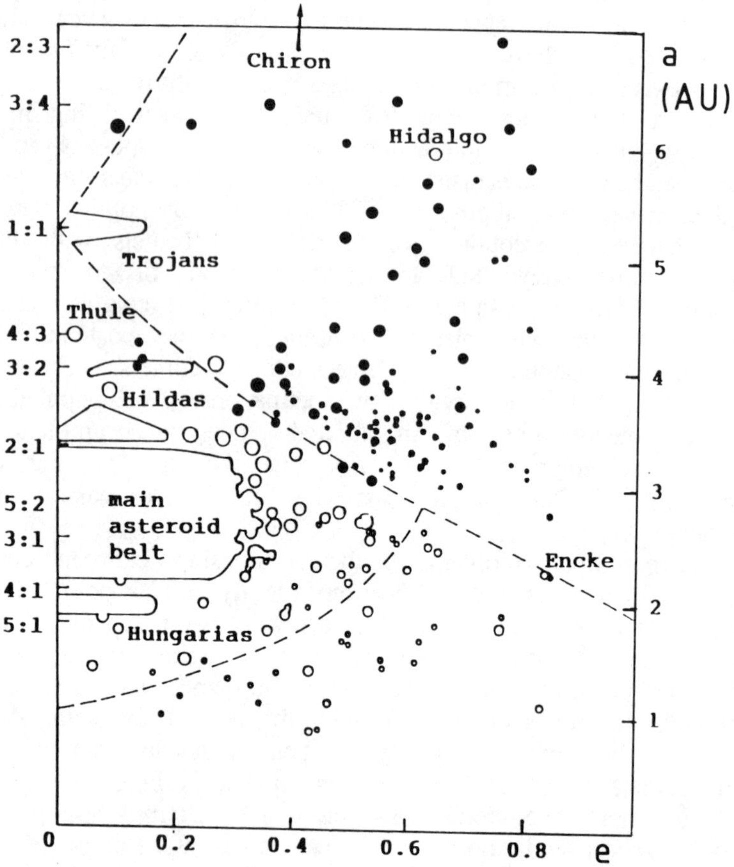

Fig. 14.2. The distribution of small solar system bodies in the plane of semimajor axis and eccentricity. On the vertical axis, the main mean motion commensurabilities with Jupiter are indicated. The dashed lines correspond to orbits which, at perihelion or aphelion, get close to the orbit of Jupiter and (the lowest one) to the orbit of the earth. While the asteroid (full circles) and comet (open circles) populations are generally well separated, it is worth noting that the short-period comet Encke has an orbit very similar to that of an earth-crossing asteroid, while a few 'asteroidal' objects like Hidalgo and Chiron have typical cometary orbits.

coma and the tail. The gas molecules are mostly ionized (and made luminous) by the interaction with solar radiation and the solar wind; the solid debris can become meteoroids strewn along the comet's orbit, affected by the Poynting-Robertson and radiation forces (see Sec. 15.4) and occasionally hitting the earth (thus producing *meteors* and *fireballs*). Reaction of the nucleus to the ejection of this material provides an explanation for the non-gravitational effects in the motions of comets. The model correctly predicts stronger outward radial components of the non-gravitational forces (the ejected material being most abundant on the sun-facing side of the nucleus) and significant transverse components, due to the comet's rotation and a thermal lag between the direction of maximum mass ejection and the subsolar meridian (see Sec. 15.4). The nucleus must be rather weak, since several cases of comet splitting and breakup have been observed; in many cases the volatile component is probably covered by an outer shell of dark, refractory material with gaseous jets from localized fractures or spots, sometimes causing sudden outbursts. Remarkably, the model has been recently confirmed by *in situ* observations and measurements by *Giotto*, the European probe which in March 1986 had a very close fly-by with Halley's comet.

14.5 INTERPLANETARY MATERIAL AND METEORITES

The interplanetary space contains a large number of bodies and particles too small to be individually visible by telescopic observation. These bodies, whose motion is frequently affected by non-gravitational forces and whose orbital lifetime is generally much shorter than the age of the solar system (see Sec. 15.4), very probably originate from the two populations of small and relatively primitive solar system bodies previously discussed: comets and asteroids. Among the asteroids, the number of bodies rapidly grows with decreasing size (see eq. (14.1)), and there is no reason to believe that this trend is not continuing at sizes smaller than those allowing telescopic detection (a few tens of km). On the other hand, cometary activity is characterized by the emission not only of gas but also of dust and solid particles; at the end catastrophic disintegration of the nucleus into a swarm of solid debris frequently occurs.

Our knowledge of these bodies – collectively called *meteoroids* – is observationally very limited. Apart from the sporadic (and quite inaccurate) spacecraft measurements of micrometeoroids during interplanetary missions, we are confined to those particles which cross the earth's orbit and enter the atmosphere, becoming *meteors* that flash momentarily across the sky. Thus most available information regards the near-earth flux in different mass ranges and its direction distribution as a function of time. Double or multi-station observations make it occasionally possible to determine the orbits of individual particles, but the accuracy is much lower than that attainable for larger bodies. Most meteors are far too small to sustain the heating due to atmospheric drag and to reach the ground; they "burn" at altitudes between 75 and 100 km. Occasional, large meteors called *fireballs* are very bright and spectacular, but often explode before hitting the ground, an indication of their fragility. A low density is also inferred from the observed decelerations due to atmospheric friction. Since the 1960's, rockets and balloons have collected micrometre-sized particles believed to be meteoroids or fragments of them. They are irregular glassy silicate and metallic aggregates of smaller dust grains, and are thought to represent samples of the dust component present in comet nuclei.

The flux as a function of time shows a prominent diurnal variation. This is controlled by the direction of heliocentric motion of the earth with respect to the zenith of the observing site. The sweeping effect of the earth's motion, combined with the higher energy of head-on collisions, tends to increase the influx on the morning side of the earth (leading in the orbital motion), where smaller particles of higher geocentric velocity are generally observed. On the other hand, the prevailing prograde motion of meteoroids on orbits with average semiaxes larger than that of the earth, favours the (trailing) evening side, partially compensating the velocity effect. For *meteorite falls* (i. e., when rocky samples are recovered), the net result is a definite overcompensation, because lower geocentric velocities yield a better chance for the survival of solid remnants after crossing the atmosphere and hitting the ground.

The permanent meteor flux, due to the so-called *sporadic meteors*, is such that on an average night about 3 meteors per hour can be seen before midnight and about 15 meteors per hour after midnight. But on certain dates each year (i. e., always at the same heliocentric longitude of the earth), *meteor showers* of 60 or more meteors per hour can be seen, all radiating from one direction in the sky. The most widely known example is the *Perseid shower*, which shows up every year around August 12th, with bright meteors streaking across the sky every few minutes from the direction of the constellation of Perseus. The annual recurrence is indicative of streams of particles dispersed all around a central elliptical orbit. Their duration – a few days to some weeks – are indicative of stream widths ranging from 10^6 to 10^8 km at the crossing point.

There is a definite connection between meteor showers and comets. In 1866 G. Schiaparelli discovered that the Perseid meteor shower appears whenever the earth crosses the orbit of comet Swift-Tuttle, implying that the shower is spread out along the orbit of that comet. Many other relationships were subsequently found between specific meteor showers and specific comets. In 1983, the Infrared Astronomy Satellite (IRAS) detected the thermal emission from a swarm of meteor dust spread along the orbit of comet Temple 2. Even the orbits of individual sporadic meteors, tracked photographically, often resemble long- or short-period cometary orbits. The conclusion that most meteors are cometary debris is not at odds with the fact that for the most conspicuous annual shower – the December Geminids – an asteroid-looking parent object was discovered by IRAS. The orbit of this earth-crossing object, of very high eccentricity (0.89), suggests that in fact it might be an extinct cometary nucleus, whose volatile component is no longer present on the surface.

The smallest interplanetary particles are microscopic dust grains, with typical sizes between 10 and 100 μm, spread out along the plane of the solar system and more concentrated toward the sun. They can barely be seen by looking westward to the evening twilight (or eastward before sunrise) as a faint, glowing band of rediffused sunlight – the *zodiacal light* – extending up from the horizon and along the ecliptic plane. It is not clear which fraction of these particles, with short orbital decay times with respect to Poynting-Robertson drag (see Sec. 15.4), are meteor-type cometary material and which fraction is due to the collisional comminution of fragments in the asteroid belt. A significant contribution from asteroidal collisions is suggested by IRAS' discovery of a few prominent zodiacal dust bands in the asteroidal belt.

An asteroidal origin is generally accepted for the meteorites, with the possible exception of the rare *carbonaceous chondrites*, made of a volatile-rich and low density material which might be abundant both in cometary nuclei and in outer belt asteroids. Their orbits do not exhibit any definite streaming pattern, and the few accurately determined orbits resemble closely those of the Apollo asteroids. Their cosmic rays exposure ages, typically of the order of 10^7 or 10^8 y, are not much shorter than the typical lifetime of earth-crossing objects with respect to impacts to the earth. This lifetime can be estimated as follows. Since for mutually inclined orbits crossings can occur only with a suitable orientation of the nodal line, and the typical revolution period for the nodal lines of inner solar system orbits is $\approx 10^4$ y, about once in 10^4 y the orbits of the meteorite and of the earth do actually cross. Then the impact occurs only if the meteorite is in the correct portion of its orbit, and the probability of this coincidence is of the order of $(1\ R_\oplus)/(1\ \mathrm{AU}) \cong 4 \cdot 10^{-5}$. It follows that the lifetime of a meteorite (or an earth-crossing asteroid) against collisions with planets is of the order of $10^4/4 \cdot 10^{-5}$ y $\cong 2 \cdot 10^8$ y. The actual average lifetimes are somewhat shorter, since many meteorites are probably shattered by small-scale collisions with other debris on their route to the earth.

On the other hand, the formation and solidification ages of meteorites, determined from isotopic abundance ratios, sharply cluster around $4.55 \cdot 10^9$ y, providing a strong indication that they formed in the solar nebula from the condensation of refractory material in the early stages of solar system formation. Meteorites have been classified in several types with respect to their mineralogical properties: volatile-rich primitive materials which never underwent melting or differentiation, like *carbonaceous chondrites*; types more chemically evolved, containing iron and sometimes collisionally brecciated, like the so called *ordinary chondrites* (that account for $\approx 80\%$ of the whole available meteorite inventory); igneous, basaltic-type rocks indicating extensive vulcanic activity on their parent bodies (*achondrites*); iron-nichel alloys, either mixed to basalt rocks (*stony-irons*) or almost pure (*iron meteorites*), which presumably come from the metallic cores of differentiated bodies. The vast majority of meteorites are believed to represent fragments of asteroids, ejected from them as a consequence of high-velocity impacts and then inserted into unstable resonant orbits leading to encounters with the inner planets. As already discussed in Sec. 14.3, though the properties of various meteorite types and asteroid taxa often appear to fit closely each other, there are still major uncertainties in the unequivocal determination of the parent bodies of most meteorites.

Every year some 10^7 to 10^9 kg of meteoritic material hit the earth, a significant part of it as small particles. However, metre-sized objects hit the ground (and form significant craters) roughly annually, and for larger sizes the typical interval between falls is roughly proportional to the square of size, consistently with the size distribution (14.1) with a characteristic exponent $q \cong 5/3$. This relationship appears to approximately hold up to catastrophic impacts of earth-crossing asteroids several km across, which occur every 10^7 to 10^8 y and may cause global ecological catastrophes. The counting of craters on the lunar surface has shown that this influx rate has been almost constant in the last few billion years, while a much more intense bombardment occurred before $4 \cdot 10^9$ y ago. This primordial heavy bombardment was probably due to the sweeping by the major planetary bodies of the residual solid material left out from accumulation of planetesimals (see Ch. 16).

14.6. PLANETARY RINGS AND THE ROCHE LIMIT

In the last decades our knowledge of the ring systems of the outer planets has vastly grown. The detection of nine narrow rings about Uranus from stellar occultation data in 1977; Voyager 1's discovery in 1979 of a faint ring circling Jupiter; the high-resolution Voyager observations of the fine structure of Saturn's rings (known since the 17th century); have shown a complex and puzzling picture. Voyager 2's images of Neptune's ring system, obtained in 1989 and displaying strong azimuthal brightness variations, have shown a complex and puzzling picture. The geometrical properties of the four ring system are shown in Table 14.6.

Since the 19th century, it is well known that planetary rings are "comprised of an indefinite number of unconnected particles" (using the words of J.C. Maxwell, who in 1857 first demonstrated that this is the only possibly stable configuration). Apart from this common feature, however, ring systems show a great variety of properties. The optical depth ranges from at most 10^{-5} for the main Jovian ring and the diaphanous Saturn's E and G rings, to almost unity for Saturn's B ring and Uranus' ε ring. The size of particles range from micrometers for the Jovian ring material to meters in Uranus' and Saturn's system (which,

(A) Jovian rings

Planetary equator, R_J	71,900
Halo, inner edge	≈100,000
Halo, outer edge	≈122,000
Main ring, inner edge	≈122,000
Main ring, outer edge	129,100
Gossamer ring, outer edge	≈225,000

(B) Saturn's rings

Planetary equator, R_S	60,330
D ring, inner edge	≈66,000
C ring, inner edge	73,200
B ring, inner edge	92,200
B ring, outer edge, and Cassini division, inner edge	117,500
A ring, inner edge, and Cassini division, outer edge	121,000
Encke division (≈200 km wide)	133,200
A ring, outer edge	136,200
F ring	≈140,000
G ring	≈170,000
E ring, inner edge	≈210,000
E ring, maximum	≈230,000
E ring, outer edge	≈300,000

(C) Uranus' rings

Planetary equator, R_U	25,700
1986U2R ring	38,300
6 ring	41,850
5 ring	42,280
4 ring	42,600
α ring	44,750
β ring	45,670
η ring	47,180
γ ring	47,630
δ ring	48,310
1986U1 ring	50,030
ε ring	51,000

(D) Neptune's rings

Planetary equator	24,750
Inner broad ring	42,000
Inner narrow ring	53,000
Plateu, outer edge	58,000
Outer narrow ring	63,000

Table 14.6. (Previous page) Geometry of the ring systems, with the radius in km.

however, also include zones where microscopic dust is abundant). The material albedo ranges from a few hundredths in Uranus' system to almost unity in Saturn's rings, whose particles are made of slightly contaminated water ice. The radial structure of the rings is complex and diverse: Jupiter's main ring appears as a smooth band; Saturn's rings extend over some 70,000 km with variations of optical depth over all length scales down to the resolution of the available data; Uranus' rings

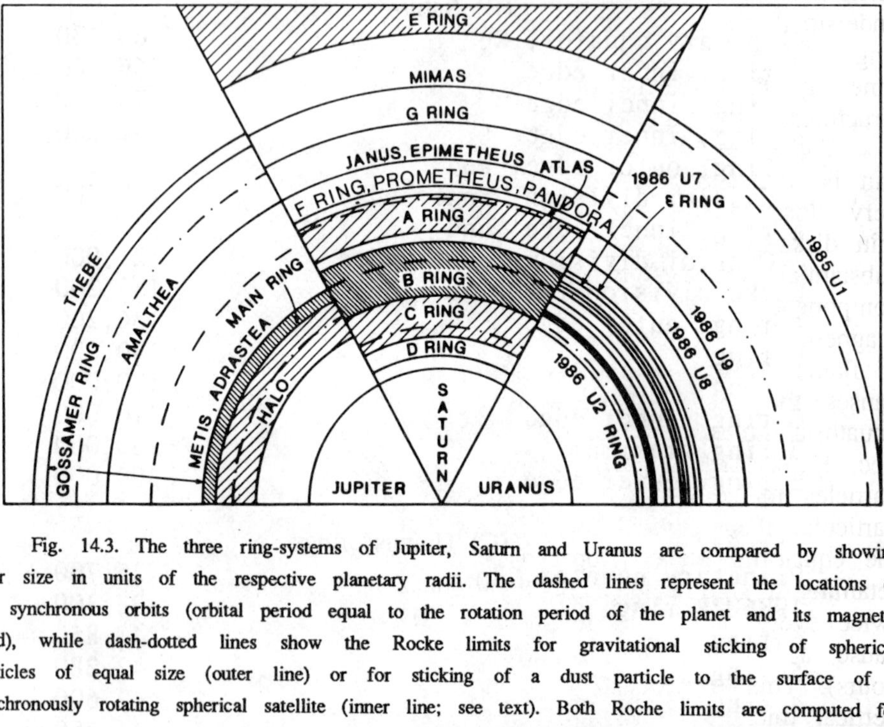

Fig. 14.3. The three ring-systems of Jupiter, Saturn and Uranus are compared by showing their size in units of the respective planetary radii. The dashed lines represent the locations of the synchronous orbits (orbital period equal to the rotation period of the planet and its magnetic field), while dash-dotted lines show the Rocke limits for gravitational sticking of spherical particles of equal size (outer line) or for sticking of a dust particle to the surface of a synchronously rotating spherical satellite (inner line; see text). Both Roche limits are computed for a particle density of 1 g/cm^3. The orbits of the small inner satellites of these planets are also shown. In the case of Uranus, 6 additional small satellites (not shown in the figure) orbit between 1986U9 and 1985U1. Some satellites (Metis and Adrastea for Jupiter, Atlas for Saturn) orbit near the outer edges of the rings; in other cases (Prometheus and Pandora for Saturn's F ring, 1986U7 and 1986U8 for Uranus' ε ring) the satellites play the role of "shepherds", constraining ring material to narrow zones.

are very narrow (≈ 1 to 100 km wide), eccentric bands with sharp edges; and Neptune's system has both narrow rings with azimuthal brightness variations and broader bands. A great variety of dynamical processes has been proposed to explain the available data: resonances with satellites external to the rings (the outer edges of Saturn's B and A rings are, respectively, located at the 2:1 mean motion resonance with

Mimas and at the 7:6 resonance with Janus); close gravitational interactions with small *shepherding moons*, constraining ring material and generating narrow ringlets and abrupt edges; mass loss from moonlets embedded in the rings; spiral density waves triggered by resonances and vertical bending waves driven by inclined satellites; interactions with magnetospheric plasma; electromagnetic forces affecting the motion of microscopic grains; collisions among ring particles and with interplanetary micrometeoroids. Although individual phenomena and features have been understood, there is still no general theory of the origin and evolution of rings. Their very age – i. e., whether they formed at about the same time as the planets or are much younger and, possibly, transient structures – is at present under intense debate.

There are, however, two common features of planetary rings which can be understood on the basis of simple considerations. First, they lie very close to the equatorial planes of their planets, in general forming thin disks or bands (An exception is the Jovian halo ring, which has a substantial vertical thickness; probably this is a result of a non-equatorial component of the Lorentz force, produced by the Jovian misaligned magnetic dipole field.) Any initial, thick and/or non-equatorial ring tends to flatten swiftly to a thin equatorial disk because the planet's oblateness causes the orbital planes of ring particles to precess relative to the equatorial plane at a rate dependent on their orbital radius (see Sec. 11.6). The gradual relative drift of the orbital planes allows the particles to collide, reducing at each collision the relative velocities, in particular their vertical components. This damping of orbital motions into the equatorial plane happens rapidly and continuously since any particle retaining a significant inclination would pass through the equatorial plane twice every orbit; and for nearly opaque rings like Saturn's, this would cause a collision during almost every passage (i.e., once every few hours). This process is very effective, as shown by the fact that the vertical thickness of Saturn's ring system is a tiny fraction, of the order of 10^{-6}, of its width. However, this out-of-plane damping never produces a perfectly thin "monolayer" state. Low-velocity, interparticle collisions occur in a thin ring also because of the gradient in the keplerian velocity along any radial direction. These collisions, as well as the scattering of particles by one another's gravitational fields (if particles at least 10 m across are present), give rise to small and random relative velocities, producing a ring at least several particles thick.

The second common feature of the ring systems is that they all lie close to their planets, inside the orbits of the major satellites. This is due to the fact that in the neighbourhood of a planet the strong gravity gradient can prevent the accretion of sizable satellites and even lead to disruption of existing bodies, provided their tensile strength is small enough. The basic mechanism of this phenomenon can be understood with elementary gravity considerations. Consider (a) the formation of a satellite by gravitational accretion of small orbiting particles, and (b) retention of loose material on the surface of an existing spherical satellite (see Fig. 14.4).

(a) The net attraction between two small spherical particles of mass m, radius r and density ρ_s, orbiting at distance a from the centre of a planet of mass M, radius R and density ρ_p, is balanced by the planetary gravity gradient when:

$$\frac{Gm^2}{(2r)^2} = \frac{GMm}{(a - r)^2} - \frac{GMm}{(a + r)^2} \approx \frac{GMm}{a^2}\frac{4r}{a} . \qquad (14.2)$$

We have assumed that the particles do not rotate, and are in contact and aligned with the radial direction; were they oriented in a different way, the limit on a/R we shall derive would be larger. Using the relation

$$\frac{M}{m} = \frac{\rho_p}{\rho_s}\frac{R^3}{r^3} , \qquad (14.3)$$

we conclude that the two particles will stick together and a larger body may form only if

$$\frac{a}{R} > \left(16\frac{\rho_p}{\rho_s}\right)^{1/3} \cong 2.52\left(\frac{\rho_p}{\rho_s}\right)^{1/3} . \qquad (14.4)$$

(b) Let us consider now a spherical satellite of mass m, radius r, orbital distance a and rotation rate equal to its mean motion n (i. e., synchronous.) The gravitational and centrifugal forces acting on a particle located on the satellite's surface, in the position facing the planet along the radial direction (where disrupting forces are maximized), balance if:

$$\frac{Gm}{r^2} + n^2(a - r) = \frac{GM}{(a - r)^2} \cong \frac{GM}{a^2}\left(1 + 2\frac{r}{a}\right) . \qquad (14.5)$$

Substituting M/m from eq. (14.3) and n^2 from Kepler's third law we find that the satellite can retain the surface particle only if:

$$\frac{a}{R} > \left(\frac{3\rho_p}{\rho_s}\right)^{1/3} \cong 1.44\left(\frac{\rho_p}{\rho_s}\right)^{1/3} . \qquad (14.6)$$

It is interesting to compare these results with those of Sec. 4.5, where we discussed the equilibrium shape of a hypothetical "liquid" satellite. We obtained there that for decreasing orbital distances the satellite becomes more and more elongated in the radial direction, until at some minimum distance (the so-called *Roche limit*) no shape can fulfil any more the equilibrium conditions. Using the results listed in Table 4.2 (for the p = 0 case, i. e., negligible satellite-to-planet mass ratio), it is easy to show that the equilibrium of a small liquid satellite requires that

$$\frac{a}{R} > 2.45\left(\frac{\rho_p}{\rho_s}\right)^{1/3}. \qquad (14.7)$$

The numerical coefficients in eqs. (14.4), (14.6) and (14.7) are different because they were derived from different assumptions about the geometrical configuration of the orbiting bodies. However, the concept of Roche limit is a very general and useful one, whenever we deal with the behaviour of self-gravitating bodies orbiting in the vicinity of a primary. In particular, as shown by Fig. 14.3, the main parts of the planetary ring systems actually lie inside or between the different limits derived above.

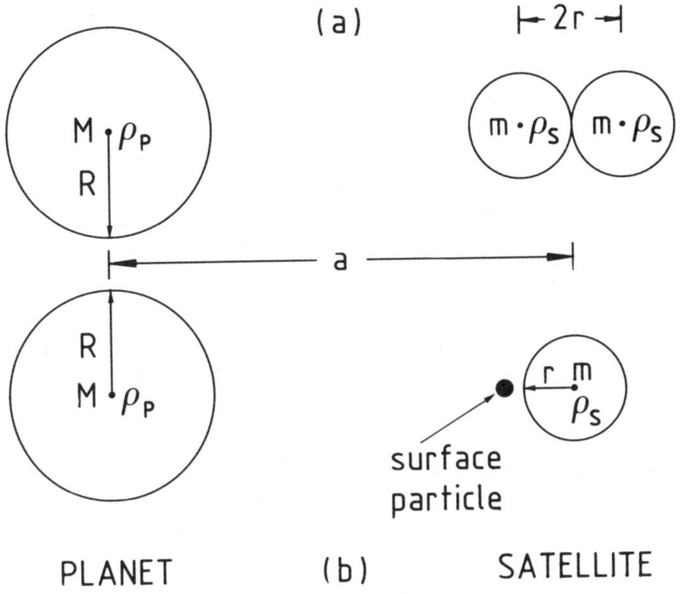

Fig. 14.4. The configurations considered to derive two different values of the Roche limit.

*14.7. FROM BOULDERS TO PLANETS

Due to some basic properties of two fundamental physical interactions – the electromagnetic force giving rise to the solid state strength and the gravitational force – an important transition occurs among the minor bodies of the solar system. For sizes smaller than a critical threshold, the shape of the bodies can be defined as "irregular", and depends essentially on their origin and history. On the other hand, beyond this critical size self-gravitational forces can over long times overcome the solid state rigidity of the material, and if we neglect minor surface elevations the shapes fit the figures of gravitational equilibrium.

As discussed in Chs. 3 and 4, these figures are different, depending on whether the object can be viewed as isolated or as a satellite distorted by the tides of its primary, and of course depend on the internal density distribution of the body. While for a star like the sun the latter factor is very important (since the central density is $\approx$ 100 times larger than the mean density), for most small solar system bodies the internal density gradient is low. Gravitational compression of the interior is limited and, even in cases of mineralogically differentiated interiors, the density cannot decrease by more than a factor 2 or 3 from the core to the outer layers. Thus, as a first approximation, we can apply the classical theory of the equilibrium shapes of homogeneous bodies, which specifies the shape parameters once the density and the spin rate of the object are known. In all the cases of interest for us, the equilibrium figures are ellipsoids; if $a \geqslant b \geqslant c$ are the three semiaxes, the shapes are determined unambiguously by the two ratios ($\leqslant 1$) c/a and b/a.

For isolated objects, the relevant equations have been derived in Sec. 3.4. We recall that as the angular momentum increases, the initial, almost spherical shape becomes more and more flattened. A threshold is reached beyond which the preferred equilibrium shape is no more an axisymmetric, but a triaxial ellipsoid. Finally, still higher angular momenta of rotation are not consistent with equilibrium of a single body, and fission into a binary or multiple system is expected to occur. As we shall discuss later, these results can be applied to understand the shapes of the asteroids.

The shape of small natural satellites, on the other hand, can be studied by using the theory of Roche ellipsoids summarized in Sec. 4.5. If the rotation of the satellites is synchronized with their orbital motion (this is the case for all but the very distant moons), both the spin rate and the angular momentum of rotation depend on the orbital radius. Therefore, for decreasing orbital radii we get shapes more and more elongated in the radial direction until, beyond the Roche limit, no equilibrium figure is any more possible (see Sec. 14.6)

For small bodies the solid state strength is often large enough to prevent relaxation toward equilibrium. In these cases, the shapes can retain an important record of the previous history of the objects, in particular when dominated by large collisional events. Indeed, it is well known that most bodies of the solar system have undergone a heavy bombardment by different populations of projectiles, so that the surviving small bodies are probably the outcome of a process of collisional fragmentation. The expected resulting shapes are rugged and irregular, like those of the fragments which can be produced in the laboratory by high-velocity impact experiments.

The critical size for the transition described above can be estimated by defining the *strength* S (Sec. 1.3) as the stress at which the material constituting the body is subject to rupture or deformation, and requiring that the maximum height h_{max} of the topographic relief on a celestial body is such that the corresponding local gravitational stress is of the order of S. We have then

$$S \cong \rho g h_{max} = 4\pi G \rho^2 R h_{max}/3 \ , \qquad (14.8)$$

where G is the gravitational constant, ρ, g and R are the density, surface gravity and radius of the body, assumed to be nearly spherical and homogeneous. Let us define a "regular" body as one whose surface topography is lower than, say, 10% of its radius. This implies that at the transition $h_{max} = R/10$, hence the critical radius is

$$R_{cr} = \sqrt{(15S/2\pi G\rho^2)}, \qquad (14.9)$$

which of course depends on the density and strength of the material constituting the object. It is difficult to give a reliable theoretical estimate of S for the materials forming the small bodies of the solar system, because it depends on the detailed microscopic structure and the electrostatic forces keeping together the molecules. The strength has the dimensions of an energy density.

An upper estimate of S is provided by the latent heat of fusion of the material, that is the energy needed to destroy the lattice arrangement at constant volume; since by supplying this energy the material becomes liquid and flows easily, it is plausible that a lower energy is sufficient to cause either disruption or "fluid" behaviour over long time scales (i. e., with very high viscosity). The heat of fusion is of the order of kT_m/m, where T_m is the melting temperature and m the molecular mass; hence $S < kT_m\rho/m$. For materials of geophysical interest (e.g., ice, metals and silicates) $kT_m/m \approx 10^9$ to 10^{10} erg/g. The laboratory values of S for terrestrial rocks, ice and meteorites vary between 10^5 and 10^9 dyn/cm^2. With densities of 1 to 3 g/cm^3, from eq. (14.9) we can only conclude that R_{cr} must be in the range from 10 to 1000 km. Probably values close to the upper limit are not realistic, because small solar system objects probably consist of weakly consolidated aggregates accumulated under conditions of small gravitational self-compression. Moreover, for km-sized and larger objects the strength might be substantially lower than for small laboratory samples of similar composition, owing to the presence of large defects or cracks caused by differential thermal contraction or collisions. The extreme case is that of a "pile of rubble", namely a body formed by a collection of small fragments held together only by their mutual gravitational attraction. In this case the strength is very low and large departures from equipotential surfaces are not possible.

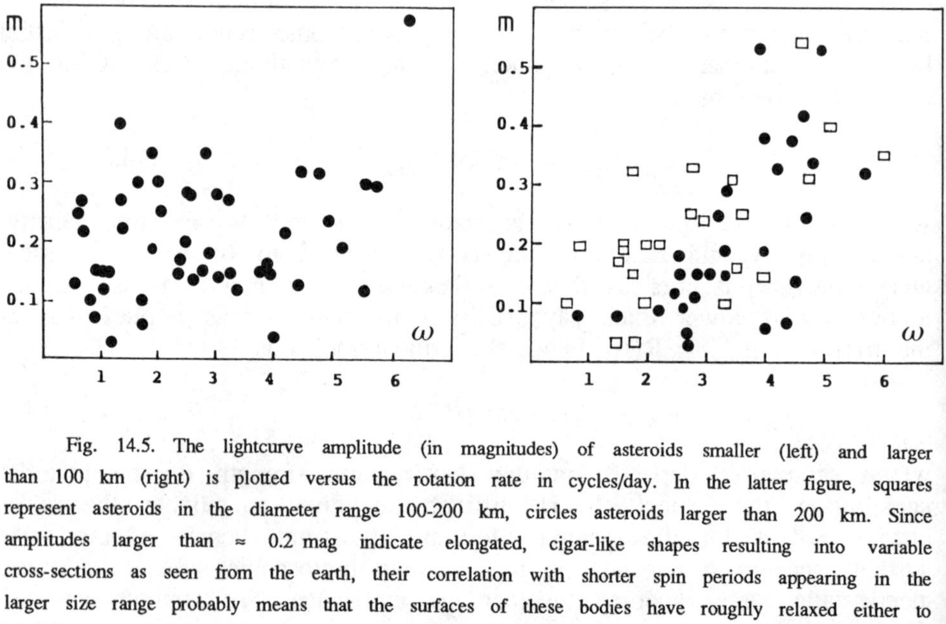

Fig. 14.5. The lightcurve amplitude (in magnitudes) of asteroids smaller (left) and larger than 100 km (right) is plotted versus the rotation rate in cycles/day. In the latter figure, squares represent asteroids in the diameter range 100-200 km, circles asteroids larger than 200 km. Since amplitudes larger than ≈ 0.2 mag indicate elongated, cigar-like shapes resulting into variable cross-sections as seen from the earth, their correlation with shorter spin periods appearing in the larger size range probably means that the surfaces of these bodies have roughly relaxed either to biaxial or to triaxial equilibrium shapes, depending on their angular momentum of rotation.

What is the available evidence? The relevant data for asteroids come mainly from photometric lightcurves, whose amplitude is a good indicator of the departure of the asteroid shape from axial symmetry (see Fig. 14.5). An analysis of the available data has shown that the transition from "irregular" fragments to nearly-equilibrium shapes (including both axisymmetric and triaxial ellipsoids) probably occurs in the diameter range 100-200 km. However, it is likely that this reflects the increased proportion of "rubble piles" in this size range (as a consequence of catastrophic impacts against asteroids having a significant self-gravitational binding – see Sec. 14.3), rather than the intrinsic strength of asteroidal materials. As for small satellites, the situation is

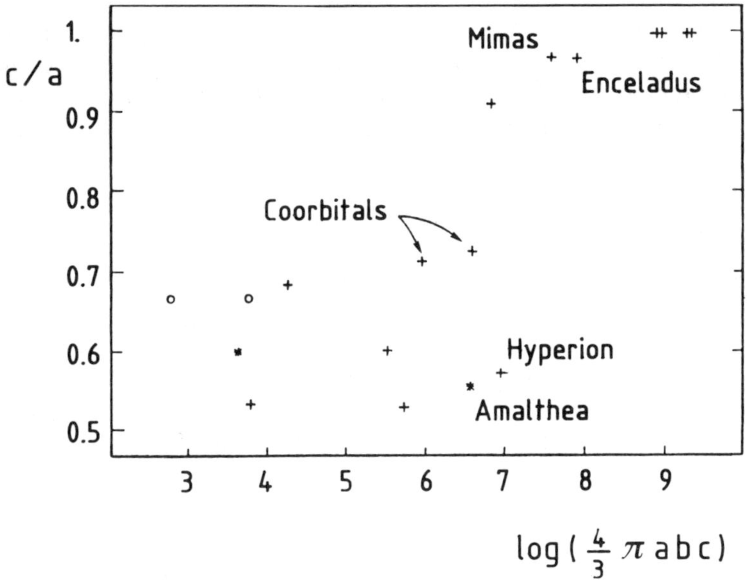

Figure 14.6. The axial ratio c/a for various satellites of Mars (o), Jupiter (*) and Saturn (+), as a function of the logarithm of their volume (in km^3), computed assuming ellipsoidal shapes. The satellites close to the transition from "irregular" to nearly spherical shapes are explicitly indicated. Titan and the four galilean satellites, which have nearly spherical shapes and volumes larger than 10^{10} km^3, are not plotted in the figure. The two "coorbitals", Janus and Epimetheus, have almost the same semimajor axis.

summarized in Fig. 14.6, where we have plotted the axial ratio c/a versus the logarithm of the volume for the bodies of the Martian, Jovian and Saturnian systems whose available images provide some estimate of the overall shape. It is clear that for volumes larger than about 10^7 km^3, corresponding to mean radii of the order of 150 km, we have only nearly-spherical objects. The largest bodies below this threshold, namely Hyperion, Phoebe, Amalthea and the Saturnian coorbitals, lie in an interesting transition region, since some of them may have been distorted by tidal effects or have had a peculiar collisional history. Inverting eq. (14.8) with $R_{cr} \approx 150$ km, we obtain that $S \approx 10^7$ dyn/cm^2, consistent with the previous discussion about the most likely values of S. At the same time, it is interesting to note that for small satellites the transition size is comparable to that obtained for asteroids, although the collisional history and the composition of the two classes of objects are certainly quite different.

Another useful approach is the following. If we can measure the maximum height of the topographic relief on a given body, this gives a lower bound to the strength of the material forming the surface layer of the object. This lower bound is close to the real strength when the surface shows no sign of endogenic evolution (e. g., tectonic and volcanic activity, melting phenomena, etc.) capable of affecting substantially the topographic features, and we can exclude any significant creep deformation which would cause slow relaxation to the equilibrium figure. For a viscous body the relaxation time scale of a non-equilibrium bulge of size comparable with the radius R is

$$\tau = k\eta/G\rho^2 R^2 . \qquad (14.10)$$

Apart from the numerical coefficient k (of order unity), this formula is obtained by recalling that the viscosity η is the ratio between the gradient of pressure ($\approx \rho g \approx GR\rho^2$) and the second spatial derivative of the flow velocity ($\approx (R/\tau)/R^2 = 1/R\tau$) (see Sec. 1.4). Of course τ is longer for topographic features of wavelength smaller than R. The viscosity has a strong temperature dependence, whose form can be expressed by the empirical formula

$$\eta(T) = \eta_0 \exp(-\, bT_m/T), \qquad (14.11)$$

where η_0 and b are experimentally determined constants (b $\cong$ 18 for many materials, including water ice), while the dependence on pressure of the melting temperature T_m can be neglected up to R $\approx$ 1000 km. At temperatures significantly lower than the melting temperature, silicate materials have viscosities at least of the order of 10^{26} g/cm•s. Using eq. (14.11), it is easy to verify that cold silicate bodies up to sizes of the order of 1000 km can retain their topography against creep deformation for times comparable with, or longer than, the age of the solar system. On the other hand, for ice at $T \cong T_m$ a reasonable estimate of the viscosity is 10^{10} g/cm•s, although the uncertainty in applying this to extraterrestrial bodies is considerable. Thus, eqs. (14.10, 11) indicate that the smaller, geologically inactive icy moons can have retained substantially irregular shapes, while for larger and/or thermally evolved icy satellites we expect fairly fast relaxation of large scale topography (e.g., giant impact craters). These expectations have been confirmed by the Voyager images of Saturn's satellites.

PROBLEMS

*14.1. Fit (see Sec. 20.1) the planetary semimajor axes from Venus to Uranus (including Ceres) with a power law of the form $a_n = A + B\alpha^n$, with n = 0,...,6. For A = 0.4 AU, B = 0.3 AU and $\alpha = 2$, this yields the so-called *Titius-Bode law* (Mercury and Pluto can be accomodated by arbitrarily choosing the values $n = -\infty$ and n = 7, respectively; Neptune does not fit anyway.) Derive the best-fit values of the parameters A, B and α; repeat the procedure for all the nine planets (excluding Ceres). The same fit can be done for the satellite systems of Jupiter, Saturn and Uranus.

14.2. Both for the solar system and for the satellite systems of the outer planets, estimate the fraction of the total angular momentum that is accounted for by the rotation of the central body.

14.3. Assume that in the asteroid belt there are 5000 bodies capable of shattering a target asteroid 200 km across. If the volume of the belt is $\approx 10^{26}$ km^3 and the average relative velocity of asteroids (due to their eccentricities and inclinations) is ≈ 5 km/s, what is the lifetime of the target asteroid against collisional fragmentation?

14.4. Estimate the Roche limit for two sticking spherical particles aligned in the radial direction and synchronously spinning about their centre of mass, as a function of the ratio of their sizes.

14.5. In order to obtain another estimate of the transition radius R_{cr} (see Sec. 14.7), assume that the deepest non-collapsing hole in a quasi-spherical homogeneous body has a depth d_{max}, and require that the strength S is equal to the hydrostatic pressure on the spherical surface of radius $(R - d_{max})$, with $d_{max} \cong 0.9 \cdot R$.

14.6. A comet in Oort's cloud has a circular orbit with radius of 20,000 AU. Assume that, as a consequence of a stellar passage, the orbital velocity suffers a random deviation, keeping the magnitude unchanged. What is the probability that the perihelion distance becomes smaller than 15 AU?

FURTHER READINGS

For a general overview of the current knowledge on the bodies forming the solar system, see J. Kelly Beatty, B. O'Leary and A. Chaikin (eds.), *The New Solar System*, Cambridge University Press, Cambridge (1981). A basic textbook with an original, comparative approach to planetary science is W.K. Hartmann, *Moons ans Planets* (2nd edition), Wadsworth, Belmont (1983). The *Space Science Series* published by University of Arizona Press (T. Gehrels, general editor) includes several comprehensive and updated volumes on various individual planets and subsystems of the solar system.

15. DYNAMICAL EVOLUTION OF THE SOLAR SYSTEM

The solar system is like a large natural laboratory, where many dynamical experiments are taking place simultaneously. They range from slow orbital evolution by tiny, accumulating changes due to gravitational perturbations acting over billions (for the satellites, even thousands of billions) of revolutions; to locking phenomena and protection mechanisms caused by resonances; to drastic, essentially unpredictable orbital changes associated again with resonances and with close orbital encounters; to steady evolution driven by non-gravitational interactions (e. g., tides and radiation forces), going on until the bodies are disrupted, removed from the system, or locked into a stable end state. In this Chapter we shall describe in some detail the most important such mechanisms, although often skipping over the rigorous derivations and technicalities that characterize many well established methods of celestial mechanics and dynamical astronomy.

15.1. SECULAR PERTURBATIONS AND THE STABILITY OF THE SOLAR SYSTEM

One of the oldest and most fundamental problems of celestial mechanics is the stability of the solar system, that is whether the planets have always orbited around the sun in more or less the present orbits and always will in the future. Here "always" has to be intended in a physically meaningful sense, namely since the formation of the solar system (about $4.5 \cdot 10^9$ y ago) and for a comparable time in the future (limited by the lifetime of the sun as a main sequence star). The problem is not just an abstract one. For instance, we know that life, present on the earth since more than $3 \cdot 10^9$ y ago, depends critically – via the energetic balance of the biosphere – on the intensity of solar illumination, which in turn is a function of the earth's orbital parameters. On shorter time scales – 10^4 to 10^8 y – substantial climatic changes are known to take place (see Sec. 7.1), probably correlated with changes in the eccentricity of the earth's orbit and the relative orientation of its perihelion and polar axis. Similar paleoclimatological problems have been raised by the images of Mars, showing erosional structures due to ancient flows of liquid water on the presently frozen surface. Another important issue is related to the current theories on the origin of the solar system (see Ch. 16), which rely on the basic assumption that the planets have not significantly changed their orbits and have formed not very far from their present locations. It is therefore frustrating that up to now no definite answer has been found about the stability of the solar system; although, as we shall see, fast and powerful computers have broken a long lasting standstill on this problem.

The motion of the planets is usually studied in the framework of the so-called gravitational N-body problem, in which the only forces are the newtonian point-mass attractions among the various bodies. Usually, N is about 10, neglecting the influence of minor bodies like asteroids, satellites and comets (their effect is indeed very small, the main exception being the earth's moon). This simple approach is possible because other forces (non-gravitational interactions, corrections due to the finite sizes of the bodies, relativistic effects, etc.) are smaller by many orders of magnitude. For instance, the aspherical shapes of the planets are not important because no approach closer than $\approx 10^4$ radii occurs; general relativistic effects are of the order of the ratio between the sun's gravitational radius ($\cong$ 1.4 km) and the orbital distances, and they do not reach 10^{-7} times the solar attraction, even for Mercury; the mass of the sun has a relative change (due to the emission of electromagnetic radiation and solar wind) not exceeding about 10^{-13} per year.

A further simplification of the problem comes from the fact that the mass of the sun is much larger (by a factor $> 10^3$) than the mass of any planet. As discussed in Chs. 10 and 11, this allows to define an approximate, zero-order solution in which every planet moves around the sun in a keplerian orbit. Such a solution is represented by six orbital elements $c_{k,i}$ (k = 1,..., 6; i = 1,..., N − 1) for every planet:

$$c_{k,i} = (a_i, e_i, I_i, \omega_i, \Omega_i, M_i), \qquad (15.1)$$

that is, semimajor axis, eccentricity, inclination, argument of perihelion, longitude of the ascending node and mean anomaly (see Ch. 11). In the following we shall drop the index k. When the eccentricity and/or the inclination are small, it is better to use another set of orbital elements, which are non-singular for vanishing values of e and I (see Sec. 11.4):

$$\begin{aligned} &a_i, && \lambda_i = M_i + \omega_i + \Omega_i, \\ h_i &= e_i \sin(\omega_i + \Omega_i), & k_i &= e_i \cos(\omega_i + \Omega_i), \qquad (15.2) \\ P_i &= \tan I_i \sin\Omega_i, & Q_i &= \tan I_i \cos\Omega_i. \end{aligned}$$

As we have already discussed in Sec 11.5, for time scales not much longer than the periods of revolution, the keplerian elements are approximately constant (apart from M_i or λ_i, whose time derivatives are the mean motions n_i and functions of the semimajor axes only). If we exclude the occurrence of close approaches between planets, the ratios of the mutual gravitational forces between planets and the forces due to the sun are of the same order of magnitude as the planet-to-sun mass ratios μ_i. Neglecting order-of-magnitude differences between the planetary masses, we can then set up a formal approximation using $\mu = O(\mu_i)$ as the smallness parameter (note that this is consistent with the definition of μ (11.50)). Using perturbation theory in Lagrange's form (Sec. 11.4) we obtain for the elements of the j-th planet (collectively indicated with c_j) their change due to the i-th planet:

$$dc_j/dt = \mu_i F_{ji}(a_j, e_j, \ldots, M_j, a_i, e_i, \ldots M_i). \qquad (15.3)$$

Therefore the unperturbed keplerian elements can be expected to be constant within $O(\mu)$ for times less than $O(1/\mu)$ periods of revolution. When these "zero-order" orbital elements are substituted in the right-hand side, the solution of eq. (15.3) differs from the exact solution by terms $O(\mu^2)$. In principle, since the right-hand side is a known function of time and of the initial conditions, the first-order equations can be solved by quadratures. As outlined in Sec. 11.5, the procedure can then be iterated: substituting the first-order solutions into the right-hand side of (15.3), a new solution can be found to $O(\mu^2)$, and so on. But the quadratures are be performed by Fourier and Taylor expansions; and since to work out the n-th iteration a solution of order (n – 1) must be used to compute the right-hand sides for all the elements and all the planets, the number of terms very soon becomes intractably large. A complete, third-order solution has never been obtained, and even second-order solutions must be produced by computer algebra techniques. Unfortunately, in order to obtain information about the stability of the solar system over time scales of the order of 10^9 y (that is $\approx 1/\mu^3$ periods of revolution), we would need at least a third-order theory; and even in this case, in general all the coefficients of the terms $O(\mu^4)$ are not of order unity.

As mentioned in Sec. 11.5, an important simplification of this analytical approach is the *averaging method.* Instead of the complete solution as a function of time, we seek only the average value of the orbital elements over the relevant orbital periods (*secular perturbations*). These running averages are slow functions of time and their series expansions are much simpler. To compute the averaged values of the orbital elements, we average the differential equations (15.3) over time. To lowest order, F_{ij} changes only (periodically) through the mean anomalies M_i; hence we can write

$$\frac{d\langle c_j\rangle}{dt} = \frac{1}{(2\pi)^2}\int_0^{2\pi} dM_j \int_0^{2\pi} dM_i \mu_i F_{ji}(\langle c_j\rangle, M_j, \langle c_i\rangle, M_i). \qquad (15.4)$$

The mean elements satisfy the equations obtained by summing over $i \neq j$, to take into account all the perturbing planets. For this purpose, we can just average the perturbing function $\mathbf{R}_{ji}$ (as given by eq. (11.36)) and use it in the Lagrange equations for the secular perturbations. Note, by the way, that according to eq. (11.36) $m_j\mathbf{R}_{ji}$ is not the interaction energy between the i-th and the j-th planet and is not symmetric in the two indices, due to an additional (*indirect*) term corresponding to the motion of the sun produced by the i-th acting planet. However, it can be easily shown that this indirect term does not give rise to first-order secular effects. According to eq. (11.36), if S is the heliocentric angle between the perturbed body (at $\mathbf{r}_1$) and the perturbing planet (at $\mathbf{r}_2$), the average of the indirect term of $\mathbf{R}_{12}$ is proportional to

$$\left\langle \frac{r_1 \cos S}{r_2^2} \right\rangle = \frac{1}{(2\pi)^2}\int_0^{2\pi} d\theta_1 \int_0^{2\pi} d\theta_2 \, \frac{r_1 \cos S}{r_2^2} \, \frac{r_2^2}{a_2\sqrt{(1-e_2^2)}} \, \frac{r_1^2}{a_1\sqrt{(1-e_1^2)}} =$$

$$= \frac{1}{(2\pi)^2 a_2\sqrt{(1-e_2^2)}\,a_1\sqrt{(1-e_1^2)}} \int_0^{2\pi} r_1^3 \, d\theta_1 \int_0^{2\pi} d\theta_2 \cos S \, ,$$

where $\theta = \omega + f$; we have used the angular momentum $h = r^2 d\theta/dt$ to change the integration variables from M_1, M_2 to θ_1, θ_2 (see Sec. 10.2). The last integral is equal to zero.

A first important result apparent from the averaged equations is the so-called *Lagrange's theorem* on the stability of the solar system: the averaged semimajor axes $\langle a_i\rangle$ in the first order approximation are constant (as noted in Sec. 11.5, this follows simply from the fact that da_i/dt is proportional to $\partial\mathbf{R}/\partial M_i$). Of course, this classical result does not solve the problem of the stability of the solar system. Apart from the fact that the first order solution can be expected to be accurate only for times much shorter than the age of the solar system, the theorem only refers to semimajor axes; the averaged eccentricities do undergo significant secular perturbations, which might lead to collisions even with no change in semimajor axes. However, over time scales of $O(1/\mu)$ periods of revolution, that is 10^3 to 10^4 y, the approximation of constant semimajor axes is fairly good.

The computation of the secular perturbations in eccentricity and inclination faces one more technical difficulty: the integrals contained in the averaged equations for e, I, ω and Ω are not elementary; in fact, the functional relationship between elements and cartesian coordinates (needed to compute **R** as a function of the elements) contains the transcendental Kepler's equation (10.34). However, we can also expand **R** in powers of the eccentricities and the inclinations (or, rather, the non-singular elements h, k, P, Q). Indeed, it is a basic property of the solar system that they are small and, in fact, roughly of the order of $\sqrt{\mu}$. Only the two extreme planets, Mercury and Pluto, make an exception, but they are small and do not affect significantly the orbits of the other planets. We can therefore confidently adopt a formal perturbation method in the single smallness parameter μ.

For this purpose, we need the averaged perturbing function $\langle \mathbf{R}_{ji} \rangle$ up to quadratic terms in h, k, P, Q; then we apply the perturbation equations (11.48) to determine the secular evolution of the non-singular elements. Some important conclusions about the structure of $\langle \mathbf{R}_{ji} \rangle$ can be drawn by noting that, as we have shown above, $m_j \langle \mathbf{R}_{ji} \rangle$ is the average potential energy of the gravitational interaction between the two planets – or between two equivalent rings of matter resulting from spreading the planetary masses along their orbits. We can thus consider the interaction between eccentric and slightly inclined rings which can be labelled by the same indices j and i used earlier for the planets. Since $m_j \langle \mathbf{R}_{ji} \rangle = m_i \langle \mathbf{R}_{ij} \rangle$, the averaged perturbing function is (apart from a constant factor) symmetric with respect to the interchange of the planets i and j. The contribution linear in h_j and k_j covers the case in which j is eccentric and i is circular and coplanar; it must vanish, because such an interaction is certainly independent of the sign of the eccentricity e_j (recall that changing the sign of the eccentricity is equivalent to interchanging pericentre and apocentre). Similarly, the contribution linear in P_j and Q_j, corresponding to the interaction between a circular ring i and another circular and inclined ring j, vanishes because it must be independent of the sign of the relative inclination. Only quadratic terms can contribute and therefore we have an averaged perturbing function of order μ. One can also see that the cross terms of the type eccentricity times inclination are not present, since the action of a circular ring upon another eccentric and inclined ring (and *viceversa*) cannot depend upon the sign of the inclination. Finally, cross terms of the types $h_i k_j$ and $P_j Q_i$ are also absent; indeed, the interaction of a circular ring and another circular and inclined ring depends only upon the (squared) mutual inclination

$$I_{ij}^2 = I_i^2 + I_j^2 - 2I_iI_j\cos(\Omega_i - \Omega_j) =$$

$$= P_i^2 + P_j^2 + Q_i^2 + Q_j^2 - 2(P_iP_j + Q_iQ_j); \qquad (15.5)$$

similarly, the interaction between an eccentric ring and another eccentric and coplanar ring can depend only upon the scalar product of their Lenz vectors (see Sec. 10.2)

$$\mathbf{e}_i \cdot \mathbf{e}_j = e_ie_j\cos(\omega_i - \omega_j + \Omega_i - \Omega_j) = h_ih_j + k_ik_j. \quad (15.5_2)$$

From the previous considerations, the averaged (and truncated to terms of second degree) perturbing functions $\langle\mathbf{R}_{ji}\rangle$ and $\langle\mathbf{R}_{ij}\rangle$ depend on the eccentricities and the inclination just through

$$\Sigma_{ij}[f(e_i^2 + e_j^2) + g\mathbf{e}_i \cdot \mathbf{e}_j + \ell I_{ij}^2], \qquad (15.6_1)$$

where the coefficients f, g and ℓ depend on the semimajor axes a_i and a_j and the masses of the planets. Introducing the non-singular elements, we can express the interaction energy between the two planets (apart from a constant) as

$$W_{ij} = m_j\langle\mathbf{R}_{ji}\rangle = [B_{ij}(h_ih_j + k_ik_j) + C_{ij}(P_iP_j + Q_iQ_j)]; \qquad (15.6_2)$$

the matrixes B_{ij} and C_{ij} are symmetric and of order of magnitude $- Gm_im_j/a$ (a being the largest between a_i and a_j).

Since the semimajor axes in our approximation are constant, Lagrange's perturbation equations (11.48) are linear and homogeneous:

$$\frac{dh_j}{dt} = \frac{1}{m_jn_ja_j^2}\Sigma_i B_{ij}k_i,$$
$$\frac{dk_j}{dt} = -\frac{1}{m_jn_ja_j^2}\Sigma_i B_{ij}h_i, \qquad (15.7)$$

and similarly for the P, Q pair. From eqs. (15.7) it follows that

$$\Sigma_j m_jn_ja_j^2(h_j^2 + k_j^2) = \Sigma_j m_jn_ja_j^2e_j^2 = \text{constant} \qquad (15.8)$$

and therefore, provided the masses and the semimajor axes are comparable, eccentricities and inclinations are bounded. Moreover, if we use the $(N - 1)$-dimensional vector notation $\mathbf{h}$ for h_i, etc. and redefine

$$h_i' = h_i\surd(m_in_ia_i), \quad k_i' = k_i\surd(m_in_ia_i),$$

$$B'_{ij} = B_{ij}/\surd[(m_j n_j a_j^2)(m_i n_i a_i^2)], \text{ etc.},$$

we can write the Lagrange equations (11.48) in the form:

$$\frac{d\mathbf{k}'}{dt} = -\mathbf{B}' \cdot \mathbf{h}', \quad \frac{d\mathbf{h}'}{dt} = \mathbf{B}' \cdot \mathbf{k}', \qquad (15.9)$$

and similarly for **P'** and **Q'**. Of course, analogous equations can be written for the other pair of non-singular elements. The matrixes ***B'*** and ***C'*** are symmetric and have entries depending only on the masses and the semimajor axes of the planets. From their definition, it is easy to see that they have the dimension of a frequency and their order of magnitude is the product between μ and the mean motions. From eqs. (15.9) we can see that **h'**, **k'**, **P'** and **Q'** fulfil the equation of the harmonic oscillator

$$\frac{d^2\mathbf{k}'}{dt^2} = -\mathbf{B}'^2 \cdot \mathbf{k}', \text{ etc.} \qquad (15.10)$$

Since the matrix $\mathbf{B}'^2$, being the square of a symmetric matrix, is also symmetric, it can be diagonalized to real and positive eigenvalues. Hence the general solution of eqs. (15.9) has the form :

$$k_j = \Sigma_s R_{js} U_s \cos(f_s t + \delta_s), \; h_j = \Sigma_s R_{js} U_s \sin(f_s t + \delta_s), \qquad (15.11_1)$$

$$Q_j = \Sigma_s S_{js} V_s \cos(g_s t + \epsilon_s), \; P_j = \Sigma_s S_{js} V_s \sin(g_s t + \epsilon_s). \qquad (15.11_2)$$

U_s, V_s, δ_s, ϵ_s are constants of integration depending upon the initial conditions, while the frequencies f_s and g_s are of the order of μ times the mean motions, typically corresponding to periods of $10^{5\pm1}$ y. The solution requires the amplitudes U and V to be small, formally $O(\surd\mu)$; it is called a *Lagrange*, or *linear solution.*

To understand its qualitative features, a simple geometric description is useful: in a fixed reference plane the vectors of components (k_j, h_j) and (P_j, Q_j) undergo epicyclic motions: their motion is the vector sum of circular uniform motions with as many independent frequencies as there are planets (One of the frequencies for (P, Q) is actually zero, owing to the constraint provided by the conservation of the total angular momentum **L** of the system; indeed, the component of **L** in the ecliptic is a linear combination of the inclinations.) Were the behaviour of each planet dominated by a single perturbing body, then each epicyclic motion would be essentially a circular motion with one dominant frequency f_1, superimposed to small epicycles with different frequencies; in this case all the perihelia (or the nodes) would be circulating in an almost uniform way with periods $2\pi/f_i$, and one could identify each frequency with the single planet whose perihelion is dominated by that frequency. For the actual solar system, Jupiter's dominance is not very strong; hence there are several cases of perihelia (and nodes) which do not revolve with a well defined frequency, an example being the earth's perihelion. As a consequence, the earth's eccentricity changes in a quasi periodic, but complex way, thus contributing to the complex variations of the solar constant in paleoclimatology. Another recently discovered case in which the simple description of the secular motions of a planet by means of a main frequency fails, is that of Uranus. Uranus' perihelion is "locked" to Jupiter's perihelion, i. e. they share the main epicyclic frequency; thus the angle $\omega + \Omega$ of Uranus minus the same angle for Jupiter "librates" around 180^0 with an amplitude of about 70^0 and a period of $1.1 \cdot 10^6$ y; Uranus' eccentricity oscillates with the same period.

How can the Lagrange solution be improved, taking into account the terms which have been neglected? Firstly, we could include the terms of degree 4 in the eccentricities and inclinations in the perturbing potential; and second, since e^2, $I^2 = O(\mu)$, we must at the same time develop a second-order theory by substituting the complete first-order solution in the right-hand side of the perturbation equations. Such a theory (order 2 and degree 4) has indeed been computed both for the planets and for the asteroids; however, the number of Fourier components used to describe the motion of the perihelia and nodes is very large, since the equations substituting (15.9) now have a right-hand side of degree 3 in the h, k, P, Q variables and the result will contain all the combinations of three fundamental frequencies. Only some of these terms have significant amplitudes, but it is impossible to know *a priori* which ones. For a third-order and sixth degree theory, the difficulties become unsurmountable, as the number of Fourier harmonics with combinations of 5 fundamental frequencies grows to hundreds of thousands. A large number of small divisors arises whenever some of the combinations of frequencies is close to zero; therefore, even the ordering of the terms by their magnitude is not easy, since the amplitudes of some third-order terms are larger than many second-order terms. Moreover, the advantage of having constant semimajor axes is lost. As a conclusion, we can state that secular perturbation theories can give a good approximation to the orbits of the planets for times up to 10^5 or 10^6 y, but not much longer.

Given the difficulties faced by analytical theories, a more direct approach by means of numerical integration of the equations of motion is attractive. The procedure may seem very simple: a finite difference approximation to the differential equations of motion is constructed, with some given step size (either fixed or variable) and initial conditions matching the observed orbits. However, there are at least three classes of problems that must be addressed before starting such a numerical computation:

1. Can a numerical integration covering a time span not much shorter than the age of the solar system (i. e., at least 10^9 y) be completed in a reasonable time, even by the fastest computers available?

2. Can the numerical error over such a time span remain small enough to keep the required level of accuracy?

3. Can the output of a numerical integration be compared with analytical theories, or be anyway translated from strings of numbers (orbital elements as functions of time) into expressions containing the qualitative informations that are really interesting?

The latter question follows from the fact that the important issue is not to know the position of the planets at any time, but whether their

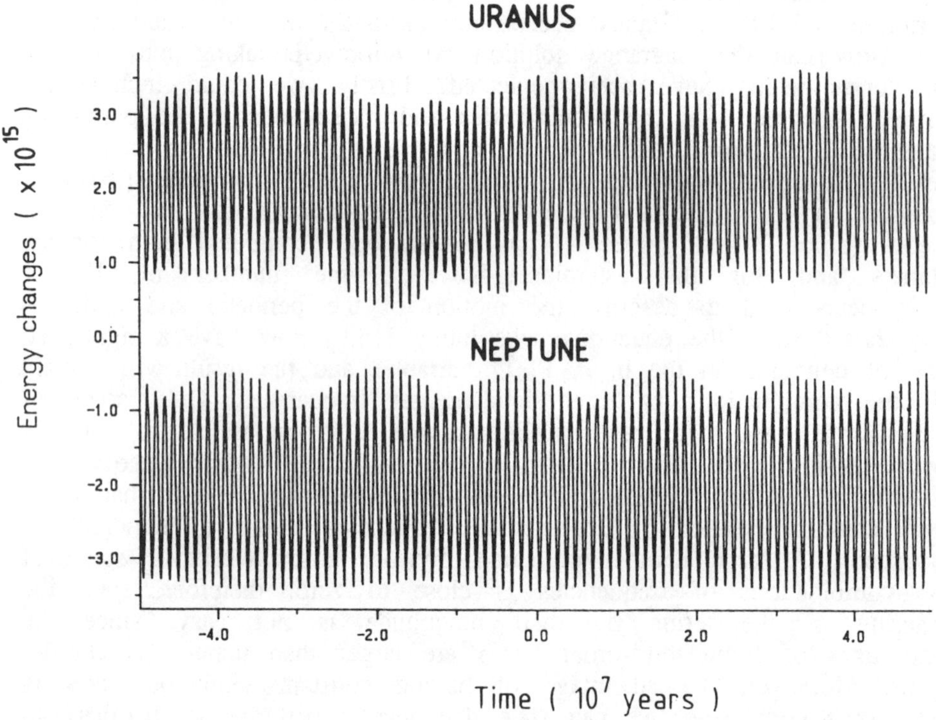

Fig. 15.1. Relative changes in the orbital energies of Uranus and Neptune over ≈ 10^8 y, as computed from a numerical integration (LONGSTOP 1B) of the five outer planets, with short periodic terms removed by a digital filter. The main periodic term has a period of about $1.1 \cdot 10^6$ y. The oscillations for the two planets being out of phase, there is an energy exchange between the two orbits, whose period is the same as the period of the locked libration of the perihelia of Uranus and Jupiter. In this plot, energies are measured in solar masses by $(AU)^2/(day)^2$; in this unit the average energies of Uranus and Neptune are of the order of $(-2 \cdot 10^{-10})$.

orbital elements vary in a quasi-periodical way, which are the longest periods involved and, in particular, whether the available secular perturbation theories correctly predict all the most important periods or there are new dynamical features. To answer these questions, one can generate a so-called "synthetic" secular perturbation theory, that is a model of the long-term evolution of planetary orbits obtained from the positions computed numerically, neglecting irrelevant, short-period effects. The needed procedure is based on *digital filtering*: a low frequency pass filter has to be designed, adapted to the available output and to the predicted spectrum of the data, and applied to the time series of the orbital elements as obtained from the numerical integration.

For eccentricities and inclinations, the computed variations are frequently in quite good agreement with analytical secular perturbation theories. As we have seen, the changes in eccentricity and inclination are controlled by first-order effects, and there is no reason to doubt the reliability of theories to $O(\mu)$. At higher order the situation is more complicated. An example is the behaviour of semimajor axes. To first order they should be constant, apart from short-period effects; therefore in this case the filter is essential because the second-order changes are much smaller than the first-order, short-period changes. The results are interesting: for instance Uranus and Neptune exchange some 10^{-5} of their orbital energies (proportional to the inverse of the semimajor axes), with exactly the same $1.1 \cdot 10^6$ y periodicity of the relative oscillation of Jupiter's and Uranus' perihelia (see Fig. 15.1). The other planets show similar effects, which were unpredicted by most analytical theories.

The main difficulty in addressing the second problem does not arise from the inaccuracy inherent in the replacement of the differential equations with a finite difference scheme, that is from the *truncation error*, well understood and predictable in a reliable way. The main problem arises from the fact that computers do not perform arithmetic operations between real numbers; they actually can remember, hence manipulate, only integer numbers (normally up to 2^{63}). Thus real numbers are approximated by keeping a limited number of decimal figures, i.e. there is a so-called *rounding-off error*. Since usually decimal figures are just dropped out, the error is systematic, and may cause problems after a very large number of operations. For example, if the rounding-off error in an integration step of length h has a relative size η, after N_s steps the accumulated error is of order $N_s^2 \eta h$: indeed, a systematic error in one of the mean motions at each step will cause a longitude shift accumulating quadratically with the elapsed time. In an orbital period about 100 steps are used; if, for example, $\eta \approx 10^{-48} \simeq 3 \cdot 10^{-15}$, the integration fails after $1/(100\sqrt{\eta})$ periods, a few million years for the outer planets and about 10^5 y for the inner planets. Some integrations for the outer planets spanning $\approx 10^8$ y have been carried out, but they have required either special methods of handling the equations to minimize errors, or the use of *ad hoc* computers.

Finally, we can now meaningfully address question (1), that is: are our computers fast enough? The reply is that such very great computational tasks, as the very long integrations of planetary orbits quoted above, have become technically possible in the 80's thanks to the availability of vectorial supercomputers. To give an idea of the needed resources, we recall that a 10^8 y integration of the five outer planets (the so-called *LONGSTOP* project) required a few days of CPU time on a Cray 1S, the fastest supercomputer available for academic research in 1986. But this estimate is likely to become out of date soon.

*15.2 RESONANCES AND CHAOTIC BEHAVIOUR

For a generic dynamical system, a *resonance* occurs when one of its periods is nearly matched by some external, periodic driving force. The simplest such case – the forced harmonic oscillator – shows that the amplitude of the response of the system can grow much more than in non-resonant situations, and is limited only by dissipation and/or by the onset of non-linear behaviour. If the evolution of a system of bodies like planets, asteroids or satellites depends on a linear combination of the proper frequencies with small integers as coefficients (*commensurability*), the same configuration is periodically repeated; then the mutual gravitational perturbations play the role of the driving resonant force and their effects are strongly enhanced. This may have either a stabilizing effect, constraining any further evolution of the system to the dynamical routes which preserve a *resonant locking* (like the Neptune-Pluto mean motion resonance, the behaviour of Hilda and Trojan asteroids, many satellite pairs or even triplets, the spin-orbit resonant coupling of Mercury); or, alternatively, may lead to instabilities, chaotic behaviour, close encounters and even collisions. The latter case applies to any asteroid or small particle that originally lied in, or was subsequently injected into, the *Kirkwood gaps* (Sec. 14.3), i.e., the *secular resonances* in the asteroid belt, as well as to ring particles formed in such sites as Cassini's division between the main Saturn's rings (Sec. 14.6). Instability mechanisms related to resonances for asteroidal orbits are especially important because they can explain the transport towards the terrestrial planets of meteorites and of Aten-Apollo-Amor asteroids.

We now discuss a model of an important, but simple orbit-orbit resonance; it is suitable to describe the resonance between the two saturnian satellites Titan and Hyperion, but many of its features are more general. Two interacting bodies have coplanar orbits about a primary; the outer body (Hyperion) has a negligible mass, but a significant eccentricity ($\approx$ 0.1); the inner body (Titan) has a circular orbit. We have, therefore, a planar, restricted three-body problem. Most important, the orbital periods are commensurate; having in mind the Titan and Hyperion pair, we assume a 4/3 ratio, that is, while Titan completes four revolutions, Hyperion makes three. As a consequence, in exact resonance, the conjunctions – when the two bodies are aligned with the primary on the same side – repeat at the same planetocentric longitude every three orbital periods of Hyperion. This is important because the gravitational pull of Titan is strongest near the conjunctions.

If the conjunction occurs while Hyperion is moving from pericentre to apocentre (see Fig. 15.2) and, therefore, has an outward velocity component, the radial component of Titan's attraction pulls inward; hence energy and angular momentum are removed from Hyperion's orbit and its semimajor axis and period decrease. The effect is enhanced by the fact that the point of closest approach, where the pull is strongest, occurs somewhat earlier than the conjunction, when Titan, which moves faster, still trails behind. As a consequence of Hyperion's speeding up,

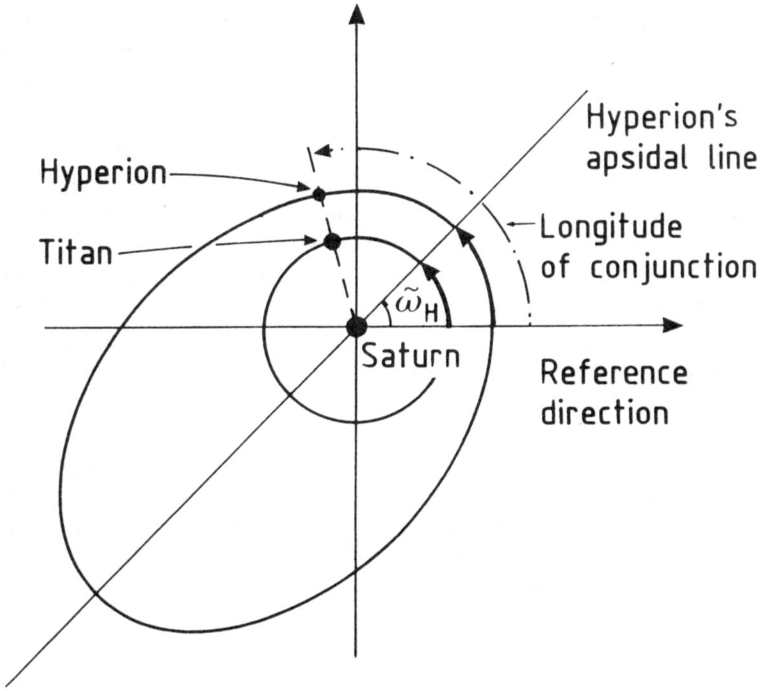

Fig. 15.2. A simplified model for the analysis of the stable resonance locking between Titan and Hyperion. Titan's orbit (the inner one) about Saturn is nearly circular. Here the conjunction between the two satellites is assumed to occur after Hyperion's passage at pericentre. Resonant perturbations from Titan cause the longitudes of subsequent conjunctions to shift in such a way that they oscillate about the direction of Hyperion's apocentre.

the conjunction moves towards the apocentre. If the conjunction occurs while Hyperion is moving from apocentre to pericentre, the opposite occurs: Titan's pull increases Hyperion's energy, slowing it down and moving the conjunction towards the apocentre.

Thus, if the initial ratio of the orbital periods is close enough to the rational value 4/3, we have a stable, pendulum-like oscillation of the conjunction position about the apocentre that tends to preserve the 4/3 ratio against such destroying effects as the differential tidal evolution of the two orbits. Like in a pendulum, if at the stable equilibrium configuration the ratio of periods is far enough from commensurability, conjunctions could in principle circulate through 360° instead of oscillating. As we shall see later, this latter behaviour is dangerous, because the conjunction is no more constrained to occur near the longitude at which the perturbing influence of Titan is weakest, and the resonance does not provide any more as *protection mechanism* against close encounters. As far as the real Titan-Hyperion pair is concerned, we indeed observe a libration of the conjunctions about apocentre with an amplitude of 36° and a period of 1.75 y.

In reality the problem is more complicated because Titan affects not only the mean motion of Hyperion, but also its eccentricity and pericentre. A quantitative model, based on Lagrange's perturbation equations, is called for. The disturbing function **R** is expanded in a Fourier series, having as arguments linear combinations of the mean longitudes λ_T and λ_H, with coefficients given by appropriate functions of the semimajor axes a_T and a_H and the eccentricity e_H. In this expansion we can retain only the resonant term, which has a very low frequency because of the commensurability of mean motions (see Sec. 11.2). This *critical argument* is

$$\sigma = 4\lambda_H - 3\lambda_T - \omega_H \ , \tag{15.12}$$

where in this planar problem ω_H denotes the longitude of the pericentre from a fixed direction. The combination (15.12) is invariant under a change in the origin of angles, as it should be.

Note that at conjunction, where $\lambda_H = \lambda_T$, σ is the angular distance of the two bodies from the pericentre; when the resonance is exact, σ does not change with time and defines the conjunction longitude at all times. The relevant part of the perturbing function must be proportional to Hyperion's (small) eccentricity, without which there is no effect; this is contrasted with the case of secular perturbations (Sec. 15.1) in which the terms linear in the eccentricities are absent and quadratic contributions are dominant. The main term in the Fourier expansion of the perturbing function is a linear combination of $\sin\sigma$ and $\cos\sigma$; since we expect a stable equilibrium configuration when the conjunction is at apocentre ($\sigma = 180°$), the perturbing function must be of the form

$$\mathbf{R} = (Gm_T/a_H)\cdot F(a_T/a_H)\cdot e_H\cdot \cos\sigma \ . \qquad (15.13)$$

$F(a_T/a_H)$ is a positive function of order unity; m_T is the mass of Titan. The semimajor axes are assumed to be of the same order of magnitude, so that $\mathbf{R}$ is of the order of the interaction energy.

We can now use Lagrange's equations (11.46) in the limit $e_H \to 0$:

$$\frac{de_H}{dt} = -\frac{1}{n_H a_H^2 e_H}\frac{\partial \mathbf{R}}{\partial \omega_H} = -\mu_T n_H F \sin\sigma \ , \qquad (15.14_1)$$

$$\frac{d\omega_H}{dt} = \frac{1}{n_H a_H^2 e_H}\frac{\partial \mathbf{R}}{\partial e_H} = \frac{\mu_T n_H}{e_H} F\cos\sigma \ , \qquad (15.14_2)$$

$$\frac{dn_H}{dt} = -\frac{3}{a_H^2}\frac{\partial \mathbf{R}}{\partial \lambda_H} = 12 e_H \mu_T n_H^2 F \sin\sigma \ . \qquad (15.14_3)$$

μ_T (= 0.00024)is the Titan-to-Saturn mass ratio; we have used Kepler's third law. From Eqs. (15.14) we obtain

$$\frac{d\sigma}{dt} = 4n_H - 3n_T - \frac{d\omega_H}{dt} = 4n_H - 3n_T - \frac{\mu_T n_H}{e_H} F\cos\sigma \ ; \qquad (15.15)$$

differentiating again

$$\frac{d^2\sigma}{dt^2} = 48 e_H \mu_T n_H^2 F \sin\sigma - \frac{\mu_T^2}{e_H^2} n_H^2 F^2 \sin\sigma\cos\sigma +$$

$$+ \frac{\mu_T}{e_H} n_H F \frac{d\sigma}{dt} \sin\sigma - 12\mu_T^2 n_H^2 F^2 \cos\sigma\sin\sigma \ . \qquad (15.16)$$

We note first that the last term in the right-hand side is negligible with respect to the second and can be dropped. The pendular oscillations are governed by the first term, corresponding to the frequency

$$\omega_r = n_H\sqrt{(48 F e_H \mu_T)} \ . \qquad (15.17)$$

If that term is prevailing, we have a stable equilibrium at $\sigma = 180^0$ and an unstable one at $\sigma = 0$; in this case $d\sigma/dt$ is of the order of ω_r, and so is the resonance offset $4n_H - 3n_T$. The other two terms are much smaller if

$$\omega_r >> \mu_T/e_H ,$$

that is to say, when

$$e_H >> (\mu_T/48)^{1/3} = 0.02 ,$$

amply satisfied in the Titan and Hyperion case. Notice that $e_H << 1$, otherwise our treatment would be inconsistent. This solution depends on the positive sign of F; the situation would be reversed if the massless body were the inner one. We have thus recovered the conclusion previously inferred from qualitative arguments: thanks to a sizable eccentricity of Hyperion's orbit, Titan's perturbation adjusts n_H in such a way to preserve the resonance and to cause the apocentric libration of the conjunction.

In the opposite limit, when e_H is sufficiently small, the dominant terms in eq. (15.16) are the last two. They correspond to coupled variations in the eccentricity and the pericentre: for a nearly circular orbit the line of apses is easily reoriented, as shown by the fact that $d\omega_H/dt \propto 1/e_H$ (see eq. (15.14)). On the other hand, $dn_H/dt \propto e_H$ and therefore the variations of n_H in this case are unimportant. The dependence on σ of the second term in (15.16) shows that we have now stable equilibria both at $\sigma = 0$ and at $\sigma = 180°$. Thus pericentric libration is also possible, but close approaches are, anyway, ruled out by the very small value of the eccentricity. The resonant behaviour can also be fruitfully studied in this case by considering the pair of variables $h_H = e_H \cdot \cos\sigma$, $k_H = e_H \cdot \sin\sigma$ and noticing that

$$dh_H/dt = - k_H (4n_H - 3n_T) , \qquad (15.18_1)$$

$$dk_H/dt = h_H(4n_H - 3n_T) - \mu_T F n_H . \qquad (15.18_2)$$

Since for very small values of e_H, n_H is nearly constant, the solution of eqs. (15.18) in this case is

$$h_H = A\cos[(4n_H - 3n_T)t + \delta] + B , \qquad (15.19_1)$$

$$k_H = A\sin[(4n_H - 3n_T)t + \delta], \qquad (15.19_2)$$

where A and δ are constants depending on the initial conditions and

$$B \equiv \mu_T n_H F/(4n_H - 3n_T) .$$

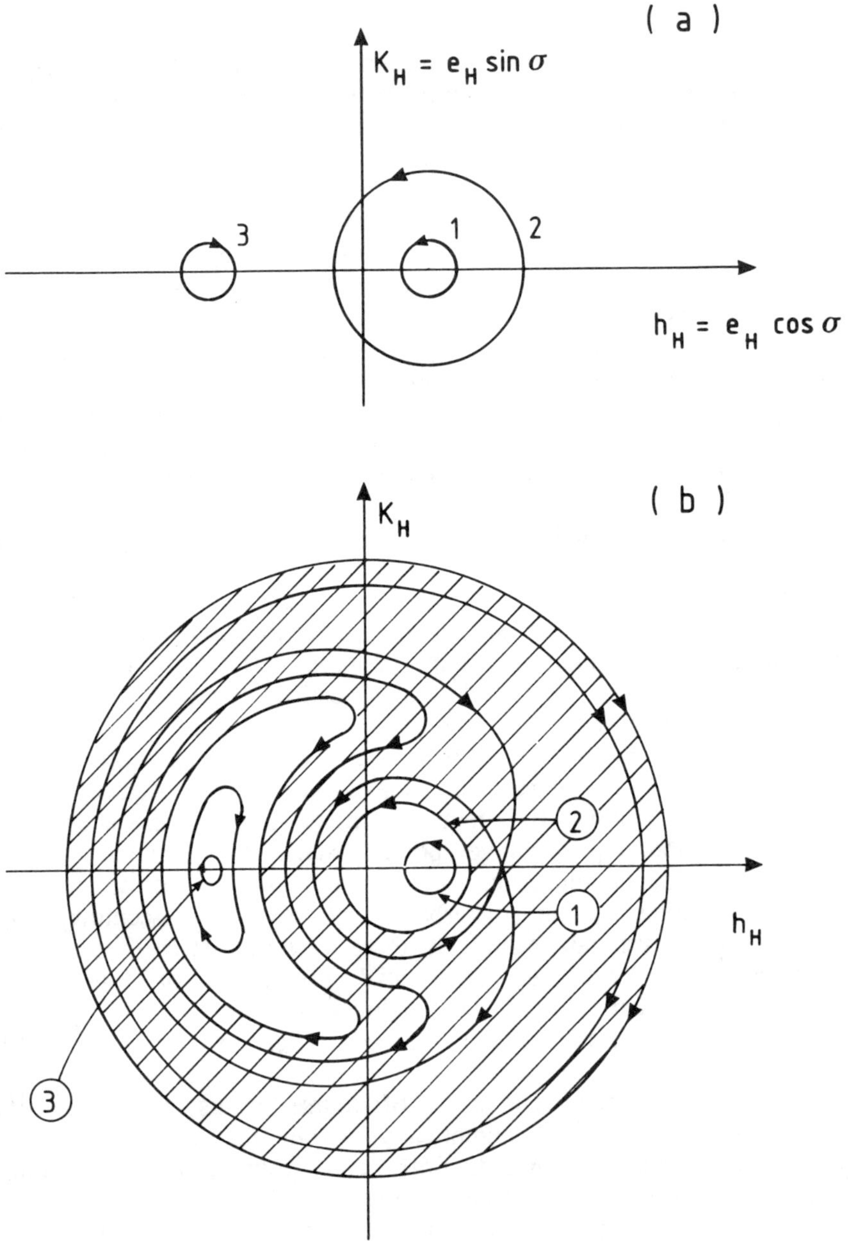

Fig. 15.3 The small eccentricity resonance mechanism (a) gives rise in the (h_H, k_H) plane to three possible types of trajectories: librations about $\sigma = 0$ (1) and about $\sigma = 180°$ (3), and circulations. For larger eccentricities (b), these trajectories are preserved only when $e_H \sin\sigma$ stays small (cases (1), (2) and (3) in (b)). Otherwise, the only stable oscillations of sizable amplitude can occur about $\sigma = 180°$, consistently with the analysis of the larger eccentricity mechanism. Both in (a) and in (b) the different curves correspond to different initial conditions of the massless body. The shaded region in (b) corresponds to initial conditions resulting into chaotic behaviour.

Thus, in the (h_H, k_H) plane (corresponding to the polar coordinates e_H and σ; see Fig. 15.3(a)), the motion occurs on a circle of radius A about the point (B, 0), at the rate ($4n_H - 3n_T$). For $A > |B|$, σ circulates; for $A < |B|$, σ librates about 0 (if $B > 0$, namely $n_H > 3n_T/4$), or about 180° (if $B < 0$). Notice that the small-eccentricity mechanism cannot work close to the exact resonance, because in this case $|B|$ is very large and e_H is not small. The transition between the small and the large eccentricity modes occurs as follows: if A is very small, so is $e_H \sin\sigma$ and thus dn_H/dt is negligible; but if A is large, even when $e_H \sin\sigma$ starts near zero, it increases (or decreases) as the system moves along its circular path on the (h_H, k_H) plane. Then n_H will vary, gradually shifting the position of the centre specified by B; it can be shown that the trajectories become distorted into a pattern of crescent-like oscillations and circulations as those shown in Fig. 15.3.

In fact, the situation is even more complicated. If we plot in the (**h**, **k**) plane of Fig. 15.3(b) a point representing the orbit every conjunction, we can obtain two clearly distinct behaviours. In the unshaded regions, the successive points tend to form the smooth curves shown in the figure, indicating that the orbit is regularly evolving in one of the two resonant modes described earlier. However, if the initial conditions are such that the orbit starts in the shaded zone (e.g., if we try to generate high-amplitude apocentric librations), the successive points tend to scatter in an irregular manner throughout the shaded area. Even if they initially follow the regular crescent-like curves shown in the figure, when the representative point reaches the extreme of such a quasi-libration, the conjunction occurs too close to pericentre ($\sigma = 0$), and the resulting strong disturbance destroys the seemingly regular behaviour. The essential feature of these *chaotic orbits* is that they are always associated with an exponential divergence of initially nearby trajectories. In other words, the evolution of the system is extremely sensitive to initial conditions, and since initial conditions can never be known with unlimited accuracy, after some time the evolution becomes completely unpredictable. The same of course occurs when the equations of motions are integrated numerically within the chaotic zone of the phase space, due to rounding-off and truncation errors that are unavoidably present in such computations.

The fact that the phase space is divided into a regular and a chaotic domain is a common feature of many dynamical systems, dissipative and conservative as well. In the latter case, at least two degrees of freedom are required to have chaotic behaviour, because for systems with one degree of freedom the energy integral ensures stability with respect to the initial conditions. For systems with more than one degree of freedom, both the regular and the chaotic domains in the phase space may be present, unless as many independent integrals are available as the number of degrees of freedom. Chaos results from time-dependent perturbations and non-linear couplings between the variables, and is commonplace especially in the presence of resonances, i. e. (in this more general framework), of small-integer ratios between two or more of the fundamental frequencies associated with the motion along each degree of freedom. Chaotic situations cannot be studied by analytical techniques based on an expansion of the perturbing function (which in these cases often diverges), nor by averaging methods, for which an obvious precondition is a regular, quasi-periodic behaviour. In order to investigate the extent and the properties of chaotic zones for various problems of physical interest, there are two techniques:

[1] *Direct numerical integrations*. Apart from the problems, quoted in Sec. 15.1, that arise from numerical errors, in many cases of interest numerical integrations are extremely time consuming. Owing to the degeneracy of Kepler's problem, which has three degrees of freedom but a single fundamental frequency, and the smallness of the perturbations by other planets (or satellites), many fundamental frequencies (e.g., $d\omega/dt$ or $d\Omega/dt$) are often so small that 10^5 or 10^6 revolutions must be integrated before the full structure of the phase space becomes apparent. This is not the case only if chaos is associated with frequently repeated close orbital encounters, which on the other hand require that special tricks – like changing step size – be adopted in the integration procedure.

[2] *Mappings*. The exact differential equations describing the evolution of the system are replaced by an algebraic mapping obtained by substituting in the averaged equations the resonant interaction with a periodic sequence of delta functions ("impulsive kicks"), which produce discontinuous changes in the orbital elements. The new equations can thus be integrated analytically both between the delta functions and across them, providing a map of the phase space into itself that qualitatively mimics the real behaviour of the system, but allows a much faster computation algorism. Consequently, for many problems more systematic investigations over longer spans of time become possible, though one has to assume that the mapping can well represent the behaviour of the original differential equations. This assumption is based on plausibility arguments, but has also been amply verified by comparisons with the results of conventional numerical integrations.

Our treatment of the Titan-Hyperion case can be immediately applied to other cases, e. g., to asteroids orbiting in the 2/1, 3/2 and 4/3 mean motion commensurabilities with Jupiter. However, there are a variety of other resonant mechanisms which have been shown to explain the dynamical behaviour of various subsystems of the solar system. Although differing in many details, they all share many typical features of the simple case discussed above (i. e., stable librations of suitably defined critical arguments and protection mechanisms on the one hand, and the possible onset of chaos on the other). A schematic list includes:

[1] *Secular resonances.* The method outlined above has another interesting application in the case of the asteroids which are far from commensurabilities and thus are not affected by mean motion resonances. These asteroids can be considered as massless bodies perturbed (predominantly) by Jupiter over time scales much longer than the orbital periods. If we again restrict ourselves to the planar problem (no mutual inclination), we can use a perturbing function of the form (15.6_1):

$$\mathbf{R} = -(Gm_J/a)[f_1 e^2 + f_2 e e_J \cos(\omega - \omega_J)], \qquad (15.20)$$

where the orbital elements without an index refer to the asteroid and f_1, f_2 are dimensionless functions of the ratio a/a_J. Then Lagrange's equations (11.46) yield:

$$de/dt = -f_2 n e_J \sin(\omega - \omega j) ,$$

$$d\omega/dt = -n[2f_1 + f_2(e_J/e)\cos(\omega - \omega_J)].$$

The quantity $\varphi = \omega - \omega_J$ fulfils

$$\frac{d\varphi}{dt} = \frac{d\varphi_0}{dt} - nf_2 \frac{e_J}{e} \cos\varphi , \qquad (15.21)$$

where

$$\frac{d\varphi_0}{dt} = \left(\frac{d\omega}{dt}\right)_0 - \frac{d\omega_J}{dt}$$

results from two contributions, the former being the precession rate of the asteroid, were Jupiter's orbit exactly circular (due to the f_1 term in eq. (15.20)), while the latter is the precession rate of Jupiter due to the perturbations by the other planets, as discussed in Sec. 15.1. One can easily see that the quantities $h = e\cdot\cos\varphi$, $k = e\cdot\sin\varphi$ fulfil the equations

$$dh/dt = -k d\varphi_0/dt , \qquad (15.22_1)$$

$$dk/dt = h d\varphi_0/dt - f_2 n e_J , \qquad (15.22_2)$$

of the same structure as eq. (15.18) for the case of Titan and Hyperion. Their solutions for small eccentricities are the same as (15.19), to wit,

$$h = e_0\cos(\varphi_0 + \delta) + f_2 n e_J/(d\varphi_0/dt) \qquad (15.23_1)$$

$$k = e_0 \sin\varphi_0 . \qquad (15.23_2)$$

Thus we have a *forced eccentricity* (the last term in eq. (15.23_1)) proportional to e_J, which shifts the centre of the circular path in the (h, k) plane away from the origin. The radius e_0 of the circle is the *proper eccentricity* and depends on the initial conditions. When $d\varphi_0/dt \to 0$ the forced eccentricity diverges and we have a *secular resonance.* Here the time scale of resonant behaviour (e. g., librations and circulations, as well as chaotic behaviour) is longer by a factor 10^3 or 10^4 than for mean motion resonances, since this is the typical ratio between mean motions and orbital precession rates. Similar relations can be derived for inclinations and nodal longitudes in the non-planar problem. For asteroids far from mean motion and secular resonances, the proper eccentricities and inclinations are almost constant (this is of course true only for small eccentricities and inclinations, consistently with our lowest-degree truncation of $\mathbf{R}$). These *proper elements,* which can include also the semimajor axis (once freed from short-periodic perturbations), are stable over times much longer than the corresponding osculating elements, and are therefore useful to study the long-term evolution of asteroid orbits and to classify asteroids into *dynamical families* (see Sec. 14.3).

[2] *Coupled* mean motion and secular resonances, with alternate librations of the arguments σ (referring to Jupiter) and φ, are observed for several asteroids. They can be analysed by applying Lagrange's equations to a perturbing function including both the secular and the longitude-dependent resonant terms (problem 15.1).

[3] Higher-order resonances, with ratios p/q between the mean motions such that the order $|p - q| > 1$. The critical arguments can be defined in a way similar to the previous one, but the corresponding terms in the perturbing function are proportional to higher powers (at least, $|p - q|$ powers) of the eccentricities and/or inclinations, so that these resonances are "weaker" than those of order 1. This is due to the fact that conjunctions occur at $|p - q|$ different longitudes, and therefore the accumulation of perturbing effects is less powerful. However, in the asteroid belt clear gaps are observed at the 3/1, 2/5 and 3/7 resonances with Jupiter (see Fig. 14.1); numerical experiments have shown chaotic behaviour there, leading to sudden increases of the eccentricity and possible encounters with planets.

[4] Resonances between bodies of similar mass, so that "restricted" models are not applicable. This is the case for instance of the pair of saturnian moons Enceladus and Dione. In this as in other cases, moreover, the precession of apsidal and/or nodal lines due to the planet's oblateness cannot be ignored.

[5] Inclination-related mechanisms, as for the pair of saturnian satellites Mimas and Tethys. In many cases, these resonances provide an effective protection mechanism, as the stable location of the conjunctions is far from the mutual node of the orbits.

[6] Resonances between bodies whose orbits are not well separated in heliocentric distance. This is the case for the Trojan asteroids, whose stable librations about the triangular Lagrangian points of the Sun-Jupiter system can be analysed by assessing the stability of the equilibrium positions in the three-body problem (see Sec. 12.3). More difficult is the case of Pluto, whose perihelion distance is well inside the orbit of Neptune. Pluto is protected by a 3/2 mean motion commensurability with Neptune, with conjunctions librating about Neptune's aphelion with a period of about $2 \cdot 10^4$ y. This is somewhat similar to the Titan-Hyperion mechanism, but Pluto's eccentricity is so large that strong interactions with Neptune are possible even far from the conjunction. Another protection mechanism is actually at work, related to Pluto's sizeable inclination: Pluto's argument of perihelion librates about 90° with a period of $3.8 \cdot 10^6$ y, implying that when conjunctions occur, Pluto is always far away from the plane of Neptune's orbit.

[7] Three-body resonances. The most famous case, already investigated by Laplace in 1829, involves the three Galilean satellites of Jupiter Io, Europa and Ganymede. For this *Laplacian resonance* the critical argument

$$\sigma = \lambda_I - 3\lambda_E + 2\lambda_G$$

is locked at 180°; $n_I - 3n_E - 2n_G = 0$ to within nine significant digits! Since $\sigma = (2\lambda_G - \lambda_E) - (2\lambda_E - \lambda_I)$, the conjunctions of adjacent satellite pairs must always occur 180° apart, and the three satellites can never line up on the same side of Jupiter.

[8] Spin-orbit resonances. We could include in this case all the satellites whose rotation has been synchronized with the orbital period by the action of tides (see Sec. 15.3), like the earth's moon. More interesting is the case of Mercury, whose "day" ($\simeq$ 59 days) lasts 2/3 of the orbital period. This ensures that at perihelion, where the tidal interaction with the sun is most effective, the longer axis of Mercury's equator is always aligned with the radial direction, though two opposite sides of the planet alternately face the sun. Even more peculiar is the rotational behaviour of Hyperion, which in the process of being tidally synchronized has apparently fallen into a state of chaotic tumbling. The reason lies in the strongly irregular shape of this satellite, and in the torque exerted by Saturn, which changes in time because of the eccentricity of Hyperion's orbit. Thus Hyperion's rotational state is essentially unpredictable over time scales as short as one year.

What is the origin of resonances? Probably there is more than a single answer. Whenever tides are effective – as in the evolution of satellite rotations and orbits – they provide a dissipative mechanism that can gradually change the spin and the orbital periods, until capture into a stable resonant state does occur. Alternatively, the origin of some resonances might date back to the origin of the solar system. Resonant orbits might have favoured the accumulation of bodies at some preferred locations, or might have provided protection mechanisms resulting into the natural selection of bodies in "lucky" dynamical configurations. Finally, in the primordial solar nebula additional dissipative mechanisms due to gas drag and interparticle collisions could in some cases have played the same role of tides, albeit on much shorter time scales. On the other hand, the relative lack of resonant bodies in some otherwise crowded regions of the solar system, like the main asteroid belt or Saturn's rings, can probably be explained by the onset of chaotic behaviour. Whenever sudden and irregular orbital changes can occur (e.g., with large jumps in eccentricity), the resonant bodies can be eliminated by close encounters or by collisions, clearing up gaps in correspondence with the resonance zones.

15.3 TIDAL EVOLUTION OF ORBITS

As we discussed in Sec. 4.3, owing to dissipation the tidal bulge raised by the moon on the earth is slightly misaligned with respect to the earth-moon direction. This misalignment causes an accelerating torque on the lunar orbit, as the tidal wave (approximately) facing the moon exerts a stronger attraction on the moon itself than that on the opposite side of the earth. This perturbation leads to a secular increase of the lunar semimajor axis and, in order to conserve the total angular momentum of the system, gradually slows down the earth's rotation (see Sec. 3.2.) The same mechanism has already synchronized with the orbital period the rotation of the moon as well as that of most other natural satellites in the solar system. It is possible to draw several interesting conclusions about the final state just on the basis of conservation laws.

If we make the simplifying assumptions that the lunar orbit is circular, with radius r, and lies in the earth's equatorial plane, the total angular momentum and the energy of the system are

$$L = I_\oplus \omega_\oplus + \frac{\sqrt{G M_\oplus} m}{\sqrt{(M_\oplus + m)}} \sqrt{r}, \qquad (15.24_1)$$

$$E = \tfrac{1}{2} I_\oplus \omega_\oplus^2 - \frac{G M_\oplus m}{2r}, \qquad (15.24_2)$$

where $M_\oplus$, $I_\oplus$, $\omega_\oplus$ are the earth's mass, momentum of inertia and rotation rate, while m is the mass of the moon. We have neglected the small contribution to L and E from the rotation of the moon, already synchronized with the orbital period and therefore much slower than $\omega_\oplus$ (the lunar momentum of inertia is also much smaller.) Owing to the tidal friction inside the earth, E decreases with time. On the other hand, L is conserved, so that there is a gradual transfer of angular momentum from the earth's rotation to the orbital motion. Eqs. (15.24) can be rewritten in the form

$$\omega_\oplus(\ell) = \frac{1}{I_\oplus}(L - \ell) , \qquad (15.25_1)$$

$$2E(\ell) = \frac{(J - \ell)^2}{I_\oplus} - \frac{\alpha}{\ell^2} , \qquad (15.25_2)$$

where ℓ is the orbital angular momentum of the system, equal to the second term in eq. (15.24_1) and $\alpha \equiv G^2 m^3 M_\oplus^3/(M_\oplus + m)$. Any dissipation process drives the system to the state of minimum energy, at which

$$dE/d\ell = \ell^4 - L\ell^3 + I_\oplus\alpha = 0. \qquad (15.26)$$

At this point the system is "completely synchronized", i. e., the earth's spin rate becomes equal to the orbital mean motion of the moon; indeed, by Kepler's third law, this condition is equivalent to saying that $\omega_\oplus = \alpha/\ell^3$ which, substituted in eq. (15.25_1), reproduces (15.26). In general, depending upon the values of L and of the product $\alpha I_\oplus$, we have three different cases:

(A) $L > L_{min} = 1.75(I_\oplus\alpha)^{1/4}$: eq. (15.26) has two positive real solutions $0 < \ell_1 < \ell_2$, which correspond to two orbital radii at which the complete tidal synchronization is possible. It is easily verified that E has a maximum at ℓ_1 (unstable) and a minimum at ℓ_2 (stable).

(B) $L = L_{min}$: there is only one real solution for $\ell = 3L/4$, which is easily shown to be unstable.

(C) $L < L_{min}$: no real roots.

These three qualitatively different possibilities are represented in Figure 15.4, where we can see that in the cases (B) and (C) tidal friction unescapably drives the satellite to fall onto the planet (or to disruption inside the Roche limit – see Sec. 14.6). Indeed neither the moon nor any other natural satellite of the solar system has $L \leq L_{min}$. In case (A) the evolution is more complex, and depends on the initial conditions. We have again three possibilities:

(A1) If initially $\ell = \ell_0 > \ell_1$, the system evolves toward the stable synchronization state, with ℓ (and r) increasing and $\omega_\oplus$ decreasing with

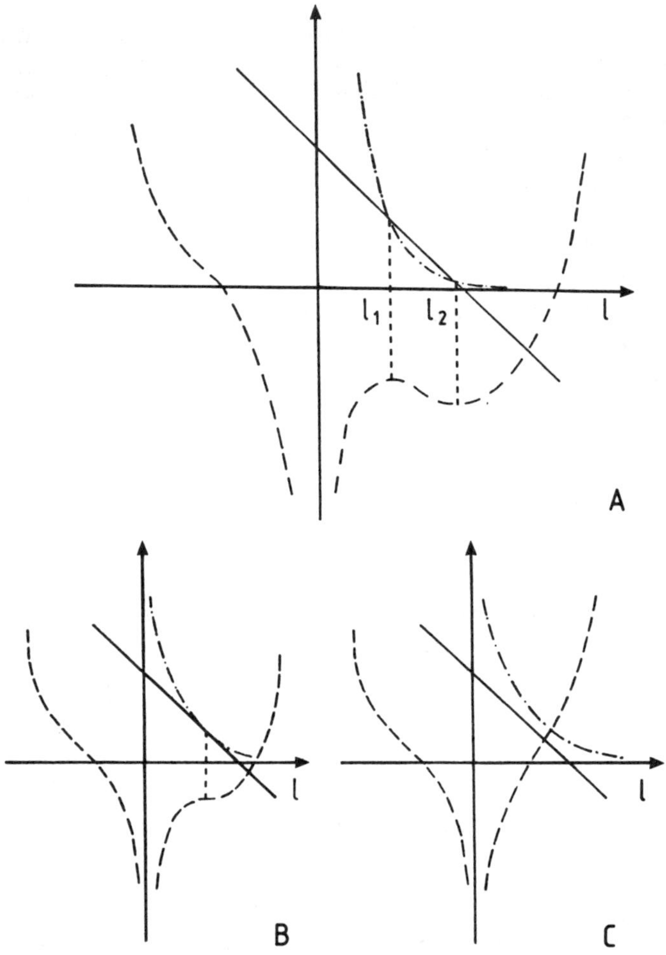

Figure 15.4. For the three cases (A), (B) and (C) discussed in the text, the various lines represent: the conservation of angular momentum, expressed by the equation $\omega_\oplus = (J - \ell)/I_\oplus$ (solid line); the total energy of the system $E(\ell) = (J - \ell)^2/(2I_\oplus) - \alpha/(2\ell^2)$ (dashed line); and the synchronization condition $\omega_\oplus = \alpha/\ell^3$ (dashed and dotted line).

time if $\ell_0 < \ell_2$, and *vice versa* (however, for no natural satellite we have $\ell > \ell_2$).

(A2) If $\ell_0 < \ell_1$, the satellite falls towards the planet while $\omega_\oplus$ gradually increases.

(A3) If $\ell_0 < 0$, i. e., if the initial orbit is retrograde (with respect to the planet's rotation), again tidal friction causes the satellite to fall onto the planet while $\omega_\oplus$ is slowed down.

For the moon and most other natural satellites, the present orbits imply that the case (A1) applies. The moon's stable synchronization distance is of about 88 $R_\oplus$ (to be compared with the present semimajor axis of about 60 $R_\oplus$), and this corresponds to an orbital period of about 48 d, which will also be the final rotation period of the earth (of course, this result is no longer true if we take into account also solar tides, which can cause a further evolution of the system.) Only in one actual case in the solar system, the Pluto-Charon pair, the stable synchronization state has already been reached. On the other hand, two satellites – Mars' Phobos and Neptune's Triton – are evolving like in the cases (A2) and (A3), respectively, approaching their planets towards their final disruption.

What is the time scale of this evolution? As discussed in Sec. 4.2 (see eqs. (4.21) and (4.15)), the gravitational potential generated by the tidal bulge at the moon's distance r is

$$U_T = \left(\frac{R_\oplus}{r}\right)^3 \left(-\kappa\frac{GmR_\oplus^2}{2r^3}\right)[3\cos^2(\psi - \epsilon_T) - 1], \qquad (15.27)$$

where κ is *Love's number* ($\cong$ 0.30 for the earth), ψ is the angle between the considered position and the lunar direction and ϵ_T is the angular lag of the moon with the axis of the earth's tidal bulge. The factor $(R_\oplus/r)^3$ is due to the fact that U_T is generated by a quadrupole deformation and hence behaves like a second-order harmonic. The tidal torque N_T on the moon is

$$N_T = -m\left(\frac{\partial U_T}{\partial \psi}\right)_{\psi=0} = \frac{3\kappa}{2}\frac{Gm^2R_\oplus^5}{r^6}\sin 2\epsilon_T, \qquad (15.28)$$

Equating N_T to the time derivative of the orbital angular momentum of the moon $m\ell$, see eq. (15.24_1), we obtain the differential equation:

$$\frac{dr}{dt} = 3\kappa\cdot\sin 2\epsilon_T\,\frac{m\sqrt{[G(m + M_\oplus)]}}{M_\oplus}\,\frac{R_\oplus^5}{r^{11/2}}\,. \qquad (15.29)$$

For the moon, with $\epsilon_T \cong 3^\circ$, this equation yields at present an increase in the semimajor axis $dr/dt \cong 3.5$ cm/y; from conservation of angular momentum (eq. (15.24_1)) we get $d\omega_\oplus/dt = -N_T/I_\oplus \cong -6\cdot10^{-22}$ rad/s^2, corresponding to an increase of the length of the day of about 2 ms per century. These values have been recently confirmed by laser ranging of the moon and of artificial satellites and are consistent with the historical record of ancient eclipses. If we integrate eq. (15.29) from r_1 to $r \gg r_1$, we find that the corresponding time interval τ_T is insensitive to r_1 and is given by

$$\tau_T \cong 2r/(13dr/dt). \tag{15.30}$$

For the earth-moon system, the observational values quoted above yield a typical timescale τ_T for the tidal expansion of the lunar orbit (say, from just outside Roche's limit, at $r_1 \approx 3R_\oplus$) as short as $1.5 \cdot 10^9$ y. As we shall see in a moment, since ϵ_T is inversely correlated with tidal dissipation, this means that the current rate of dissipation in the earth probably is higher than it has been on the average in the history of the system. A possible reason for this is the present (anomalously?) large abundance of shallow seas at the margins of the currently fragmented continental blocks.

If we now consider the very early phase of tidal despinning of the moon's rotation, according to eq. (15.28) (where we have to exchange the roles of the earth and the moon, indicated with a subscript m) this process was due to a torque

$$N_T' = N_T(M_\oplus/m)^2(R_m/r_\oplus)^5 .$$

Since $d\omega_m/dt = - N_T'/I_m$, we have

$$d\omega_m/dt = (M_\oplus/m)^3(R_m/R_\oplus)^3 d\omega_\oplus/dt .$$

This shows that the corresponding timescale is some 10^4 times shorter than the time needed to synchronize the earth's rotation, so that in a few million years the moon became tidally locked.

How much energy is currently dissipated by tidal friction in the earth? From eq. (15.24_2) we have

$$\frac{dE}{dt} = \frac{d}{dt}\left[\tfrac{1}{2} I_\oplus \omega_\oplus^2 - \frac{GM_\oplus m}{2r}\right] =$$

$$= I_\oplus \omega_\oplus \frac{d\omega_\oplus}{dt} + \frac{GmM_\oplus}{2r^2}\frac{dr}{dt} = N_T\cdot(n - \omega_\oplus), \tag{15.31}$$

where n is the orbital mean motion. Using the value given above for dr/dt (or $d\omega_\oplus/dt$) to estimate N_T, one obtains a dissipation rate of about $3 \cdot 10^{12}$ watts, comparable to that estimated from data on tidal currents. By analogy with the case of the damped harmonic oscillator (see also Sec. 3.6), we can now define the *tidal quality factor* Q as the inverse of the ratio between the energy dissipated during a tidal cycle (about 1 day) and the total energy stored in the tidal bulges. We have then

$$-\frac{dE}{dt} \approx \frac{2\rho g h^2 R_\oplus^2 \omega_\oplus}{2\pi Q} \approx \frac{\omega_\oplus N_T}{\epsilon_T Q} , \tag{15.32}$$

where $(\rho g h^2 R_\oplus^2)$ is of the order of the potential energy stored in a bulge of height $h \approx R_\oplus(m/M_\oplus)(R_\oplus/r)^3$ (see eq. (4.18)) and we have used eq. (15.28). Since $\omega_\oplus >> n$, the comparison between (15.32) and (15.31) shows that Q is of the order of $1/\epsilon_T$, for the earth about 20 (of course, this holds for the low forcing frequencies associated with tides).

A similar argument can be used to estimate the tidal dissipation in a satellite whose rotation is already synchronized with the orbital period, but whose orbit is not circular. This case is important because the surfaces of several satellites of the outer planets display widespread marks of volcanism and/or resurfacing events, probably associated with a melted interior, and tidal friction appears as the most plausible energy source for procucing such melting processes. Usually, dissipation in the satellites tends to reduce the orbital eccentricity, but in some cases a complete circularization of the orbit is prevented by resonant interaction with other satellites, which gives rise to a forced eccentricity term (see Sec. 15.2). This is, for instance, the case for Io, the innermost of the three Jovian satellites involved in the Laplace resonance. Images from the Voyager probes have shown that Io's surface morphology is indeed dominated by volcanism, and several ongoing eruptions have even been directly detected on this satellite. An eccentric orbit causes the tidal bulge of the satellite to oscillate back an forth by an amount $\delta\ell \approx (P_T/\mu) R_m$, where R_m is the satellite's radius, μ is the rigidity of its crust, $P_T = (GMmR_m/r^3)/R_m^2$ is the pressure due to the tidal force (a similar expression was used in Sec. (3.6) to derive eq. (3.44)) and M is the planet's mass. The displaced mass is $\delta m \approx \rho R_m^2 \delta\ell \approx GMm^2/\mu r^3 R_m$ (ρ is the satellite's density) and the corresponding kinetic energy is $\delta m(enr)^2$, where e is the orbital eccentricity and nr the orbital velocity. If we now assume that friction in the satellite's interior dissipates during a cycle (i. e., an orbital period) a fraction 1/Q of this tidal energy, we get

$$-\frac{dE}{dt} \approx \frac{\delta m(enr)^2 n}{Q} \approx \frac{\rho^2 R_m^7 n^5 e^2}{Q\mu} . \tag{15.33}$$

Assuming for Io a crustal rigidity of the order of 10^{12} dyne/cm^2 and a quality factor $Q \approx 20$ (of the same order as those of the earth and the moon), and using the forced eccentricity $e \cong 0.004$ induced by the Laplacian resonance, we obtain a dissipation rate

$$- dE/dt \approx 10^{13} \text{ watts} . \tag{15.34}$$

It corresponds to a dissipation rate per unit mass of $\approx 10^{-7}$ erg/(g s), higher than the radioactive heating rate; it is sufficient to cause extensive melting of Io's interior (For another explanation of Io's volcanism, see Sec. 13.4.)

15.4. DYNAMICS OF DUST PARTICLES

While the orbital evolution of "large" solar system bodies like planets, satellites and even asteroids can be described to a very high degree of accuracy as being solely due to gravitational interactions, this approximation is no longer valid for small solid particles up to ≈ 1 m in size. Interplanetary space is not empty, but contains photons of solar radiation and solar wind particles, both of which carry momentum and, upon being absorbed or scattered by interplanetary dust particles, impose a net force on them. Circumplanetary environments – as those sometimes populated by planetary ring particles – are even more complex, owing to the presence of magnetospheres and of radiation from the planets. Generally speaking, these non-gravitational forces – both radiation forces and drag – are roughly proportional to the cross-section of the particles (for radiation this holds only if the size s is larger than the wavelength of light), so that the corresponding accelerations are proportional to the inverse of the size. This is the reason why these effects are most important for small solid particles, frequently making their motion quite complicated and puzzling. In this section we review the most important types of forces and their consequences on the long-term orbital evolution, basing on the work by J.A. Burns and coworkers.

The force on a perfectly absorbing particle due to solar radiation can easily be derived by simple arguments (see also Sec. 18.1). The energy of a radiation beam intercepted per unit time by an absorbing particle of cross-section S is ΦS. If the particle is moving relative to the source of the beam (e. g., the sun) with a velocity $\mathbf{v}$, Φ must be replaced by

$$\Phi' = \Phi(1 - v_r/c), \tag{15.35}$$

where c is the speed of light and $v_r = \mathbf{v} \cdot \mathbf{e}_r$ is the component of the velocity along the radial unit vector $\mathbf{e}_r$. The factor in parentheses is due to the Doppler effect, which changes the incident energy flux by shifting its frequency. The momentum flux transferred to the particle, producing the *radiation pressure force*, is then

$$\mathbf{f}_{RP} = (\Phi' S/c)\mathbf{e}_r. \tag{15.36}$$

Momentum may also be also exchanged by the radiation emitted by the moving particle. If the radiation is isotropic (small particles are effectively isothermal), in the rest frame there is no net force thereby exerted on the particle. However, in the frame of reference of the sun the emission is anisotropic and a net momentum loss results, giving rise to the *Poynting-Robertson drag*

$$\mathbf{f}_{PR} = -\,(\Phi' S/c^2)\,\mathbf{v}. \qquad (15.37)$$

This follows from the fact that radiation, emitted at the rate $\Phi' S$, is equivalent to a mass loss rate $-\,\Phi' S/c^2$ from the moving particle; this, multiplied by $\mathbf{v}$, gives the momentum loss in the solar frame. To first order in v/c, the total radiation force $\mathbf{f}_R$ reads

$$\mathbf{f}_R = (\Phi S/c)[(1 - v_r/c)\mathbf{e}_r - \mathbf{v}/c] \qquad (15.38)$$

and, decomposing it along the radial and transversal directions, such that $\mathbf{v} = v_r\mathbf{e}_r + r(d\theta/dt)\mathbf{e}_t$,

$$\mathbf{f}_R = (\Phi S/c)[(1 - 2v_r/c)\mathbf{e}_r - (r/c)(d\theta/dt)\mathbf{e}_t]. \qquad (15.39)$$

For real particles the situation is more complicated, because they absorb (and re-emit) only a fraction of the incident light and scatter the remaining fraction in an anisotropic and often complex pattern. As a consequence, the efficiency of momentum transfer depends on the size, shape and optical properties of the particles, as well as on the wavelength distribution of the radiation flux. However, these effects are roughly accounted for by replacing in eqs. (15.35, 36) S with $S\kappa_r$, the effective cross-section. κ_r is a sensitive function of the ratio between the size s of the particle and the typical wavelength λ_R of the radiation. Of order unity for $s \gg \lambda_R$ (when geometrical optics apply), it decreases rapidly for $s < \lambda_R$, so that smaller particles become almost "transparent" to radiation and are less affected by radiation forces.

In order to estimate the importance of these forces, it is useful to consider the ratio β between the quantity $(\Phi S\kappa_r/c)$ and the sun's gravitational attraction. Since $\Phi = \Phi_0(1\ \mathrm{AU}/r)^2$ ($\Phi_0 \cong 1.4\cdot10^6$ erg/(cm^2 s) is the energy flux at 1 AU from the sun, the *solar constant*, see Sec. 7.1), for a spherical particle of radius s, mass m and density ρ, we get (in cgs units)

$$\beta \simeq 7.7\cdot10^{-5}\ \kappa_r S/m = 5.8\cdot10^{-5}\ \kappa_r/\rho s \qquad (15.40)$$

(see also the Problem 10.6). As expected, this force ratio is proportional to 1/s, and if $\kappa_r \approx 1$, $\rho \approx 1$ g/cm^3, it becomes of order unity for particle radii of about 0.5 μm. The solar spectrum is peaked at $\lambda_R \cong 0.5$ μm, and since κ_r decreases faster than s for smaller sizes, in the range 0.1 to 1 μm radiation forces really compete with gravitation. We recall that the zodiacal dust particles are typically larger, their typical size being in the range from 10 to 100 μm, so that for them $\beta \approx 10^{-2}$. It is also remarkable that β is independent of the heliocentric distance. As a consequence, if we restrict ourselves to the main radial term in eq. (15.39) (of order zero in v/c), the orbits remain keplerian conic sections, with a smaller "effective" solar mass. Of course, if $\beta > 1$, so that the outward radiation pressure exceeds gravity, particles move on unbound hyperbolic orbits.

In addition, even for $\beta < 1$, if a small particle originally residing on a larger body is released (e. g., by a collision, or in a comet jet), it will initially have a velocity close to that of its parent, but will feel a weaker solar gravity. Provided the velocity exceeds the escape velocity with respect to this weaker gravity, the particle can be expelled from the solar system. With the two-body formulae it is easily seen that at pericentre, where this mechanism is most effective, ejection occurs for particles with $\beta \geqslant (1-e)/2$. For long-period comets the eccentricity e can approach unity (often within $\approx 10^{-4}$, see Sec. 14.4) and therefore particles of larger sizes can be eliminated in this way. Indeed, micrometeoroid detectors aboard spacecraft have identified a class of particles streaming away from the sun which could have been produced in the inner solar system by collisions or rapid sublimation. Of course, the orbital evolution can become complex if the particle properties, in particular their size, change with time.

For absorbing particles in planetocentric orbits the solar radiation pressure has different effects. Since the perturbing acceleration has essentially a constant magnitude and is directed away from the sun, it performs no work when averaged over one orbit, and therefore no secular or long-period effect in semimajor axis arises (see Sec. 11.1). On the other hand, Gauss' equation (11.6) for de/dt shows that at every revolution e suffers a change of the order of $\beta M_\odot r^2 / M_p r_p^2$, the ratio between the perturbing force and the planet's gravity (M_p and r_p are the planet's mass and orbital radius, respectively and r is the planetocentric orbital radius of the particle.) This eccentricity change builds up (or down) for a time of the order of 1/4 of the orbital period of the planet, which controls the orientation of $\mathbf{f}_R$. Therefore the eccentricity can approach unity, causing escape from the system or collision with the planet, when β exceeds a critical value of the order of $4v/v_p$; here v/v_p is the ratio between the particle's planetocentric velocity and the planet's velocity about the sun (we have used Kepler's third law), e. g., ≈ 0.03 at the orbit of the earth's moon.

Although smaller than the radiation pressure term by a factor of the order of v/c, Poynting-Robertson drag causes important orbital evolution effects for dust particles. Since it acts in the direction opposite to the particle's velocity, this force takes away energy and angular momentum from the orbital motion, decreasing both the semimajor axis and the eccentricity (compare with the gas drag problem discussed in Sec. 11.3.) For particles in heliocentric orbits having eccentricities significantly smaller than unity, Gauss' equations (11.4,6) yield the characteristic decay time for a and e:

$$\tau_{PR} = mc^2/2\Phi\kappa_r S \approx 10^7 \ \rho s r^2/\kappa_r \ y, \qquad (15.41)$$

where ρ and s are again given in cgs units and r in AU. As it could be expected, τ_{PR} is of the order of the time it takes for a particle to be struck by its own equivalent mass in solar radiation. The Poynting-Robertson decay occurs on a time scale not exceeding 10^3 y for μm-sized particles, and even metre-sized bodies can be lost from 1 AU in the solar system's lifetime. A similar computation for the semimajor axis decay of a planetocentric orbit yields the same value of τ_{PR}, provided for r we take the planet's orbital distance from the sun. Small particles, however, can be lost earlier due to changes in eccentricity induced by radiation pressure (which, as discussed above, have the period of the planet's orbit), than through Poynting-Robertson decay. Moreover, in the case of planetary rings this decay time may not apply, because the particles can interact with each other collisionally and gravitationally, and rings probably contain some big bodies that are unaffected by radiation forces.

Another type of radiation force which can influence the dynamical evolution of small particles occurs if the surface of a particle is not isothermal, producing a radiative recoil. The surface temperature depends on the size, the thermal properties and the rotational state (both spin rate and direction of the polar axis), as well as on the distance from the sun. In general, the hemisphere facing the sun is warmer, producing an extra force roughly directed outward in the radial direction. The order of magnitude of this force is

$$f_{ATR} \approx 4\pi s^2 (\sigma T^4/c)(\Delta T/T), \qquad (15.42)$$

where T is the temperature of the particle, ΔT the typical temperature difference and σ the Stefan-Boltzmann constant (see Sec. 7.1). Since $\sigma T^4 \approx \Phi$, the order of magnitude of f_{ATR} is in general that of the radiation pressure, reduced by a factor $4\Delta T/T$.

A fast rotation induces the same temperature along each parallel and the "summer" hemisphere is warmer; the resulting force is directed along the spin axis and depends on the angle ξ it makes with the direction of the sun. We are interested, in particular, on the drag force acting on a particle with a circular orbit. f_{ATR} vanishes at the "equinoxes" ($\xi = \pi/2$) and is maximum at the "solstices", where however it has no tangential component. In between, such a component is present. If the body has a mirror symmetry with respect to the equatorial plane, this force does not change sign when ξ is replaced with its supplementary, which occurs after half an orbital period; in this case, therefore, the drag averages out over one orbit.

When the rotation is not fast enough, compared with the thermal relaxation time, the "evening" side is warmer than the "morning" side and a net force in the equatorial plane arises. Consider, for simplicity, a body spinning around an axis perpendicular to the orbital plane. If the rotation is prograde, besides a radial force, there is a transversal component in the same sense as the motion, and *vice versa*, similar to the recoil force due to outgassing from cometary nuclei (Sec. 14.4). This is the *Yarkovsky effect* and results in complex, long-period orbital changes, whose magnitude and sign depend on the magnitude and the orientation of the spin vector. The latter is not likely to remain constant, owing to collisions and other effects; moreover, the temperature asymmetry ΔT depends on the thermal properties and the spin rate itself. For real particles in the solar system the information available on these matters is very scarce. However, from the discussion above and eq. (15.37) it is easy to infer that the Yarkovsky force exceeds in magnitude the Poynting-Robertson drag whenever $4\Delta T/T \succ v/c$, that is, when ΔT is larger than ≈ 0.01 K.

The solar wind, like the solar radiation, produces a radial force and a drag. Since the velocity of the solar wind V is approximately constant (Sec. 13.2), its momentum flux $\rho_{sw}V^2$ is proportional to $1/r^2$, like the radiation. The radial force is obtained from eq. (15.36) by replacing Φ'/c with $\rho_{sw}V^2$, giving a value $3 \cdot 10^{-4}$ times smaller for the typical values $\rho_{sw} \approx 10^{-23}$ g/cm^3 and V = 400 km/s. On the other hand the drag, given by the radial force times the factor $v/V \approx 0.1$, is only about 0.3 times the Poynting-Robertson drag, for which the corresponding factor is $v/c \approx 10^{-4}$. For particles smaller than ≈ 0.1 µm, however, the coefficient κ_T which determines the radiative cross-section becomes much smaller than unity and the orbital decay due to the solar wind prevails over the effect of drag.

Finally, particles (and spacecraft!) in space acquire an electric charge, whose sign and magnitude are determined by the competition between the ionization produced by the solar UV radiation (which contributes a positive charge) and the differential flow of electrons and ions in the plasma. Since the ratio of their thermal velocities is of the order of the square root of the mass ratio, a body of size s in a plasma charges negatively, until an electrostatic potential arises at its boundary which restores the equilibrium situation of no net current. This potential V_e extends to a distance of the order of the Debye length (see Sec. 1.7, eq. (1.78)) and is of the order of kT/e, in the magnetosphere typically a few volts. The corresponding charge, of order V_e/s, produces a Lorentz force in an external magnetic field. It turns out that in interplanetary plasma the corresponding acceleration, proportional to $1/s^4$, is relevant only for microscopic sizes (≈ 1 μm and smaller).

PROBLEMS

*15.1. Consider an asteroid for which the very small eccentricity resonance mechanism described in Sec. 15.2 works in conjunction with secular perturbations, and find the solutions of Lagrange's equations when both the resonant and the secular terms are retained in the perturbing function.

15.2 Which fraction of the total angular momentum of the earth-moon system is presently accounted for by the moon's orbital motion? And which are the corresponding fractions for Jupiter's Galilean satellites and for Pluto's Charon? (See Ch. 14 for the relevant parameters of the planet-satellite systems.)

15.3 How long it will take for the earth-moon system to reach the complete tidal synchronization state? (Neglect tides due to the sun.) How long did the same process last for the Pluto-Charon pair?

*15.4 Write down Gauss' equations for da/dt and de/dt of an orbiting particle subjected to Poynting-Robertson drag, and show that $C = ae^{-4/5}(1 - e^2)$ is a constant of motion. Find the explicit relationships $a(t)$ and $e(t)$ in the approximation $e \ll 1$, and show that the characteristic decay time for the eccentricity is 4/5 times that for the semimajor axis.

15.5 Compute the recoil due to anisotropic thermal emission for a spherical particle with a dipole temperature distribution. How large is this force compared to that due to direct radiation pressure?

15.6. Estimate the radiative recoil produced by the sun on LAGEOS (Sec. 20.2) and its secular effects in absence of eclipses.

FURTHER READINGS

For the classical treatment of secular perturbations, see D. Brouwer and G.M. Clemence, *Methods of Celestial Mechanics*, Academic Press, New York (1961) (especially Ch. XVI). Other treatments of this problem are given by A.E. Roy, *Orbital Motion*, Hilger, Bristol (1978), and J.G.

Williams, *Secular Perturbations in the Solar System*, Ph. D. Dissertation, Univ. of California, (1969) (with a particular emphasis on the case of asteroids). As regards long-term numerical integration of planetary orbits, see A.E. Roy *et al.*, *Project LONGSTOP*, Vistas in Astronomy, **32** (1988). On orbital resonances, see S.J. Peale, *Orbital Resonances in the Solar System*, Annual Reviews of Astronomy and Astrophysics (1976), and R. Greenberg, *Orbit-orbit Resonances in the Solar System: Varieties and Similarities*, Vistas in Astronomy, **21**, (1977). On tidal evolution, see K. Lambeck, *The Earth's Variable Rotation*, Cambridge Univ. Press, Cambridge (1980), Ch. 10, and A.M. Nobili, *Secular Effects of Tidal Friction on the Planet-Satellite Systems of the Solar System*, The Moon and the Planets **18**, pp. 203-216 (1978). On radiation forces on dust particles, see J. A. Burns, P.L. Lamy and S. Soter, *Radiation Forces on Small Particles of the Solar System*, Icarus **40**, pp. 1-48 (1979), and F. Mignard, *Radiation Pressure and Dust Particle Dynamics*, Icarus **49**, pp. 347-366 (1982).

A brilliant exposition of the theoretical aspects of some of these problems is in V. Béletski, *Essais sur le Mouvement des Corps Cosmiques*, Editions Mir, Moscow (1986).

16. ORIGIN OF THE SOLAR SYSTEM

About 4.6 billion years ago, a cloud of interstellar gas and dust, perhaps triggered by a nearby supernova explosion, began to collapse. The central part of the cloud contracted under its own gravity and heated, until the temperature became so high that thermonuclear reactions were initiated and the early sun began to release growing amounts of energy. But the peripheric part of the cloud, owing to its increasing rotation rate, took the shape of a flattened disk with a density maximum on its equatorial plane. A fraction of this material, prevented from falling towards the centre by angular momemtum conservation, gradually aggregated in a swarm of small lumps of solid matter gravitationally bound to the sun. These bodies underwent a process of further accumulation and coalescence, and in the outer part of the nebula the biggest ones could also accrete a substantial amount of gas. Through a complex sequence of mutual interactions and disturbances, both gravitational and non-gravitational, this process eventually gave rise to planets, satellites, asteroids and comets, the variety of bodies which we presently observe in the solar system. In this chapter we shall discuss, mainly with order-of-magnitude estimates and computations, several features of this process, as it is presently understood on the basis of a growing body of evidence. While the basic ideas of this so-called *nebular theory* for the origin of the solar system go back to Kant and Laplace, the quantitative contemporary version of it has been worked out in the 50's and the 60's by O. Yu. Shmidt, V. S. Safronov and coworkers. However, it has to be noted that important uncertainties still affect our knowledge and it is quite possible that significant revisions will be needed as new data will constrain the most speculative portions of the theory.

16.1. MASS AND STRUCTURE OF THE SOLAR NEBULA

The current understanding of collapse and fragmentation processes of interstellar clouds is not refined enough to yield plausible estimates, even of the gross properties, for the primeval solar nebula. As regards the mass of the nebula, however, we can use the present masses and overall

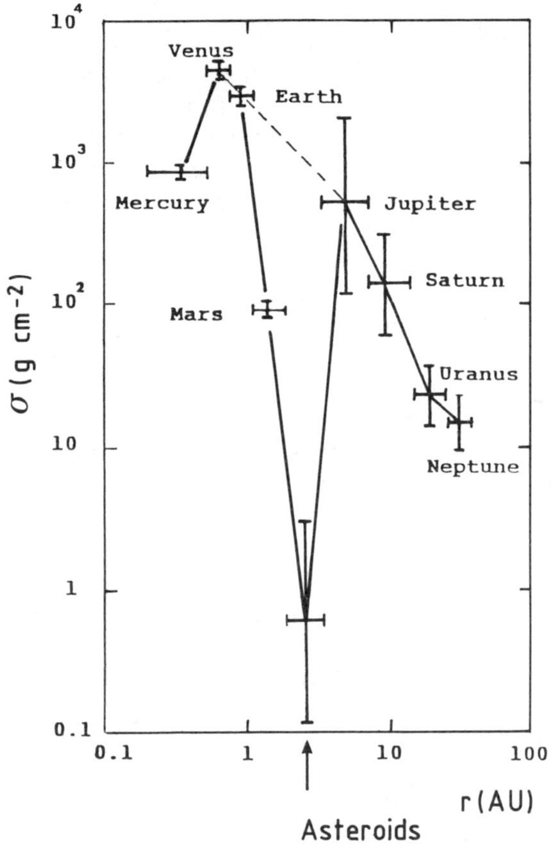

Fig. 16.1. The surface density of the primordial solar nebula vs. heliocentric distance, obtained by restoring the planets to solar composition and spreading the resulting masses through contiguous zones surrounding the present orbits. The zone boundaries are taken as the arithmetic means of adjacent orbits (for Mercury and Neptune, the zones are assumed to cover equal distances inward and outward from their orbits). The horizontal bars in the figure show the resulting zone widths, and the vertical error bars reflect the estimated uncertainties in planetary compositions. For the asteroids, an original mass ranging between the present mass ($\approx 5.10^{-4}$ $M_{\oplus}$) and 25 times this value has been assumed. (Figure adapted from Weidenschilling (1977)).

compositions of the planets as the input data to derive a lower bound. The distribution of element abundances in the planets of different groups in general does not coincide with that of the sun, but allows us to complement the planetary masses in such a way as to reconstitute the solar abundances (see Table 13.1). As for the terrestrial planets, their major constituents are heavy elements like Mg, Si, Fe and their present mass is about 10^{-5} $M_{\odot}$; if we add enough light elements (H, He) to obtain the present solar composition, we obtain an original mass some 300 times the present mass. For Jupiter and Saturn the situation is different, since, although enriched in heavy elements with respect to the sun, they still retain a substantial amount of H and He. As a consequence, it can be estimated that their present mass of about $1.3 \cdot 10^{-3}$ $M_{\odot}$ has to be increased by a factor of the order of 10. Uranus and Neptune are probably an intermediate case, since they are quite depleted in H and He; in fact, their present mass of about 10^{-4} $M_{\odot}$ should be probably increased by a factor of the order of 50. Since the contribution of other bodies (Pluto, satellites, etc.) is negligible, our inventory leads to the conclusion that the total initial mass of nebula was at least of the order of a few times 10^{-2} $M_{\odot}$. The reconstructed density distribution in the nebula is shown in Fig. 16.1.

In carrying out these computations, we have implicitly assumed that the missing mass, mostly light elements in gaseous form, was ejected from the system soon after the sun began its life as a star. Indeed, young solar-type stars are known to undergo short phases of violent activity (or *T-Tauri phases*, from the name of a well-known star of this class), with strong stellar winds and high mass loss rates. The reason why the typical composition of planets of the three groups quoted above is so markedly different is probably related both to their different distances from the sun (implying different condensation sequences in the nebula and different efficiency of light gas ejection by T-Tauri winds) and to the timing of planet formation versus T-Tauri activity (Jupiter, Saturn and, to a lesser extent, Uranus and Neptune could accrete gas as they probably reached the minimum mass for making this process effective before the whole gaseous component was lost from the nebula, while terrestrial planets never grew enough for that.)

Given the present rotation rate of the sun, it is easy to see that while it contains about 99.9% of the mass of the solar system, 98% of the total angular momentum of the system is provided by the orbital motion of the planets. Such an uneven distribution, of course, had to apply also to the proto-sun and the nebula from which the planets formed; this puzzling situation has been for a long time a convincing argument against the nebular theory itself, and at the same time a motivation of the search for alternative theories explaining more naturally why most angular momentum is concentrated in the planets (e.g., the once popular hypothesis that the planets formed from a spray of solar material ejected during a close encounter with another star). Today, we know that at least two plausible physical mechanisms might have effectively slowed down the rotation of the sun by coupling it with that of the inner part of the nebula: (1) *magnetic* coupling, due to the freezing of the magnetic field lines of the proto-sun into the plasma component of the inner nebula; (2) *turbulent* coupling, due to the effective turbulent viscosity in the inner nebula, which damps differential rotation. Both these mechanisms are complex and it is difficult to model them in detail, but the anomalously slow rotation of the sun is no longer seen as a fundamental difficulty of the nebular theory. Moreover, it is known that many solar-type stars have similarly slow rotations; this provides indirect evidence that strong angular momentum transfer to an external medium is common for them.

Let us now choose a typical location in the disc-shaped nebula, say about at the heliocentric distance of Jupiter, and let us approximately estimate the main physical quantities there. For a distance from the sun $a \approx 8 \cdot 10^{13}$ cm and a mass of the disc inside it of (say) 10^{-2} $M_\odot$, the surface density σ is given by

$$\sigma \approx 0.01\ M_\odot/\pi a^2 \approx 10^3\ \text{g cm}^{-2}\ , \qquad (16.1)$$

comparable with the column density through the earth's atmosphere. As for the thickness of the disc, we shall assume that the component g_z of the gravity acceleration along the nebular rotation axis z was balanced by the acceleration due to the gas pressure gradient in the same direction (as in the hydrostatic equilibrium condition (1.6)):

$$\frac{dP}{dz} = -\rho_g g_z \approx -\rho_g \frac{GM_\odot}{a^2}\frac{z}{a} = -\rho_g n^2 z\ . \qquad (16.2)$$

P and ρ_g are the gas pressure and density, n is the keplerian mean motion at distance a from the sun ($\approx 2 \cdot 10^{-8}$ s^{-1}); we have assumed that the contribution to g_z due to the nebula itself can be neglected.

From the equation of state of perfect gases (1.12) we obtain

$$P/\rho_g = kT/m = v_T^2/3, \qquad (16.3)$$

where m is the mean molecular mass and v_T is the r.m.s. thermal velocity, $\approx 10^5$ cm/s for hydrogen at a temperature of 100 K (reasonable if the nebula was in radiative equilibrium with the sun; see eqs. (7.9, 10)). If v_T does not depend on z, we can solve eq. (16.2) in terms of $\rho_g(z)$:

$$\rho_g(z) = \rho_0\ \exp[-\ 3n^2 z^2/(2v_T^2)]. \qquad (16.4)$$

The effective thickness of the disc can be defined as

$$H \equiv \sigma/\rho_0 = \frac{1}{\rho_0}\int_{-\infty}^{+\infty} \rho_g(z)dz = \sqrt{(2\pi/3)}v_T/n\ ; \qquad (16.5)$$

we obtain $H \approx 10^{13}$ cm, $\rho_0 = \sigma/H \approx 10^{-10}$ g cm^{-3} and the pressure in the equatorial plane $P_0 = \rho_0 v_T^2/3 \approx$ 0.3 dyne/cm^2 $\approx 3 \cdot 10^{-7}$ atmospheres. Thus the ratio H/a is of the order 10^{-1} at the distance of

Jupiter, changing slowly with heliocentric distance. Since for $z \approx H$ the self-gravity of the disc produces an acceleration along z of the order of $2\pi G\sigma$ (the value corresponding to a plane sheet of mass from Gauss' theorem), its ratio to the solar gravity g_z is

$$2\pi G\sigma/n^2 H \approx 2\pi\rho_0 a^3/M_\odot, \qquad (16.6)$$

i. e., of the order of the ratio between the nebular density and the density of the sun spread in a spherical volume of radius a. For the solar nebula, this is of the order of 0.1; hence the neglect in eq. (16.2) of the nebular gravity is correct.

The fact that this ratio is less than 1 has another important consequence: unless the total mass of the nebula was much larger than the lower bound discussed earlier (and in fact approaching 1 $M_\odot$), planets could not form directly from gravitational instability of the nebular matter causing it to break up into big, separate blobs. This can be seen by using the concept of Roche limit introduced in Sec. 14.6: a planet (or satellite) can accrete material only if its orbital radius is larger than a minimum value or, equivalently, its density exceeds the critical value

$$\rho_{cr} \approx \left(\frac{1.44\ R_\odot}{a}\right)^3 \rho_\odot \approx \frac{M_\odot}{a^3} \qquad (16.7)$$

(compare with eq. (14.6)). Here we have considered a hypothetical "protoplanet" at a distance a from the sun, whose radius and mean density are $R_\odot$ and $\rho_\odot$. As we shall see in Sec. 16.3, this stability criterion is correct only if the nebular material is cold enough. At Jupiter's distance, ρ_{cr} is about $4 \cdot 10^{-9}$ g cm^{-3}, that is, almost two orders of magnitude larger than the density of our minimum-mass solar nebula. Thus separate protoplanets could not arise from gravitational collapse of the nebular gas, unless we can identify some process which, after their formation, depleted their gas content by a factor of $\approx 10^3$ in mass. Although this possibility has been debated (intense mass loss phenomena have been observed in many star-forming regions), most researchers today favour a nebular mass not larger than $\approx 10^{-1}$ $M_\odot$, implying that no large-scale gravitational instability occurred.

In addition to a vertical pressure gradient, in the gaseous nebula a radial pressure gradient contributed to hydrostatic equilibrium and therefore affected the orbital motion of the gas. If n_g is the orbital angular velocity of the gas, the radial equilibrium condition is

$$a n_g^2 = \frac{GM_\odot}{a^2} + \frac{1}{\rho_g}\frac{dP}{da}. \qquad (16.8)$$

With the approximation $dP/da \approx P/a$ and using eq. (16.3), we get

$$n_g^2 = n^2 - v_T^2/3a^2 \quad . \tag{16.9}$$

For our model nebula, at Jupiter's distance $v_T/na \approx 0.1$, so that, to lowest order, we obtain $(n - n_g)/n \approx v_T^2/(6n^2 a^2)$. Since $v_T/na \approx 1/10$, the deviation from keplerian motion is small.

16.2. GROWTH AND SETTLING OF SOLID GRAINS

The solar nebula, after its separation from the protosun, gradually cooled. As a consequence, the vapour pressure of a number of constituents rapidly decreased and eventually fell below their partial pressure. At this stage, the condensation of small solid particles (*grains*) began: first, at a temperature of ≈ 1200 K, metals like iron and nickel could condensate; then, for decreasing temperatures, other materials became dominant: first silicates, then carbon-rich minerals, and finally water and water-ammonia ices. Since the temperature was lower in the outer part of the nebula, a compositional gradient arose in the solid component, which is reflected today in the very different compositions of planets, satellites and minor bodies at various heliocentric distances (see Ch. 14).

Once condensation occurred, grains continued to grow by collecting material still in the vapour phase. If we assume that every impinging molecule sticks to a (spherical) target grain of radius r and mass $m = 4\pi\rho_s r^3/3$, the growth rate of the grain is:

$$\frac{dm}{dt} \approx (4\pi r^2)(\alpha\rho_g)\left[\frac{v_T}{\sqrt{A}}\right] \quad , \tag{16.10}$$

where α and A are the mass fraction and the molecular weight of the condensate. We get the equation:

$$\frac{dr}{dt} = \alpha \frac{\rho_g}{\rho_s} \frac{v_T}{\sqrt{A}} \quad . \tag{16.11}$$

With the estimates of the typical nebula properties made in Sec. 16.1, and choosing a grain density $\rho_s \approx 1$ g/cm³, $A \approx 20$ (ice) and $\alpha \approx 0.01$ (from the solar abundance of elements other than H and He), we have $dr/dt \approx 2\cdot10^{-8}$ cm/s ≈ 0.6 cm/y. Thus the grains became macroscopic particles in a short time, their growth occurring at the remarkably constant rate expressed by eq. (16.11).

Since the solid particles were denser than the gaseous component, they tended to sink toward the equatorial plane of the nebula. We shall now estimate the time-scale of this *settlement* process and the size the grains reached during this time. The equation of motion of grains in the z-direction is:

$$\frac{d^2 z}{dt^2} = - g_Z + \frac{F_D}{m} \quad , \tag{16.12}$$

where F_D is the gas drag force. Under the assumption (justified later) that the inertial term in the left-hand side is negligible, gravity and drag balance each other. For the small particles with which we are concerned, the mean free path of the gaseous molecules ($\approx 10^2$ cm, see eq. (1.77)) is larger than the particle size and Stokes' drag law does not apply. On the other hand, we shall verify below that the grain velocity dz/dt is much smaller than the gas thermal speed v_T. Therefore each molecular impact, on the average, takes away a momentum given by the molecular mass times dz/dt; hence the drag force is given, in a free-flow regime, by the *Epstein formula*

$$F_D = 4/3(\pi r^2)(\rho_g v_T)\,|dz/dt| \quad , \tag{16.13}$$

where the factor 4/3 holds for a fully absorbing (or a specularly reflecting) sphere. Thus from eq. (16.12) we obtain

$$\frac{F_D}{m} = \frac{\rho_g}{\rho_S}\,\frac{v_T}{r}\,\frac{dz}{dt} = - n^2 z \tag{16.14}$$

and using eq. (16.11) to account for the changing grain size, we get the solution

$$z(t) = z(0) \exp[- \alpha n^2 t^2/(2\surd A)] \tag{16.15}$$

The time-scale for settlement to the equatorial plane is

$$\tau_S = \left(\frac{2}{\alpha}\right)^{\frac{1}{2}} \frac{A^{\frac{1}{4}}}{n} \quad , \tag{16.16}$$

of the order of 50 y; after this time, $r \approx 30$ cm. We can verify that $|dz/dt| \approx H/\tau_s \approx v_T/n\tau_s << v_T$ (consistently with the assumption we made to estimate F_D) and that $|d^2 z/dt^2| \approx 2H/\tau_s^2 = \alpha n^2 H/\surd A \approx$

$(\alpha/\sqrt{A})g_z << g_z$. Thus in a short period of time the grains could sink through the gas to the equatorial plane and form there a flat and thin layer of high density, somewhat resembling Saturn's rings.

This conclusion is not very much modified if the nebula cooled so fast that a substantial fraction of the condensable component was converted into grains in a time smaller than τ_s. In this case the grain growth was mainly due to *coagulation* with other grains during the settling process, caused by surface forces; its rates were:

$$\left(\frac{dm}{dt}\right)_{coag} \approx (\pi r^2)(\alpha\rho_g)\left|\frac{dz}{dt}\right| , \tag{16.17}$$

$$\left(\frac{dr}{dt}\right)_{coag} \approx \frac{\alpha}{4}\frac{\rho_g}{\rho_s}\left|\frac{dz}{dt}\right| , \tag{16.18}$$

taken instead of (16.10,11). Solving eq. (16.18) we get

$$r(z) = r_0 + \alpha(\rho_g/\rho_s)(z_0 - z)/4, \tag{16.19}$$

implying that during the settlement process the grains reached a size

$$r \approx \alpha(\rho_g/\rho_s)H/4 \approx \alpha(\sigma/\rho_s)/4 , \tag{16.20}$$

that is, a few cm at our assumed location in the nebula. Using again eqs. (16.13,14) for the gas drag, it is easy to verify that also in this case the settling process was fast, its time scale being of the same order as τ_s.

During the settlement process, since the motion of the gas was somewhat slower than keplerian, the grains were subjected also to a drag force acting in the tangential direction, which made them spiral in toward the sun. What is the time scale of this effect? Let us consider a grain on a circular orbit of radius a, and define the frequency $\chi \equiv \rho_g v_T/\rho_s r$. Then the tangential drag acceleration, by analogy with (16.14), is $-\chi(n_p - n_g)a$, where n_p is the mean motion of the particle, intermediate between that of the gas n_g and the keplerian value n. We can now apply to the grain motion Gauss' perturbation equation (11.4), which yields

$$\dot{a} = -2\chi(n_p - n_g)a/n. \tag{16.21}$$

Another equation is provided by the equilibrium condition in the rotating reference frame bound to the grain (see eq. (16.14) and note that the inertia is negligible):

$$n_p^2 a = n^2 a + \chi \dot{a} \; . \tag{16.22}$$

Eqs. (16.21, 22) can easily be solved, by assuming $(n_p - n) \ll n$, to derive the orbital decay time:

$$\tau_D \equiv |a/\dot{a}| \cong \frac{1 + (\chi^2/n^2)}{2\chi} \frac{n}{(n - n_g)} \cong \frac{(\chi^2 + n^2)}{\chi} \frac{3a^2}{v_T^2}, \tag{16.23}$$

where we have used eq. (16.9). Since $\chi \propto 1/r$, $\tau_D \propto r$ for $\chi \ll n$ (big bodies), while $\tau_D \propto 1/r$ for $\chi \gg n$. The minimum value of τ_D occurs for the grain radius r_D which discriminates between the two regimes. We easily get $r_D = \rho_g v_T/\rho_s n \approx 5$ m and $\tau_{D,min} = 6na^2/v_T^2$. Comparing the latter expression with eq. (16.16), we obtain $\tau_{D,min} \approx 30\ \tau_s$; therefore, the orbital decay during settlement was small, and the use of $|dz/dt|$ as the appropriate grain-to-gas relative velocity is justified. If the settling process gave rise very quickly to formation of big planetesimals via gravitational instability (see Sec. 16.3), we can also conclude that no important radial transport of solid material took place in the nebula due to gas drag.

It must be stressed that the picture presented in this Section is oversimplified. For instance, if in the gas component there were turbulent motions with decay time longer than τ_s, the settling process would have been disturbed and the grain layer would have retained a thickness depending on particle size (since smaller particles were dragged more by gas motions.) Moreover, the cooling rate of the nebula may have been slower than the settling process, implying a gradual condensation of grains from gas and an early chemical fractionation. The above-mentioned computations, however, prove that the formation of a thin layer of solid particles was a fast process, much faster than either the collapse of the nebula from interstellar gas or the growth of sizable solid bodies.

16.3. GRAVITATIONAL INSTABILITY AND FORMATION OF PLANETESIMALS

In Sec. 16.1 we have used the Roche limit criterion (16.7) to estimate the critical density of the disk which causes its fragmentation in separate lumps via a gravitational instability mechanism. In fact, we concluded that this instability did not arise in the gaseous component of the nebula, unless its mass was almost 2 orders of magnitude larger than the minimum mass estimated in Sec. 16.1. However, in Sec. 16.2 we have seen that due to grain settlement to the equatorial plane, a flat layer of high density gradually develops. When this layer reaches the critical density $\rho_{cr} \approx M_{\odot}/a^3$, its thickness is (see eq. (16.5)):

$$H_{cr} \approx \alpha\sigma/\rho_{cr}; \qquad (16.24)$$

since at Jupiter's distance $\alpha \approx 0.01$, $\rho_{cr} \approx 4 \cdot 10^{-9}$ g/cm³ $\approx 40\ \rho_g$, we obtain $H_{cr} \approx 2 \cdot 10^{-4}$ H $\approx 3 \cdot 10^{-5}$ a (i. e., a thickness-to-radius ratio of the same order as in Saturn's rings). According to eq. (16.5), this stage is reached only if the velocity dispersion of the grains, due to collisions or to random motions in the gas, did not exceed

$$\sqrt{3}nH_{cr} = \sqrt{3}\frac{n\alpha\sigma}{\rho_{cr}} = \sqrt{3}\frac{\alpha G\sigma}{n} \approx 50 \text{ cm/s}, \qquad (16.25)$$

where we have used the analogue of (16.5), with an additional $\sqrt{(2\pi)}$ factor to account for the gravity of the grain layer which, at the critical density, is about 2π times that of the sun (see eq. (16.6)).

Roche's limit criterion provides a good order of magnitude estimate of the critical density. However, it does not take into account that not only differential rotation of the disc but also random motions in the material tend to oppose the collapse. As a consequence, this criterion does not allow us to estimate the size of the condensations arising from the instability, which was also influenced by the fact that the grain layer had a very small thickness-to-radius ratio. For these reasons, a more accurate treatment is needed. Since this theory has many other interesting astrophysical applications, we shall discuss it in some detail.

The *gravitational instability*, discovered by J. Jeans and often called after him, arises when in a localized disturbance the effect of self gravity prevails over the thermal pressure, so that collapse ensues. This is the basic process which gives birth to the stars and to galaxies (though in the latter case it is contrasted by the general expansion of the universe.) It also played a crucial role in the formation of condensations in the primeval nebula. In a medium of mass density ρ_0 the gravitational instability is characterized by *Jeans' frequency*

$$\Omega_J = \sqrt{(4\pi G\rho_0)}, \qquad (16.26)$$

easily obtained on dimensional grounds (see also Sec. 5.3). If the gas is composed of n particles of mass m per unit volume, the quantity $-\Omega_J^2$ is exactly what one gets from the plasma frequency (8.11) by replacing e^2 with $-Gm^2$; this replacement indeed transforms the electrostatic repulsion into gravitational attraction between two bodies (see Gauss' theorem (1.3)). Although this is only a plausibility argument, it suggests that the gravitational analogues of plasma waves are indeed unstable.

The normal mode approach, already used in Sec. 8.2, cannot be applied to Jeans' instability in a straightforward manner because the equilibrium situation is never homogeneous: a gravitating system is necessarily limited and one must make the further approximation that the size of the disturbance is much smaller than the size of the whole system. Here we follow a simpler approach, which has the advantage of allowing an extension to the non-linear regime. The main purpose of this analysis is the derivation of the largest wavelength λ_{cr} of a disturbance beyond which self gravity prevails and instability ensues.

Consider a disturbance whose characteristic length λ is much smaller than the size of the system. If a localized and small displacement $x \rightarrow x + \xi(x)$ takes place in one dimension x, it produces a density perturbation

$$\delta\rho(x) = -\rho_0 d\xi(x)/dx \ . \qquad (16.27)$$

This is easily derived in the lagrangean formulation of the conservation of mass, similarly to what was done in Sec. 1.2 to get eq. (1.21). From Poisson's equation (1.23) the extra self-gravitational force per unit mass is proportional to the displacement:

$$\delta F_g(x) = 4\pi G\rho_0\xi(x) = \Omega_J^2\xi(x) \ . \qquad (16.28)$$

There is also an extra pressure force per unit mass produced by the density gradient:

$$\delta F_p = -\frac{1}{\rho_0}\frac{dP}{dx} = -\frac{c_s^2}{\rho_0}\frac{d\delta\rho}{dx} = c_s^2\frac{d^2\xi}{dx^2} \ , \qquad (16.29)$$

where c_s is the speed of sound. Suppose now that $\xi > 0$ displaces in the positive sense a small lump of fluid of size 1/k; the new force acting on it will tend to increase ξ if

$$0 < \delta F_p + \delta F_g = \Omega_J^2\xi + c_s^2 d^2\xi/dx^2 \approx (\Omega_J^2 - c_s^2k^2)\xi. \qquad (16.30)$$

The last expression estimates $d^2\xi/dx^2$ as $-k^2\xi$, valid at the worst point, where ξ is largest. In the opposite case the force will tend to restore the fluid element back at its previous position. When the factor in the brackets is small, we can expect that the acceleration is small and the inertia plays little role; therefore we get the critical wave number below which we have instability:

$$k_{cr} = 2\pi/\lambda_{cr} = \Omega_J/c_s \ . \tag{16.31}$$

It is easy to include the inertia in this model: in lagrangean variables the motion of the fluid element whose original position is x is described by

$$\frac{\partial^2 \xi}{\partial t^2} = \Omega_J^2 \xi - c_s^2 \frac{\partial^2 \xi}{\partial x^2} , \tag{16.32}$$

Note that we are working in the linear approximation of small displacements, so that the argument x of ξ – a label of the fluid element – can be replaced with the point $x + \xi$ where we sit in the eulerian description. Since the velocity v is just the time derivative of ξ, the previous equation is equivalent to the eulerian equation

$$\frac{\partial^2 v}{\partial t^2} = \Omega_J^2 v - c_s^2 \frac{\partial^2 v}{\partial x^2} \ . \tag{16.33}$$

In this form we can use the normal mode approach (see Sec. 8.2, eq. (8.17)) and assume for v the dependence $\exp[i(\omega t - kx)]$. This yields at once the exact dispersion relation

$$\omega^2 = c_s^2 k^2 - \Omega_J^2 \ . \tag{16.34}$$

When $\Omega_J \to 0$ we get acoustics; but gravity modifies the acoustic waves. In particular, when $k < k_{cr}$, the "frequency" ω becomes imaginary and we have an exponentially growing solution, with growth rate

$$\sqrt{(\Omega_J^2 - k^2 c_s^2)} . \tag{16.35}$$

This result can easily be understood by noting that the collapse of a region of size $1/k$ releases a specific potential energy of order $4\pi G\rho/k^2 = \Omega_J^2/k^2$, but requires a specific energy input of order c_s^2 to overcome the pressure forces. By the same argument we can guess the effect of rotation of the medium. For an orbital angular velocity Ω of the disc material, the collapsing lump has an intrinsic angular velocity of the same order and an additional specific energy of order Ω^2/k^2 is required to overcome centrifugal forces. Thus we expect that Jean's criterion is modified by replacing Ω_J^2 with $(\Omega_J^2 - \Omega^2)$. A more precise result can be obtained by applying the same lagrangean approach that was used before to a rotating fluid, with arbitrary angular velocity $\Omega(r)$. In the case of cylindrical symmetry, let us assume that at a distance r from the rotation axis we have an additional mass coming from a small radial displacement $\xi(r)$ (equal at all azimuths.) The total mass $\mu(r)$ up to the distance r per unit height is then changed by $\delta\mu = -2\pi r\rho\xi$ and according to Gauss' theorem the gravitational force (per unit mass) $F_g = -2G\mu/r$ is changed by $\delta F_g = 4\pi G\rho\xi$ (compare with eq. (16.28)). To include the centrifugal force, we must determine how the orbital angular velocity Ω' of the displaced element is changed. Note that with a purely radial motion the gravitational force produces no torque, so that the orbital angular momentum per unit mass of the element $\ell = \Omega' r^2$ is constant during the displacement. The centrifugal force (per unit mass) felt by the element at r, is then $r\Omega'^2 = \ell^2/r^3$, where ℓ is the value of the angular momentum before the displacement, at $(r - \xi)$; hence the additional centrifugal force (per unit mass) is

$$\delta F_c = -\xi \frac{d\ell^2}{r^3\,dr} = -2\xi\Omega \frac{d(\Omega r^2)}{r\,dr} .$$

With the same argument as before we arrive at the dispersion relation

$$\omega^2 = c_S{}^2k^2 - \Omega_J{}^2 + \frac{2\Omega}{r}\frac{d(\Omega r^2)}{dr} . \tag{16.36}$$

For a uniform rotation $(\Omega_J{}^2 - 4\Omega^2)$ replaces $\Omega_J{}^2$; for keplerian motion $\Omega = n \propto r^{-3/2}$ and we get

$$\omega^2 = c_S{}^2k^2 - \Omega_J{}^2 + n^2 . \tag{16.37}$$

We see that, provided (Ωr^2) increases with r, rotation always decreases the critical wave number below which we have instability. However, rotation does not affect the propagation of longitudinal waves along directions parallel to the rotation axis; in this case we still have instability below the critical wave number (16.31). It can be shown that the rotation induces another mode and that the system is unstable below k_{cr} for all directions of propagation, except those lying in the (x, y) plane.

Our problem, however, is different because of the finite vertical thickness. It is impossible, in general, to define a wave vector k if along its direction the medium changes over a scale H of order 1/k or less: the linearized equations cannot be made algebraic. In our case we have a system of ordinary differential equations in z of the fourth order. Rather than going through this complex analysis, we follow the previous order-of-magnitude method. A disturbance with a wave number k in the (x, y) plane is expected to vary along z with a scale not larger than H, so that the gravitational potential energy per unit mass is of order

$$-\ 4\pi G\rho/(k^2 + 1/H^2)$$

or less. (In fact, if the mass density changes from ρ to $\rho + \delta\rho$ over a volume V, a gravitational disturbance is produced with potential

$$\delta U = -\ 4\pi G\delta\rho/(k^2 + 1/H^2);$$

It interacts with the background with a potential energy of order $\rho V\delta U$ which, for a unit displaced mass $V\delta\rho = 1$, gives the previous expression.) Being interested in the criterion for gravitational instability, we stick to this upper limit. We expect marginal stability if the three relevant energies balance:

$$\frac{4\pi G\rho}{k^2 + 1/H^2} = c_S{}^2 + \frac{\Omega^2}{k^2} . \tag{16.38}$$

When $Hk >> 1$ we get, apart from numerical coefficients, the marginal stability condition (16.37)

$$\Omega_J{}^2 - \Omega^2 = k^2 c_S{}^2 .$$

When H is finite we have a biquadratic equation for k:

$$c_S{}^2 k^4 + k^2(\Omega^2 - \Omega_J{}^2 + c_S{}^2/H^2) + \Omega^2/H^2 = 0 . \tag{16.39}$$

In order to have positive values for k^2 we must require

$$\Omega_J{}^2 - \Omega^2 - c_S{}^2/H^2 \geqslant 0$$

and the reality condition

$$\Omega_J{}^2 - \Omega^2 - c_S{}^2/H^2 \geqslant 2\Omega c_S/H, \tag{16.40}$$

which is sufficient. If it holds, only a finite range of wave numbers is unstable. As one sees from eq. (16.38), for small k the stabilization is ensured by rotation; for large k it is the pressure that overcomes gravity. The minimum value ρ_{cr} of the density for which the system is unstable corresponds to the equality sign in eq. (16.40), to wit

$$4\pi G\rho_{cr} = (\Omega + c_S/H)^2 ; \tag{16.41}$$

for $\rho = \rho_{cr}$ there is only one unstable wave number

$$c_s k_{cr}^2 = \Omega/H \ . \qquad (16.42)$$

In our case $\Omega = n$ and, in order of magnitude, $c_s \approx nH$; hence we recover Roche's criterion

$$4\pi G\rho_{cr} \cong n^2 \qquad (16.43)$$

and

$$Hk_{cr} \cong 1 \ . \qquad (16.44)$$

For $\rho > \rho_{cr}$, an initial perturbation of size of the order of H collapses in a time scale of the order of the reciprocal of Jeans' frequency, that is to say, an orbital period. The grain disc breaks up into separate blobs of mass

$$m_{pl} = H^3\rho_{cr} \ . \qquad (16.45)$$

When $\rho_{cr} = 2 \cdot 10^{-9}$ g/cm³ and $H = 10^{10}$ cm, this gives a mass $2 \cdot 10^{21}$ g, corresponding to a size of the order of 100 km for the density of water ice. In the region of the terrestrial planets the product $H\rho$ was comparable, but the density was about 100 times larger; thus the gravitational instability could form here bodies of size $\approx$ 5 km. These bodies are called *planetesimals* and were the building blocks for further planetary growth.

A necessary condition for the occurrence of gravitational instability is that the random velocities of the grains did not exceed the value given by (16.25); otherwise, the grain disk could not become thin enough to achieve the critical density. It is uncertain whether this condition was satisfied in the inner part of the nebula, where the maximum allowed velocity dispersion was very small and, on the other hand, turbulent motions in the gaseous component were likely to arise, as well as other sources of non-gravitational forces on the grains, due to interaction with the young sun. An alternative possibility is that in this zone the grains grew gradually by collecting material in the vapour phase and sticking at low-velocity collisions. This gradual growth could occur at a rate of the order of 1 cm/y (see Sec. 16.1) and with an efficiency depending upon the microscopic properties of the grain surfaces.

*16.4. COLLISIONAL ACCUMULATION OF PLANETESIMALS

Planetary growth by collisional accumulation of planetesimals depended in a critical way on the average relative velocity in the swarm of accumulating bodies. Had this velocity been too high, collisions would have caused catastrophic fragmentation instead of accumulation (such a disruptive process is actually occurring in the present asteroid belt; see Sec. 14.3.) On the contrary, too small relative velocities would have diminished orbital intersections and the outcome of the process would have been a large number of small "planets". The critical value of the relative velocity v_r is the characteristic escape velocity v_e of the planetary embryos. If $v_r \gg v_e$, fragmentation is the typical outcome of a collision, since the delivered kinetic energy is larger than the gravitational binding energy of the impacting bodies; in this case further accumulation is prevented. We shall show with a simple argument that the growth process adjusted itself in such a way that v_r was always of the order of v_e, in spite of the fact that v_e, approximately proportional to the size of the body, increased by three orders of magnitude in the mean time. We shall neglect the effects of gas drag.

Let us consider the planetesimals at a given orbital radius a. If e is their mean eccentricity (and mean inclination), only the bodies at a distance less than ea from the given orbit can interact. Their relative velocity, of order $v_r = ena$, is increased by elastic collisions (i. e., orbital encounters without impact) and decreased by those inelastic collisions (i. e., impacts) which result in the coalescence of the two colliding bodies. As long as v_r is less than, or comparable with, v_e, this is indeed the usual outcome of impacts. We now discuss these two processes in the approximation – initially valid – in which all the bodies have a comparable mass m_{pl} and a comparable radius r_{pl}. If σ is the local surface mass density, the number density of the planetesimals is then $\sigma n/(m_{pl}v_r)$.

An orbital encounter with an impact parameter b changes the relative velocity v_r approximately by

$$\delta v_r = \frac{Gm_{pl}}{b^2} \frac{2b}{v_r} = \frac{2Gm_{pl}}{bv_r} , \tag{16.46}$$

the product of the relative acceleration and the interaction time, of order $2b/v_r$. The cumulative effect of the encounters is a diffusion in velocity space or, equivalently, an increase of the average inclination and eccentricity. This statistical process is crucial in the relaxation to thermal equilibrium of a gas whose particles interact with forces inversely proportional to the square of the distance (e.g., a plasma or a stellar cluster). Like every diffusion process, it can be understood on the basis of the theory of random walk (see Problem 7.13).

There are three intervals for the impact parameter b. When

$$r_{pl} < b < Gm_{pl}/v_r^2 \equiv b_0$$

the deflection is substantial (strong encounter). When

$$b > (m_{pl}/3M_{\odot})^{1/3} a \equiv d \tag{16.47}$$

the two bodies never enter the confinement region, as defined by the zero-velocity curve passing through the Lagrangian point L_2; see Sec. 12.2. One can say, the effect of the sun is prevailing and no effective interaction takes place. Introducing the mass density of the planetesimals ρ_{pl}, we have

$$d = (\rho_{pl}/3\rho_{\odot})^{1/3}\ r_{pl} a/R_{\odot}\ .$$

Since $\rho_{pl}/3\rho_{\odot} \approx 1/4$ and, at Jupiter's distance, $a \cong 10^3\ R_{\odot}$, we have $d \cong 600\ r_{pl}$. The bulk of the effect comes from the intermediate range $d > b > b_0$, where we have many weak interactions; but the strong interaction range $b_0 > b > r_{pl}$ can be treated in the same way. Let us consider the effect of those encounters whose impact parameter lies in the interval $(b, b + db)$. A typical body experiences in a time dt $2\pi b db v_r dt \cdot \sigma n/(m_{pl} v_r)$ encounters of this kind, $\sigma n/(m_{pl} v_r)$ being the number density of planetesimals. Each encounter will produce a small, random change in the relative velocity; the sum of these changes has a zero mean, but its mean square is obtained by a "stochastic" summation, that is to say, by multiplying the square of the elementary change (16.46) by the number of encounters. To obtain the total change, we must sum this mean square for all the relevant values of the impact parameters:

$$\left(\frac{dv_r^2}{dt}\right)_{enc} = \int_{r_{pl}}^{d} 2\pi b v_r \frac{\sigma n}{m_{pl} v_r} \left(\frac{2Gm_{pl}}{b v_r}\right)^2 db.$$

The integral produces a logarithmic dependence from the limits of integration (see also Sec. 13.3); choosing for them an accurate value therefore is unimportant. We can express the result in terms of the escape velocity $v_e = \sqrt{(2Gm_{pl}/r_{pl})}$ and get:

$$\left(\frac{dv_r^2}{dt}\right)_{enc} = 2\pi \frac{\sigma n}{m_{pl}} \frac{v_e^4\ r_{pl}^2}{v_r^2} \ln\left(\frac{d}{r_{pl}}\right) . \tag{16.48}$$

At the same time, relative velocities are decreased by impacts resulting into coalescence of the two bodies. As we shall see later, in the velocity range $v_r \approx v_e$ the impact cross-section is of the order of the geometrical value πr_{pl}^2. Since in an impact the velocity loss is substantial, i. e., $\approx - v_r$, the mean decrease in the relative velocity is

$$\left(\frac{dv_r}{dt}\right)_{imp} = -\pi r_{pl}^2 \frac{\sigma n}{m_{pl}} v_r \ . \qquad (16.49)$$

The combination of the two effects gives

$$\frac{dv_r}{dt} = \pi \frac{\sigma n}{m_{pl}} r_{pl}^2 v_r \left[\left(\frac{v_e}{v_r}\right)^4 \ln\left(\frac{d}{r_{pl}}\right) - 1\right] . \qquad (16.50)$$

The characteristic time-scale

$$\tau_{vel} = m_{pl}/\pi r_{pl}^2 \sigma n \qquad (16.51)$$

in most cases turns out to be shorter than the time needed for a substantial growth of a planetary embryo (which we shall obtain later); we can therefore solve this equation keeping m_{pl} and r_{pl} (and hence v_e) constant. We see that in a time of order τ_{vel} a steady state is reached with $v_r \approx v_e$ (Problem 16.4).

This simplified discussion neglects any difference in mass and size among the planetesimals and gives the correct result only if the mass of the largest growing body is just a small fraction of the total mass in the planetesimal swarm. However, we shall show in the next Section that the largest bodies grow faster than the smaller ones, not only because of their larger size, but also due to their stronger gravity; a *runaway* process ensues, in which one body becomes predominant in the ring. In general, as shown by Safronov with realistic (power law) mass distributions, the mean relative velocity adjusts itself to the escape velocity of the bodies that dominate the system in terms of mass. When there is just one big body (of mass m_1 and radius r_1) comprising a large fraction of the total mass, Safronov's theory gives

$$v_r = Gm_1/\theta r_1 \ . \qquad (16.52)$$

Here θ is an adimensional parameter, nearly constant during the this type of accumulation process and typically ranging from 1 to 10 (the highest values are obtained in the presence of gas, whose friction decelerates the relative motion of planetesimals.) As a consequence, the width of the ring where planetesimals interact with the embryo (proportional to the average eccentricity v_r/na) is proportional to the escape velocity of the latter; it is from this ring that the growing planetary embryo draws its material.

16.5. ACCRETION

The *accretion* process, by which a gravitating body in a medium of gas or particles grows in mass and size, has a fundamental role in astrophysics, especially when it is enhanced by the gravitational field of the accreting body.

Let us consider the capture of a particle by a growing planetary embryo in a two-body approximation. We assume that initially, at a large distance from the embryo, the particle travels in a straight line with relative velocity v_∞ and impact parameter b. For a grazing impact (when b is equal to the radius r_{pl} of the embryo), conservation of angular momentum and energy gives

$$v_\infty b = \sqrt{[v_\infty^2 + (2Gm_{pl}/r_{pl})]}\, r_{pl} ,$$

where the expression in square brackets is the velocity at distance r_{pl}. It follows that impacts do occur when the impact parameter is less than

$$b_i = r_{pl}\sqrt{(1 + v_e^2/v_\infty^2)} , \tag{16.53}$$

in terms of the escape velocity v_e. The gravitational capture cross-section is then πb_i^2, and is significantly larger than the geometrical cross-section πr_{pl}^2 whenever $v_e \succ v_\infty$. The fact that it becomes infinite when v_∞ approaches zero is due to the long-range nature of the gravitational interaction. However, this singularity would disappear in a more realistic three-body approach (i. e., when $b \succ d$; see eq. (16.47)).

Using (16.53) with $v_\infty = v_r$, we can see that in the first phase of the accumulation process, when the largest embryo accounts just for a small fraction of the total mass of the planetesimal swarm and v_r does not depend on its size, its rate of mass accretion is roughly proportional to the fourth power of the radius. If we consider two embryos with different masses, the mass ratio between the larger and the smaller one increases with time, that is, the larger embryo "runs away" from the smaller, and we have the so-called *runaway growth* process. On the other hand, when the largest embryo comprises most of the mass in the swarm and v_r tracks its escape velocity according to eq. (16.52), the accretion rate of the dominant embryo is

$$dm_1/dt = \pi r_1^2 \alpha\rho_g v_r (1 + 2\theta) , \tag{16.54}$$

where

$$\alpha\rho_g = \left(\frac{3}{2\pi}\right)^{\frac{1}{2}} \frac{n}{v_r}\, \alpha\sigma_0 \left(1 - \frac{m_1}{m_{tot}}\right) \tag{16.55}$$

is the density of solid material in the "feeding zone" of the planetary embryo; we have used here eq. (16.5) and have taken into account the fact that the original surface density $\alpha\sigma_0$ of the disk is decreased as the mass of the largest embryo m_1 approaches the total available mass m_{tot}. Notice that in this case the increase in v_r with growth of the largest body ensures that its gravitational cross-section can never get much larger than its geometrical size. Inserting (16.55) into (16.54) and defining

$$z \equiv \left[\frac{m_1}{m_{tot}}\right]^{1/3}, \quad \tau_0 \equiv \frac{4\sqrt{(2\pi/3)}}{(\alpha\sigma_0)n(1 + 2\theta)}\left[\frac{m_{tot}\rho_s^2}{36\pi}\right]^{1/3}, \quad (16.56)$$

we obtain the differential equation

$$3\tau_0 dz/dt = 1 - z^3 \quad . \quad (16.57)$$

When z (the ratio of the embryo's radius to the final radius) is $\ll 1$, it increases linearly with time at the rate $1/3\tau_0$. For terrestrial planets, $\tau_0 \approx 10^7$ y, and the rate of increase of their size was initially of $\approx$ 20 cm/y. On the other hand, in the final phase of the accumulation, when the remaining mass fraction $(1 - z^3)$ was $\ll 1$, it was decreasing exponentially with a time constant τ_0. We can also explicitly solve eq. (16.57) by separation of variables:

$$\frac{t}{\tau_0} = \sqrt{3}\arctan\left[\frac{1 + 2z}{\sqrt{3}}\right] + \frac{1}{2}\ln\left[\frac{1 - z^3}{(1 - z)^3}\right] - \frac{\pi\sqrt{3}}{6} \quad . \quad (16.58)$$

The total accumulation time can be estimated by noting that $z = 0.99$ (i. e., 97% of the total mass is accumulated into one planet) when $t \cong 6\tau_0$. For terrestrial-type planets, we conclude that they reached their final mass within at most 10^8 y. Notice that τ_0 increases fast for increasing heliocentric distances.

How big was the second largest planetesimal (of mass and radius m_2, r_2) during the final stage of the accumulation process? In analogy with (16.54), we have

$$\frac{dm_2}{dt} = \pi r_2^2 \alpha\rho_g v_r\left[1 + 2\theta\,\frac{r_2^2}{r_1^2}\right] \quad (16.59)$$

and, therefore,

$$\frac{dm_1/m_1}{dm_2/m_2} = \frac{r_2}{r_1} \frac{1 + 2\theta}{1 + 2\theta r_2^2/r_1^2}. \tag{16.60}$$

At the beginning of the growth process m_1 and m_2 were comparable; since $\theta \gg 1$, one gets $2\theta r_2^2/r_1^2 \gg 1$ and the ratio (16.60) was approximately proportional to r_1/r_2. Therefore

$$dm_1/dm_2 = r_1 m_1/r_2 m_2 \gg 1 ,$$

that is to say, the largest body grew faster than the second, further increasing the ratio dm_1/dm_2. Hence, later in the process $2\theta(r_2/r_1)^2$ became smaller than 1. When the size ratio r_1/r_2 approached 2θ, the right-hand side of eq. (16.60) became of order unity; from then on, the ratio of the two masses remained approximately constant at $(2\theta)^3$.

Although this steady-state value of the mass ratio may have not been necessarily reached, we can conclude that the bodies impacting the planets during the final phase of their growth had masses ranging from 0.1% to 10% of the planetary mass. These huge impacts probably had several important consequences. In general, they contributed a random component to the angular momenta of rotation of the planets, giving rise to the observed obliquities with respect to the orbital planes (Uranus is the most extreme such case.) In the case of terrestrial planets, (i) they caused a strong heating of a thick outer layer in the newly formed planets; (ii) they produced inhomogeneities in the chemical composition and temperature of this layer, possibly resulting into the formation of differentiated geological units (e.g., continents); (iii) perhaps, finally, they could eject from the earth a spray of material which subsequently accumulated in a geocentric orbit and formed the moon.

During the process, each growing embryo was sweeping an annular feeding zone of width $\approx 2v_r/\sqrt{(GM_\odot/a^3)}$, which, since $v_r \approx v_e$, increased proportionally to the radius of the embryo. Therefore, at the beginning there were many small bodies and the feeding zones were narrow; as the planetesimals grew, so did the average relative velocities, and the feeding zones gradually broadened and overlapped, causing coalescence of different embryos. At the end, the population of small bodies was completely exhausted and the distance between adjacent embryos had become so large that mutual gravitational perturbations were unable to cause orbital crossings even over very long spans of time. This physical mechanism can well explain the spacings between adjacent planetary orbits. Accordingly, the so-called *Titius-Bode law* (i. e., the nearly geometric progression of planetary semimajor axes; see Sec. 14.1) is not due to some mysterious physical principle, but is a natural outcome of the process of collisional accumulation of planets from planetesimals.

This conclusion has been confirmed by numerical simulations of the accumulation process, carried out since the 70's at increasing levels of sophistication and detail.

16.6. FORMATION OF GIANT PLANETS AND MINOR BODIES

In the case of the giant planets, the growth process was complicated by a number of additional factors like the overlapping of the feeding zones, the ejection of bodies to unbound orbits, the dissipation of gas by solar wind particles and the accretion of light gases (hydrogen and helium) by growing embryos. Moreover, we have the difficulty that for Uranus and Neptune the characteristic accumulation time τ_0 given by eq. (16.56) for $\theta \approx 5$ is of the order of 10^{11} y, that is, much longer than the age of the solar system. To solve this problem, one should assume either much lower relative velocities among the planetesimals (i. e., values of θ of the order of 100) or much larger amounts of solid material inside these zones. The former possibility seems more likely, since on the one hand it is not easy to find a reasonable physical mechanism capable of ejecting from the solar system large amounts of solid material, and on the other hand θ was probably increased by gas drag and by the fact that, when the average relative velocity (16.52) became sufficient to eject the bodies from the solar system, a cut-off appeared on the typical velocity of the accumulating planetesimals (see later). The alternative possibility that the giant planets were formed by gravitational instabilities of gaseous condensations is not very attractive, both because of the big mass excess that again would be implied and because the composition of the giant planets, especially of Uranus and Neptune, does not match well that of the sun (and no chemical segregation mechanism can be easily identified).

In an initial stage, when the masses of the embryos were small, their feeding zones did not overlap. The growth formulae derived in Sec. 16.4 allow us to estimate the time after which overlapping occurred, the criterion being that the average relative velocity (16.52) had to become sufficient for the planetesimals to reach the neighbouring planet at aphelion. For Jupiter, this occurred for an embryo's mass of the order of 1 $M_\oplus$; for a surface density of solid material of $\approx$ 10 g/cm^2, the corresponding accumulation time is of the order of 10^8 y. Since inner planets grew faster, the zones of Jupiter and Saturn overlapped first; after some time, planetesimals from the Jupiter zone acquired a relative velocity high enough to allow escape beyond Neptune's heliocentric distance and out of the solar system.

Ejection became substantial when the average relative velocity of planetesimals v_r reached the critical value

$$v_{ej} = (\sqrt{2} - 1)\sqrt{(GM_{\odot}/a)}, \tag{16.61}$$

corresponding to a value of θ

$$\theta_{ej} = \frac{1}{3 - 2\sqrt{2}} \frac{m_1}{M_{\odot}} \frac{a}{r_1} , \tag{16.62}$$

that is, about 100 for all the outer planets in the final phase of their accumulation. Comparing (16.61) with (16.46), it can be seen that in this phase a large fraction of the close encounters led to the ejection of planetesimals, so that the ejection process was much more effective than the accumulation of mass by the embryo. A small, but not unsignificant fraction of bodies were also inserted into highly elongated elliptical orbits, with aphelion distances not much smaller than the distance to the nearest stars. Due to random encounters with passing stars, these bodies gradually spread into a huge "cloud" surrounding the solar system, the so called *Oort's cloud*; when some of them fall back to the inner solar system (owing again to gravitational disturbances by passing stars), we can observe it as a *comet*, releasing volatiles which were embedded in these outer solar system planetesimals before their ejection into the Oort's cloud (see Sec. 14.4).

Since the planetesimals were preferentially ejected in the prograde direction, planetary embryos experienced a recoil leading to a decrease of their angular momentum and orbital radius. If δm_{ej} is the ejected mass, conservation of angular momentum yields

$$\delta a/a \approx - \delta m_{ej}/m_1 ; \tag{16.63}$$

this implies that the total ejected mass was at most of the order of the present planetary masses. The corresponding orbital evolution of the growing planets might have had important dynamical consequences, due to sweeping of mean motion resonances through vast regions of the primordial solar system. Some bodies (e.g., Pluto) might have been trapped into stable resonant orbits with the outer planets; on the other hand, sweeping Jovian resonances through the asteroid belt might well have caused disturbances leading to increased eccentricities and inclinations and eventually to ejection. Today, the most intense Jovian mean motion resonances correspond indeed to clear gaps in the semimajor axis distribution of asteroids (see Secs. 14.3, 15.2.)

As shown by Fig. 16.1, if the surface density of the solar nebula was a smooth function of radius, a strong depletion of solid material (by a factor ≈ 1000) must have occurred in the region of the asteroid belt, and to a lesser extent a similar process might have limited the growth of Mars. At the same time, the present relative velocities of asteroids (resulting from their average eccentricities and inclinations, about 0.15 and 8° respectively) amount to about 5 km/s. This is approximately 10 times higher than the escape velocity of the largest asteroid, Ceres, and causes disruptive collisions to occur rather than accumulation. The unavoidable conclusion is that planetary growth in the asteroid zone was stopped at an intermediate stage – the accumulation of Ceres must have taken less than $\approx 10^7$ y according to (16.56) – by some mechanism which stirred up the asteroid orbits and at the same time caused the mass loss. Besides sweeping resonances, another such mechanism has been proposed: the interaction with planetesimals scattered by Jupiter. If a planetesimal is scattered in the direction opposite to the orbital motion at the velocity v_{ej} given by (16.61), its subsequent orbit has a perihelion distance of ≈ 0.21 times the orbital radius of the scattering planetary embryo (Problem 16.4). Since in the final growth phase the planetesimal swarms probably had a sharp cutoff in the velocity distribution at v_{ej}, each planet scattered as many bodies inward up to $\approx 0.21 \cdot a$, as outward to escape. In the case of Jupiter, the bodies scattered inward crossed the asteroid belt and reached Mars' zone; the most massive ones possibly swept many newly formed asteroids by direct collisions and/or increased their random velocities by close encounters.

Coming back to the formation of giant planets, another important process – the accretion of the gaseous component of the nebula – affected their growth, starting at a critical size that probably was close to that of the earth. In fact, gaseous molecules can be retained by a planetary body when the escape velocity – that grows approximately proportionally to the planetary radius – is several times larger than the thermal velocity (see Sec. 8.4), which in this region of the nebula was of the order of 1 km/s (see Sec. 16.1.) On the contrary, the effect of gravity upon gaseous accretion is unimportant when the escape velocity is smaller than the thermal velocity, a condition soon violated for the giant planets. The problem of the radial flow of a gas toward a spherically symmetric body at rest is essentially identical to that of the solar wind (see Sec. 13.2): indeed, the equations are invariant under a change in sign of the velocity. With the assumption of a polytropic flow, the density and the velocity are determined by the conservation of mass (13.8) and Bernoulli's equation (13.11); however, the boundary conditions are different because now we require the velocity, not the pressure, to vanish at infinity.

Referring to Figs. 13.2 and 13.3, we can have either the case A) (the two curves of Fig. 13.13) do not intersect and the solution is of type c, subsonic); or the case B) (the two curves have a tangent intersection and the solution is the separatrix a). Note, however, that Fig. 13.2 refers to the isothermal case; in an adiabatic flow the solution is only quantitatively different. As N.L. Balazs has shown (*Mon. Not. R.*

astr. Soc. **161**, 217 (1973)), the first possibility is prevented by an arbitrarily small viscosity, so that the flow becomes supersonic below the critical point. The accretion rate is then determined by the conditions at the critical point r_c given by eq. (13.11), and on turn is related to the conditions at infinity by the simple expression

$$\frac{dM}{dt} = 4\pi r_c^2 \rho_c v_c = \pi f(\gamma) \frac{G^2 M^2}{4 c_\infty^3} \rho_\infty , \quad (16.64)$$

The solution exists only if the polytropic index γ lies between 1 and 5/3; the function

$$f(\gamma) = \left[\frac{2}{5 - 3\gamma}\right]^{(5-3\gamma)/2(\gamma-1)} \quad (16.64')$$

changes very little in this interval, from 1 at $\gamma = 5/3$ to $e^{3/2} \cong 4.5$ when γ tends to unity.

For Jupiter's embryo in the solar nebula, accretion of gas became possible when r_c exceeded the physical radius r_1 of the embryo, that is to say at

$$r_1 = r_{accr} = \left[\frac{3}{\pi(5 - 3\gamma)G\rho_1}\right]^{\frac{1}{2}} c_\infty , \quad (16.65)$$

where ρ_1 is the embryo's density. For $\rho_1 \approx 1$ g/cm³, $c_\infty \approx 10^5$ cm/s, $\gamma \approx 1.5$, we obtain $r_{accr} \approx 5000$ km, corresponding to a mass of $\approx 0.1 M_\oplus$. In the inner part of the nebula, gas accretion was hampered both by the higher temperature and by the fact that the available amount of solid material was not enough to allow any embryo to substantially exceed the critical radius r_c. On the other hand, for Jupiter the accretion time scale (see eq. (16.64))

$$\tau_{accr} = \frac{M}{dM/dt} = \frac{4 c_\infty^3}{\pi f(\gamma) G^2 M \rho_\infty} \quad (16.66)$$

is very short. With $c_\infty \approx 10^5$ cm/s, $\rho \approx 10^{-9}$ g/cm³, we get $\tau_{accr} \approx 1000(M_\oplus/M)$ y. In fact, the accretion could be so fast only where the planet's gravity was dominant with respect to that of the Sun, that is only in a fraction of the feeding zone; and diffusion of the gas toward the centre of the zone was hampered by angular momentum conservation. However, eq. (16.66) shows that gas accretion was a rapid process both with respect to the accumulation of solid material and with respect to the dissipation of the nebula by primordial solar wind. Hence, some 90% of the original gaseous component of the nebula had to dissipate from the zones of Jupiter and Saturn by the beginning of accretion (i. e., some 10^8 y after the formation of planetesimals).

Finally, as far as the formation of satellites is concerned, it is important to note that they probably represent a very heterogeneous set of bodies; all the main theories, proposed for instance to explain the origin of the earth's moon, might be correct for some particular category of satellites. Let us review briefly these theories.

(1) *Capture*. This appears as the most likely hypothesis for the small outermost satellites of Jupiter, Saturn and Neptune, whose physical properties resemble very much those of outer-belt and Trojan asteroids and whose "irregular" (high eccentricity and/or inclination) orbits suggest some peculiar dynamical history. Two other cases of possible capture are those of the Martian moonlets Phobos and Deimos, which have "regular" orbits but are very small ($\approx$ 10 km across) and asteroid-like, and of Triton, which on the contrary is a large body but moves around Neptune on a retrograde orbit (due to the large satellite-to-planet mass ratio, Triton's orbit might have tidally evolved from an original quasi-parabolic one; see Sec. 15.3). Capture from an unbound orbit, however, is a dynamically unlikely phenomenon even in presence of three-body solar perturbations or tidal effects; therefore, it has been suggested that capture was aided by collisions or by gas drag in extended primordial gaseous envelopes surrounding the planets. Both these effects could indeed cause rapid and strong dissipation of planetocentric orbital energy for small bodies, but are unlikely to work for big objects like the earth's moon.

(2) *Rotational fission*. This theory has been proposed for high mass ratio ($\succ$ 0.01) planet-satellite pairs like the earth-moon and the Pluto-Charon systems. It assumes that rotational instability occurred during the final phase of planetary accumulation, when the spin angular momentum of the planet exceeded the critical value for fission (see Sec. 3.4), either due to collisions or owing to the formation of a high density core. For the earth-moon case, the latter idea is attractive because it can explain in a natural way why the moon is much more deficient in iron than the earth, while the lunar density and composition are very close to those of the earth's mantle alone (from which, according to this theory, the lunar material would in fact come). After the fission, the moon would have tidally evolved outward from the Roche limit (see Secs. 4.5 and 14.6) to its present orbit. However, for the earth-moon pair (but not for the Pluto-Charon one) the fission theory meets the serious objection that the present total angular momentum of

the system is only about half of the critical value for a spinning proto-earth, and it is very difficult to find out a reasonable mechanism to get rid of such a large amount of angular momentum during the subsequent history of the system.

(3) *Circumplanetary accumulation.* This theory applies in a natural way to the "regular" satellite systems of Jupiter, Saturn and Uranus, which mimic on a much smaller scale the structure of the whole solar system. The basic idea is that in the course of accumulation of a planet from planetesimals, a swarm of debris arises in planetocentric orbit, initiated by collisions between planetesimals in the vicinity of the growing planet and then further replenished by collisions of planetesimals with bodies previously trapped in the swarm. The evolution of these circumplanetary swarms is very complex, being governed by a balance between mass inflow from outside and infall of material toward the planet, caused by gas drag and by the growing central mass. When sizeable satellites are accumulated in the swarm, they also spiral inward unless, as for the earth's moon, there are counteracting tidal effects; for Mercury and Venus, the present lack of satellites could also be due to tidal evolution, having different outcomes than for the earth because of the different rotational states of these planets. For the outer planets, the existing satellites are instead just the last generation of bodies accumulated in the swarms, and account only for a small fraction of the total trapped mass, most of which fell into the planets. Again, for our moon this theory meets with a serious problem: that of explaining why the moon's iron abundance is so low, if one assumes that the satellite formed from the same basic material as the earth.

(4) *Megaimpact.* During the final phase of planetary accumulation, planetary embryos were struck by large secondary bodies (see Sec. 16.3). It is possible that such impacts, especially if occurred in a "grazing" mode, ejected from the proto-earth a spray of mantle material which subsequently reaccumulated outside the Roche limit and formed the (iron-deficient) moon. The estimated mass of the impactor ($\approx 0.1\ M_\oplus$, close to the mass of Mars) is however in the very upper limit of the range predicted by the planetary accumulation theory (implying $\theta \approx 1$), and the spin angular momentum provided to the earth by the impact would be such that only by chance the earth's rotational axis could remain fairly close to the pole of the ecliptic. The impact theory, on a much smaller scale, is of course a very plausible way to explain the origin of some suspected double asteroids, like 216 Kleopatra and 624 Hektor, whose binary nature has been inferred from the very large amplitude of their light-curves. For them, after a catastrofic collision with another (smaller) asteroid, the resulting angular momentum might simply have been too high to allow reaccumulation of the fragments into a single body. On the other hand, collisional disruption of pre-existing satellites could explain the formation of planetary rings and the presence,

in the inner part of the satellite systems of the outer planets, of many small, fragment-like moonlets.

PROBLEMS

16.1. Confirm, in order of magnitude, eq. (16.13).

*16.2. Using the fluid dynamics equations in the adiabatic approximation, prove the dispersion relation (16.34).

16.3. Show that if the ejection velocity (16.61) is added in the retrograde sense to a circular orbit, the resulting pericentre distance is about 0.21 times the initial orbital radius.

16.4. Solve the differential equation (16.50) for the function v_r, keeping everything close to constant. From eq. (16.25), the appropriate initial condition is $v_r \ll v_e$.

16.5. Compute the total angular momentum of the earth-moon system and compare it with the critical value for rotational instability of a homogeneous proto-earth (see Sec. 3.4).

16.6. Evaluate the ratio (16.6) as function of the distance from the sun, assuming a constant temperature, on the basis of Fig. 16.1.

16.7. Prove eq. (16.53) and calculate the accretion rate (16.54).

16.8. Solve the accretion equation (16.54) in the case when v_r is constant and $\ll v_e$ (i. e., with $\theta = v_e^2/2v_r^2 \gg 1$). Show that his yields the maximum rate of embryo growth, because any increase in v_r due to stirring by the embryo or to planetesimal growth results in a reduced growth rate.

FURTHER READINGS

The basic book summarizing the main aspects of the contemporary nebular theory is V.S. Safronov, *Evolution of the Protoplanetary Cloud and Formation of the Earth and the Planets*, Nauka, Moscow (1969) (English translation by the Israel Program for Scientific Transations, Jerusalem (1972).) Two other more recent and useful books are: A. Brahic, ed., *Formation of Planetary Systems*, Cepadues, Toulouse (1982); S.F. Dermott, ed., *The Origin of the Solar System*, Wiley, New York (1978). See also R. Greenberg, *Planetary Accretion*, in S.K. Atreya and

J.B. Pollack, eds., *Origin and Evolution of Planetary Atmospheres*, University of Arizona Press, Tucson (1988).

Fig. 16.1 has been adapted from S.J. Weidenschilling, *The Distribution of Mass in the Planetary System and Solar Nebula,* Ap. Spoce Sci., **51**, 153 (1977).

17. RELATIVISTIC EFFECTS IN THE SOLAR SYSTEM

The large amount of optical observations of the planets and the moon available have provided an excellent confirmation of the newtonian laws of celestial mechanics. However, a minute, but important discrepancy, was known since the second half of the 19th century: the perihelion of Mercury undergoes an unexplained advance of 43" per century. In 1916 Einstein's discovery of general relativity pointed out an entirely new way to understand gravitation: this was suggested not by the anomaly of Mercury, but by the equality between inertial and gravitational mass – the *Principle of Equivalence* – and was brought to a complete mathematical formulation on the basis of theoretical simplicity. Within the theory of general relativity one can predict small corrections to the two-body problem; the only secular correction is an advance of the periastron, which for the case of Mercury agrees with the observed discrepancy. In recent years the relativistic corrections of gravitational motion in the solar system and in a binary pulsar have been tested with a much greater accuracy and Einstein's predictions have been confirmed. At present these corrections are an essential ingredient of celestial mechanics and space navigation.

In this chapter we review the theory of general relativity and its approximation to the description the motion of bodies and photons in the solar system. This is done without excessive mathematical formalism, having in view the physical aspects and the actual measurements. In this approximation special relativistic effects have the same general structure as those due to general relativity and are included in this review; however, special relativity is assumed to be known.

Some remarks on notation: in this chapter greek letters range from 0 to 3 and label the space-time variables; latin indices, with values 1, 2, 3, denote the space coordinates. Covariant (downstairs) and controvariant (upstairs) indices, if repeated in a product, are understood to be summed over ("summation convention"). $\partial_\mu = \partial/\partial x^\mu$ is the ordinary, partial derivative. The velocity of light c is unity. The masses are measured in terms of their gravitational radii $m = GM/c^2$. Sometimes we shall denote four-vectors by an ordinary letter without any index and their scalar product by a dot (•).

17.1. CURVATURE OF SPACE-TIME

In newtonian mechanics space is taken to be endowed with an absolute structure, namely, euclidean geometry; and time is a given, absolute variable, uniformly flowing for all observers. In special relativity we have a four-dimensional manifold, space-time, whose coordinates x^μ are determined to within translations and Lorentz transformations. Given an infinitesimal displacement dx^μ, the quadratic form

$$d\tau^2 = \eta_{\mu\nu} dx^\mu dx^\nu \quad (\eta_{\mu\nu} = \mathrm{diag}(1, -1, -1, -1)) \qquad (17.1)$$

is unchanged under these transformations and yields the proper time interval $d\tau$ when $d\tau^2 > 0$ and the space interval $\sqrt{(-d\tau^2)}$ when $d\tau^2 < 0$. $d\tau = 0$ for two events lying on the world line of a photon.

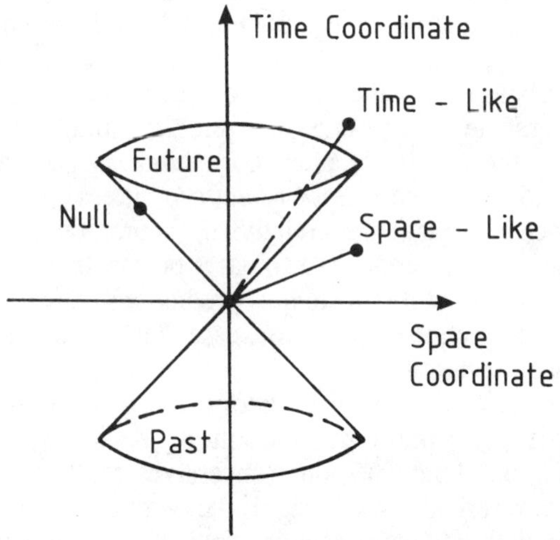

Fig. 17.1. The light cone structure.

In general relativity cartesian coordinates are meaningless; the coordinates x^μ are *arbitrary* labels of events, not the result of a measurement; generally their choice is dictated by mathematical simplicity. They can be replaced with regular, arbitrary functions $x'^\mu(x^\nu)$ of themselves, so that their differentials transform according to a general, linear transformation:

$$dx^\mu \to dx'^\mu = \frac{\partial x'^\mu}{\partial x^\nu} dx^\nu. \qquad (17.2)$$

All laws of physics are invariant under this replacement (*general covariance*). Instead of the Minkowski line element (17.1) we now have a generic quadratic form, the *metric*

$$ds^2 = g_{\mu\nu}(x)dx^\mu dx^\nu \text{ (with } g_{\mu\nu} = g_{\nu\mu}). \qquad (17.3)$$

This endows space-time with a rich geometric structure and makes it a *pseudo-riemannian manifold* (in a *riemannian* manifold the metric has a diagonal form (1, 1, 1, 1)). The *metric tensor* $g_{\mu\nu}$ also defines the scalar product $g_{\mu\nu}u^\mu v^\nu$ between two vectors u and v. Under a coordinate transformation the new metric at the point of new coordinates x'^μ is given by

$$g_{\mu\nu}'(x') = g_{\rho\sigma}(x(x')) \frac{\partial x^\rho}{\partial x'^\mu} \frac{\partial x^\sigma}{\partial x'^\nu}; \qquad (17.4)$$

in other words, the metric is a *symmetric, covariant* tensor of the second rank. Of course, it is always possible to choose the transformation in such a way that *at a given point* the metric reduces to a diagonal form, with one of the numbers 1, − 1 or 0 on the diagonal; for consistency with special relativity (eq. (17.1)), we require this diagonal to be (1, − 1, − 1, − 1). Therefore a light cone is defined at any event.

The question then arises, is it possible to reduce a given metric to the diagonal form (17.1) globally? If so, clearly space-time is equivalent to a Minkowski space-time and one could not expect anything different from special relativity, except for the general coordinates. The answer to this question is given in terms of a fourth rank tensor $R_{\mu\nu\rho\sigma}$ − called after *Riemann* − which can be expressed in terms of the metric in a way to be indicated later. Its vanishing is the necessary and sufficient condition in order that a coordinate transformation exists that reduces the metric to the form (17.1); if this happens, the space-time is said to be *locally flat.* Otherwise space-time is *curved* and the laws of euclidean geometry do not hold; in particular, the sum of the internal angles of a triangle is not equal to π. The difference between this sum and π is the *excess angle* ϵ.

Since we are interested only in the solar system, we confine ourselves to riemannian manifolds and to coordinate transformations such that the metric deviates little from its flat form (17.1):

$$g_{\mu\nu} = \eta_{\mu\nu} + h_{\mu\nu} \ , \tag{17.5}$$

with $|h_{\mu\nu}| \ll 1$ (*weak field approximation*). This also says that we have adopted coordinates which are nearly minkowskian; it is convenient to attribute to them the dimension of a time (the same as τ and s) and to have the h's dimensionless, like η. Expressed in a formal way, this means that, besides the coordinate transformations, we also consider the scale transformations

$$x^\mu \Rightarrow x'^\mu = kx^\mu, \tag{17.6}$$

which leave the metric unchanged and changes s into s' = ks. The choice of the same unit for proper time, coordinate time and distance is in agreement with the current metrological use (see Sec. 3.1.) In the approximation (17.5) it is now possible to work with the special relativistic formalism; in particular, the indices are lowered and raised with the metric (17.1).

With the introduction of a generic metric (17.3), albeit at any given point it is formally equivalent to the Minkowski metric (17.1), we have opened the way to a complex and subtle geometric content, which goes under the general name of curvature. According to Einstein's theory, gravitation is described by this new geometric structure.

In the approximation (17.5) the *Riemann* (or *curvature*) *tensor*, to within linear terms in the h's, reads:

$$R_{\mu\nu\rho\sigma} = \tfrac{1}{2}(\partial_\mu\partial_\sigma h_{\nu\rho} + \partial_\nu\partial_\rho h_{\mu\sigma} - \partial_\mu\partial_\rho h_{\nu\sigma} - \partial_\nu\partial_\sigma h_{\mu\rho}) . \tag{17.7}$$

Its components have dimension $[1/T^2]$. This tensor fulfils three symmetry properties :

(1) antisymmetry in each of the two pairs of indices:

$$R_{\mu\nu\rho\sigma} = - R_{\nu\mu\rho\sigma} = - R_{\mu\nu\sigma\rho} \ ; \tag{17.8}$$

(2) symmetry with respect to the interchange of the two pairs:

$$R_{\mu\nu\rho\sigma} = R_{\rho\sigma\mu\nu} \ ; \tag{17.9}$$

(3) the cyclic symmetry:

$$R_{\mu\nu\rho\sigma} + R_{\mu\sigma\nu\rho} + R_{\mu\rho\sigma\nu} = 0 \ . \tag{17.10}$$

These properties reduce from $4^4 = 256$ to 20 the number of independent components.

The curvature tensor is related to the *gaussian curvature* of the two-dimensional sections of space-time. At a point P of an ordinary surface Σ, consider a section with a plane π through its normal. Let $R(\pi)$ be the radius of the osculating circle to this section. As π rotates around the normal, $R(\pi)$ changes and goes through a maximum R_1 and a minimum R_2 (except cases of degeneracy). The gaussian curvature

$$K = 1/R_1 R_2 \tag{17.11}$$

gives an overall measure of the curvature of Σ at P. It is positive if the tangent plane to Σ at P does not intersect the surface, negative if otherwise (like for a saddle point). It can also be shown that the excess angle ϵ of any infinitesimal triangle of area A is given by

$$\epsilon = A|K| . \tag{17.11'}$$

In this case the Riemann tensor has only one component R_{1212}; the invariant $R_{\mu\nu\rho\sigma}g^{\mu\rho}g^{\nu\sigma}$ is the gaussian curvature K. In more than two dimensions the components of the Riemann tensor determine the gaussian curvature of two-dimensional sections of arbitrary orientation.

17.2. GEODESICS

In special relativity free motion is described by straight world lines (the coordinates are linear functions of proper time.) They can be defined in a synthetic way by saying that the proper interval

$$T = \int_A^B |d\tau| \tag{17.12}$$

on a world line joining the two events A and B is an extremum. (In euclidean geometry we would say, a minimum; in Minkowski geometry the quantity (17.12) is a minimum on a space-like straight line, but a *maximum* on a time-like world line. It is an extremum in all cases.) Since we care only for the property of extremality, we can drop in eq. (17.12) the absolute value. These world lines are termed *geodesics*.

This variational principle can obviously be generalized to a riemannian geometry with metric (17.3): the action

$$S = \int_A^B ds \tag{17.13}$$

is an extremum on the geodesic world line joining A and B. If $x^\mu(\lambda)$ are the parametric equations of the world line, they are solutions of the Euler's equations for the lagrangian function

$$L = \surd[g_{\mu\nu}(x^\rho(\lambda))\dot{x}^\mu\dot{x}^\nu] , \qquad (17.14)$$

to wit,

$$\frac{\partial L}{\partial x^\rho} = \frac{d}{d\lambda}\frac{\partial L}{\partial \dot{x}^\rho} . \qquad (17.15)$$

A dot indicates here the derivative with respect to the parameter λ. The value of the action

$$S = \int_A^B L d\lambda$$

does not change if λ is replaced with another parameter λ'; in other words, if $x^\mu(\lambda)$ is a solution, so are the functions of λ' that can be expressed as $x^\mu(\lambda(\lambda'))$. This follows from the circumstance that L is homogeneous of degree 1 in the velocities (Problem 17.13). We can remove this arbitrariness by stipulating that the parameter be the proper time s. This, of course, is impossible for a null line, on which $L = ds = 0$; in that case, anyway, eq. (17.15) loses its meaning. To encompass every case it is necessary to define the geodesic lines with a slightly different variational principle based upon the Lagrange function L^2:

$$S' = \int_A^B d\lambda L^2 = \int_A^B d\lambda g_{\mu\nu}(x^\rho(\lambda))\dot{x}^\mu\dot{x}^\nu . \qquad (17.13')$$

This principle holds also when L vanishes. A straightforward calculation from Euler's equation for L^2 shows that the quantity L itself is a constant along any world line which extremizes S'. When S' is varied we can write, therefore,

$$\delta S' = 2\int_A^B d\lambda L\delta L = 2L\int_A^B d\lambda\delta L .$$

Hence a curve which is an extremum for S' is also an extremum for S;

but the new principle is universally valid, also for a null line. Moreover, while in the previous formulation the parameter λ was not specified, now from L = const. we conclude that the parameter is the proper time, unless the world line is null; in this case we have a new, special variable called the *affine* parameter. In the generic case the proper time is determined within a linear substitution $s \Rightarrow s' = as + b$.

The theory of the geodesics based upon the action S' is formally equivalent to the theory of a mechanical system whose hamiltonian function contains only the kinetic energy, and the mass is an arbitrary, tensorial function of the configuration variables. This kinetic energy is a constant of the motion.

17.3. DYNAMICS AS GEOMETRY

The previous section points to an obvious generalization of the free motion of a particle in special relativity based upon a pseudo-riemannian structure; we ask the question, what kind of dynamics does it describe? As an epistemological remark, note the peculiar way this reasoning goes: first one invents, on the basis of simplicity and mathematical beauty, a generalization of a previous formal scheme; next he looks for physical phenomena described by it. This typically deductive argument has met with extraordinary success in the theory of general relativity.

A dynamical theory succinctly described by the variational principle (17.13') is a geometrical theory; in other words, the motion does not depend on the nature of the particle (its mass, shape, etc.), but only on the underlying geometry. The variational equations (17.15), being of the second order, determine the unknown functions $x^\mu(\lambda)$ if the initial derivatives and the initial point are specified (Problem 17.10): through an event we have one integral curve for any direction.

This characteristic property is not generally true in newtonian dynamics, but only when *the force is proportional to the inertial mass*. In that case, in fact, the motion depends only on a vector function of space and time, the acceleration per unit mass. This, of course, is certainly true for the apparent forces, which arise in newtonian mechanics from the inertial term when a moving frame is adopted; but, as far as we know, to a very good approximation, it is also true for the gravitational forces. This fundamental property of gravitation can be described quantitatively. From the accelerations suffered by different bodies in the presence of the same external force (e. g., a mechanical force) we can measure their inertial mass m_i. From the inertial force (inertial mass times acceleration) experienced by bodies placed at the same event in a gravitational field we can measure the ratio of the gravitational forces to which they are subjected. In general these forces vary from body to body at the same event; one assumes that this force is the product of a vector $\mathbf{g}$ function of space, time and possibly the velocity – the gravitational acceleration – and a scalar m_g which depends on the nature of the body. Therefore in this general formulation the equation for gravitational dynamics reads

$$m_i \ddot{\mathbf{r}} = m_g \mathbf{g} \ . \tag{17.16}$$

The equality of m_i and m_g goes under the name of *Weak Equivalence Principle* and is the basis of the geometrical description of gravitational dynamics. Of course, such a statement must be phrased within an appropriate system of units. One should really say, having chosen arbitrarily $m_i = m_g$ for a given reference body, the same equality holds for all the other bodies; in other words, the ratio of the inertial masses of two bodies is equal to the ratio of their gravitational masses. The strength of this requirement most clearly appears when looking at electromagnetism: there in place of $\mathbf{g}$ we have the force per unit charge – the electric field; the analogue of m_g is the electric charge. We can say that the "gravitational charge" is the inertial mass. The Weak Equivalence Principle has been experimentally verified with great accuracy, to about one part in 10^{12}.

We can consider also a third mass, the *active gravitational mass* m_{ag}. It is obtained by measuring the gravitational accelerations produced on a given test body by two different sources placed at the same place. In newtonian theory

$$\mathbf{g} = -\ \mathbf{R}Gm_{ag}/R^3 \ .$$

In classical physics the three masses are equal; in general relativity the definition of m_{ag} is more complicated because the gravitational force depends also on the velocity of the source: the concept of active gravitational mass is really inappropriate.

According to the Principle of Equivalence apparent forces and gravitational forces are locally indistinguishable: both of them are geometrical in nature, that is to say, are defined by a family of curves in space-time, one for each event and for each direction. In general relativity we assume that they follow from the same variational principle (17.13), with the only difference that when one has only apparent forces the metric can be transformed into the Minkowski form (17.1) with a suitable change of coordinates. This is what happens in a freely falling elevator: the motion of test bodies in its interior is straight and uniform and shows no trace of the presence of the earth. The apparent force generated by the fall cancels exactly the gravitational force. Conversely, within an accelerated rocket in free space, bodies experience a uniform acceleration, exactly like on the surface of the earth. One can say, with *local* experiments performed with a single test body it is impossible to discover the presence of a true gravitational field; in other words, bodies fall because one observes them in a "wrong" frame of reference, subject to external forces (the molecular forces of the ground which support an earth laboratory or the thrust of the rocket.) Weight is an illusion, not a force.

However, with two mass points (making thereby the experiment not quite local), in general gravitational forces cannot be exactly compensated by apparent forces. For example, in an accelerated rocket in empty space the acceleration of a point mass is uniform in space, while in a laboratory at rest on the surface of the earth the directions of the accelerations of two masses are not exactly parallel, but converge to the centre of the earth. The presence of gravitation can be ascertained only through the *relative motion* of bodies.

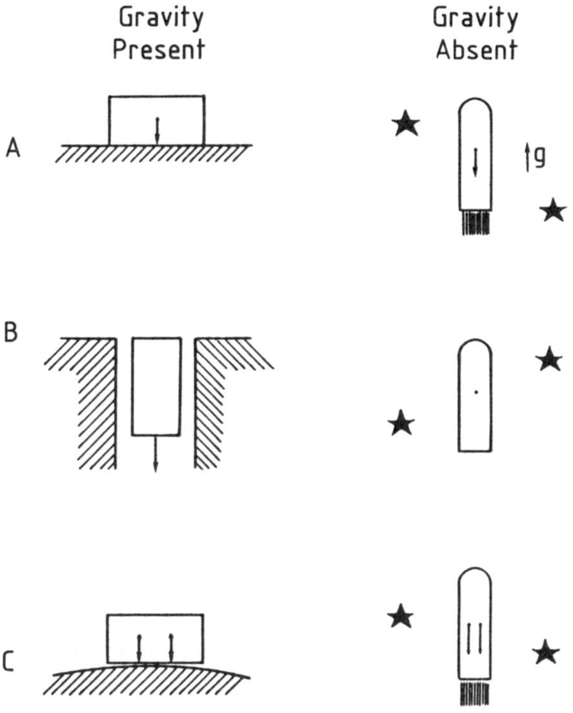

Fig. 17.2. The Weak Equivalence Principle and the measurement of the gravitational field. With a single particle experiment one cannot distinguish between the dynamics in a ground laboratory and in a rocket (A and B); to do this it is necessary to compare the motion of two particles (case C).

The accuracy of compensation of gravitational forces by apparent forces in a region of size r is contained in a theorem of riemannian geometry due to E. Fermi. Given a geodesic world line ℓ, it is always possible to construct in its neighbourhood a frame of reference in which the time axis coincides with ℓ: ℓ is at rest. In this frame the metric tensor has the form:

$$g_{\mu\nu} = \eta_{\mu\nu} + O(Kr^2). \tag{17.17}$$

r is the (proper) distance to ℓ and K indicates the order of magnitude of the curvature tensor. One could say, the metric is in a form which is, in the neighbourhood of ℓ, as near as possible to the flat form (17.1). Fermi's construction is invariantly defined: the time coordinate is the proper time s along ℓ and the simultaneity hypersurfaces s = const are defined by means of spatial geodesic lines orthogonal to ℓ. On these hypersurfaces we have three invariant coordinates r^i (i = 1, 2, 3) and at ℓ a corresponding basis of three unit and orthogonal vectors. Note that in this way we can also define in an invariant way the components of any vector on ℓ and we can give an invariant meaning to the *constancy* of a vector, or, more generally, a tensor, on a geodesic ℓ: it is constant if so are its Fermi components. This important construction of riemannian geometry takes the name of *parallel transport* and establishes in a general way a linear mapping between the vectorial spaces at two points. Fermi's coordinates and the parallel transport construction can also be generalized to a line ℓ which does not possess the geodesic character.

In the Fermi frame the gravitational acceleration of a point P in the neighbourhood of ℓ is of order Kr; this is in agreement with the fact that, since ℓ is at rest, this acceleration is the relative acceleration of P with respect to ℓ and hence is of *tidal* character, proportional to the distance r. It is important to note that the coefficients of the metric corrections in eq. (17.17) are linear combinations of the curvature tensor; if there is gravity, it is impossible to find a frame of reference in which they all vanish. One could say, the geometrical properties of space-time and hence the structure of the true gravitational field show up in the relative accelerations.

It is illuminating to write down, even without proof, the dynamical equation which describes the motion in Fermi's frame. If $\mathbf{r}(s)$ is a point lying on a nearby geodesic, it fulfils the equation of *geodesic deviation*

$$\frac{d^2 r^i}{ds^2} + R^i{}_{0j0}\, r^j = 0 , \tag{17.18}$$

where

$$R^i{}_{0j0} = \eta^{ih} R_{h0j0} = - R_{i0j0}$$

is a 3·3 matrix, function of s only, constructed with the curvature tensor and the (minkowskian) metric at ℓ. This is the generalized tidal equation; it has the same formal structure as in newtonian mechanics (see Sec. 4.2), where, however, the gradient of the force is expressed in terms of the gravitational potential energy U:

$$R_{i0j0} = \frac{\partial^2 U}{\partial r^i \partial r^j} . \qquad (17.19)$$

The equivalence principle applies to photons as well and requires that the spectral lines received from a source in a gravitational field be redshifted. This corresponds to the fact that if a photon moves to a region of greater gravitational potential energy, it loses energy and hence frequency. According to the equivalence principle (case A of Fig. 17.2) this corresponds to a photon moving upward in an accelerated rocket; at the moment of reception at O the system moves faster than at the moment of emission at S and hence less wavelengths per unit time are received than emitted (Fig. 17.3). Because of the equivalence principle, the same frequency shift is observed when there is a difference (gh) in the gravitational energy per unit mass between the observer and the source:

$$\Delta\nu/\nu = U_S - U_0 = gh. \qquad (17.20)$$

In the case of light emitted from the sun, for example, this number is $2 \cdot 10^{-7}$ and hence barely observable. The effect becomes drastic in the case of black holes, but then eq. (17.20), which requires for its validity

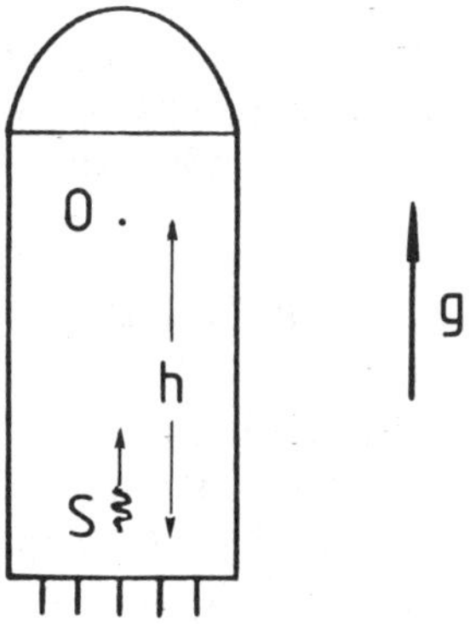

Fig. 17.3. Doppler shift in an accelerated rocket. During a photon transit time h/c the receiver O increases its velocity by gh/c and detects a fractional frequency shift gh/c^2.

a small fractional frequency shift, does not apply and a more rigorous treatment is needed. A more complete discussion of the Doppler effect will be given in Sec. 17.5; see also Sec. 3.1.

17.4. SLOW MOTION, RELATIVISTIC DYNAMICS

Relativistic dynamics replaces the variational principles of classical dynamics with the geodesic principle. The comparison between them will clearly show under which approximation newtonian mechanics is recovered from general relativity and which corrections have to be taken into account for a more precise description. The classical motion of a point mass moving in a gravitational potential energy (per unit mass) U is governed by the action

$$S_n = \int dt L_n(\mathbf{r}, \mathbf{r}) = \int dt(\tfrac{1}{2}v^2 - U). \qquad (17.21)$$

The requirement that S_n be stationary with respect to all infinitesimal variations of the trajectory, which leave the end positions unchanged, leads to the newtonian equations of motion.

To write the general relativistic action in an appropriate form it is convenient to treat the time coordinate t and the space coordinates x^i separately; factoring out dt, (17.13) reads

$$S = \int_A^B \surd[g_{\mu\nu}dx^\mu dx^\nu] = \int dt \surd[g_{00} + 2g_{0i}v^i + g_{ij}v^iv^j].$$

$v^i = dx^i/dt$ is the ordinary velocity. Besides the weak field approximation, we now assume that the motion is slow, to wit, $v \ll 1$; moreover, we are interested in situations in which the potential and the kinetic energies are of the same order of magnitude. In terms of an infinitesimal parameter V

$$O(v^2) = O(U) = V^2 . \qquad (17.22)$$

For the earth $V \approx 10^{-4}$. We also assume – to be confirmed later – that the metric corrections $h_{\mu\nu}$ are of the same order of magnitude or smaller; indeed, it can be shown that the expansion in powers of V has only even components for h_{00} and h_{ji} and odd components for h_{0i}. This allows us to expand the Lagrange function in powers of V^2:

$$S = \int dt[1 - \tfrac{1}{2}v^2 + \tfrac{1}{2}h_{00} + O(V^4)]. \qquad (17.23)$$

Since the first term does not contribute to the variation, the relativistic variational principle is equivalent, to this order, to the newtonian counterpart (17.21), provided that

$$h_{00} = 2U \ . \tag{17.24}$$

This shows that the whole of classical gravitational dynamics, in the slow motion approximation, is contained in the geodesic principle and reduced to geometry!.

In this (3 + 1)-dimensional formulation, gravitation is described by three dimensionless potentials: a scalar h_{00} (with respect to spatial transformations, of course), a vector h_{0i} and a symmetric tensor h_{ij}. To lowest order, the scalar potential, according to eq. (17.24), is the sum of newtonian contributions from all bodies present; a mass M at a distance r contributes

$$h_{00} = -2m/r \quad (m = GM/c^2 = 1.48\text{km}\ (M/M_\odot)). \tag{17.24'}$$

This determination places a condition upon all the other metric deviations. Note that under a Lorentz transformation $h_{\mu\nu}dx^\mu dx^\nu$, being the difference between two invariant quantities, is itself an invariant; hence $h_{\mu\nu}$ behaves as a tensor under a Lorentz transformation. We expect that a gravitating body contributes to the h's, to lowest order, with terms proportional to the dimensionless scalar quantity m/r. If the body has a (special relativistic) velocity

$$\bar{u}^\mu = \frac{dx^\mu}{d\tau} = \frac{dx^0}{d\tau}(1,\ u^m) = (1,\ \mathbf{u})/\surd(1 - u^2), \tag{17.25}$$

we can construct only two second rank tensors of this kind, proportional to $\eta_{\mu\nu}$ and $\bar{u}_\mu\bar{u}_\nu$:

$$h_{\mu\nu} = 2(\gamma\eta_{\mu\nu} + \gamma'\bar{u}_\mu\bar{u}_\nu)m/r.$$

γ and γ' are arbitrary, dimensionless coefficients. When the body is at rest ($u_0 = 1$, $u_i = 0$), we must recover eq. (17.24'); this leads to $\gamma' = -1 - \gamma$ and

$$h_{\mu\nu} = 2[\gamma\eta_{\mu\nu} - (1 + \gamma)\bar{u}_\mu\bar{u}_\nu]m/r. \tag{17.26}$$

To lowest order we have, besides (17.25), a vector component

$$g_{0i} = h_{0i} = -2(1 + \gamma)u_i m/r \tag{17.27}$$

and a tensor component

$$h_{ij} = -\,2\gamma\delta_{ij}m/r. \qquad (17.28)$$

We see that the vector component is of order V^3; it enters in the Lagrange function via its scalar product with the velocity of the particle, exactly like the vector potential in electromagnetism. For this reason this is called a gravitomagnetic term (see Sec. (17.7).)

To get the relativistic corrections to the newtonian motion we must go beyond the approximation (17.23) and compute the Lagrange function up to $O(V^4)$. To do this the approximations (17.27) and (17.28) are sufficient, but we need h_{00} to $O(V^4)$. For a single gravitating body at rest the obvious choice is a correction proportional to $(m/r)^2$:

$$h_{00} = -\,2\frac{m}{r} + 2\beta\left(\frac{m}{r}\right)^2 + \ldots \qquad (17.29)$$

For several sources the situation is more complicated because of the interaction terms. For the motion of planets and interplanetary navigation, however, the only important relativistic corrections are those of the sun, which can be taken at rest. It is therefore important to study the geodesic motion for the spherically symmetric metric

$$ds^2 = (1-2m/r+2\beta(m/r)^2)dt^2 - (1+2\gamma m/r)\delta_{ij}dx^i dx^j. \qquad (17.30)$$

Higher order terms in m/r are neglected.

The dimensionless parameters β and γ must be determined from the field equations, whose detailed derivation goes beyond our scope. The scalar component (17.24) must fulfil, to lowest order, Poisson's equation (1.23), or

$$\nabla^2 h_{00} = 8\pi G\rho\ . \qquad (17.31)$$

We need ten equations for the ten unknowns $g_{\mu\nu}$; they can be expected to include several different novel features. First of all they must be invariant under general coordinate transformations; in other words, if $g_{\mu\nu}(x)$ is a solution, so must be (17.4). Secondly, they are not linear: the field produced by different sources is not the sum of the fields of the single sources. As indicated by eq. (17.26), the velocity of the sources will come into play. Also, as expected in any relativistic theory, we must expect equations of hyperbolic character and hence propagation phenomena (while Poisson's equation is elliptic). The energy-momentum tensor of the matter and the fields (ten components!) is the appropriate source in the field equations. It turns out that, as a consequence of general covariance, from the field equations a constraint is imposed upon the energy-momentum tensor: this is a set of four differential equations, which lead to the correct equations of motion. Contrary to other field theories, in general relativity we do not need to postulate separately the dynamical equations for the sources; they follow from the field equations themselves.

As stated above, here we need only the metric solution for the exterior of a spherical body at rest (at the origin). This is the *Schwarzschild solution;* it depends on the radial variable r only through the combination m/r. Expanding the Schwarzschild metric with respect to this small quantity we get the expressions (17.29) and (17.28), with $\beta = \gamma = 1$.

Of course it is possible to think of different field equations leading to different values of these parameters. An important task of experimental gravitation, discussed in Ch. 20, is to measure them and thereby to test Einstein's theory of gravitation. In any case, we have laid down the theoretical tools needed to describe the relativistic effects in the solar system in a general metric framework in accordance with the equivalence principle.

17.5. THE DOPPLER EFFECT

Since most measurements in the solar system are performed by means of electromagnetic radiation, it is important to discuss the relativistic effect on the propagation of photons. In special relativity a photon is described by a null vector p, whose time component is equal to its energy $h\nu$. With respect to an observer $x^\mu(\tau)$ the energy is the scalar product $p \cdot \dot{x} = \eta_{\mu\nu} p^\mu \dot{x}^\nu$. The components orthogonal to $\dot{x}^\mu$ give the momentum.

It is beyond our scope to discuss in full electromagnetic propagation in a curved space; let it suffice to say that the photon four-momentum p is proportional to $\dot{y}^\mu(\lambda)$, where $y^\mu(\lambda)$ is a null solution of the variational principle (17.13'), function of the affine parameter λ:

$$p^\mu = h\nu \dot{y}^\mu(\lambda) \ . \tag{17.32}$$

Here $h\nu$ is a constant, with the dimension of an energy, which characterizes that particular photon, of frequency ν. The propagation is described by Euler's equations for that variational principle. The energy of the photon with respect to an observer $x^\mu(\lambda)$ is the (riemannian) scalar product $g_{\mu\nu} p^\mu \dot{x}^\nu$.

Consider now a source $x_1{}^\mu(s_1)$ and a receiver $x_2{}^\mu(s_2)$ on the path of the photon. The exact expression of the frequency ratio is then

$$\frac{\nu_2}{\nu_1} = \frac{[g_{\mu\nu}p^{\mu}\dot{x}^{\nu}]_2}{[g_{\mu\nu}p^{\mu}\dot{x}^{\nu}]_1} . \tag{17.33}$$

This formula, whose intuitive meaning is obvious, can be rigorously derived from the theory of null geodesics.

Let us first recover and generalize the gravitational frequency shift (17.20). Consider a stationary space-time in which the components of the metric are independent of the time x^0; and the source and the observer are at rest with respect to this time coordinate. Because of the definition of the proper time s

$$(\dot{x}^0 = 1/\sqrt{(g_{00})}, \quad \dot{x}^i = 0) . \tag{17.34}$$

We do not need to solve the variational principle (17.13') explicitely for the motion of the photon; it is enough to use the fact that, since the Lagrange function L^2 does not contain the time, its canonically conjugate quantity

$$\frac{\partial L^2}{\partial \dot{y}^0} = 2g_{0\mu}\dot{y}^{\mu} \tag{17.35}$$

is a constant along the ray (and so is $g_{0\mu}p^{\mu}$). But this is just what is needed to compute (17.33):

$$\frac{\nu_2}{\nu_1} = \frac{[g_{0\mu}p^{\mu}/\sqrt{(g_{00})}]_1}{[g_{0\mu}p^{\mu}/\sqrt{(g_{00})}]_2} = \left[\frac{g_{00}(2)}{g_{00}(1)}\right]^{\frac{1}{2}} . \tag{17.33'}$$

This is exact; in the linear approximation (17.24) we recover (17.20).

From the general formula (17.33) we can also recover the frequency shift in special relativity (3.3). There the four-momentum of the photon p is proportional to (1, **k**), where **k** is the (constant) unit vector along the direction of propagation; and the metric is given by eq. (17.1). We also have

$$\dot{y}^0 = 1/\sqrt{(1 - v^2)} \quad , \quad \dot{y}^i = v^i/\sqrt{(1 - v^2)} . \tag{17.36}$$

Hence

$$g_{\mu\nu}p^{\mu}\dot{y}^{\nu} = p^0\dot{y}^0 - \delta_{ij}p^i\dot{y}^j = (1 - \mathbf{k}\cdot\mathbf{v})/\sqrt{(1 - v^2)}$$

and

$$\frac{\nu_2}{\nu_1} = \frac{1 - \mathbf{k} \cdot \mathbf{v}_2}{1 - \mathbf{k} \cdot \mathbf{v}_1} \left[\frac{1 - v_1^2}{1 - v_2^2}\right]^{\frac{1}{2}} . \qquad (17.37)$$

The first factor represents the ordinary Doppler effect; the second describes the second-order, transversal correction.

It is important to observe that it is not possible to distinguish between the Doppler effect due to a gravitational field and that due to a velocity difference. Upon transforming to an accelerated frame, a relative velocity shows up in the scalar product as an energy difference due to the apparent force. The whole effect is fully described by eq. (17.32), which is invariant and covers all cases.

This discussion shows clearly that, as expected, there is no global and privileged time variable: we only have independent clocks on their own world lines and the means to compare their frequencies with electromagnetic signals (eq. (17.33)). It does not make sense to say that two clocks show the same time; we can only say that the rates of two "well functioning" clocks are related by eq. (17.33). There is, however, an important case of global synchronization of practical interest. Consider a set of clocks at rest on the (rigidly) rotating geoid of the earth. In this rotating frame (neglecting the influence of other cosmic bodies) the metric is still time-independent and the clocks are at rest. The potential U appearing in eq. (17.24) is, up to $O(V^2)$, the sum of the gravitational potential and the centrifugal potential and is a constant on the geoid; hence, because of eq. (17.33'), these clocks do not show, to this order, any frequency shift when compared electromagnetically.

17.6. RELATIVISTIC DYNAMICAL EFFECTS

Consider first the motion of material bodies in the small velocity approximation. As explained in Sec. 17.4, it is sufficient to consider the metric field of the sun (17.30), corresponding to the Lagrange function

$$L = [1 - 2m/r + 2\beta(m/r)^2 - (1 + 2\gamma m/r)v^2 + ..]^{\frac{1}{2}}. \qquad (17.38)$$

The radicand has terms $O(V^2)$ and $O(V^4)$; it is necessary to expand the square root up to $O(V^4)$. Neglecting an irrelevant additive constant and an irrelevant sign, we get:

$$L = \tfrac{1}{2}v^2 + \frac{m}{r} - \beta\left(\frac{m}{r}\right)^2 + \tfrac{1}{8}\left(v^2 + 2\frac{m}{r}\right)^2 + \gamma \frac{m}{r} v^2 . \qquad (17.39)$$

The canonical momentum (per unit mass)

$$\boldsymbol{\pi} = \frac{\partial L}{\partial \mathbf{v}} = \mathbf{v}\left[1 + \tfrac{1}{2}\left(v^2 + 2\frac{m}{r}\right) + 2\gamma\frac{m}{r}\right]$$

has small correction $O(V^3)$ with respect to the ordinary value $\mathbf{v}$. To get the hamiltonian, we solve this (in the V^2 expansion) for $\mathbf{v}$ and express in terms of $\mathbf{r}$ and $\boldsymbol{\pi}$ the function

$$H_0 + H_1 = \boldsymbol{\pi}\cdot\mathbf{v} - L =$$

$$= \tfrac{1}{2}\pi^2 - \frac{m}{r} + \beta\left(\frac{m}{r}\right)^2 - \tfrac{1}{8}\left(\pi^2 + 2\frac{m}{r}\right)^2 - \gamma\frac{m}{r}\pi^2 . \qquad (17.40)$$

We see that H_1, the correction to the newtonian hamiltonian function $H_0 = p^2/2 - m/r$, is of order H_0V^2; hence we can expect that every orbital element suffers in a period a fractional change of order V^2. For example, the distance from the earth to the sun differs from its newtonian value by a variable term of the order of $V^2\cdot 1\text{AU} = 1.5$ km, the gravitational radius of the sun. This is not much, but is well observable with current instrumentation. The corresponding angular deviation, of order 10^{-8} rads = 2 mas, cannot be seen with ordinary telescopes. It is then important to assess whether there are secular effects.

The Lagrange function (17.39) describes a plane motion, a consequence of the fact that $\partial L/\partial\mathbf{v}$ is a vector parallel to $\mathbf{v}$. Introducing polar coordinates (r, φ) in the orbital plane, we see that

$$\frac{\partial L}{\partial\dot\varphi} = r^2\dot\varphi\left[1 + \tfrac{1}{2}(\dot r^2 + r^2\dot\varphi^2) - (1 + 2\gamma)\frac{m}{r}\right] , \qquad (17.41)$$

the relativistic angular momentum, is a constant of the motion. It differs by $O(V^2)$ from the ordinary expression $r^2\dot\varphi$. The energy (17.40) is also a constant.

The perturbation energy H_1 depends only on three of the six orbital elements: the mean anomaly M, the semimajor axis a and the eccentricity e. The secular effects are obtained by replacing H_1 with its average over an orbital period in terms of the eccentric anomaly U (see Chs. 10 and 11). This corresponds to the averaging operation

$$\frac{1}{2\pi}\int_0^{2\pi} dU(1 - e\cos U) \cdot \ldots$$

on a function of $r = a(1 - e\cos U)$, expressed in terms of the eccentric anomaly U (see Fig 10.2).

This kills the dependence on the mean anomaly M; the dependence of $\langle H_1 \rangle$ on the semimajor axis a produces only a change in the mean anomaly, corresponding to a slightly different Kepler's third law. This secular effect is difficult to observe because of the error in the mass m, which is not directly measured. We therefore conclude that the only relevant secular effect is a motion of the argument ω of the periastron according to the equation (11.46_5), or

$$\frac{d\omega}{dt} = \frac{\sqrt{(1 - e^2)}}{na^2 e}\frac{\partial H_1}{\partial e} , \qquad (17.42)$$

The perturbation hamiltonian (17.40) is a quadratic form in 1/r :

$$H_1 = -\frac{m^2}{2a^2} + (1 + \gamma)\frac{m^2}{ra} + (\beta - 2 - 2\gamma)\frac{m^2}{r^2} . \qquad (17.43)$$

Here we have used the energy conservation law

$$p^2 = m\left(\frac{2}{r} - \frac{1}{a}\right)$$

(eq. (10.26)). The first two terms yield a contribution which does not depend on the eccentricity; to evaluate the last term we need the integral

$$\frac{1}{2\pi}\int_0^{2\pi} dU/(1 - e\cos U) = \frac{1}{\sqrt{(1 - e^2)}} .$$

We finally get the advance of the periastron in a period

$$\delta\omega = \frac{2\pi}{n}\left(\frac{d\omega}{dt}\right)_{sec} = (2 - \beta + 2\gamma)\frac{2\pi m}{a(1 - e^2)} . \qquad (17.44)$$

In Einstein's theory the numerical expression in the bracket equals 3.

The relativistic effects on the motion of a photon are governed by the Lagrange function L^2, with

$$L^2 d\lambda^2 = (1 - 2m/r)dt^2 - (1 + 2\gamma m/r)dx^i dx^j \delta_{ij} . \qquad (17.45)$$

The affine parameter λ is obtained from the constancy of $\partial L^2/\partial(dt/d\lambda)$; setting this constant equal to unity (so that λ coincides with the coordinate time t at infinity), we have

$$d\lambda = (1 - 2m/r)dt . \qquad (17.46)$$

We now note that the magnitude v of the ordinary velocity of the photon dx^i/dt differs very little (indeed, by $O(V^2)$ from unity; hence the Lagrangean L^2 is also $O(V^2)$ and the difference between λ and t can be neglected. We have, therefore,

$$L^2 = 1 - v^2 - 2(\gamma + 1)m/r + O(V^4) . \qquad (17.47)$$

This is essentially Lagrange's function of the newtonian gravitational theory in the field of a point mass $(\gamma + 1)m$ (2m in general relativity). The motion is hyperbolic with an angle between the asymptotes

$$\delta = 2(1 + \gamma)m/b , \qquad (17.48)$$

where b is the impact parameter (see Sec. 10.4, Problem 17.2 and Fig. 17.4). This effect moves the stars away from the sun by the angle δ and increases the transit time of a photon between two points on the opposite side of the sun.

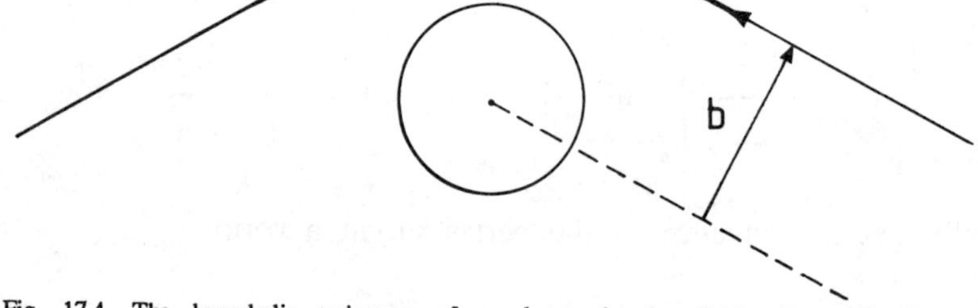

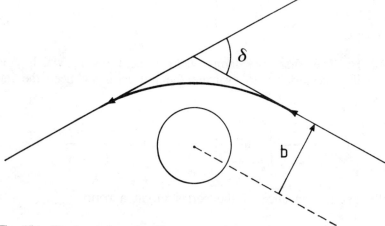

Fig. 17.4. The hyperbolic trajectory of a photon in the field of a point-like body. δ is the deflection.

*17.7. GRAVITOMAGNETISM

Contrary to newtonian theory, relativistic gravitation produces an additional force between two bodies which depends on their velocities. This is already apparent in the linear approximation (17.26), which contributes to the Lagrange function the term (eq. (17.21'); the sign is taken to agree with (17.21)):

$$L_{gm} = -\, m' h_{0i} v^i = -\, m' \mathbf{v} \cdot \mathbf{h}_0 \, . \qquad (17.49)$$

This is the term arising from the time and space components of the metric; we have written it, for convenience, for a test particle of mass m' and velocity **v**. Using the explicit expression (17.27) we get

$$L_{gm} = 2mm'(1 + \gamma)\mathbf{v} \cdot \mathbf{u}/r \, . \qquad (17.49')$$

There is an obvious, formal analogy with magnetism which we now proceed to fully exploit. The magnetic term in the Lagrange function for a charge q' in a vector potential **A** (with c = 1) is

$$L_m = q'\mathbf{v} \cdot \mathbf{A} \, . \qquad (17.51)$$

The vector potential produced by another charge q at a distance r and moving with a velocity **u** is

$$\mathbf{A} = q\mathbf{u}/r \, . \qquad (17.51')$$

Note, however, that this expression holds in the Lorentz gauge. In the electromagnetic lagrangian only the sum $(-\, V + \mathbf{A} \cdot \mathbf{u})$ is gauge-invariant; with a change of gauge the velocity dependence of **A** (eq. (17.50) can be altered at the expense of adding to the electrostatic potential V a term quadratic in the velocity **u**. It turns out, however, that the correspondence with magnetism in the Lorentz gauge leads to the correct results. For a detailed treatment we refer the reader to the book by Will.

From the three previous equations we see that the gravitomagnetic lagrangean function between two massive bodies is formally equivalent to the magnetic lagrangian for a charge

$$q = \surd[-\, 2(1 + \gamma)]m \, . \qquad (17.51)$$

(Of course, since only the product of two charges appears, the imaginary quantity is of no consequence.) For example, as a consequence of this effect the relative motion of an earth satellite with respect to the earth feels the gravitomagnetic effect of the sun, which in this relative frame is moving around. The gravitomagnetic force which arises in this way has an effect on the orbital plane of the satellite which is exactly like the lunisolar precession (Sec. 4.1): the normal to the orbital plane precesses around the pole of the ecliptic. Its rate (to which contributes also another term in the lagrangian) amounts to about 19 mas/y.

We now discuss in detail another important consequence of gravitomagnetism on rotating bodies. Just like moving charges produce a magnetic dipole moment

$$\mathbf{d} = \tfrac{1}{2}\int dq(\mathbf{r} \times \mathbf{u}) \ , \qquad (17.51)$$

moving masses produce a "gravitomagnetic" moment proportional to the angular momentum **L**

$$\mathbf{d}_{gm} = \tfrac{1}{2}\surd[-2(1+\gamma)]\int dm(\mathbf{r} \times \mathbf{u}) = \tfrac{1}{2}\surd[-2(1+\gamma)]\mathbf{L}. \quad (17.52)$$

The corresponding gravitomagnetic potential and field are

$$\mathbf{A}_{gm} = -\,\mathbf{d}_{gm} \times \nabla(1/r) = -\,\tfrac{1}{2}\surd[-2(1+\gamma)]\cdot[\mathbf{L} \times \nabla(1/r)], \quad (17.53_1)$$

$$\mathbf{B}_{gm} = \nabla \times \mathbf{A}_{gm} = -\,\tfrac{1}{2}\surd[-2(1+\gamma)]\nabla \times [\mathbf{L} \times \nabla(1/r)]. \quad (17.53_2)$$

Consider now another body with angular momentum **L'** = **IS** expressed in terms of the moment of inertia I and the angular velocity **S**. The gravitomagnetic torque acting on it is

$$I d\mathbf{S}/dt = \mathbf{d'}_{gm} \times \mathbf{B}_{gm} = \tfrac{1}{2}(1+\gamma)I\mathbf{S} \times \nabla \times (\mathbf{L} \times \mathbf{r}/r^3) \ .$$

Comparing this equation with the precessional equation corresponding to an angular velocity **Ω**, to wit,

$$d\mathbf{S}/dt = \mathbf{\Omega} \times \mathbf{S} \ , \qquad (17.54)$$

we get

$$\mathbf{\Omega} = \tfrac{1}{2}(1+\gamma)\ \nabla \times (\mathbf{L} \times \mathbf{r}/r^3) \ . \qquad (17.55)$$

This precession was discovered by Lense and Thirring and has never been tested experimentally in a direct way. Its order of magnitude is easily obtained using the value (1.16) of the angular momentum per unit mass; for the earth it is

$$\Omega \simeq 3\cdot 10^{-14}(R_{\oplus}/r)^3\,\mathrm{rad/s} \simeq 0.2\ (R_{\oplus}/r)^3\,''/\mathrm{y}. \qquad (17.56)$$

PROBLEMS

17.1. Evaluate the frequency shift (17.33) in a newtonian potential U when the observer and the source move slowly, taking into account the corrections of the order of $v^2 \approx U$.

17.2. Calculate the angle between the asymptotes of a keplerian hyperbola with an impact parameter b.

17.3. Describe quantitatively the motion in the sky of the image of a star as the sun moves by it.

17.4. Find the exact value of the Lense-Thirring precessional frequency (17.55), averaged over a circular polar orbit.

17.5. Find the order of magnitude of the geodetic precession using the gravitomagnetic force produced by the moving sun in a frame of reference where the earth is at rest.

*17.6. Study the effect of retardation on the gravitational force. (Just like in electromagnetism, gravity propagates with the velocity of light and Poisson's integral (1.24) should be replaced with the corresponding retarded expression. In the slow motion approximation this leads to a correction to Newton's potential of a mass m at a distance r of order $m d^2 r/dt^2$.)

17.7. Compare the relativistic advance of the perihelion of Mercury with the classical advance due to the oblateness of the sun (see Sec. 11.6). The quadrupole moment of the sun is not well known, but one can assume, as a lower estimate, the value corresponding to a uniform rotation, with an angular velocity equal to the observed equatorial velocity (see Sec. 20.6.)

*17.8. Clearly a difference between the inertial and the (passive) gravitational mass has no effect on the two-body problem. To study the possible violations of the Principle of Equivalence, write the equations of motion for the moon relative to the earth in the presence of the sun, assuming that their inertial and gravitational masses are different (see eq. (18.24_2)), and study the induced perturbations.

17.9. Calculate the excess angle of a curvilinear triangle on a sphere with a vertex at the pole and two vertices on the equator and verify eq. (17.11').

17.10. Find from the variational principle (17.15) the geodetic equations.

17.11. Describe the effect on a geoid clock of a) a vertical displacement with respect to the geoid; b) the sun; c) the moon.

17.12. Find the elapsed time difference between the ground and an aircraft flying once eastward or westward around the world on the equator. This time difference has in fact been measured.

17.13. Show that the hamiltonian function corresponding to a Lagrange function homogeneous of degree one in the velocities is constant.

FURTHER READING

The standard textbook on the theory of general relativity, with a particular emphasis on the slow motion approximation, is C.W. Misner, K.S. Thorne and J.A. Wheeler, *Gravitation*, Freeman (1973). Another old, but useful book is C. Møller, *The Theory of Relativity*, Clarendon (1972). Experimental gravity and its theoretical foundations are extensively treated in C.M. Will, *Theory and Experiment in Gravitation Physics*, Cambridge University Press (1981). For an update, C.M. Will, in *Three Hundred Years of Gravitation*, S.W. Hawking and W. Israel, eds., Cambridge University Press (1987), p. 80.

18. ARTIFICIAL SATELLITES

Since 1957 several thousands of artificial objects have been flown in earth and interplanetary orbits, with a variety of functions, like communications, scientific missions, meteorology, navigation, military reconnaissance, surveillance and warning. The extensive use of the space around the earth for these and other purposes is very likely to continue during the next decades and an understanding of the way these spacecraft move and are operated, while frequently representing a straightforward application of the physical concepts and methods discussed in this book, is interesting and instructive *per se*. In particular, scientific space missions related to gravimetry, geodesy, geophysics and relativity can provide unique contributions to these disciplines. In this chapter we shall concentrate on the motion of spacecraft, including launch and interplanetary trajectories. The thermal and mechanical structure of a satellite, its attitude control systems, the power sources and the interfaces with the experimental payloads largely belong to engineering and are not treated in detail here.

18.1. PERTURBATIONS

The forces acting upon an artificial satellite fall into four categories. First we have gravitational forces produced by bodies in the solar system (the effects of stars and galaxies are negligible); they obey the principle of equivalence and are characterized by an acceleration field. Secondly, the forces due to the environment, mainly the radiation pressure and the drag due to impact of particles; they are generally proportional to the cross section of the spacecraft. Thirdly, any change in momentum due to the propulsion system and, in general, to the internal dynamics, to be discussed in the following section. Finally, there are apparent forces – e. g., Coriolis forces – due to the rotation of the frame of reference; indeed, the accuracy with which an inertial frame can be defined is finite and in principle apparent forces are always present. To decide whether a given force is negligible is not straightforward, because forces may be neglected not in themselves, but with respect to other forces and/or to the achievable (or required) accuracy in the orbit determination and control. The latter is also a function of the accuracy of the observations of the spacecraft position and velocity. It is precisely because of the remarkable increase in accuracy of the measurements of distance and velocity that the need has recently arisen for a very refined analysis of the orbital perturbations, especially when carrying out geodetic studies of the earth from space.

We consider three typical examples of today's (1989) technology in satellite geodesy. Satellite Laser Ranging (SLR) directly measures the distance from a ground station to a satellite by the two-way travel time of a laser pulse (see Sec. 20.2). The instantaneous keplerian orbit of the geophysical satellite LAGEOS, for example, can be determined to within an accuracy of a few cm; and this will further improve in the future. On the other hand, when the orbit is predicted (usually a few months in advance for LAGEOS), these predictions fail by an amount – typically tens of metres – much larger than the scatter of the measured distances due to the accidental errors of the observing apparatus. This means that the accuracy of the model of the orbit fails to cope with the accuracy of the observations. The main physical cause of these prediction errors is the presence of tiny non-gravitational forces, which cannot be modelled well enough.

The Global Positioning System (GPS) is a network of satellites and ground stations of the U.S. Navy. Every satellite broadcasts a coded radiowave beam; comparing the signals from several satellites, the position of an aircraft or a ship can be determined within a few metres. This has obvious military and navigational applications, but in addition the GPS can be used to measure the vector between two ground stations with greater accuracy, down to a few cm. However, this does not allow geodetic positioning in a global reference frame with the same accuracy, since the satellites wander around their computed orbits owing to unmodellable non-gravitational perturbations. Thus again the identification, understanding and quantitative modelling of the perturbations is crucial.

Radar altimeter satellites, such as the U.S. spacecraft SEASAT flown in 1978 and the European project ERS-1, can provide measurements of the height of the satellite above the mean sea surface with an accuracy of a few tens cm. Updated radar technology could give even better accuracy, but it would be very difficult to exploit it, because these satellites emit very powerful radiowave beams, resulting into significant recoil forces, and have large orientable solar panels and radar antennas: these surfaces with their variable attitudes produce relatively large and quite unpredictable non-gravitational perturbations, and again the orbit cannot be determined very accurately even if the best tracking methods are used. Since accuracies of a few cm in the mean sea surface height would represent an important improvement both in gravimetry (through measurement of the geoid) and in oceanography (by measuring the *sea surface topography*, i. e. the difference between the mean sea surface and the geoid), two solutions are being studied. One is the *drag-free* technology, to be discussed later. The second solution is based on the idea of tracking the altimeter satellite in an essentially continuous way, thus relying very little on the computation of the orbit based on a complex force model. Both methods essentially abandon the idea of modelling non-gravitational perturbations; however, the technologies involved are complex and expensive.

In the future the techniques for precise measurements of distances, angles and velocities in space will certainly improve dramatically. We mention the use of very precise frequency standards for Doppler measurements and of orbiting VLBI antennas (QUASAT project); optical interferometry in space; the upgrading of radio telecommunication systems, in particular with the use of X- and even K-band. For LAGEOS-type earth satellites, sub-centimetre accuracies in distance are probably achievable. It is clear, therefore, that the uncertainty in the forces acting on the satellite, which soon deteriorates an initial observational accuracy, will play an important role in future space experiments.

In interplanetary orbits the main non-gravitational acceleration is due to the solar radiation pressure. This problem has been discussed for natural orbiting particles in Sec. 15.4; for a spacecraft we also must take into account the dependence of the effective cross section S on the attitude, and on the complex way reflection and diffusion may occur on the surface. The situation is further complicated by the deterioration of the optical properties of a surface in space due to the solar UV radiation and to energetic particles. It is thus difficult to model the radiation acceleration for an ordinary spacecraft (both earth-orbiting and interplanetary) with an accuracy better than $\approx$ 10%. For artificial earth satellites, on the other hand, the main non-gravitational force is the atmospheric drag, already discussed in Sec. 11.3; but, bisede solar radiation pressure, there is also the pressure due to radiation diffused, reflected and thermally re-emitted by the earth and the force resulting from the electromagnetic interactions with the magnetosphere.

One way to get rid of non-gravitational forces is to use a *drag-free* spacecraft. Such a system has a small proof mass m' freely moving near the centre of a cavity which shields it completely from external perturbations, aside from gravitation. The cavity is centreed at the centre of mass of the spacecraft. The relative motion of m' with respect to the cavity is determined by the difference between the (gravitational) acceleration of m' and the total (gravitational plus non-gravitational) acceleration of the spacecraft. Accurate measurements of the relative motion allow the determination of the non-gravitational force. This information can be used to control suitable thrusters which correct the orbit of the spacecraft, so as to compensate the non-gravitational force; or can produce in software the corresponding orbital corrections. In these ways accuracies in the compensation of non-gravitational forces better than 10^{-8} cm/s^2 have been reached and even better performances are possible. Note, however, that at present forces at the level of 10^{-10} cm/s^2 are relevant for the orbitography of LAGEOS. It is also interesting to note that the performance of a drag-free system is limited by the accuracy with which we can evaluate and/or eliminate the forces that the body of the spacecraft exerts on the proof mass (in particular the gravitational force; see Problem 18.6).

Of course, a more direct way to reduce non-gravitational forces is to decrease the area to mass ratio of the spacecraft. For a fixed density, this ratio is inversely proportional to the radius, and this is the reason why sizeable natural satellites are practically drag-free. A spherical shape is also important, because it makes the non-gravitational force independent of attitude and much better modellable; both these advantages are used for LAGEOS.

In the rest of this Section we confine ourselves to earth's satellites. In Table 18.1 we list all relevant forces, with specific data concerning two important types of satellites LAGEOS and geosynchronous satellites. The second column gives an order of magnitude formula for the acceleration and the third column lists the parameters which must be known in advance (or solved for) to obtain an accurate dynamical model, together with their current or estimated values. These values are then inserted in the formulae to calculate the accelerations "felt" by the two satellites. (Their semimajor axis (a) and area to mass ratio (S/m) are given in the first row.) The fourth column of the Table shows the relative, estimated uncertainties of the different parameters; the sixth and eighth columns display the corresponding accelerations for the two satellites, giving an idea of the achievable levels of accuracy and of the most critical modelling problems arising when an accurate orbital propagation is needed.

The specific comments that follow are numbered as in the Table.

(1) The earth's monopole term $GM_\oplus/r^2$ is, of course, the largest force on the satellite. The corresponding acceleration can be taken as the fundamental quantity to which the various perturbing effects must be compared in evaluating how much they displace the satellite's motion from a keplerian orbit on a timescale comparable with the orbital period (the long term behaviour of the perturbation is determined by other factors, as discussed in Ch. 11). Thus, e. g., a perturbing acceleration of 10^{-6} cm/s^2 on LAGEOS will cause a fractional change in the satellite semimajor axis of $10^{-6}/(3 \cdot 10^2) \approx 3 \cdot 10^{-9}$, about 4 cm. The determination of $GM_\oplus$ requires, by Kepler's third law, simultaneous measurements of the orbital period and of the semimajor axis. While the former is

straightforward and very accurate, the latter is not. Thus, the uncertainty for $GM_\oplus$ given in the Table is directly connected with the uncertainty in ranging: 3 cm over 12,000 km corresponds to a fractional uncertainty of $2.5 \cdot 10^{-9}$ in semimajor axis and, from Kepler's third law, an uncertainty 3 times larger in $GM_\oplus$. This error does not have serious consequences on the orbit determination, because in the differential correction procedure an error in $GM_\oplus$ can be compensated almost perfectly by an error in the orbital semimajor axis. Note also that it is impossible with satellites to measure separately the mass $M_\oplus$ and the gravitational constant G; this requires weighing the earth with a non-gravitational force.

(2) The perturbations due to departure of the earth from spherical symmetry, as explained in Ch. 2, can be expanded into series of spherical harmonics. For each value of ℓ (the *order* index) we have $\ell + 1$ harmonics which (for $\ell > 2$) are of the same order of magnitude. To evaluate the acceleration caused by *all* the terms of a given order we multiply the effect of a single harmonic by $\sqrt{(\ell + 1)}$ (assuming that the various terms are uncorrelated and add up like in a random walk). The indicated coefficients correspond to normalized harmonics and are dimensionless (see eqs. (2.8, 20)). To estimate the perturbing accelerations, we have neglected the angular dependence of the force, involving the Legendre functions and their derivatives. The uncertainties shown in the Table are consistent with the estimated errors of global geopotential models like GEM-T1 (J. Geoph. Res., **93**, 6169-6215 (1988)). In Sec. 11.3 we have already discussed the most important effects of the J_{20} term, due to the oblateness of the earth, which causes a secular motion of the perigee and the node of the orbit.

(3) The gravitational perturbations due to other celestial bodies (the moon, the sun and the planets) are not determined by the full gravitational attraction of these objects, but only by the corresponding *tidal* terms, i.e., by the difference between the force on the earth and that on the spacecraft. As we can see from the Table, the present accuracy of the planetary ephemerides is greater than needed for satellite geodesy purposes, since for the sun and the moon the main problem is the mass ratio with respect to the earth. Planetary perturbations are small – Venus providing the largest contribution – and cause no significant uncertainy in acceleration.

(4) The corrections to the Newtonian equations of motion introduced by general relativity are equivalent, in order of magnitude, to an acceleration given by the earth's monopole acceleration times the ratio between the gravitational radius of the earth $GM_\oplus/c^2 = 0.44$ cm, and the satellite's distance (see Ch. 17); the relative order of magnitude of these corrections, therefore, decrease with the distance.

(5) The orbital perturbations due to atmospheric drag have been discussed in Sec. 11.4. Although their qualitative features are easily

TABLE 18.1

Accelerations of spacecraft in earth orbit. The estimates and the relative uncertainties are given for a typical geosynchronous satellite and LAGEOS. M, R and $J_{\ell m}$ are the mass, the radius and the geopotential coefficients of the earth; r is the average satellite–earth distance; the subscripts $m, \odot$ and v refer to the moon, the sun and Venus; S/m, C_D and V are the satellite's area-to-mass ratio, drag coefficient and velocity relative to the atmosphere (of density ρ); Φ is the solar constant; A the earth's albedo; α and $\Delta T/T$ are the satellite's absorption coefficient and the fractional temperature difference between parts of its surface.

	Origin of acceleration	Formula	Parameters (in CGS units)	Relative uncertainty in the parameters	Geosynchronous satellite $a = 42,160$ km, $S/m = 0.1$ cm^2/g in cm/s^2		LAGEOS $a = 12,270$ km, $S/m = 0.007$ cm^2/g in cm/s^2	
					Acceleration	Uncertainty	Acceleration	Uncertainty
1.	Earth's monopole	GM/r^2	$GM = 3.986 \cdot 10^{20}$	$5 \cdot 10^{-9}$	22	$2 \cdot 10^{-7}$	280	$3 \cdot 10^{-6}$
2.	Earth's oblateness	$\frac{3GMR^2}{r^4} J_{20}$	$J_{20} = 4.84 \cdot 10^{-4}$ $R = 6.378 \cdot 10^8$	$6 \cdot 10^{-7}$	$7.4 \cdot 10^{-4}$	$4.5 \cdot 10^{-10}$	0.1	$6 \cdot 10^{-8}$
2.	Low-order geopotential harmonics ($\ell = m = 2$)	$\frac{3GMR^2}{r^4} J_{22}$	$J_{22} = 2.81 \cdot 10^{-6}$	$4 \cdot 10^{-4}$	$4.3 \cdot 10^{-6}$	$2 \cdot 10^{-9}$	$6.0 \cdot 10^{-4}$	$2 \cdot 10^{-7}$
2.	*Ibidem*, ($\ell = m = 6$)	$\frac{7GMR^6}{r^8} J_{66}$	$J_{66} = 2.36 \cdot 10^{-7}$	$7 \cdot 10^{-3}$	$4.4 \cdot 10^{-10}$	$3 \cdot 10^{-12}$	$8.6 \cdot 10^{-6}$	$6 \cdot 10^{-8}$
2.	High-order geopotential harmonics ($\ell = m = 18$)	$\frac{19GMR^{18}}{r^{20}} J_{18\,18}$	$J_{18\,18} = 7 \cdot 10^{-9}$	0.3	negligible	negligible	$2.7 \cdot 10^{-10}$	$1 \cdot 10^{-10}$
3.	Perturbations due to the Moon	$\frac{2GM_m r}{r_m^3}$	$GM_m = GM/81.3$ $r_m \approx 3.8 \cdot 10^{10}$	10^{-5} 10^{-9}	$7.3 \cdot 10^{-4}$	$7 \cdot 10^{-9}$	$2.1 \cdot 10^{-4}$	$2 \cdot 10^{-9}$
3.	Perturbations due to the Sun	$\frac{2GM_\odot r}{r_\odot^3}$	$GM_\odot = 3.33 \cdot 10^5 GM$ $r_\odot \approx 1.5 \cdot 10^{13}$	$3 \cdot 10^{-8}$ 10^{-8}	$3.3 \cdot 10^{-4}$	negligible	$9.6 \cdot 10^{-5}$	negligible
3.	Perturbations due to other planets (e.g., Venus)	$\frac{2GM_v r}{r_v^3}$	$GM_V = 0.82 \cdot GM$ $r_V \geq 4 \cdot 10^{12}$	10^{-4} 10^{-7}	$4.3 \cdot 10^{-8}$	negligible	$1.3 \cdot 10^{-8}$	negligible
4.	General relativistic corrections	$\frac{GM}{c^2 r} \frac{GM}{r^2}$	$\frac{GM}{c^2} = 0.44$	$5 \cdot 10^{-9}$	$2.3 \cdot 10^{-9}$	negligible	$9.5 \cdot 10^{-8}$	negligible
5.	Atmospheric drag V = volume, S = surface	$C_D \frac{S}{2m} \rho V^2$	$C_D = 2$–4 $\rho = \rho(r)$	1 1	negligible	negligible	$3 \cdot 10^{-10}$	$3 \cdot 10^{-10}$
6.	Solar radiation pressure	$\frac{S}{m}\frac{\Phi}{c}$	$\Phi = 1.38 \cdot 10^6$	S: 0.01–0.3 Φ: 0.001 Opt. coeff.: 0.01–0.1	$4.6 \cdot 10^{-6}$	$9 \cdot 10^{-7}$	$3.2 \cdot 10^{-7}$	$6 \cdot 10^{-9}$
7.	Radiation pressure due to the albedo of the Earth A	$\frac{S}{m}\frac{\Phi}{c}\left(\frac{R}{r}\right)^2 A$	$A \approx 0.4$	Opt. coeff.: 0.01–0.1 A: 0.1–1 S: 0.01–0.3	$4.2 \cdot 10^{-8}$	10^{-8}	$3.4 \cdot 10^{-8}$	$8 \cdot 10^{-9}$
8.	Recoil from thermal emission	$\frac{4}{9}\frac{S}{m}\frac{\Phi}{c}\alpha\frac{\Delta T}{T}$	$\alpha \simeq 0.1$–0.7 $T \simeq 300$ K [illegible]	0.01–0.1 0.01 0.1	$3 \cdot 10^{-8}$	$6 \cdot 10^{-9}$	$5 \cdot 10^{-9}$	$1 \cdot 10^{-9}$

understood, quantitative predictions in the case of real satellites are often difficult, since they require the use of a model of the upper atmosphere, i.e., the density ρ of the resisting medium as a function of height. Even the best atmospheric models, however, are inadequate to account for the large temporal variations of ρ, depending on the level of solar and geomagnetic activity. As a matter of fact, the procedure currently used is based on the determination from the tracking data itself of a few solve-for parameters, e.g., those appearing in the formula given in the Table.

(6) We have assumed a 2% accuracy in the radiation pressure model for a nearly spherical satellite like LAGEOS, and a 20% accuracy for a geosynchronous satellite. Very significant long-term perturbations in longitude are caused by radiation pressure when the satellite is three-axis stabilized or equipped with a large, despun telecommunication antenna, since in these cases there is a resonance effect between the perturbing force and the orbital period. Moreover, the modelling of eclipses, when the satellite crosses the earth's shadow, is not straightforward and introduces further uncertainties. Notice also that, at a lower level, there is an unpredictable change in the radiation pressure due to fluctuations in the solar luminosity.

(7) For the albedo A of the earth, far from uniform and constant in time because of meteorological and seasonal effects, we have assumed a 25% uncertainty; however, the closer the satellites, the greater these effects.

(8) The thermal re-emission of the radiation absorbed by the satellite (α is the average absorption coefficient of its surface) usually is anisotropic, due to anisotropies of shape, emissivity and surface temperature. The corresponding recoil acceleration is proportional to $\Delta T/T$, in order of magnitude the fractional difference of temperature between significant parts of the spacecraft (see Sec. 15.4). The values of α and ΔT shown in the Table are, of course, different for a rapidly spinning "passive" satellite (like LAGEOS) and for a stabilized, "active" spacecraft, with an internal energy conversion system (generally used to emit radio waves) and equipped with solar panels. The availability of an accurate thermal model of the satellite can reduce these uncertainties. In satellites with radio transmitters the emitted beams also produce a recoil acceleration; when the emitted energy is got from solar panels, it is approximately given by the absorbed energy times the efficiency ϵ of the conversion in electrical energy, and the order of magnitude of the acceleration is ϵ times the direct radiative acceleration.

There are other physical processes which perturb the motion of a satellite, but are less relevant than those of Table 18.1 in relation to the present accuracy of the observations. We can quote (in brackets an estimate of the corresponding acceleration of LAGEOS is given):

(a) Drag due to micrometeorite impacts ($\approx 10^{-11}$ cm/s^2) (This is the mean acceleration; of course, exceptionally large fragments can at times produce larger perturbations.)

(b) Poynting-Robertson effect ($\approx 10^{-11}$ cm/s^2) (see Sec. 16.5).

(c) Additional, general relativistic terms ($\approx 10^{-12}$ cm/s^2).

(d) Emission of gravitational waves ($\approx 10^{-35}$ cm/s^2) (negligible).

We next discuss the effects of tides on the motion of a spacecraft (see Ch. 4), relevant in three ways: (i) they cause periodic changes in the shape of the earth, hence in the positions of the tracking stations (kinematical effect); (ii) they produce a time variation of the geopotential that affects the satellite orbit (dynamical effect); (iii) they perturb the rotation of the earth, thus affecting the reference systems that is commonly adopted when tracking data are used (reference system effect).

Formula	Parameter (cgs units)	Relative uncertainty	Geosyn. sat.	LAGEOS
1. Kinematic solid tide				
$\delta h_T \left[\dfrac{2\pi}{T_{syn}/2}\right]^2$	$\delta h_T \cong 30$	0.03	$5.8 \cdot 10^{-7}$	$5.8 \cdot 10^{-7}$
2. Kinematic ocean loading				
$\delta h_L \left[\dfrac{2\pi}{T_{syn}/2}\right]^2$	$\delta h_L \cong 5$	0.2	$1 \cdot 10^{-7}$	$1 \cdot 10^{-7}$
3. Dynamic solid tide				
$3\kappa \dfrac{GM_m}{r_m} \left[\dfrac{R}{r_m}\right]^2 \dfrac{R^3}{r^4}$	$\kappa = 0.3$	0.01	$2.7 \cdot 10^{-8}$	$3.7 \cdot 10^{-6}$
4. Dynamic oceanic tide				
$\approx$ 10% of dynamic solid tide		0.1	$2.7 \cdot 10^{-9}$	$3.7 \cdot 10^{-7}$
5. Reference system: non-rigid earth nutation (14 d term)				
Nutation coefficient	0.002" in 14 d	0.1	$1.2 \cdot 10^{-9}$	$3.5 \cdot 10^{-10}$

Table 18.2. Tidal effects on the orbits of earth's satellites (the last two columns give accelerations in cm/s^2.) M_m, r_m and T_{syn} are, respectively, the mass and the distance of the moon, and the synodic period. The last two columns refer to a geosynchronous satellite (a = 42,160 km) and LAGEOS (a = 12,270 km).

In order to compare the tidal effects among themselves and with the other perturbations discussed earlier, we have transformed all of them into equivalent accelerations and listed them in Table 18.2.

(1) The main tidal term produces a symmetric deformation of the earth along the direction of the perturbing body; therefore the period of the corresponding tidal wave is one half the synodic period of the perturbing body. Considering only the moon, whose tidal effect on the earth is about twice that of the sun, a tracking station on the ground would oscillate in height with period $T_{syn}/2 \approx 4.5 \cdot 10^4$ s and amplitude δh_T together with the solid body of the earth. The *synodic period* of the moon T_{syn} is given in terms of the lunar mean motion n by

$$\frac{2\pi}{T_{syn}} = n - \frac{2\pi}{1 \text{ year}} . \qquad (18.1)$$

Were this effect ignored in the determination of satellite orbits, the station acceleration would appear as a residual acceleration on the satellite. The amplitude δh_T depends upon the elastic response of the earth to the perturbing potential; the relevant response coefficient (Love number) is fairly well known, giving uncertainties of the order of 1 cm in δh_T (see Sec. 4.3).

(2) Tidal waves raised in the oceans also affect positions of the station sites. Displacements of nearby water masses produce a time dependent elastic response of the continental shelves (*oceanic loading*) and this changes the station position with respect to the theoretical tidal displacement corresponding to an oceanless, solid earth. Of course, the size of this effect depends strongly on the geographical situation of the station; the corresponding displacements can be either as large as 12 cm on a peninsula protruding into an ocean (such as Cornwall) or smaller than 1 cm in the middle of a continental plate. Even though some modelling is possible, this effect can be measured with ground based tidal instruments, with uncertainties of the order of 1 cm. The equivalent acceleration of the kinematical effects and their uncertainties, as listed in Table 18.2, may appear large with respect to the other forces listed in Table 18.1; however, they affect the satellite tracking data at well defined tidal frequencies (see Sec. 4.4) and do not have secular or long period effects. It is therefore possible to separate the kinematical effects from other dynamical effects, in such a way that their uncertainty does not affect other geophysically relevant parameters. Of course, when station positioning is the main goal, the tidal displacements must be properly modelled.

(3) As seen in Secs. 4.2 and 4.3, the main term of the tidal

perturbation to the geopotential is obtained by computing the corresponding (i.e., with the same frequency) term of the perturbing gravitational potential generated by the other body, say the moon, times a Love number κ which measures the ratio between the response of the real earth and the theoretical response of a perfect fluid. Since the main contribution is the quadrupole term (due to a symmetric elongation of the earth in the direction of the perturbing body), the corresponding acceleration felt by the satellite is proportional to $1/r^4$. The Love number κ is currently determined from geodetic satellite orbits with a fairly good accuracy from its long term effects. An additional source of model errors is the truncation of the tidal potential expansion to a finite, and usually small, number of terms.

(4) Tides appear in the oceans as waves that move on the surface, forced by the gravitational perturbations of the moon and the sun; however, waves cannot propagate without friction: they dissipate energy and interact with the bottom (especially in shallow waters) and with the shore line. As a result, tidal waves in the seas are not in phase with the perturbing potential. The displaced water masses also give a contribution to the time variation of the geopotential, which affects the orbit of the satellite together with the solid tides; since it is difficult to separate them, the uncertainty arising from the oceanic tides limits the accuracy in the modelling of thc solid tides as well.

(5) The earth does not respond to external torques as a rigid body and its rotation axis undergoes forced nutations different from the theoretical case of a rigid earth. The most important difference lies in the fortnightly term (with argument twice the mean anomaly of the moon) and has an amplitude of about 2 mas. Were this effect ignored, the satellites would appear to oscillate by that amount, as a result of an incorrect definition of the reference system in which the equations of motion are written. VLBI measurements and models of the elastic response of the earth have improved by about one order of magnitude in the accuracy of nutation tables in the last decade.

We now come to the last kind of forces, due to the non-inertial character of the frame of reference. Since satellites are tracked from the earth, it is convenient to use a body-fixed cartesian frame (assuming the earth to be rigid). As explained in Ch. 3, the angular velocity of the earth

$$\boldsymbol{\omega}(t) = \boldsymbol{\omega}_0 + \Delta\boldsymbol{\omega}(t) \tag{18.2}$$

is the sum of a constant, nominal part along the z axis (ω_0 = $7.29 \cdot 10^{-5}$ rad/s) and a small change $\Delta\boldsymbol{\omega}(t)$ which accounts for polar wobble and variations of the length of the day (LOD). Keeping only first order terms in $\Delta\omega$, the perturbing acceleration in an earth-fixed reference frame is

$$\mathbf{F} = \mathbf{r} \times \Delta d\boldsymbol{\omega}/dt + 2\mathbf{v} \times \Delta\boldsymbol{\omega} + 2(\boldsymbol{\omega}_0 \cdot \Delta\boldsymbol{\omega})\mathbf{r} +$$

$$- (\Delta\boldsymbol{\omega} \cdot \mathbf{r})\boldsymbol{\omega}_0 - (\boldsymbol{\omega}_0 \cdot \mathbf{r})\Delta\boldsymbol{\omega}, \qquad (18.3)$$

where **r** is the position vector of the satellite in the earth-fixed frame and $\mathbf{v} = d\mathbf{r}/dt$ is the velocity vector. Each of the five terms of eq.(18.3) has an effect on the satellite with a characteristic signature, that is different for a $\Delta\boldsymbol{\omega}$ parallel to $\boldsymbol{\omega}_0$ (i.e., ΔLOD) or perpendicular to it (i.e., polar wobble). We list in Table 18.3 an order of magnitude estimate of these apparent forces for a geosynchronous satellite (in equatorial orbit) and for LAGEOS. We have assumed a variation in the pole's position and a ΔLOD of about 0.2 arcsec and 1 ms, respectively, over a time-scale of 1 y, obtaining

$$\begin{aligned}
&\Delta\omega_X \approx \Delta\omega_Y \approx 0.2\ \omega_0/2\cdot10^5 \approx 7.3\cdot10^{-11}\ \text{rad/s},\\
&\Delta\omega_Z \approx 10^{-3}\ \omega_0/8.6\cdot10^4 \approx 8.5\cdot10^{-13}\ \text{rad/s},\\
&\Delta d\omega_X/dt \approx \Delta d\omega_Y/dt \approx 2\pi\ \Delta\omega_X/3\cdot10^7\ \text{s} \approx 1.5\cdot10^{-17}\ \text{rad/s}^2,\\
&\Delta d\omega_Z/dt \approx 2\pi\ \Delta\omega_Z/3\cdot10^7\ \text{s} \approx 1.7\cdot10^{-19}\ \text{rad/s}^2.
\end{aligned} \qquad (18.4)$$

Term	Accelerations (in cm/s²) of geosyn. sat. due to polar wobble	geosyn. sat. Δ(LOD)	LAGEOS polar wobble	LAGEOS Δ(LOD)
$(\Delta\boldsymbol{\omega} \cdot \mathbf{r})\boldsymbol{\omega}$	$2.2\cdot10^{-5}$	$4.5\cdot10^{-9}$	$2.0\cdot10^{-6}$	$7.0\cdot10^{-8}$
$2\mathbf{v} \times \Delta\boldsymbol{\omega}$	$7.8\cdot10^{-7}$	$5.0\cdot10^{-10}$	$7.7\cdot10^{-5}$	$2.0\cdot10^{-7}$
$\mathbf{r} \times d\Delta\boldsymbol{\omega}/dt$	$1.0\cdot10^{-9}$	$7.5\cdot10^{-10}$	$1.7\cdot10^{-8}$	$6.8\cdot10^{-11}$
$2(\boldsymbol{\omega} \cdot \Delta\boldsymbol{\omega})\mathbf{r}$	–	$5.2\cdot10^{-7}$	–	$1.5\cdot10^{-7}$
$(\boldsymbol{\omega} \cdot \mathbf{r})\Delta\boldsymbol{\omega}$	$3.9\cdot10^{-7}$	$4.5\cdot10^{-9}$	$6.2\cdot10^{-6}$	$7.2\cdot10^{-8}$

Table 18.4. Apparent accelerations due to polar wobble and variations in the length of the day (LOD). The following orbital elements were assumed: geosynchronous satellite: a = 42,160 km, e ≈ 0.001, I ≈ 1°; LAGEOS: a = 12,270 km, e ≈ 0.004, I ≈ 110°.

The uncertainty in the knowledge of the pole position has recently been reduced by LAGEOS' orbit analysis, by VLBI and by lunar laser ranging techniques down to a few milliarcseconds, at least for variations with a period longer than a few days, with a relative uncertainty of 1%. The relative uncertainty in ΔLOD amounts to a few percents. Therefore, if the data obtained in this way are used as an *a priori* model, the uncertainties of the orbital effects become correspondingly smaller.

18.2. LAUNCH

In order to push off the ground and put into an earth orbit a satellite, we need to carry it to an altitude $(r - R_\oplus)$ (at least ≈ 150 km) such that the atmospheric drag does not cause it to fall back very quickly, and there impart to it a velocity **v** such that

$$\frac{v^2}{2} - \frac{GM_\oplus}{r} = - \frac{GM_\oplus}{2a} \tag{18.5}$$

(see eq. (10.26)). Here r is the distance from the earth's centre and a is the semimajor axis of the orbit. The direction of **v** can be specified by the angle γ with the horizontal direction. By conservation of angular momentum we have

$$rv\cos\gamma = \sqrt{[GM_\oplus a(1 - e^2)]} = \sqrt{[GM_\oplus r_p(1 + e)]}, \tag{18.6}$$

where e is the orbital eccentricity and r_p is the perigee distance. From eq. (18.5) we see that v must be at least of the order of $\sqrt{(GM_\oplus/R_\oplus)} \approx 8$ km/s; this is $1/\sqrt{2}$ the earth's *escape velocity* v_{esc} = 11.2 km/s, which is needed to permanently leave the neighbourhood of the planet. The velocity required for insertion into orbit is somewhat decreased if the launch occurs eastwards and close to the equator, thus exploiting the earth's rotational speed (0.465 km/s at the equator). On the other hand, v must be higher if the initial orbit of the satellite is eccentric (since in this case r must be close to r_p and lower than a, with a small value of γ).

The usual way to achieve such velocities is by putting the satellite on the tip of a rocket, a device where thermal combustion occurs, producing through a nozzle a jet of gas having an exhaust velocity v_{ex} (of the order of its thermal velocity) with respect to the rocket itself and resulting into a mass loss rate dm/dt (< 0). In order to obtain the equation of motion of a rocket, let us neglect for the moment the external forces and apply the conservation of linear momentum to the [rocket + exhaust gas] system. In absence of external forces, the rate of change in the momentum of the rocket (mv) is equal and opposite to the rate at which new momentum of the expelled gas is created:

$$d(mv)/dt - (v - v_{ex})dm/dt = 0 \ , \tag{18.7}$$

that is $dv/dt = - v_{ex}dm/(mdt)$. We can define the *specific impulse* $I_{sp} \equiv v_{ex}/g$, with the dimension of time, and write, in general

$$m d\mathbf{v}/dt = \mathbf{F}_P + \mathbf{F}_G + \mathbf{F}_D \ , \tag{18.8}$$

where $\mathbf{F}_p$ is the rocket propulsion "force" due to the loss of momentum, of magnitude $- v_{ex}dm/dt = - I_{sp}g dm/dt$, while $\mathbf{F}_G$ and $\mathbf{F}_D$ are the gravitational and aerodynamic drag forces, respectively. Assuming for simplicity a vertical, one-dimensional motion, we obtain for the velocity

$$v(t) = \int_{t_0}^{t} \frac{F_P}{m} dt - \int_{t_0}^{t} \left[\frac{F_D}{m} + g \right] dt \ , \tag{18.9}$$

where t_0 is the launch time. For a more realistic, two-dimensional trajectory, we should take into account the variable direction of gravity with respect to **v**, the fact that the drag force is not necessarily along the track (owing to the motion of the atmosphere and to aerodynamic lift effects) and the possible misalignment between **v** and the axis of the nozzle. Provided I_{sp} is constant, the first integral in eq. (18.9) gives the *propulsion velocity increment* at time t:

$$\Delta v_P = - g I_{sp} \int_{t_0}^{t} \frac{dm}{m} = g I_{sp} \ln \frac{m_0}{m(t)} \ , \tag{18.10}$$

where m_0 is the initial mass. In general, the main parameter which characterizes a propulsion system is the total change Δv_p in velocity it produces in absence of other forces. The second integral in eq. (18.9) can be seen as the sum of two *velocity losses*, due to the drag and to gravity, which depend on the properties of the rocket and on the trajectory. If they are optimized, the order of magnitude of these velocity losses is not higher than 1.5 or 2 km/s, significantly less than the final velocity of (at least) ≈ 8 km/s. (The gravity velocity loss is about $g\Delta t \cdot \sin\gamma \approx g\Delta t/2$, where Δt is the rocket boost time, of the order of a few minutes.) Since with current chemical fuels I_{sp} is at most of the order 300 s, corresponding to exhaust velocities of ≈ 3 km/s, one sees that the initial mass of the rocket must be much higher than the mass of the *payload* (i. e., the satellite and the rockets needed for further orbital changes). The initial mass must include, besides those of the payload (m_s) and of the fuel (m_f), the tank, the guidance system and the other structural parts of the rocket. With current technology, these parts weigh km_f, with $k \approx$ 10-15%. From eq. (18.10) we see that, even if the mass of the satellite were negligible, a one-stage rocket could reach a maximum velocity given by

$$\Delta v_{max} = gI_{sp}\ln\left(\frac{1 + k}{k}\right) \approx 2.2\ gI_{sp} \approx 6\ \text{km/s} \qquad (18.11)$$

(for $k \cong 0.12$), which is not enough for insertion into orbit.

This problem can be overcome by using multi-stage rockets, which abandon the empty, expendable tanks of the successive stages. In this case eq. (18.10) becomes

$$\Delta v_p = \sum_{i=1}^{n} gI_{sp,i}\ln\left(\frac{m_{o,i}}{m_i}\right) , \qquad (18.12)$$

where $m_{o,i}$ and m_i are the masses of the rocket at the beginning and at the end of the burning time of the i-th stage. Since when the (i + 1)-th stage is fired, the structural parts of the i-th stage are abandoned, we have $m_{o,i+1} = m_i - k_i m_{f,i}$. Now let us assume that (a) the rocket is made of n stages of similar technology (i. e., having the same values of k and I_{sp}); (b) each stage provides 1/n of the final propulsion velocity Δv_p; (c) each stage has a mass x times the previous one ($0 < x < 1$). We have

$$m_{o,i} = (1 + k)m_{f,i} + xm_{o,i}, \quad m_i = km_{f,i} + xm_{o,i} . \qquad (18.13)$$

Eliminating $m_{f,i}$ one obtains

$$m_i/m_{o,i} = (k + x)/(1 + k) , \qquad (18.14)$$

independent from i. The overall mass ratio m_o/m_s of the complete rocket to the satellite is $1/x^n$, that is, recalling eq. (18.12)

$$m_o/m_s = [(1 + k)\exp(-\Delta v_p/ngI_{sp}) - k]^{-n} . \qquad (18.15)$$

For I_{sp} = 300 s, k = 0.12, Table 18.4 shows the values of m_o/m_s attainable with different numbers n of stages for a total velocity Δv_p of 9.5 and 13 km/s, such as needed for a low-orbiting satellite and an interplanetary probe, respectively.

	Δv_p = 9.5 km/s	Δv_p = 13 km/s
n = 1	Impossible	Impossible
n = 2	94	$1.1 \cdot 10^5$
n = 3	56	390
n = 4	48	251
n = 5	45	211

Table 18.4. Ratio of the total mass of the rocket to the pay-load mass for different number of stages.

We see that the total mass to payload ratio decreases by a significant factor going from n = 2 to n = 3, but much more slowly for further increases of n. Since the complexity of the rocket (hence its cost) grows rapidly with the number of stages, usually the best trade-off is for n = 2 to 4. In this case the payload can weigh 1% or 2% of the rocket mass for low earth orbits and a fraction of a percent for interplanetary missions. Better performances are possible only by using cryogenic propellants, which have higher values of I_{sp} but cannot be stored for a long time and are much less safe.

The problem of optimizing the ascent trajectory is complex. Usually the gravitational velocity losses are the largest. Therefore the flight path is quickly bended away from the vertical after launch, in such a way that only a small component of the gravitational acceleration slows down the ascent. Due to the atmosphere, however, the bending is somewhat delayed, lest the rocket builds up high speeds in the lower and denser atmospheric layers (the drag force is proportional to the atmospheric density and the square of the velocity.) The trade-off between the two types of losses leads, for the required final values of v and γ, to the optimization of the path; this may be either computed in advance and stored in the memory of the guidance system or, if a better accuracy is needed, obtained in real time by the rocket's computer using the data provided by an accelerometer on board about the non-gravitational forces.

Once outside the atmosphere, the spacecraft is placed in a low circular orbit in order to test its functioning. If a higher, circular orbit is needed, the transfer is obtained with a velocity increment Δv_1 along the trajectory, which places the spacecraft on an elliptical orbit; the place where this change takes place is the perigee. At apogee the velocity is again increased by Δv_2, to transform the elliptical orbit in a circle. The required values of the velocity increments are easily found on the basis of energy conservation, as functions of the radii of the initial and final orbits (Problem 18.10). It can also be shown that this *transfer orbit* requires the least energy.

18.3. SPACECRAFT

A space vehicle must generally include subsystems for the following functions: (1) mechanical and thermal performance; (2) manoeuvring; (3) attitude determination and control; (4) telecommunications; (5) designed experimental (or other) activity; (6) power generation. Since a detailed description of all these functions and subsystems is beyond the scopes of this book, we confine ourselves here to a qualitative discussion, which may be of interest to those who want to plan a space mission for scientific purposes. All the functions depend on the basic choice for the spacecraft, whether it is spinning or not. In absence of external torques the spin axis, if it is aligned along a principal axis of inertia, is fixed; the rotation will then distribute the solar heat along the equator of the spacecraft, according to the thermal conductivity of its parts. In this case the radio communications with the earth can be ensured in two ways: for interplanetary travels one can align the spin axis with the earth direction and the high gain antenna with the spin axis; otherwise, one can build receiving and transmitting circuitry on the equatorial belt and use them in suitable phase arrays to achieve the required gain. From the experimental point of view, an instrument on a spinning spacecraft scans a section of the sky around a parallel; its output is modulated with the rotation period. A spinning spacecraft must have an accurate monitoring of its rotational vector; this is usually accomplished with electromagnetic sensors of celestial bodies, like the earth, the sun or a star. The spin can be changed by actuating suitable thrusters. The dynamics of a spinning satellite is governed by the same equations already given for the earth (Sec. 3.5).

A spacecraft rotationally stabilized along three axes may become necessary when two directions are to be kept: for example, the earth and a planet. Its structure and control are more complicated. In both cases the stabilization can use, besides thrusters, also moving parts inside the spacecraft and their own angular momentum. For example, the precessional state of an axially symmetric spacecraft (see Secs. 3.5 and

3.6) has an energy larger than a purely rotational state and tends to be damped away by any dissipative dynamical process; hence special containers with moving fluids can be used to kill the precessional motion. One can also use wheels driven by special motors to control the angular momentum of the spacecraft body. A solution intermediate between a spinning and a fully stabilized spacecraft is provided by despun parts (e. g., antennas or instrument platforms), which are directed along a constant azimuth on a rotating structure.

The power needed by a scientific spacecraft usually depends in a crucial way on the communication bit rate (see Ch. 19). An exception is provided by the use of radar techniques from space, which bring the energy requirements into a different order of magnitude and may call for nuclear energy generation. In imaging systems the very large bit rates may require special processing computers on board.

For spacecraft at a great distance (r) from the sun, the solar energy flux, proportional $1/r^2$, may be inadequate and radioactive thermoelectric generators can be used; they are based upon the thermoelectric effect produced by radioactive heat. But in general the obvious energy source is the use of photovoltaic solar cells powered by solar radiation. Assuming a 10% efficiency, at 1 AU from the sun we need a cross section of 1 m^2 to obtain 140 Watts.

Special problems have to be faced by scientific satellites with particular requirements of "cleanliness", in a general sense. There is a chemical contamination, produced by the exhaust gases from the engines and other leaks; the radiation they emit, either directly or diffused from the sun, can be obnoxious for infrared detectors. There is an electromagnetic cleanliness problem, due to the intense, high frequency voltages on board. Some experiments involving very precise measurements of velocity and distance require a dynamical cleanliness, which may be disturbed not only by external forces, but also by moving parts on the spacecraft, including the astronauts. A very accurate pointing also requires handling this problem.

A scientific space mission must be planned as an integrated whole, including the tracking from the ground. The orbit is the major point of interaction between the scientific requirements and the overall structure of the spacecraft; in particular, its weight, crucial for the launch operations, depends not only on the payload, but also on the amount of fuel needed for the manoeuvres and on the duration of the mission. The latter, of course, is related to its scientific aims, the required accuracy of the observations, the reliability of the components of the spacecraft and the resources needed to operate the satellite and to analyze its data. For earth spacecraft the lifetime of large structures in space is greatly affected by *space debris*. They are mostly fragments of other spacecraft produced by normal space operations like launch and explosions; more than 7000 objects of size larger than about 2 cm have been identified and are at present in orbit. The smaller fragments are soon drawn down by the atmospheric drag; but the other ones remain in orbit for a very long time, especially at altitudes between 500 and 1000 km, and the high relative velocity of impacts between orbiting bodies makes them a positive danger for large structures.

The orbit of an earth spacecraft must take into account three main time scales: the day (LOD), the orbital period $2\pi/n$ and the period of the node in its equatorial motion $2\pi/(d\Omega/dt)$ (see Sec. 11.2). We discuss in the following Section the effects of the important resonance LOD = $2\pi/n$; more generally, when

$$n = 2\pi s/\mathrm{LOD} = s\omega \ (s \text{ integer}) , \tag{18.16}$$

a keplerian satellite, in absence of perturbations, passes over the same point of the earth after s revolutions. The largest possible value of s is 16, at a heigth of 239 km. Of course, a resonant orbit of this kind is very much affected by the inhomogeneities of the earth.

The orientation of the orbital plane is affected by the secular advance of the node, with rate $d\Omega/dt$. It may be useful to have this orbital plane always aligned with the direction of the sun; in this way the spacecraft will cross the equator at noon every orbital period and will be able to monitor continuosly the earth-sun relationship, in particular the radiative budget. This is the *heliosynchronous orbit*; it is determined by the relationship (see eqs. (11.15) and (11.57_2))

$$\frac{2\pi}{1y} = \frac{d\Omega}{dt} = -\frac{3J_2}{2}\, n\, \frac{R_\oplus^2}{a^2}\, \frac{\cos I}{(1-e^2)^2} . \tag{18.17}$$

We see that the orbit must be retrograde ($\cos I < 0$) and that the inclination depends on semimajor axis and eccentricity; for a vanishing eccentricity and $a \cong R_\oplus$, we have

$$\cos I = -\frac{2}{3}\, \frac{1}{16\cdot 365\ J_2}, \tag{18.17'}$$

that is, about 96° in inclination.

A space mission requires also a precise orbit determination. This is accomplished by tracking systems; appropriate software integrates the equations of motion and uses least square fits to determine the unknown parameters, in particular the initial position and velocity (see Sec. 20.1). We can list six tracking systems:

– Optical observation of low satellites with special telescopes. (The angular accuracy, determined by the local seeing, is a few arcsec.)
– Laser tracking using retroreflectors on board (for LAGEOS the current accuracy in distance is a few cm).
– Radar from the ground.
– Radio signals received from the satellite; one can measure the relative velocity with a coherent beam or the range with modulated signals.
– Accelerometers on board.
– Optical sensors on board, capable of measuring the relative position of planets and stars.

18.4 GEOSTATIONARY SATELLITES

A *geostationary satellite* has, by definition, a *geosynchronous orbit*, whose period is very close to one sidereal day (s = 1 in eq. (18.16)). In addition, the orbital eccentricity and inclination to the earth's equatorial plane are small. From Kepler's third law, the corresponding semimajor axis $a = (GM_\oplus/\omega^2)^{1/3}$ is 6.61 $R_\oplus \cong$ 42,160 km and the satellite is about 35,800 km high above the earth's surface.

An "ideal" geostationary satellite always retains the same position in the sky, as seen from the earth. Moreover, it is able to image always the same part of the earth's surface, spanning $180^\circ - 2\sin^{-1}(R_\oplus/a) \cong 160^\circ$ both in latitude and in longitude, that is, somewhat less than a whole hemisphere. These two features make geostationary satellites very suitable for a variety of applications, ranging from telecommunications, to meteorology and earth resources data collection, to early warning of nuclear missile attacks.

A real geosynchronous satellite, with finite eccentricity and inclination, does not remain at an exactly fixed position in the sky, but oscillates about an average position with a daily frequency and with an angular amplitude approximately given by the eccentricity and/or by the inclination (Problem 18.2). Apart from this short-period motion, a slow, but accumulating longitude drift is caused by the fact that the earth is not exactly axially symmetric. (Note that the polar flattening just causes a small shift of the geosynchronous orbital distance with respect to the value given by Kepler's third law.) Since the earth's equator, as a result of the $\ell = 2$, $m = 2$ geopotential coefficients (of order 10^{-6}, see Ch. 2 and Sec. 18.1), deviates from a perfect circle by an amount of the order of 100 m, the satellite feels a small force along the track towards the equatorial "bulges", slowly changing its semimajor axis and longitude. This effect is important because, if the satellite is constrained by operational or other requirements to remain inside a small longitude window (usually, 0.1° to 1°), orbital station-keeping manoeuvres must be planned and carried out.

If the orbit is nearly circular and equatorial, and φ is the satellite longitude (in the earth-fixed rotating frame, $\varphi = f - \omega t$, f being the true anomaly), from eq. (2.20) we obtain for the along-track component T of the perturbing acceleration:

$$T = -\frac{\partial U}{a\partial\varphi} = -\frac{GM_\oplus}{a^2}\sum_\ell\sum_m\left(\frac{R_\oplus}{a}\right)^\ell mJ_{\ell m}P'_{\ell m}(0)\sin[m(\varphi - \varphi_{\ell m})]. \quad (18.18)$$

The non-vanishing contributions are those with $m \neq 0$ and $(\ell - m)$ even, since the latitude is assumed to be zero and the associated Legendre functions fulfil $P'_{\ell m}(0) = 0$ for all odd values of $(\ell - m)$. Therefore the most important perturbative term is $\ell = 2$, $m = 2$, with an amplitude $J_{22} \simeq 2.8\cdot10^{-6}$:

$$T \cong -\frac{2GM_\oplus}{a^2}\left(\frac{R_\oplus}{a}\right)^2 J_{22}P_{22}(0)\sin[2(\varphi - \varphi_{22})] = K\sin\lambda; \quad (18.19)$$

$$K = (2GM_\oplus/a^2)(R_\oplus/a)^2 J_{22}P_{22}(0) \cong 5.6\cdot10^{-6}\ \text{cm/s}^2$$

is the acceleration constant; we have also introduced the angle

$$\lambda = 2(\varphi - \varphi_{22}) + \pi\ .$$

According to Gauss' equation for a circular orbit (see Sec. 11.1), $\dot{a} = 2T/n$. Then, from the relationship $\ddot{\varphi}/n = -3\dot{a}/2a$ (derived from Kepler's third law), we get

$$a\frac{d^2\lambda}{dt^2} = 2a\frac{d^2\varphi}{dt^2} = -3n\frac{da}{dt} = -6T = -6K\sin\lambda\ . \quad (18.20)$$

This is the well-known pendulum equation for the variable λ. Note that, as explained in Sec. 11.1, because of the "perverse" nature of gravity, it would be wrong to equate directly the force T to $a\ddot{\varphi}$. The two equilibrium positions $\lambda = 0$ and $\lambda = \pi$ correspond to two pairs of equilibrium longitudes, the former stable (for $\varphi = \varphi_{22} \pm \pi/2$) and the latter unstable (for $\varphi = \varphi_{22}$ and $\varphi = \varphi_{22} + \pi$). Therefore, the stable positions correspond to the longitudes at 90° from the major axis of the earth's equator, where the perturbing potential energy (proportional to $\cos\lambda$) has two maxima. The period of the small oscillations about the two stable equilibrium longitudes is

$$P = 2\pi \left[\frac{a}{6K} \right]^{\frac{1}{2}} \approx 800 \text{ days} , \qquad (18.21)$$

while the order of magnitude of the angular acceleration $\ddot{\varphi}$ is 10^{-3} degrees/(day)^2 (provided the satellite is not too close to an equilibrium position). Since $a = a_0 + \Delta a$ differs very little from a constant value a_0, eq. (18.20) integrates to

$$2a_0 d\varphi/dt = 3n\Delta a + \text{const};$$

if at $t = 0$, $\dot{\varphi} = 0$ and $\Delta a = 0$, we have

$$2a_0 d\varphi/dt = 3n\Delta a . \qquad (18.22)$$

Therefore the plane $(\varphi, \Delta a)$ is the phase space of the motion.

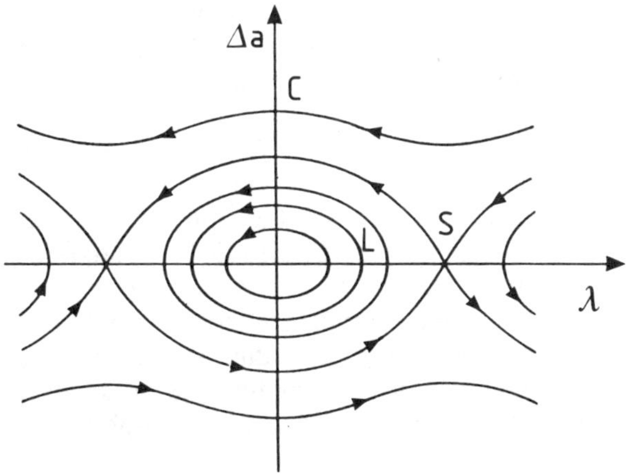

Fig. 18.1. The figure illustrates with the energy level curves the longitude and semimajor axis drifts of nearly-geostationary satellites due to $\ell = 2$, $m = 2$ geopotential terms. The longitude origin is chosen at a stable equilibrium position, about which librational (L) motions occur. Farther away from the "ideal" geosynchronous distance ($\Delta a = 0$), the satellite circulates all around the earth's equator (C-type paths), and a "separatrix" (S) orbit lies at the boundary between the two types of motion.

For small longitude changes, we can approximate $\ddot{\varphi}$ (or T) with a constant, so that in this case φ depends quadratically upon time. If a satellite has to stay inside a longitude window of width δ, centreed at a longitude φ_0, we can readily estimate the time Δt needed for the round trip from $\varphi_0 - \delta/2$ to $\varphi_0 + \delta/2$ and back:

$$\Delta t \cong 4 \left[\frac{\delta}{2\ddot{\varphi}(\varphi_0)}\right]^{\frac{1}{2}} . \qquad (18.23)$$

If $\ddot{\varphi}(\varphi_0) = 10^{-3}$ degrees/(day)^2 and $\delta = 1^\circ$, Δt is about 3 months. This shows that station-keeping manoeuvres must be performed routinely during the lifetime of a geostationary satellite.

18.5. INTERPLANETARY NAVIGATION

Interplanetary navigation presents an orbital problem different from ordinary planetary dynamics, in which there is a main field, due to the sun or the planet, and perturbation theory can be used to study the orbital evolution. An interplanetary spacecraft must first escape the attraction of the earth and acquire a keplerian heliocentric orbit; later on, it can be inserted in a bound orbit near another planet. In both these situations – and in the missions to the moon as well – we must match together two keplerian orbits relative to different centres of force. This matching is crucial to determine the manoeuvre strategy and the fuel requirements.

A quantitative solution to this problem – which is in fact a part of the three-body problem, see Ch. 12 – can be obtained only hy numerical calculations; however, the concept of the *sphere of influence* is a useful introduction. We must consider two different descriptions of the motion of the space probe: with respect to the sun, with relative coordinates **R**; with respect to the planet, with relative coordinates **r**. The first description is suitable when the force due to the planet, written on the right-hand side of the equations of motion, is a small perturbation; and *viceversa* for the second description. We therefore have in the two cases

$$\frac{d^2\mathbf{R}}{dt^2} + \frac{GM_\odot}{R^3}\mathbf{R} = GM\left(-\frac{\mathbf{r}}{r^3} - \frac{\mathbf{R}-\mathbf{r}}{|\mathbf{R}-\mathbf{r}|^3}\right) ; \qquad (18.24_1)$$

$$\frac{d^2\mathbf{r}}{dt^2} + \frac{GM}{r^3}\mathbf{r} = GM_\odot\left(-\frac{\mathbf{R}}{R^3} + \frac{\mathbf{R}-\mathbf{r}}{|\mathbf{R}-\mathbf{r}|^3}\right) . \qquad (18.24_2)$$

The last terms in the round brackets are the accelerations produced by the planet (of mass M) on the sun and *vice versa* (see eqs. (11.35)).

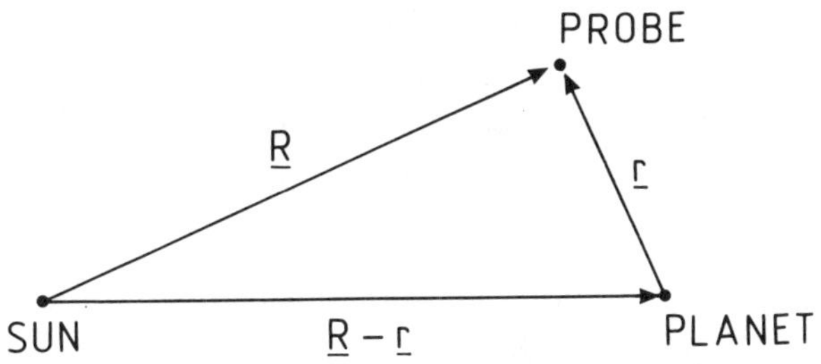

Fig. 18.2. The geometry of interplanetary navigation.

The ratio of the perturbing to the main acceleration (in modulus) is in the two cases the required smallness parameter:

$$\epsilon_R = \frac{R^2 M}{M_\odot} \left| \frac{\mathbf{r}}{r^3} + \frac{\mathbf{R} - \mathbf{r}}{|\mathbf{R} - \mathbf{r}|^3} \right| , \qquad (18.25_1)$$

$$\epsilon_r = \frac{r^2 M_\odot}{M} \left| \frac{\mathbf{R}}{R^3} - \frac{\mathbf{R} - \mathbf{r}}{|\mathbf{R} - \mathbf{r}|^3} \right| . \qquad (18.25_2)$$

When $M \ll M_\odot$, the concept of the sphere of influence is relevant only where $r \ll R$; then the two parameters read

$$\epsilon_R = \frac{R^2}{r^2} \frac{M}{M_\odot} , \qquad (18.26_1)$$

$$\epsilon_r = \frac{r^3}{R^3} \frac{M_\odot}{M} \sqrt{(1 + 3\cos^2\alpha)} . \qquad (18.26_2)$$

Note that the acceleration of the sun produced by the earth – and the other planets as well – is negligible. The second parameter depends also on the angle α between **r** and **R**. As a first, rough approximation one could say that the first description is better than the second when $\epsilon_R < \epsilon_r$ and *vice versa*. The surface defined by

$$\epsilon_R = \epsilon_r \qquad (18.27)$$

corresponds to the transition between the two approximations, given by

$$r = R \left(\frac{M}{M_\odot} \right)^{2/5} (1 + 3\cos^2\alpha)^{-0.1} . \qquad (18.27')$$

This surface is not very different from a sphere: the ratio between the largest value (for $\alpha = \pi/2$) and the smallest value (for $\alpha = 0$) of r is $4^{0.1} = 1.15$. The *radius of influence* around the smaller body is

$$r_I = R(M/M_\odot)^{2/5} \quad . \qquad (18.27'')$$

Note also that at this radius $\epsilon_r = \epsilon_R = O(M/M_\oplus)^{1/5} < 1$, so that we do have for both descriptions a *bona fide* perturbation approximation. The concept of sphere of influence was introduced by Laplace in 1845.

More generally, for a better control of the errors, we can define an inner sphere of influence at a given level of approximation ϵ by setting $\epsilon_r < \epsilon$; and an outer sphere of influence by $\epsilon_R < \epsilon$. For the sun-earth system, if the two parameters are set at 0.01, the radii of the outer and inner spheres of influence are 0.0178 AU and 0.0027 AU; when

$$\epsilon_r = \epsilon_R = (M/M_\odot)^{1/5} = (328,700)^{-1/5} = 0.079$$

we have the single sphere with a radius of 0.0062 AU from the earth's centre. One way to construct an interplanetary trajectory is to assume a keplerian orbit inside the inner sphere of influence, to integrate numerically the orbit in the intermediate shell and to match the solution to another, solar keplerian orbit at the outer sphere of influence.

To send a space vehicle on an interplanetary flight we must first raise it from the parking orbit (of radius r_0) to the sphere of influence (of radius r_I), with a suitable hyperbolic velocity V. If $v_0 = \surd(GM_\oplus/r_0)$ is the parking velocity, from conservation of energy a velocity increment Δv will produce at r_I the velocity

$$V = \surd[2GM_\oplus/r_I + \Delta v(2v_0 + \Delta v)] \quad . \qquad (18.28)$$

Since $r_I \gg r_0$, the hyperbolic orbit will reach the sphere of influence very near the asymptote, along an almost radial direction. To match this orbit with the required interplanetary trajectory, the heliocentric velocity $V_\oplus$ of the earth (at distance R from the sun) must be taken into account. Let us consider the most favourable case, in which the two velocities are parallel. Suppose we want to reach an outer planet on a circular orbit with radius R' in the ecliptic plane; we need an interplanetary, elliptic transfer orbit in which the earth is at perihelion and the planet at aphelion. This transfer orbit is in principle identical to the one used in transfers between earth-bound orbits (see Sec. 18.2, Problem 18.10). To find out the required value of V we use the energy conservation law

$$\frac{(V + V_\oplus)^2}{2} - \frac{GM_\odot}{R} = - \frac{GM_\odot}{R + R'} \ ,$$

to obtain

$$V = \left(\frac{GM_\odot}{R}\right)^{\frac{1}{2}}\left[\left(\frac{2R'}{R + R'}\right)^{\frac{1}{2}} - 1\right] . \qquad (18.29)$$

In combination with eq. (18.28), this equation gives the necessary velocity increment from the parking orbit (Problem 18.11) to the transfer one. Once the spacecraft has reached the radius R' of the outer planet, an additional aphelion manoeuvre is required to insert it on an orbit bound to the planet. The relative velocity V_{rel} with respect to the planet is given again by the formula (18.29), but with R and R' interchanged

$$V_{rel} = \left(\frac{GM_\odot}{R'}\right)^{\frac{1}{2}}\left[\left(\frac{2R}{R + R'}\right)^{\frac{1}{2}} - 1\right] . \qquad (18.29')$$

This is the so-called *Hohmann transfer orbit.* Of course, the actual encounter must also take into account the transfer time and the relative position of the two planets; this defines a time window for the injection manoeuvre.

The velocity increments can be very large and pose great difficulties to interplanetary space travel. One way to overcome this problem is to use the gravitational force of the planets to change the direction and the magnitude of the spacecraft velocity in a *swing by* a planet. A rudimentary assessment of the swing-by capability of a planet is provided by the ratio between the lowest potential energy (– GM'/r') in the field of a planet of mass M' and radius r' and the potential energy in the field of the sun for a circular orbit of radius R' (apart from a factor of 2, this is the same as the squared ratio between the escape velocity and the orbital velocity of the planet). This ratio $M'R'/M_\odot r'$ has the following values:

Mercury	0.0039
Venus	0.043
Earth	0.069
Mars	0.022
Jupiter	10.7
Saturn	6.6
Uranus	5.1
Neptune	10.3

The outer planets are very effective in a swing-by; Mercury is useless; Venus and the earth require several passages or an eccentric orbit, less bound to the sun than the planet itself.

The swing-by problem can be solved in an approximate way with the help of the sphere of influence concept (eq. (18.27")). Within this sphere the orbit can be assumed to be hyperbolic, with an asymptotic relative velocity given in terms of the planetocentric semimajor axis a by

$$V_{rel}^2 = GM' \left(\frac{2}{r_I} + \frac{1}{a} \right) . \qquad (18.30)$$

Its periastron $r_p = a(e - 1)$ determines the eccentricity and via the latter quantity we can get the angle ψ between the outgoing and the incoming velocity (see Sec. 10.4):

$$\tan(\psi/2) = a/b = 1/\surd(e^2 - 1) . \qquad (18.31)$$

The greater ψ is, the smaller must be the periastron distance r_p, down to the minimum value which is just the radius r' of the planet. Consider, for example, a Hohmann transfer orbit from the earth to Jupiter, arriving there with a velocity parallel to Jupiter's velocity and of magnitude given by eq. (18.29'), that is $V_{rel} = 5.8$ km/s. The radius of Jupiter's sphere of influence is 0.322 AU; the radius of the planet itself is $4.77 \cdot 10^{-4}$ AU. Using eqs. (18.30,31), we find $\psi \cong 160^\circ$, a substantial deviation angle. This calculation is performed in the frame of reference where Jupiter is at rest, so that after the swing-by the heliocentric velocity of the probe can become larger, at the expense of the energy of the planet.

PROBLEMS

*18.1. Study the orbit of an ICBM (Inter-Continental Ballistic Missile) outside the atmosphere. For a given angular distance between the launch site and the target, compute the launch velocity v as a function of the initial angle from the vertical and plot the time Δt required to cross the atmosphere as a function of v.

*18.2. Show that a geosynchronous satellite having a circular orbit and a small inclination I with respect to the earth's equator moves in the sky along an eight-shaped path crossing on the celestial equator, and covering a range equal to 2I in latitude and to $I^2/2$ in longitude.

*18.3. During the oscillation of a geosynchronous satellite with period P given by eq. (18.21), we make N equally spaced and independent measurements of the semimajor axis with accuracy σ_a. Evaluate the corresponding error in the parameter J_{22} and apply the calculation to the case $\sigma_a = 50$ cm.

18.4. Estimate the force on an interplanetary spacecraft due to the solar wind.

18.5. Compute the semimajor axis of a satellite in any resonance with the rotation of the earth (see eq. (18.16)).

18.6. The proof mass of a drag-free system does not feel any gravitational force from a spacecraft made as a spherical shell; but spacecraft are more complex than shells. Evaluate the gravitational potential near the center due to two diametrically opposite point masses and a ring in their equatorial plane and show how it can be minimized.

18.7. Estimate the tidal forces on an earth satellite and on an interplanetary probe due to the nearest star and to the galaxy.

18.8. Construct the table of the radii of influence for the sun and the planets; and for the earth and the moon.

18.9. Calculate the geosynchronous radius for each planet.

FURTHER READING

The non-gravitational forces perturbing the earth's satellites are discussed in detail in A. Milani, A.M. Nobili and P. Farinella, *Non-gravitational Perturbations and Satellite Geodesy*, Hilger, Bristol 1987. On the particular problem of atmospheric drag and its models, see D. King-Hele, *Satellite Orbits in an Atmosphere. Theory and Applications*, Blackie and Son, Glasglow 1988. The latest nutation model is given by J. Wahr, *The forced nutations on an elliptical, rotating, elastic and oceanless earth*, Geophys. J. Astron. Soc. 64, 705 (1981). A course of lectures especially devoted to the engineering aspects of space technology is Centre National d'Etudes Spatiales, *Le Movement du Vehicule Spatial en Orbit*, CNRS, Toulouse 1980. Useful general textbooks are J.W. Cornelisse, H.F. Schöyer and K.F. Wakker, *Rocket Propulsion and Spaceflight Dynamics*, Pitman, London 1979, and A.E. Roy, *Orbital Motion*, Hilger, Bristol 1978; the former emphasizes technology, the latter celestial mechanics.

19. *SPACE TELECOMMUNICATIONS*

Radio communications to and from a spacecraft are its lifeline, with which commands are received from the ground control centres and information collected in space is returned. Particular electromagnetic links provide also the possibility of very precise measurements of distance and velocity and of very accurate determination of the orbit of the spacecraft and the earth. Similar techniques are also used in radar ranging to the moon, to passive artificial satellites and to the planets (both in the radio and optical band). In this chapter we discuss the energetics of radio communications in space and introduce spectral theory, a necessary tool to describe their noise performance. Particular attention is devoted to coherent propagation, a particular radio link used to measure the relative velocity, and to the effect of media upon electromagnetic propagation.

19.1. THE POWER BUDGET

The energy balance of a radio link in space depends crucially upon the directivity of the antennas on the ground and on board. This directivity is measured by the *gain* G, equal, for a given direction, to the ratio between the actual emitted energy flow and the isotropic flow from an antenna without any directivity at all. Of course, usually an antenna has a maximum gain along its axis of symmetry and is used near that direction. At a distance D, therefore, an emitted power P produces an energy flux

$$\Phi = PG/(4\pi D^2) \ . \qquad (19.1)$$

Engineers usually measure gains and attenuations – that is, ratios of powers – in a logarithmic scale:

$$g = 10 \cdot \log_{10} G; \qquad (19.2)$$

the unit of g – the *decibel, dB* – corresponds to a gain G = 1.25.

To achieve a high gain, reflecting and parabolic antennas are used, with an oscillator placed at their foci. If the source were point-like, in the limit of vanishing wavelength, the emitted spherical beam would be transformed in a parallel beam with a perfect directivity and infinite gain; but due to the finite wavelength and other effects there is always an angular spread. Physical optics demands that the main emission lobe have an angular width of the order (wavelength)/(radius of the dish); therefore the solid angle into which most radiation is emitted is about (square of the wavelength)/(area of the dish). The gain (4π times the reciprocal of this solid angle) is appropriately written as

$$G = 4\pi A_e/\lambda^2 \quad . \tag{19.3}$$

For a perfect paraboloid A_e, the effective area, is the area A of the antenna; in practice it can reach up to 0.8 A or 0.9 A. We see from eq. (19.3) that with large dishes (diameters of 70 meters have been realized) one can reach in the microwave band gains of 60 to 80 dB, corresponding to angular widths of 0.001 to 0.0001 rads. High gains place stringent requirements upon the orientation of the antenna.

It turns out that the *effective area* A_e defined by eq. (19.3) for the emission is also the cross-section in the receiving mode; thus the received power on board is

$$P' = PGA_e'/4\pi D^2 \quad . \tag{19.4}$$

A prime denotes the quantities related to the spacecraft receiver. Using the receiver gain G' it is convenient to write this as

$$P' = PGG'(\lambda/4\pi D)^2 = PGG'L. \tag{19.4'}$$

The *space loss* $L = (\lambda/4\pi D)^2$ measures the attenuation due to the geometrical spread of the beam during the propagation. It is common to write the energy budget (19.4') in a logarithmic scale, using

$$p = 10\cdot\log_{10}(P/1\text{mW}). \tag{19.5}$$

This is the power enhancement, measured in dBm (with respect to a mW), with respect to a nominal level of 1 mW. Therefore we have

$$p' = p + g + g' + \ell. \tag{19.4"}$$

ℓ – the logarithmic space loss – is negative. In this scheme we do not consider other minor losses, in particular ohmic losses in the transmitter and the receiver and the atmospheric attenuation.

Radio systems can be used also in the *radar* mode, in which the signal arriving upon a passive body – like a planet – is reflected or diffused back to the ground. This technique has many applications in planetary physics; its optical counterpart provides the very precise tracking of LAGEOS-type satellites (LAser GEOphysic Satellite; see Sec. 20.2) and of the moon.

The link budget now uses eq. (19.4) to get the power P' impinging upon the body, of geometrical cross-section A_e'. Part of this power is reflected and diffused with a gain G_1. This re-emission from a planetary body, with albedo A, occurs in a solid angle of order 2π, so that G_1 is about 2A. This gain is substantially increased if special optical systems (*retroreflectors* or *corner reflectors*) are used (like those on the moon and on LAGEOS) which, within the limits of the optical approximation, are able to reflect back the received radiation in the same direction it arrives from. The simplest version of such a system is a set of two orthogonal mirrors and works only for beams orthogonal to their intersection; in practice, more complex prisms are used. In a retroreflector the gain G_1 is again of the order of 4π times the ratio between its area and the square of the wavelength. With the same reasoning as before and a suitable gain G' we get the power received on the ground by the same transmitting antenna:

$$P_1' = P \frac{GA_e'}{4\pi D^2} G_1 G \left(\frac{\lambda}{4\pi D}\right)^2 . \qquad (19.6)$$

This is proportional to $1/D^4$, the typical dependence of radar links.

The United States use for their interplanetary satellites the *Deep Space Network*, a very large network of steerable paraboloids at three locations: Goldstone (California), Canberra (Australia) and Madrid (Spain) (Fig. 19.1). They are used mainly for telecommunication and space navigation, for radar measurements and, in a receiving only mode, as a part of a VLBI network (see Sec. 20.5).

The power flux in space is given by Poynting's flux

$$\mathbf{\Phi} = c(\mathbf{E} \times \mathbf{B})/4\pi . \qquad (19.7)$$

In empty space this is directed along the propagation vector **k** and, since electric and magnetic fields are equal in size and orthogonal,

$$\Phi = cE^2/4\pi . \qquad (19.7')$$

This momentum flux exerts a force upon a material body and contributes to the momentum balance of a spacecraft. There are two basic modes of propagation, according to the *polarization* of the beam. Any electromagnetic wave – propagating along z – is the superposition of two modes in which the electric field is aligned along x or along y (*linear polarization*); if the two orthogonal components have the same amplitude and a $\pi/2$ phase lag, *circular polarization* results. Note that the rotation of the spacecraft affects the state of polarization generated by its oscillator.

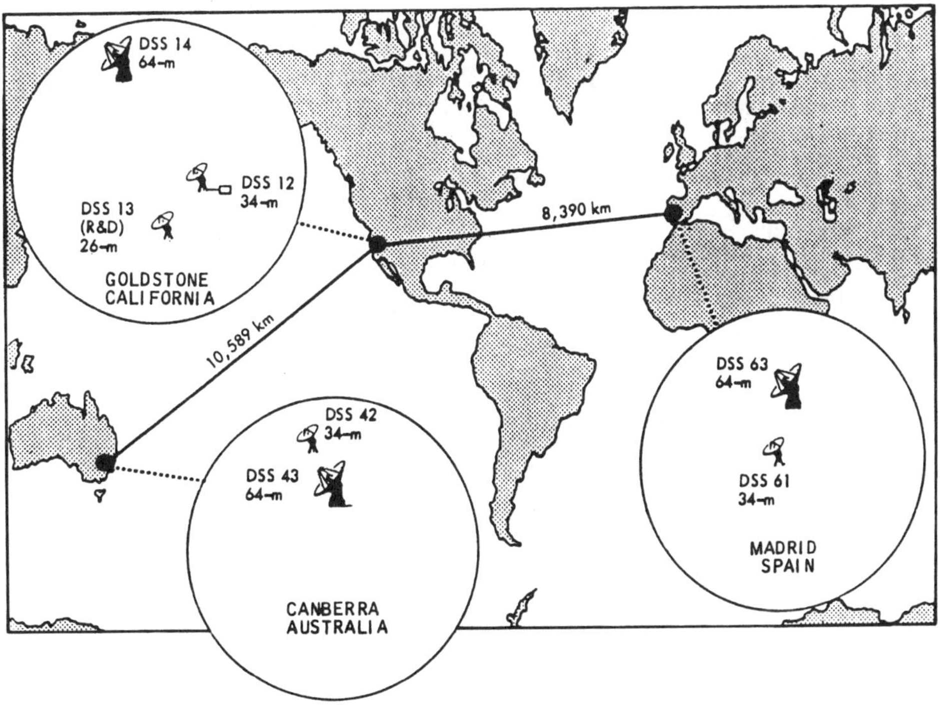

Fig. 19.1. The radio stations of NASA's Deep Space Network. (From N.A. Renzetti et al., *The Deep Space Network*, JPL Publication 82-102 (1982); the system has been improved since.)

19.2. SPECTRA

At the receiver the wave electric field E(t) is transformed into a voltage which is amplified and then recorded; but the gain of the amplification is frequency dependent, so that only a fraction of the incoming energy is measured; moreover, a noise is added in the process. It is therefore essential to discuss the decomposition of a signal into its frequency components, namely, spectral theory.

Consider first, for simplicity, a real, periodic function $a(t) = a(t + P)$ with period P. In reality we are interested in signals of indefinite duration and with no periodicity; but, by choosing P sufficiently large, we can always approximate them with periodic functions. The use of periodic functions simplifies the mathematics considerably. A periodic function has the Fourier expansion (with n ranging over all the integers)

$$a(t) = \Sigma_n a_n \cdot \exp(2\pi int/P); \qquad (19.8)$$

since a(t) is real, $a_n^* = a_{-n}^*$. The proper frequencies $\omega_n = 2\pi n/P$ are equally spaced by $2\pi/P$; hence, when P is large, in a unit frequency interval we have $P/2\pi$ frequencies. We call the *spectrum* of the periodic function the frequency dependent quantity

$$S_a(\omega) = P|a_n|^2/2\pi = S_a^*(-\omega) \ . \qquad (19.9)$$

This function, even in its argument ω and with the dimension of $a^2 \cdot$time, is here defined by its discrete realization at the frequencies ω_n.

Squaring and integrating eq. (19.8) over the period P we find

$$\mathrm{Ave}(|a(t)|^2) = \frac{1}{P}\int_0^P dt|a(t)|^2 = \Sigma|a_n|^2 = \int_{-\infty}^{\infty} d\omega S(\omega) \ ; \qquad (19.10)$$

in other words, the mean squared value of a(t) is decomposed in different frequency components according to the spectrum $S(\omega)$.

When a generic function a(t) defined over the whole real line is considered, the definition (19.9) is not applicable. We must instead start from the *correlation function*, even in its argument τ (the *lag*):

$$C(\tau) = \lim_{T\to\infty} \frac{1}{T}\int_{t_0}^{t_0+T} dt \ a(t-\tau/2)\ a(t+\tau/2) \ . \qquad (19.11)$$

We require it to exist and to be independent of the time t_0, which in the following will always be taken to be zero. These two properties characterize mathematically a signal which has no beginning nor end and is sufficiently irregular so that its behaviour at distant times is not related. Note that C(0) is the mean squared value of a(t). Since

$$0 \leqslant [a(t-\tau/2) - a(t+\tau/2)]^2 =$$

$$= a^2(t-\tau/2) + a^2(t+\tau/2) - 2a(t-\tau/2)\ a(t+\tau/2),$$

taking the average of both sides over a time T, we obtain

$$C(\tau) \leqslant C(0) \ , \qquad (19.12)$$

i.e., the correlation function has an absolute, positive maximum C(0) at the origin. The correlation can be expected to be much smaller than C(0) when the lag is so large that the values of a at two instants separated by $\tau \gg \delta\tau$ have no relationship (i. e., no correlation) between each other, so that their product has a random sign and averages to zero. This characteristic value $\delta\tau$ of the lag is called *correlation time.*

When a(t) is periodic with a period P much smaller than T, the average (19.10) can easily be computed

$$C(\tau) = \Sigma_n |a_n|^2 \cdot \exp(2\pi i n\tau/P). \qquad (19.13)$$

Replacing the sum with an integral (which is possible if P is much larger than the characteristic time scale of the correlation) and introducing the spectrum (19.9), we find that C(τ) is the Fourier transform of the spectrum :

$$C(\tau) = \int_{-\infty}^{\infty} d\omega S(\omega) e^{i\omega\tau} = \int_{-\infty}^{\infty} d\omega S(\omega) \cos(\omega\tau) = 2 \int_{0}^{\infty} d\omega S(\omega) \cos(\omega\tau). \qquad (19.14)$$

This definition applies also to the case in which a(t) is not periodic and its Fourier transform does not exist within the class of ordinary functions. The inversion of eq. (19.14) gives

$$S(\omega) = \frac{1}{2\pi} \int_{-\infty}^{\infty} d\tau C(\tau) \exp(-i\omega\tau) = \frac{1}{\pi} \int_{0}^{\infty} d\tau C(\tau) \cos(\omega\tau). \qquad (19.15)$$

This definition uses the double-sided Fourier transform, with the frequency ranging over the whole real line; sometimes one calls 2S(ω) the spectrum and uses the single-sided Fourier transform, with positive frequencies only. We stick to the double-sided form.

A function whose spectrum has the shape of a narrow line of width $\delta\omega$ around the frequency ω_0 corresponds to a correlation function whose correlation time $\delta\tau$ is about $1/\delta\omega$; in fact, a more precise definition of the two quantities leads to the *uncertainty principle*

$$\delta\tau\delta\omega \geqslant 1 \ ; \qquad (19.16)$$

the equality sign holds for a gaussian spectral profile (see Problem 19.3). This is the same principle as in quantum mechanics.

An important way to realize a narrow band spectrum is a sinusoidal function slowly modulated in phase and amplitude:

$$a(t) = A(t)\cdot\exp(i\omega_0 t + i\varphi(t)). \qquad (19.17)$$

The amplitude A and the phase φ change very little in a period $2\pi/\omega_0$ of the carrier; that is to say, the Fourier transform of the modulating function $A\cdot\exp(i\varphi)$ has components only within a frequency interval $\delta\omega \ll \omega_0$ near the origin. Therefore the Fourier transform of a(t) has components only within two bands of width $\delta\omega$ around $\pm\,\omega_0$.

The most important use of space telecommunication links is to transmit data, either uplink to control the spacecraft or downlink to carry information gathered on board. The data are transmitted in digital form, that is to say, encoded in a binary number $q = q_1 q_2 \ldots$ (where each digit is either 0 or 1). With an elementary pulse a(t) of duration P_1 (for instance, a rectangular pulse) one constructs the signal

$$s(t) = \Sigma_k q_k a(t - kP_1)\ . \qquad (19.18)$$

Its Fourier transform is easily read off from the Fourier decomposition of the pulse (19.8):

$$s_n = \Sigma_k q_k a_n \cdot \exp(-\,2\pi ikn)\ , \qquad (19.18')$$

where the index n corresponds to the frequency $2\pi n/P_1$. According to eq. (19.9) the spectrum of the signal s is

$$S_s(2\pi n/P_1) = \Sigma_{k,k'} q_k q_{k'} |a_n|^2 (P_1/2\pi)\cdot\exp[2\pi in(k' - k)];$$

or, in terms of the spectrum of a(t),

$$S_s(\omega) = S_a(\omega)\Sigma_{k,k'} q_k q_{k'} \cdot\exp[i\omega P_1(k' - k)]. \qquad (19.19)$$

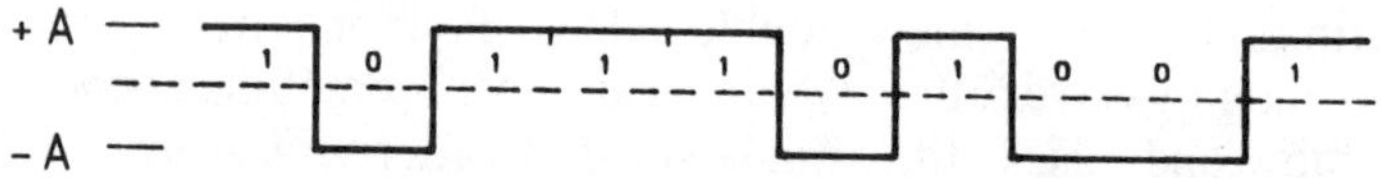

(a) TRANSMITTED SIGNAL

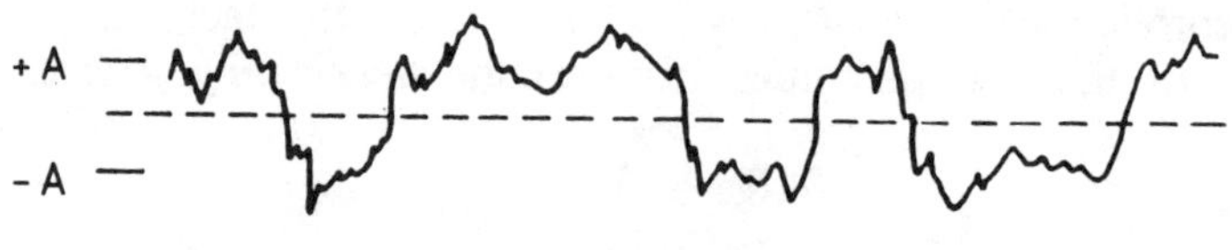

(b). RECEIVED SIGNAL

Fig. 19.2. The amplitude modulated signal corresponding to the binary number ...1011101001...

In a completely analogous way to what was done earlier for a continuous function, we define by

$$\Sigma_k q_k q_{k+h} = C_q(h) \qquad (19.20)$$

the "correlation function" of the numerical sequence q_k and by

$$S_q(r) = \Sigma_h C_q(h) \cdot \exp(2\pi i h r) \qquad (19.21)$$

its "Fourier transform", to wit, the "spectrum" of the binary number q at the "frequency" r. If the binary digits are not correlated, $C_q = 0$ unless $h = 0$ and S_q is white; if the digits are correlated over a length N, S_q has a width of order 1/N. Eq. (19.19) can now be written as

$$S_s(\omega) = S_a(\omega) S_q(\omega P_1 / 2\pi) \ . \qquad (19.22)$$

Therefore, in order to encode n uncorrelated bits per second, we need a signal whose bandwidth is equal to the bandwidth $\delta\omega$ of the elementary signal, which in turn cannot be smaller than the reciprocal of its duration P_1. We come therefore to the simple and fundamental conclusion that *the maximum channel information capacity*, in bits per second, *is the bandwidth of the signal.* If the binary number is correlated, the required bandwidth is smaller.

This result dictates much of the specifications of space communications. The need of sending back to the earth large quantities of data (i. e., when remote sensing instruments are on board) points to the need of high-frequency carriers (for practical reasons it is not convenient to have a frequency bandwidth exceeding a small and given fraction of the carrier). For this reason most space transmitters operate in S-band at about 2 GHz or even in X-band, at about 8 GHz. Higher frequencies are also being planned.

As we shall see in the next section, the requirement of a large bandwidth conflicts with the need of reducing the noise, which in general is a wide-band process.

19.3 NOISE

Since noise is inherently a statistical quantity, it requires the use of *random functions.* A random function is a set of ordinary functions, chosen in a class over which a probability distribution is defined. For example, a random periodic function in the class (19.8) is defined by assigning an infinite dimensional probability distribution for all its amplitudes $|a_n|$ and all its phases $\arg(a_n)$. A very useful way to realize a stationary random function with a given spectrum is provided by the *random phase method.* Referring to eq. (19.8), assume that $a_n = |a_n| \cdot \exp(i\varphi_n)$ is a set of random variables defined by assigning for the phases φ_n a uniform probability distribution over the circle, with no correlation. In other words, the function is made up of oscillators whose phase changes randomly from one to the next.

An angular bracket will denote the average over the set, or ensemble; $\langle a(t)\rangle$ is the mean and $\langle[a(t) - \langle a(t)\rangle]^2\rangle$ the mean square deviation of the function at time t. A random function is called *stationary* if all its statistical averages are time-independent; this is the most interesting case to which we confine ourselves. If the function is defined over the whole real line, under fairly general and reasonable assumptions, the stationary character has an important and very useful consequence. Consider the statistical average of a product of n values of the function at different times $t_1, t_2, \ldots, t_n$; it does not change if these arguments are displaced by the same amount:

$$\langle a(t_1)a(t_2)\ldots a(t_n)\rangle = \langle a(t_1 + t)a(t_2 + t)\ldots a(t_n + t)\rangle.$$

For two factors $\langle a(t_1)a(t_2)\rangle$ it depends only on the difference $(t_2 - t_1)$. It turns out that, under wide conditions, *the time average of such a product is equal to the statistical average (ergodic theorem):*

$$\lim_{T\to\infty} \frac{1}{T}\int_{-T/2}^{T/2} dt\, a(t_1 + t)\ldots a(t_n + t) = \langle a(t_1)a(t_2)\ldots a(t_n)\rangle. \tag{19.23}$$

In the left-hand side of this equation we sample the product constructed with a given realization of the random function with different time translations; in the right-hand side we keep the arguments fixed and sample over different realizations of the random function taken with the appropriate probability weight. The ergodic theorem says that if a random function is stationary, a given realization, so to speak, reproduces in time the statistical properties of the ensemble. The statistical average gives the expected mean value of the product from many different experiments performed under the same conditions at the same times; under the ergodic property it equals the time average of a single, infinitely long measurement. Under this assumption we can therefore write the correlation function (19.11) as

$$C(\tau) = \langle a(t - \tau/2)\cdot a(t + \tau/2)\rangle. \tag{19.24}$$

If a is complex, the appropriate definition is

$$C(\tau) = \langle a(t - \tau/2)\cdot a^*(t + \tau/2)\rangle. \qquad (19.24')$$

It is easy to see that in the random phase realisation

$$C(\tau) = \Sigma_n |a_n|^2 \cdot \exp(-\, i\omega_n\tau). \qquad (19.24'')$$

From eq. (19.15) we see that the spectrum is a line spectrum:

$$S(\omega) = \Sigma_n \delta(\omega + \omega_n)|a_n|^2,$$

where $\delta(\omega)$ is the Dirac delta function. With the same assumption of random phases we see that the correlation of a(t) with itself and $a^*(t)$ with itself vanish.

Any physical operation on the signal s(t) (amplification, propagation, filtering, etc.) adds a noise n(t) to it; the key problem of communication theory is the extraction of the signal s(t) from the observed superposition [s(t)+n(t)]. s(t) is an ordinary function; n(t) is a random function, whose statistical properties are determined by the instrumental and physical, poorly known perturbations. This extraction is done with the use of *linear filters*, constructed either in hardware or in software, which are mathematically equivalent to an integral transform with a suitable kernel K(t):

$$\int dt'\, K(t-t')[s(t') + n(t')].$$

The kernel is chosen in such a way to reduce as much as possible the noise contribution. For example, if the noise is known to occur at frequencies higher than the signal, a low-pass filter is needed.

A very important type of noise is the *thermal noise*, in particular the electrical thermal noise. Any resistance R at an absolute temperature T produces very short and accidental bursts of electrical carriers which travel in one sense or another: a voltmeter measures between its ends a fluctuating voltage V, whose statistical properties are entirely determined by the temperature. This effect was predicted by Nyquist in 1928 and confirmed experimentally by Johnson in 1929. Thus a signal of power P', passing through a resistor, is superimposed to a thermal noise power P_T. Nyquist has shown that the spectrum $S_T(\omega)$ of the thermal noise is flat up to frequencies of order kT/h, where k and h are, respectively, Boltzmann and Planck's constants. At this point quantum effects must be taken into account; they ensure that the spectrum is integrable and the total power is finite. The value of the (two-sided) spectrum is determined solely by the temperature:

$$S_T(\omega) = kT/4\pi \qquad (\omega h \ll kT) \qquad (19.25)$$

(the amount of power in a cycle ($d\omega = 2\pi$) is just kT.) Note that the spectrum of a power signal has the dimension of an energy. It turns out that the same expression (19.25) holds for arbitrary linear electronic devices, in particular for amplifiers. The temperature which defines their spectrum is an effective temperature T_e, not necessarily equal to the room temperature. For example, the receivers of NASA's Deep Space Network have noise temperatures of 19 to 40 K.

Any electronic device used to process the signal has a bandwidth $\delta\omega$ which limits the amount of noise introduced into the system. We call *signal-to-noise ratio (SNR)* the ratio of the signal power to the noise power in the appropriate bandwidth. Using the budget equation (19.4') we get:

$$\mathrm{SNR} = \frac{2\pi P'}{kT\delta\omega} = \frac{2\pi PGG'L}{kT\delta\omega}, \qquad (19.26)$$

which shows how the SNR depends upon the distance of the source. Generally one can detect the signal only if its power is larger than the noise power (unless special filters are used); this gives the largest admissible bandwidth and hence the largest information capacity:

$$\delta\omega = 2\pi P'/kT = 2\pi PGG'L/kT. \qquad (19.27)$$

This is proportional to the inverse square of the distance: in interplanetary missions one needs a large power on board, a short wavelength (to ensure large gains) and cooled receivers.

A digitally modulated signal is used also to measure the round trip light-time to, and hence the distance of, a spacecraft. The emitted beam is modulated in amplitude or in phase according to a random number of sufficient length; on board a repeater is able to retransmit it back with the same modulation. On the ground the uplink modulation A(t), delayed by an arbitrary amount τ, is multiplied by the incoming modulation A'(t) and the average of the product is computed

$$C_A(\tau) = \frac{1}{T}\int_0^T dt\, A(t-\tau)\, A'(t).$$

This average has a maximum when the lag equals the round trip light-time. This method is affected by several noise sources, which at present limit its accuracy, for interplanetary distances, to a few meters.

19.4. PHASE MEASUREMENTS

A better way to determine the orbit of a spacecraft is the *Doppler effect.* This uses only the central carrier (the side bands being reserved for telemetry, command and ranging), stabilized to a very high accuracy by appropriate frequency standards. More generally, measurements of phase have a great importance in the physics of the solar system and in space navigation (see Sec. 8.3 for an example).

The accuracy of a frequency standard is usually measured by *Allan's variance.* Let $\nu(t) = \omega(t)/2\pi$ be the output frequency and

$$y(t) = (\nu(t) - \nu_0)/\nu_0 \qquad (19.28)$$

its fractional change with respect to its nominal value $\nu_0 = \omega_0/2\pi$. What counts experimentally is not its instantaneous value but its mean over an integration time τ:

$$\bar{y}_\tau(t) = \frac{1}{\tau}\int_t^{t+\tau} dt'\, y(t') = \frac{\varphi(t+\tau) - \varphi(t)}{\omega_0 \tau} . \qquad (19.29)$$

A measure of stability is provided by considering this mean at two instants separated by an interval τ; Allan's variance is:

$$\sigma^2(\tau) = <[\bar{y}_\tau(t+\tau) - \bar{y}_\tau(t)]^2>/2$$

$$= <[\varphi(t+2\tau) - 2\varphi(t+\tau) + \varphi(t)]^2>/2\omega_0{}^2\tau^2 . \qquad (19.30)$$

This quantity vanishes when ω is a constant. An important case occurs when τ can be subdivided in $\tau/\Delta t$ elementary intervals of length Δt, during which y has averages y_i, uncorrelated and with the same variance σ_y. Then

$$\sigma^2(\tau) = <(\Delta t \Sigma_i y_i/\tau)^2> = (\Delta t/\tau)\sigma_y{}^2 . \qquad (19.31)$$

This case – Allan's variance inversely proportional to the square root of the integration time – corresponds to a *white frequency noise.* If there are frequency drifts or other systematic errors, $\sigma(\tau)$ becomes worse or even increases as τ increases (see Fig. 3.1). In general, Allan's variance is determined by the spectrum $S_y(\omega)$ of y (see Problem 19.6). The case in which S_y is a power of the frequency ($S_y \propto \omega^\alpha$) is particularly important. When $\alpha = 0$ we have a white frequency spectrum; $\alpha = 2$ is a white phase spectrum. It can be shown that the corresponding Allan's variance is also a power:

$$\sigma^2(\tau) \propto \tau^{-\alpha-1} \quad .$$

When $\alpha > 1$ or $\alpha < -3$, however, a cutoff is needed to ensure convergence.

The frequency standards commonly used for precision radio tracking use hydrogen masers and have a variance of a few parts in 10^{15}, for integration times of the order of an hour. This field is in rapid development, also for metrological purposes. Frequency stabilizations of a few parts in 10^{17} have been obtained with superconductive cavities in the laboratory and improvements are planned. For standard, less demanding applications frequency standards based upon resonances in cesium atoms are used, with a typical frequency stability of a few parts in 10^{14}.

A Doppler measurement requires on board of the spacecraft a transponder capable of locking the phase of the emitted signal to the phase of the incoming wave; on the ground the received frequency is compared with the standard, to get the fractional change y. Phase-locked loops over interplanetary distances can be realized in this way. The main contribution to this change is the relative velocity v between the ground antenna and the antenna on board, taken with the positive sign when the two are approaching:

$$y_D(t) = (\nu_{rec}(t) - \nu_{em})/\nu_0 = v/c. \qquad (19.32)$$

To this velocity contribute mainly: the orbital and the rotational motion of the earth; the orbital velocity of the spacecraft; the motion of the antenna on board. For higher order corrections the fully relativistic formula (3.3) must be used; in the presence of a gravitational field the effect (17.20) of the gravitational potential must also be taken into account.

We are in the position to compute, as an interesting exercise, the effect of the thermal noise of the receiver in a Doppler measurement. Let n(t) be a (small) noise added to a monochromatic signal

$$a(t) = \sqrt{P'}\cdot\exp(i\omega_0 t)$$

of power P' and frequency ω_0. This produces a fluctuation in amplitude and in phase in the signal a(t) given by

$$a + n = \sqrt{(P' + \delta P)}\cdot\exp[i(\omega_0 t + \delta\varphi)]; \qquad (19.33)$$

we get, after some algebra,

$$\delta\varphi = [n\cdot\exp(-i\omega_0 t) - n^*\cdot\exp(i\omega_0 t)]/2i\sqrt{P'} \quad (\ll 1) .$$

From the definition (19.24') we get

$$C_\varphi(\tau) = \cos(\omega_0\tau)C_n(\tau)/2P' . \tag{19.34}$$

$C_n(\tau) = \langle n(t)\cdot n^*(t + \tau)\rangle$ is the correlation function of the noise and $C_n(0) = \langle|n|^2\rangle$ its mean power. From eq. (19.34) we get

$$\sigma_\varphi^2 \approx \frac{\text{noise power}}{\text{signal power}} .$$

This relation is of immediate interpretation in the complex plane, where the signal describes a circle of radius $\sqrt{P'}$ superimposed to a random motion of size $\sqrt{\langle|n|^2\rangle}$. This produces a phase uncertainty of the order of $\langle|n|^2\rangle^{\frac{1}{2}}/\sqrt{P'}$.

Allan's variance (19.30) can be written in terms of the correlation function of the phase as

$$\sigma^2(\tau) = [3C_\varphi(0) - 4C_\varphi(\tau) + C_\varphi(2\tau)]/\omega^2{}_0\tau^2 . \tag{19.35}$$

In the case of white noise $C_n = 0$ for $\tau \neq 0$ and the mean power over a bandwidth $\delta\omega$ is $kT\delta\omega/2\pi$; hence

$$\sigma^2(\tau) = 3kT\delta\omega/(4\pi P'\omega_0{}^2\tau^2). \tag{19.35'}$$

When $\delta\omega\tau \approx 1$ this corresponds to a power spectrum for y with exponent $\alpha = 2$ (white phase).

This relationship shows that the frequency measurement is improved by increasing the received power P' and the integration time τ. It is also important to lower the bandwidth. In practice this is limited by the fact that the receiver frequency must be changed continuously in order to accommodate within its bandwidth the predicted Doppler change; this accommodation can be done with a finite accuracy and a suitable window must be kept open in order not to lose the signal.

19.5. REFRACTION AND DISPERSION

A phase measurement in space is affected by the refractive and dispersive properties of the medium. The phase φ of a wave is changed by a refraction index η_r according to eq. (8.34), or

$$\Delta\varphi = \int ds(n_r - 1) = -\omega\Delta\ell/c, \qquad (19.36)$$

The change in the optical path

$$\Delta\ell = \int ds(n_r - 1) \qquad (19.37)$$

is an integral along the ray. Its rate of change gives the equivalent relative velocity v.

An important contribution to the refraction in space comes from the plasma in the interplanetary space and the ionosphere; the fact that it is dispersive (eq. (8.23)) proves of great advantage in eliminating its effect or in using it to determine the electron density n. At high frequencies ($\omega \gg \omega_p$) we have

$$y = -\frac{d\Delta\ell}{cdt} + \frac{v}{c} = \frac{1}{2\omega^2 c}\frac{d}{dt}\int ds\omega_p{}^2 + \frac{v}{c}, \qquad (19.38)$$

where (eq. (8.11)) $\omega_p = \sqrt{(4\pi e^2 n/m_e)}$ is the plasma frequency along the ray. To eliminate the plasma contribution one can use either a very high frequency carrier (i. e., optical) or two distinct radio frequencies ω_1 and ω_2. Indeed, the quantity

$$\frac{\omega_1{}^2 y_1 - \omega_2{}^2 y_2}{\omega_1{}^2 - \omega_2{}^2} = \frac{v}{c} \qquad (19.39)$$

does not contain the dispersive contribution. *Vice versa*, the total electron content along the beam is given by

$$\frac{d}{dt}\int ds\omega_p{}^2 = -\frac{2c\omega_1{}^2\omega_2{}^2}{\omega_1{}^2 - \omega_2{}^2}(y_1 - y_2). \qquad (19.40)$$

This discussion is limited by magnetic fields and by the fact that the paths of the two beams are not quite the same and therefore are affected by a slightly different plasma content.

Another way to measure the plasma content is to use simultaneously phase measurements (eq. (19.38)) and round trip light-time measurements using a random code as described in Sec. 19.3. A wave packet propagates with the group velocity, which in an electrostatic plasma is given by (when $\omega \gg \omega_p$)

$$\frac{d\omega}{dk} = c\left(1 - \frac{\omega_p{}^2}{2\omega^2}\right). \qquad (19.41)$$

The round trip light-time $\Delta T = 2\int ds/(d\omega/dk)$, when differentiated with respect to time, gives the relative velocity and a plasma correction:

$$-\frac{1}{2}\frac{d\Delta T}{dt} = -\frac{v}{c} - \frac{1}{2\omega^2 c}\frac{d}{dt}\int ds\,\omega_p^2 \ . \qquad (19.42)$$

The negative sign corresponds to the fact that when the source is approaching ($v > 0$), the transit time decreases. We see that this correction is equal and opposite to the plasma correction for the phase. Summing these two quantities we obtain the rate of change of the plasma content along the beam. This method is also used in space navigation to calculate the relative velocity. Radio measurements of this kind have also produced excellent measurements of the electron density in the solar corona and in interplanetary space.

19.6. PROPAGATION IN A RANDOM MEDIUM

A refractive medium is often turbulent and its refractive index $n_r(\mathbf{r}, t)$ is stochastic with a correlation function

$$C_n(\xi, \tau) = \langle n_r(\mathbf{r}, t)n_r(\mathbf{r} + \xi, t + \tau)\rangle \ . \qquad (19.43)$$

In a homogeneous and isotropic medium C_n is a function only of the time lag τ and the distance ξ between the two points. The function $C_n(\xi, 0)$ describes the space correlation of the refractive index. Its characteristic decay length $\delta\xi$ is the distance beyond which the values of the function n_r are not well correlated and therefore measures the "average size" of the turbulent cells. Similarly, the characteristic decay time $\delta\tau$ of the function $C_n(0, \tau)$ measures the mean life of a turbulent cell at a given place. Usually the time $\delta\xi/c$ taken by light to traverse a turbulent cell is much smaller than $\delta\tau$; then only the function $C_n(\xi, 0)$ is relevant.

The structure of a turbulent medium depends also upon its mean velocity $\mathbf{V}$, but often this dependence arises simply from a Galilean transformation; in other words, the correlation function (19.42) (and also the higher order moments) is obtained from the correlation function for $\mathbf{V} = 0$ with a change in the argument $\mathbf{r} \rightarrow \mathbf{r} + \mathbf{V}t$. This says that the wind drags along the turbulence without change. Hence

$$C_n(\xi, \tau) \rightarrow C_n(\xi + V\tau, \tau) \ . \qquad (19.44)$$

The wind mixes up the spatial and the temporal correlations.

Let us consider a turbulent medium at rest. Although its structure is determined by complicated non-linear processes, often for a large interval of ξ it obeys a universal law derived by Kolmogorov on the basis of dimensional arguments (see Sec. 7.5). In the simplest case considered here it says that the spatial correlation function is determined by two constants:

$$C_n(\xi, 0) = C_n(0, 0)[1 - (\xi/\xi_0)^{2/3}]. \tag{19.45}$$

One, $C_n(0, 0)$ is the mean square value of the refractivity; ξ_0 measures the characteristic scale of the turbulence. This law is found to hold, for example, in the troposphere and in interplanetary space over a wide range of distances; of course it is replaced by a different law if ξ is comparable with ξ_0.

In the troposphere the turbulence is strongest near the ground, where thermal exchanges produce instabilities within one or two hundred metres of height. This has an important effect upon the quality of astronomical imaging and places an essential limitation upon ground astronomy. A surface of constant phase is distorted by this effect through the simple formulas (19.36) and (19.37). To describe it in a simple way, assume that the turbulent refractive index is made up of elementary cells with size $\delta\xi$, mean life $\delta\tau$ and amplitude δn_r, and extends over a distance L. We can evaluate the integral (19.37) in a random-walk approach by splitting it into $L/\delta\xi$ elementary contributions, each of them being about $\delta n_r\delta\xi$. They are independent random numbers and sum up stochastically to the total value

$$\Delta\varphi = k\delta n_r \delta\xi \sqrt{(L/\delta\xi)}.$$

This quantity varies transversally to the ray over the characteristic distance $\delta\xi$, so that the normal to the surface of constant phase changes by about

$$k\delta n_r \sqrt{(L/\delta\xi)}.$$

Since the unperturbed gradient of the phase has a magnitude k, the angular deviation of the ray is about

$$\delta\theta = \delta n_r \sqrt{(L/\delta\xi)}$$

and changes in time with a time scale of order $\delta\tau$.

A perfect and small optical system which takes an exposure of a source outside the troposphere for a time much larger than $\delta\tau$ will superpose elementary, point-like images and produce an extended image of size $\delta\theta$. Taking, typically, $\delta n_r = 5\cdot 10^{-7}$, $L = 200$ m and $\delta\xi = 10$ cm, we get $\delta\theta \approx 1$ arcsec, a reasonable value. Of course this depends very much upon local conditions; moreover, a telescope of area A, by averaging over $A/(\delta\xi)^2$ independent coherence regions for the phase, diminishes the effect by the factor $\delta\xi/\sqrt{A}$. The tropospheric turbulence produces also a random fluctuation in the intensity and a random displacement of the centre of the image (*seeing*). Of course, these phenomena are of crucial importance for observational astronomy, but turbulent propagation occurs also for radio waves, both in the ionosphere and in interplanetary plasma.

PROBLEMS

19.1. Using geometrical optics, show how a two-dimensional retroreflector (i. e., two orthogonal mirrors) works.

*19.2. Compare the typical Allan's variance due to the thermal noise in a deep space antenna with the variance due to the hydrogen maser instability. For example, consider the following values:

transmitted power	100 km
diameter of the ground antenna	60 m
distance	5 AU
diameter of the on-board antenna	2 m
receiver bandwidth	100 Hz
carrier's frequency (S-band)	2.3 GHz
integration time	1000 s

19.3. Prove the "uncertainty principle" $\delta\omega\delta\tau = 1$ for a gaussian correlation function.

*19.4. Using a model of the solar corona (see Sec. 13.3), find the plasma correction to the optical path as a function of the impact parameter.

19.5. Find out the number of photons received on the ground in a laser pulse to the lunar retroreflectors:

energy of the pulse	1J
telescope diameter	2.72 m
diameter of each retroreflector	3.8 cm
number of retroreflectors	300
ruby laser wavelength	$6.9\cdot10^{-5}$ cm

*19.6. Find out Allan's variance for a power frequency spectrum $S_y(\omega) \propto \omega^\alpha$. (Hint: calculate first the spectrum of $\bar{y}_\tau$ and then use it in eq. (19.30).)

19.7. What is the effect of the rotation of a planet on its radar scattering ?

*19.8. A coherent wave is sent from the earth to an interplanetary spacecraft in the ecliptic plane. Find out the frequency received on board as a function of time, neglecting the eccentricities of the earth and of the spacecraft.

19.9. Estimate the number of bits required to code a black and white planetary image.

FURTHER READINGS

In view of the very large number of books on the general topics of this introductory chapter we quote only a few related to space, in particular J.J. Spilker, *Digital Communication by Satellite*, Prentice-Hall (1977). On the operation and the performance of NASA's Deep Space Network one can look up the publications of the Jet Propulsion Laboratory (Pasadena, California), in particular J.H. Yuen, ed., *Deep Space Telecommunications Systems Engineering*, Jet Propulsion Laboratory (1982). More details about frequency measurements are in J.A. Barnes et al., *Characterization of frequency stability*, IEEE Trans. on Instrum. and Meas., IM-20, 105-120 (1971).

20. PRECISE MEASUREMENTS IN SPACE

In parallel and in close interaction with the progress of information gathering and our understanding of the physical processes in the solar system, the recent years have witnessed an impressive development of precise measurements in space. This has been made possible by the new space technologies, in particular by dedicated scientific space missions and radio communications. The extreme accuracies obtained in measuring distances, time intervals and relative velocities not only have led to new and deeper insights in the structure of planetary bodies, but have also affected our conception of space-time. It is appropriate, therefore, to review in this last chapter the mathematical and the experimental techniques used in these experiments and to survey some of their most interesting results. We do not mention here the large body of space plasma experiments, which have provided extensive and accurate information about particle distributions and electromagnetic fields in the magnetosphere and interplanetary space.

*20.1. LEAST SQUARES FIT

When large amounts of data about complex systems are available, as in space physics, sophisticated methods of data analysis are needed to determine the physical effects one is looking for. There are three main problems that must be faced. The errors in the measured quantities y_i $(i = 1, 2, ..., N)$ must be determined, on the basis of the known performance of the measuring apparatus and, if possible, a repetition of the experiment. Mathematically, this means that y_i are random variables – usually assumed to be gaussian, with a correlation function

$$<(y_i - <y_i>)(y_j - <y_j>)> = Q_{ij} \ . \qquad (20.1)$$

One can often assume that the measured quantities are uncorrelated, with the same mean square deviation:

$$Q_{ij} = \sigma_y{}^2 \ \delta_{ij} \ . \qquad (20.2)$$

The ideal situation (20.2) corresponds to purely accidental errors, in which the final error in the measurement is inversely proportional to $\sqrt{N}$; the presence of correlations essentially decreases the number of independent data and the total amount of available information. The estimate of systematic errors is, of course, the major experimental problem to be faced.

Secondly, a decision must be taken as to the unknown parameters p_r one wishes to determine, in addition to those which are the main object the experiment. This means that, for any set of p_r (r, and later s, = 1, 2,..., M), a theoretical law $y_i^t(p_r)$ is assigned for the expected result. For example, if p_r are the initial position and velocity of a spacecraft, its distance at a time t_i is obtained by integrating the equations of motion. The choice of the unknown parameters to be solved for is a delicate one: if they are too many, accuracy is lost and the data analysis becomes unnecessarily complicated; if, on the other hand, some important parameters are assumed to be known and are excluded from the set p_r, their error may unknowingly affect the result.

Finally, we have the uncertainty in the theoretical model. Often the forces acting on the system are not known exactly and it is impossible to propagate in time exactly its state; in this case, even if the state is perfectly known at the beginning, an uncertainty will develop and increase with time, which can be reduced only by other measurements. The proper way to deal with this difficulty – which will not be considered any further – is to replace the deterministic evolution of the system with a stochastic evolution, in which there are *stochastic forces*, known only statistically. This evolution is determined not by ordinary differential equations, but by a diffusion-like equation fulfilled by the probability distribution of the state vector. This approach is the basis of the *Kalman filtering procedure.*

Let us now consider the problem of determining the parameters when a set of observations y_i are available. When the theoretical values are non-linear functions of the parameters, an iteration procedure must be used. One starts with an initial determination p_r of the parameters and looks for their corrections δp_r which produce the linear changes in the observables

$$\delta y_i{}^t = \frac{\partial y_i{}^t(p_r)}{\partial p_r} \delta p_r \; . \qquad (20.3)$$

These changes are compared with the residuals

$$\delta y_i = y_i - y_i{}^t(p_r) \qquad (20.4)$$

in the measured quantities. Having determined the corrections δp_r with a least square fit, as it will be explained below, one iterates the procedure by taking $p_r + \delta p_r$ as a starting set, and so on until convergence is obtained.

We now describe the fitting procedure when the observables are linear functions of the parameters:

$$y_i{}^t = \Sigma_r A_{ir} p_r \quad (r = 1,\ldots,M;\ i = 1,\ldots,N). \qquad (20.5)$$

Consider first the case (20.2) in which the measured quantities are independent. One looks for an *estimator*, i. e., a function $p_r^*(y_i)$ from the observable space to the parameter space which is the "best estimate" of the parameters. This estimate is obtained by requiring the quadratic form

$$\Phi = \Sigma_i (y_i - \Sigma_r A_{ir} p_r)^2 \qquad (20.6)$$

to be minimum, leading to the conditions

$$\Sigma_i y_i A_{ir} = \Sigma_i \Sigma_s A_{ir}\ A_{is}\ p_s{}^* \ (s,\ r = 1,\ldots,M). \qquad (20.7)$$

When the measured quantities are correlated, one must use instead of (20.6) the quadratic expression

$$\Phi = \Sigma_{ij} (1/Q)_{ij} (y_i - \Sigma_r A_{ir} p_r)(y_j - \Sigma_s A_{js} p_s) \qquad (20.6')$$

in terms of the inverse 1/Q of the correlation matrix (20.2). The solution is obtained by inverting the M•M matrix

$$B_{rs} = \Sigma_i A_{ir} A_{is} \qquad (20.8)$$

and reads

$$p_r{}^* = \Sigma_s (1/B)_{rs} \Sigma_i A_{is} y_i\ . \qquad (20.9)$$

If the determinant of (20.8) vanishes, the parameters cannot be all determined.

The statistical meaning of this solution is seen by assuming that, because of measurement errors, the quantities y_i differ from the theoretical values (20.5) by the random variables

$$\delta y_i = y_i - y_i{}^t = y_i - \Sigma_r A_{ir} p_r .$$

This means that infinite repetitions of the same experiment with the same values of p_r would produce a distribution of estimates p_r^*. It can easily be shown from eq. (20.9) that they are *unbiased*, namely,

$$\langle p_r^* \rangle = p_r \ . \tag{20.10}$$

One can also compute their covariance matrix

$$R_{rs} = \langle (p_r^* - p_r)(p_s^* - p_s) \rangle = (1/B)_{rs}\sigma_y^2 , \tag{20.11}$$

which contains the main information about the errors in the parameter determination. It can be shown also that the estimate (20.9) makes, in a precise mathematical sense, the covariance matrix (20.11) a minimum.

In this point of view we have a repeated experiment affected by measurement errors; the parameters have fixed ("real") values p. However, in space physics we usually have a single experiment and we must find out from it the statistical distribution of the best parameter solutions p^*, in particular their uncertainties; together with the values themselves, these uncertainties are the main information we need. This is a different problem, usually solved with the method of *Bayesian estimation*. We need to establish a region in parameter space around the estimated value $\bar{p}$ in which we can assume the true parameters p to lie with a given probability (*level of confidence*) η.

The Bayesian approach consists in assuming an *a priori* probability distribution for the true parameters p. Its determination is, of course, a matter of subjective judgment, depending on previous determinations and on theoretical expectations. Let $\pi(p)dp$ be the corresponding probability density and $\Pi(\bar{p}\,|\,p)d\bar{p}$ the conditional probability density of the estimated values $\bar{p}$ when the "true" parameters have the value p. For example, *a priori* ignorance is described by a function π constant over a very large region. With the symbolic volume elements dp and $d\bar{p}$ we indicate the variables with respect to which we have a probability distribution. Usually one assumes that $\bar{p}$ are normally distributed around their means p with the correlation matrix (20.11), but the argument is general. From the theorem about the joint probability of independent events we see that

$$\Pi(\bar{p}, p)dp d\bar{p} = \pi(p)dp \cdot \Pi(\bar{p}\,|\,p)d\bar{p}$$

is the joint probability distribution density in p and $\bar{p}$. The required conditional probability density of p for a given $\bar{p}$ is then obtained simply by normalizing the function Π so that its p integral is unity:

$$\Pi'(p\,|\,\bar{p})dp = \pi(p)\Pi(\bar{p}\,|\,p)dp/\int dp'\pi(p')\Pi(\bar{p}\,|\,p') \ . \tag{20.12}$$

Consider, e. g., a two-dimensional parameter space, with a gaussian distribution and write explicitely the elements of the covariance matrix (20.11) as

$$R_{11} = \sigma_1^2, \quad R_{22} = \sigma_2^2, \quad R_{12} = R_{21} = \rho\sigma_1\sigma_2. \qquad (20.13)$$

The dimensionless quantity ρ is the *correlation coefficient*; it always lies in the interval (– 1, 1). The (conditional) gaussian probability distribution for the estimator reads

$$\Pi(\bar{p}_1, \bar{p}_2 | p_1, p_2) = \frac{1}{2\pi\sigma_1\sigma_2\sqrt{(1 - \rho^2)}} \exp(- q), \qquad (20.14)$$

where

$$q = \frac{1}{2(1-\rho^2)}\left[\left(\frac{\bar{p}_1 - p_1}{\sigma_1}\right)^2 - 2\rho\left(\frac{\bar{p}_1 - p_1}{\sigma_1}\right)\left(\frac{\bar{p}_2 - p_2}{\sigma_2}\right) + \left(\frac{\bar{p}_2 - p_2}{\sigma_2}\right)^2\right]. \qquad (20.15)$$

We have assumed that the probability distribution $\pi(p)$ is constant over the relevant domain in which the gaussian distribution (20.14) is appreciably different from zero. Then in the integral (20.12) the function π drops out; moreover, since the function (20.14) is symmetric in p and $\bar{p}$, the normalization factor in the denominator is unity. We conclude, therefore, that in this case eq. (20.14) describes not only the outcome $\bar{p}$ in a set of experiments, but also the distribution of the "true" parameter values p in a given experiment.

Consider now the elliptic region in p space

$$q \leqslant C^2; \qquad (20.16)$$

it is easy to show that the probability that the point (p_1, p_2) lies in such a region is

$$\eta = 1 - \exp(- C^2) . \qquad (20.16')$$

This is the required region within which we can expect the parameters to lie with the probability η. For example, for $\eta = 0.9$, $C = 1.52$. Note also that the value of η depends on the dimensionality of the parameter space.

If $\rho = 0$ this region is an ellipse centered at $(\bar{p}_1, \bar{p}_2)$, with the semiaxes parallel to the coordinates and magnitudes $(\sqrt{2})C\sigma_1$, $(\sqrt{2})C\sigma_2$; we can say, the uncertainties of the two parameters are $\pm (\sqrt{2})C\sigma_1$, $\pm (\sqrt{2})C\sigma_2$. In general eq. (20.16) describes an ellipse which becomes more and more elongated as the correlation coefficient ρ tends to 1 or to – 1. In the simple case in which $\sigma_1 = \sigma_2 = \sigma$, the ellipse is oriented along the bisector of the coordinate axes and has semiaxes

$$C\sigma\sqrt{[2(1-\rho)]}, \quad C\sigma\sqrt{[2(1+\rho)]}.$$

When $|\rho|$ approaches unity the ellipse degenerates into a segment, making it inappropriate to speak of the separate uncertainties; at $\rho = 1$, only the combination $(p_1 - p_2)$ enters in the probability distribution and this is the appropriate variable one really measures. Similarly, when $\rho = -1$, one in fact measures $(p_1 + p_2)$.

A large correlation means that the uncertainty in one variable is much affected by any additional knowledge we may gain or loose about the other. Thus, the result of an experiment is greatly affected if a parameter which is highly correlated with the quantity of interest is dropped from the analysis: if in a two-parameter experiment we assume p_2 to be known and equal to its estimate $\bar{p}_2$, the parameter of interest p_1 has an uncertainty $C\sigma\sqrt{(1-\rho)}$; but if we fit p_2 as well, its uncertainty, the semimajor axis of the ellipse divided by $\sqrt{2}$, has the larger value $C\sigma\sqrt{(1+\rho)}$. This shows how important it is, in a complex experiment, to understand qualitatively how the parameters affect the observables and to be aware of correlations. The result of a complicated experiment should be expressed in terms of the full correlation function (20.11) of the parameters, in order to show any substantial correlation.

We conclude by mentioning two important practical problems which may arise in complex space experiments. It is often necessary to extract from a noisy record a periodical signal, determined by its frequency ω, its amplitude and phase. This is a particular example of *spectral estimation*. It turns out that the duration T_1 of the record constrains the accuracy with which the frequency can be determined; indeed, the largest possible frequency resolution $\Delta\omega$ is essentially $1/T_1$. One can view this as the *uncertainty principle* $T_1 \Delta\omega \lesssim 1$ in data analysis.

Spectral estimation is also constrained by the sampling of the continuous signal. Usually, the experiment consists in individual measurements; for instance, let Δt be their (constant) separation. In this case the highest frequency which can be detected in the record is the *Nyquist frequency* $\omega_n = \pi/\Delta t$. This phenomenon arises because when a periodic signal of frequency ω is sampled every Δt seconds, new *alias* lines arise at the frequencies displaced from ω by multiples of the Nyquist frequency.

20.2. LASER TRACKING

The transit time method to measure distances, already discussed in Sec. 3.1, with the use of short laser pulses has reached an extraordinary level of accuracy. It is now currently used to determine the distance of the moon and of earth satellites, in particular LAGEOS (LAser

GEOdynamic Satellite). In both cases the reflector is a special corner-cube shaped optical system, capable of reflecting back in the same direction the light it receives from a ground station.

Using the radar link budget equation (19.6), one can find the energy $w_1' = P_1'\tau$ received on the ground for a pulse of duration τ and mean power P_1'. The mean number of photons received in a pulse

$$N = w_1'/h\nu \tag{20.17}$$

is often small or even less than unity (For the lunar laser ranging it is between 0.05 and 0.1.) Therefore the technique of laser tracking requires single photon detection systems, based upon photomultipliers and a timing system capable of measuring the round trip light-time T_r. A photomultiplier is characterized by an efficiency ϵ (< 1); even when $\epsilon N > 1$ an appropriate statistical analysis, using averages over many shots, is required to produce the best timing measurements. For example, if $\epsilon N = 1$ and the intensity profile of the emitted pulse is taken to be constant in a time interval τ, the uncertainty in the determination of its middle point is just τ. Since the photons are not correlated and each arrival time has a uniform probability distribution within that interval, with ϵN detected photons we have a reduction of the error to $\tau/\sqrt{(\epsilon N)}$. Therefore, it is convenient to average over many consecutive pulses. Statistical averaging is limited by the need of keeping track of any physically meaningful variation in the round trip time T_r produced, for example, by the motion of the satellite. The results of a laser-tracking experimental run, after this preprocessing stage, are presented as a time series of round trip times $T_r(t_i)$, embodying all the information contained in all the single shots within a time interval Δt around the time argument t_i. The corresponding distance is $T_r(t_i)/2c$ and the error usually quoted is the formal error in this quantity. The laser-ranging technique requires pulses as short as possible in relation to the energy needed; Q-switched lasers, producing pulses with duration of 10^{-9} s or less, are used.

LAGEOS is a passive satellite of spherical shape, with a diameter d of 60 cm and a mass M of 411 kg; to diminish the effects of non-gravitational forces like gas drag or radiation pressure it has a low ratio of the cross section S to the mass M

$$S/M = \pi d^2/4M = 0.0069 \text{ cm}^2/\text{g}.$$

A large fraction ($\cong$ 40%) of its surface is covered with 426 retroreflectors. Its orbital elements are listed in Table 20.1.

Semimajor axis	a	12270 km
Mean motion	n	6.38 revs/d = $4.65\cdot10^{-4}\,s^{-1}$
Period	$2\pi/n$	$1.35\cdot10^{4}$ s
Eccentricity	e	0.0044
Inclination	I	109.9°
Nodal rate	$d\Omega/dt$	0.343°/d

Table 20.1 - Orbital elements of LAGEOS.

LAGEOS is routinely tracked by many stations around the world: the best stations have an error of about 1 cm in the mean distance, as defined above, for averaging times of about 2 minutes. This error will decrease in the future. At this high level of accuracy, the correction in the round trip time due to tropospheric refraction must be estimated very accurately.

The main use of LAGEOS is to determine the distance AB between two ground stations A and B and its time variations, as well as polar motion. If the position P of the spacecraft is known to within, say, 1 cm (corresponding to an angular accuracy of less than 10^{-3} arcsec), from the measurement of the distances AP and BP we should be able to determine AB with a comparable accuracy. This is relevant to measure the kinematics of plate tectonics, whose velocities are of the order of a few cm/y. The distance between LAGEOS and a station is also affected by the rotation of the earth; it is thus possible to measure very accurately (with an error of a few cm) the position of the pole of rotation, but not the constant part in the rotation rate. In fact, mainly due to the oblateness of the earth, the orbital plane precesses around the polar axis by about 125°/y; the main contribution to this precession is proportional to J_2 and depends on the inclination (see Sec. 11.6). Since J_2 is currently known only to within about 10^{-6}, the corresponding uncertainty in the precession rate is of the order of one arcsec/y, far larger than the measuring accuracy.

The data analysis of LAGEOS requires the simultaneous determination of the initial keplerian elements, of the parameters of polar motion, of the positions of the stations with respect to an ideal, rigidly rotating earth and of several other parameters. To evaluate the theoretical values of the round trip times $T_r(t_i)$ as a function of these parameters, we need to know the forces acting on the spacecraft, in particular those

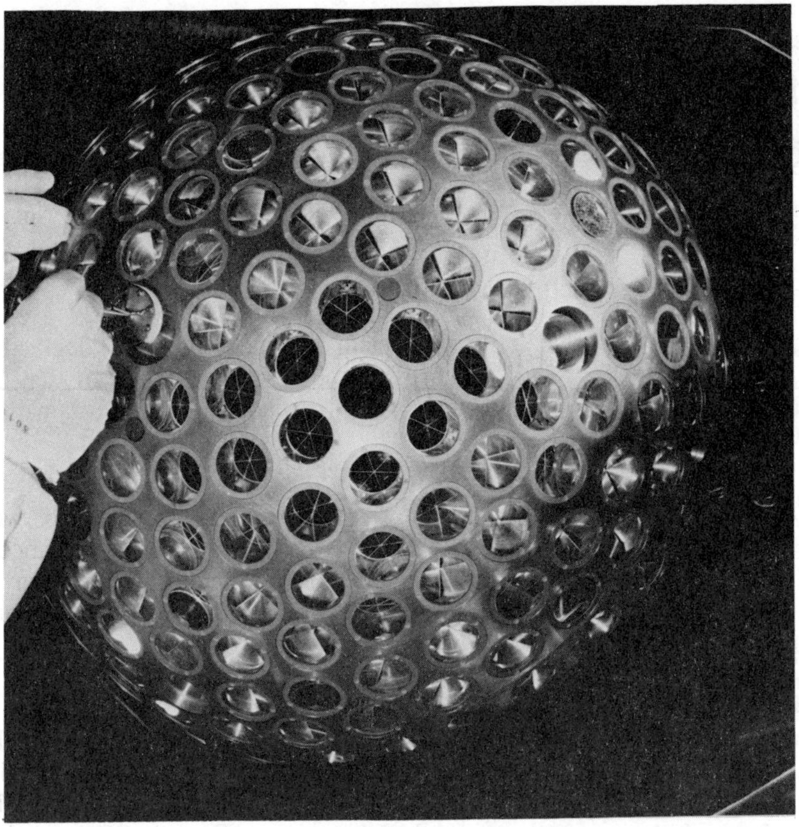

Fig. 20.1. The LAGEOS spacecraft. (Courtesy of Aeritalia Space System Group.)

which do not change much in an orbital period. As explained in Sec.18.1, this is a rather complex task; for LAGEOS the force model is good to a few times 10^{-10} cm/s^2 on time scales of the order of one month or more. An along-track acceleration T, in particular, changes the semimajor axis at the rate 2T/n (see eq. (11.4)) and, by Kepler's third law, displaces the spacecraft by $3Tt^2/2$ in a time t (see Sec. 11.1). LAGEOS's semimajor axis a has been (unexpectedly) found to decay at the rate of about 1.1 mm/day; this correponds to a drag-like T component of $\simeq -3.4 \cdot 10^{-10}$ cm/s^2, causing an acceleration along the track of about 4 cm/d^2; there are also long term changes of a of the same order. The physics of the non-gravitational forces upon LAGEOS is an interesting problem in itself. Since drag from neutral particles at LAGEOS' altitude is expected to be at least one order of magnitude smaller than the value of T quoted above, other mechanisms must be identified. A list of possible physical causes of this effect includes: the interaction with the plasma; the recoil produced by the anisotropic thermal emission from the surface of the satellite, which is differentially heated by sunlight (when the satellite is not eclipsed by the earth's shadow) and by the infrared and visible radiation from the earth; and the radiation pressure due to the earth-reflected sunlight, which depends on the variable optical properties of the earth' surface.

The analysis of the motion of the node of LAGEOS has shown that it has an anomalous deceleration, believed to be produced by a secular decrease of the quadrupole moment of the earth

$$dJ_2/dt = -\ 3{\cdot}10^{-11}\ y^{-1}\,.$$

Such a change is expected on geophysical grounds: the last deglaciation, occurred about 20,000 y ago, has displaced large amounts of ice from the poles to the oceans, producing a slow rebound of the crust which is still going on.

On the moon there are four retroreflectors placed by the Apollo and the Lunakhod missions; they are routinely tracked from the ground with large telescopes and have provided so far several thousands of "normal values" of the distance (about one per month) with an accuracy which at present is about 5 cm. These data provide important information on the motion of the moon and the polar motion and have allowed an important test of the equivalence principle (the *Nortvedt effect*; see Secs. 17.3, 20.6 and Problem 20.8.)

20.3. SPACE ASTROMETRY

The word "astrometry" refers in general to a variety of techniques which are used to determine the apparent angular positions in the sky of the stars and other celestial bodies. Ground based astrometry can be divided into two broad categories: [1] *global* astrometry, which seeks to measure large angles across the sky with the aim of determining positions and proper motions of stars in a single reference system; and [2] *local* astrometry, confined to small areas of the sky for investigations of *relative* trigonometric parallaxes, proper motions, orbits of visual binaries and local extensions to faint sources of the global astrometric catalogues. The former type of research is traditionally carried out by exploiting the diurnal rotation of the earth, i. e., with meridian instruments and accurately calibrated circles; the latter investigations are performed by taking photographic plates with astrographs or telescopes with long focal length over angular ranges of the order of one to a few degrees. A basic noise in all these ground based observational techniques is due to *scintillation*, which makes the pointlike stellar images scatter over a zone of the order of 1 arcsec across (in the best seeing condition; see Sec. 19.6.) Moreover, even when relative positions and displacements are measured with high precision by large telescopes, the accuracy of the absolute positions is still limited by the accuracy with which the global reference frame is defined in the vicinity of any small area. Current global catalogues contain only a few thousands of bright stars (of the order of 0.1 per square degree) and are based on many unrelated observations made by different instruments, which require corrections for various instrumental and refraction errors; as a consequence, the accuracy is of around 0.2 arcsec at best.

Global astrometry will be drastically improved in the next decade by space observations. Taking advantage of zero gravity, full-sky visibility, constant thermal environment and absence of a refracting atmosphere, the achievable accuracy will be of the order of a few mas for positions and a few mas/y for proper motions. One such project, under the name of *HIPPARCOS* (chosen as a tribute to the discoverer of precession and as an acronym for HIgh-precision Positions and Parallaxes COllecting Satellite) is being carried out by the European Space Agency; unfortunately the satellite, launched in July 1989, could not reach its planned (geosynchronous) orbit due to a failure of the apogee thruster, and this will cause a significant downgrading of the final results. However, we describe here the original plans for the HIPPARCOS mission, as they give most clearly an idea of the current potential of space-based astrometry.

The basic purpose of the mission is the observation by a single dedicated instrument of about 10^5 stars, all those brighter than magnitude 9 plus a selection of fainter ones, down to magnitude 12.4, with an almost isotropic distribution all over the sky. (We recall that two stars with apparent brightness ratio B_2/B_1 have a *magnitude difference* $\Delta m = -2.5 \log(B_2/B_1)$; the sun seen from a distance of 10 pc would have an apparent magnitude 4.72.) Besides producing a final, comprehensive astrometric catalogue, HIPPARCOS will accurately measure many trigonometric parallaxes (all individually significant stars up to $\approx$ 150 pc from the sun will be included in the observation list), significantly improving the status of one of the most complex and long-standing problems in astrophysics, namely the calibration of stellar absolute magnitudes and consequently of the extragalactic distance scale.

The essential feature of HIPPARCOS is the simultaneous observation in two directions, at a fixed angle $C \cong 58^\circ$ apart, by means of a beam-splitting mirror, formed by two plane surfaces which reflect into a single telescope two stellar fields $0.9^\circ \cdot 0.9^\circ$ wide (see Fig. 20.2). The sky is scanned by rotating the telescope (in fact, the whole satellite) at a rate of 11.25 revolutions per day about an axis which is nominally perpendicular to the two beams. As the instrument rotates, star images from either beam (close to the same great circle in the sky) will transit across the focal plane. Here a grid, composed of alternately opaque and transparent bands with angular period φ, modulates the incoming light from the moving sources. Since only preselected stars will be observed, they will be framed in a window of $38 \cdot 38$ (arcsec)^2, isolated by an image dissector tube; here the incoming grid-modulated signal will be sampled at a frequency of ≈ 1200 Hz, in such a way to reconstruct a light curve. A star image crosses the grid in about 20 seconds, and the image dissector can switch rapidly from one position to another in the focal plane recording several photoelectric sequences for each star. The phase difference between the light curves of any pair of stars (plus the fixed angle C if the stars belong to different fields) yields the projection of the angle between the stars on the great circle perpendicular to the spin axis. Since the aperture D = 25 cm of the telescope produces a diffraction pattern of width ≈ $\lambda/D \cong 0.4$ arcsec, comparable with the grid period, diffraction must be accounted for by a deconvolution of the light curve, based on a suitable signal model.

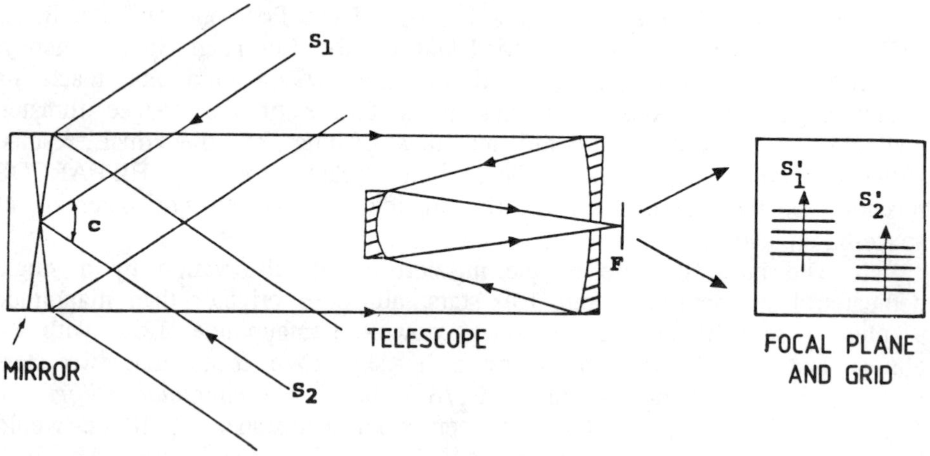

Fig. 20.2. A scheme of the HIPPARCOS telescope: light rays from two stars S_1 and S_2 of two different fields separated by an angle C are sent by a beam combiner mirror to the same focal plane, where they form the images S_1' and S_2', which are then modulated by the grid. The telescope will actually be an all-reflecting Schmidt.

The accuracy is limited by the performances of the attitude determination and control system, owing to some high-frequency, irregular jitter in the rotation of the satellite, and by the imperfections of the optical system, producing an estimated noise in the observed phases of the order of 0.1 arcsec. Since every star will be observed for a few seconds, taking $\varphi \cong 1$ arcsec (about 2700 slits in the whole focal plane), a few hundreds of light curve periods will be detected; if the error in the phase of each elementary period is of the order of 0.1 arcsec, a simple application of the square-root law (see Sec. 20.1) leads to estimate the final accuracy in the determination of the overall phase at several mas. More sophisticated error analyses have confirmed this order of magnitude estimate; of course, the accuracy depends also upon the apparent brightness of the stars (as the quality of the light curves depends on the absolute number of photons.)

Complete sky coverage will be achieved by changing smoothly the direction of the axis of rotation of the satellite. This motion must be related to the apparent motion of the sun in the sky, since on the one hand the solar panels must receive sunlight continuously, and on the other hand direct sunlight must never penetrate the telescope. The optimal scanning law is thus obtained by maintaining the spin axis at a constant angle ($\approx 45^{\circ}$) from the sun direction, with a precession rate of several revolutions per year. This law results into an epicyclic pattern of the spin axis around the ecliptic, such that any point in the sky will be scanned a few times per year (as the nominal mission lifetime is of 2.5 y, this should result in a further reduction of the random errors by a factor of ≈ 3). The instantaneous attitude of the satellite can be determined at the arcsec level from observations of transits of selected bright stars across an auxiliary grid near the edge of the field; this grid, called "star mapper", is much coarser than the main modulating grid and has some slits parallel to those of the main grid and some inclined to it, so that both coordinates can be measured. From a set of some 10^7 one-dimensional observations of relative star positions along different great circles and the attitude data (for a total amount of information of $\approx 10^{12}$ bits), a complex iterative data reduction procedure will be required to obtain the final catalogue of 5 astrometric parameters - two position coordinates, two components of the proper motion and one parallax - for some 10^5 stars.

The method of measurement based on the modulating grid requires a special reduction procedure for double stars, which will be detected as such only if their separation is not smaller than the grid period (i. e., larger than ≈ 0.1 arcsec) and their magnitude difference is not too large ($\lesssim 3$ magnitudes). It will also be impossible to include objects with non-stellar images, such as planets; however, a few tens of bright asteroids will be observed in order to link the HIPPARCOS reference system with a dynamically defined inertial system (see Sec. 3.2). Another way of referring the reference frame of the HIPPARCOS catalogue to an inertial frame is through measurements of the apparent positions (with respect to stars of this catalogue) and proper motions of extragalactic objects. These measurements will be feasible (as we shall see in a moment) by using the astrometric facilities of the Space Telescope. Choosing for this purpose a set of point-like extragalactic sources at radio and optical frequencies and observed by VLBI techniques, it will be possible to determine the rotation of HIPPARCOS' proper motions relative to an extragalactic frame.

As for *local* (or relative) astrometry, an order of magnitude increase in accuracy with respect to traditional techniques will be achievable by using the Fine Guidance Sensors (FGS) of the Space Telescope. Two of the three FGS will be routinely used to control the pointing of the telescope, while the third will be available for astrometric observations to measure relative positions of stars in its field of view, up to magnitude 17 and with an accuracy of about 2 mas. Each FGS views a 90° sector of an annulus in the outer portion of the focal plane, corresponding in the sky to a 4 arcmin width and to a 18 arcmin cord length. The objects whose relative position is to be measured must be visible in the same field, and thus be no more than a few arcmin apart. Each FGS contains an optical beam splitter, two orthogonal prism interferometers (one for each optical axis) and their associated photomultiplier tubes. Two beam deflectors, called *star selectors*, rotate to bring the light from a source anywhere in the field into the 3 arcsec wide square aperture of the FGS detector assembly; if the object is exactly on the axis of the interferometer, the difference between the signals from the two photomultipliers associated with that axis will vanish. If the source is off the axis, the difference provides a non-zero signal that measures how far from the axis the source is; in this case, the star selectors are repositioned to bring the object back to the axis, killing the signal. The relative position of two objects can then be computed from the displacements of the star selector.

The scientific output of these observations will be significant in the following fields: measurements of parallaxes of faint objects (HIPPARCOS will be limited in this respect); astrometric measurements of double stars (separations smaller than 1 arcsec are difficult to observe from the ground), to improve the available statistics and to increase the sample of directly measured stellar masses; search for planets of nearby stars, analysing the perturbations in their proper motions; studies of the motions of stars in clusters and of the orbits of planets and satellites; measurements of the gravitational deflection of light (see Sec. 17.6). All these objectives, as well as the possibility mentioned earlier of linking the VLBI reference system with the HIPPARCOS catalogue, shows that there is a lucky complementarity between the two projects, which will make space astrometry one of the most exciting areas in modern astrophysics.

One should also mention the plans to fly optical interferometers, designed to determine the position of stellar sources with the theoretical accuracy of Michelson interferometers of arm length d. If two coherent beams of wavelength λ from the same source are brought together with a path difference Δd, the intensity is modulated according to

$$I = 4I_0 \cos^2(\pi \Delta d/\lambda) = 4I_0 \cos^2(\delta\Phi/2) \quad . \qquad (20.18)$$

An angular displacement of the source $\Delta\theta$ corresponds to a path difference $\Delta d \approx d\Delta\theta$. The purpose of the measurement consists in measuring very accurately the change in intensity and hence the change in the phase difference $\delta\Phi$. The fundamental limitation to this accuracy for astronomical and space experiments is the *photon noise*, which produces an uncertainty

$$\sigma_I = I_0 / \sqrt{N} \qquad (20.19)$$

in an intensity measurement carried out with N photons. This is a simple consequence of the fact that photons obey the Poisson statistics. Differentiating eq. (20.18), we see that this corresponds to an uncertainty in phase difference

$$\sigma_\Phi \approx 1/(4\sqrt{N}) \approx 2\pi\sigma_\theta d/\lambda \ . \qquad (20.20)$$

The angular accuracy depends on the intensity of the source and the duration of the measurement. The project POINTS (Precision Optical INTerferometry in Space), under study at MIT, is a space interferometer with a baseline of two meters and is planned to reach the limiting accuracy (20.20), at the level of a few microarcsec for the angular position of a source. Another possible use of space interferometry is the detection of low frequency gravitational waves.

20.4. PLANETARY IMAGING

In the last twenty years many new results concerning the physics, geology and meteorology of planetary bodies (including satellites, rings and one cometary nucleus) have been obtained through direct close-up imaging by instruments carried on board by space vehicles.

For the earth, imaging by reconnaissance satellites has played a crucial role for a variety of purposes, ranging from military surveillance and warning to meteorology and to search and monitoring of natural resources. Low satellites (a few hundreds of km high) can sweep a narrow ribbon of the surface, about as wide as the height of the satellite; at a height of 200 km this means that in one orbital revolution (lasting about 90 minutes) a satellite can monitor about 1.5% of the earth's surface. Due to the rotation of the earth and the precession of the orbital plane, every revolution spans (in absence of resonances) a different ribbon, so that within several days a complete coverage is obtained of the part of the surface whose latitude (either northern or southern) does not exceed the satellite inclination. A helio-synchronous, low orbit with an inclination of 96° (see Secs. 11.6 and 18.3) is well suited to obtain images with homogeneous conditions of illumination. At high altitudes the attainable resolution is smaller, but the surface coverage is wider and faster. For instance, three geosynchronous satellites equally spaced over the equator can continuously monitor almost all the globe, a valuable feature for global weather monitoring and for strategic military warning. The best resolution attainable on the ground is of the order of 1 arcsec (the typical atmospheric scintillation noise) times the equivalent thickness of the atmosphere (say, 10 km), that is, about 5 cm. This limit is independent on the distance D of the observer (the satellite), but larger telescopes are needed to achieve the limiting resolution as the orbital distance increases. At an altitude of 200 km, the λ/D diffraction law requires an aperture of about 2 m.

As an outstanding example of imaging of other planets by a spacecraft, we shall describe the imaging system of the two Voyager probes, which over more than one decade have approached the four outer planets and their satellite and ring systems. The scientific objectives of the imaging experiments ranged from observation of atmospheric dynamics to identification of geological structures on the satellites, search for rings and new small satellites and low-resolution photometry over a wide range of phase angles. The imaging system of each Voyager probe consists in two television cameras, each equipped with a set of color filters (mostly in the visible band, including the sodium line and two lines of methan). Both a wide- and a narrow-angle camera were needed to obtain the highest possible resolution, while retaining the capability to study global features on the planets and the satellites. In normal photographic terms, both cameras used telephoto lenses. For the wide angle camera, a focal length of 200 mm was selected (focal number f/3.5), giving a field of view of about 3° (this field is similar to that obtained with a 400 mm telephoto lens on a normal 35 mm camera). The narrow angle Voyager camera had a focal length of 1500 mm (f/8.5) and a field of view of 0.4°. Each camera had a rotating filter wheel to select the color of the light reaching the focal surface. To create a color picture, the cameras are commanded to take, in rapid succession, pictures of the same area in blue, green and orange light. These three pictures can then be reconstructed on earth into a "true" color image. Other combinations of colors can be used to investigate particular scientific problems and to determine the spectrum of sunlight reflected from features on the planets and satellites. The detector in the cameras is not a photographic film, but the photoconductive surface of a 2.5 cm selenium-sulfur vidicon television tube. The resulting charge distribution is then scanned by an electron beam, sampled and digitized to produce an array ($800 \cdot 800$; ordinary television images have only 525 lines) of picture elements (*pixels*), whose brightness is specified by a number corresponding to eight bits. Unlike the commercial TV cameras, these tubes are designed for slow scan readout, needing 48 seconds to acquire each picture. The shutter speed can be varied according to the source luminosity, from a fraction of a second (e.g., for Jupiter) to many minutes (for search of faint features, such as aurorae on the night sides of the planets). The total content of information of an image is $8 \cdot 640{,}000 = 5 \cdot 10^6$ bits. Even at a transmission rate of one frame per 48 seconds, the needed *bit rate* is more than 10^5 bits per second. For comparison, the bit rate from the first spacecraft to Mars (Mariner 4) was about 10 bits per seconds, requiring a week to transmit to earth 21 pictures, with a total information content equivalent to a single Voyager picture. Altogether, the Voyager probes took several tens of thousands of pictures, for a total of some 10^{11} bits of information. The resolution σ of the images, of course, is proportional to the distance D from the object and given by

$$\sigma = \alpha D/N_p$$

where α is the width of the field of view and N_p is the number of pixels per line (800). For the Voyager narrow-field camera, $\alpha = 0.00740$ rads; at a distance $D \approx 10^5$ km, this yields a resolution $\sigma \approx 1$ km. Even for Jupiter's system, the improvement in resolution with respect to ground-based observations is comparable to the transition from naked-eye observations of the earth's moon to the best telescopic photographs.

Recently a new kind of technology has become available for application to astronomical imaging and, in particular, to planetary imaging from space probes. The first such application will be carried out in the frame of NASA's GALILEO project, an ambitious, two-vehicle mission launched in 1989, aimed at the exploration of Jupiter and its satellite system. Besides a probe to be dropped into Jupiter's atmosphere, GALILEO consists of an orbiter which will use satellite fly-by's to alter its Jovicentric orbit, thus carrying out a two year long satellite tour, including many close encounters with the four Galilean satellites at different geometries. Instead of the vidicon television camera of Voyager, GALILEO's imaging will use a new solid state detector called a *charge coupled device* (CCD). The CCD has a wider spectral response (from about 0.4 to 0.8 μm) and greater photometric accuracy. In addition, its increased sensitivity permits shorter exposures, so that even on very close fly-by's the pictures will not be blurred by spacecraft motion. For instance, substantial coverage of the Galilean satellites at a resolution of 100 metres will be possible, leading to a much more detailed knowledge of their geological features. A CCD chip consists of an array of hundreds to thousands of rows and columns of pixels ($800 \cdot 800$ for the GALILEO orbiter); each of them is a square of ≈ 15 μm side and consists of a light sensitive capacitor made with a metal-oxide semiconductor, capable of using the incoming light to create electric charge. Besides a wider bandwidth, they have a higher quantum efficiency (by a factor of the order of 100) than both photographic plates and vidicon cameras. This allows integration times of a few seconds, in which hundreds of thousands of pixels can be read out, each with its own accumulated charge proportional to the source brightness. The charge is then transferred to a computer for real time, automated processing. CCD technology will be extensively used in many new astronomical projects of the next decade, including the Space Telescope and sky scanning for searching faint comets and near-earth asteroids.

20.5 VERY LONG BASELINE INTERFEROMETRY

The comparison of the phases of two electromagnetic signals, besides in the planned optical interferometers in space, has a very important role in geophysics and astrophysics with the technique called Very Long Baseline Interferometry (VLBI). It provides measurements of the angular structure of celestial radio sources with milliarcsec accuracy (or even less) and of geodetic quantities (positions and distances of ground stations) with errors of, say, 1 cm.

Consider two radio antennas A and B separated by a distance d, receiving signals from a distant, extraterrestrial source at an angle ψ (see Fig. 20.2) from AB. Neglecting noises and assuming that appropriate filters generate a monocromatic source signal with frequency ω, the received signals are proportional to

$$V_A(t) \propto \cos\omega t \ , \quad V_B(t) \propto \cos\omega(t - \tau) \ , \tag{20.21}$$

where

$$\tau = (d/c)\cos\psi \tag{20.21'}$$

is the time delay. Each signal is then multiplied with a standard signal from a local oscillator, at a frequency ω_0 not far from ω, and the high-frequency part of the product is filtered out. For A we get a signal proportional to

$$V_A'(t) \propto \cos(\omega - \omega_0)t .$$

The signal at B, after low-pass filtering, is delayed by a variable amount τ' near the estimated value of τ:

$$V_B'(t) \propto \cos[(\omega - \omega_0)t + (\omega - \omega_0)\tau' - \omega\tau] \ .$$

Taking the product of V_A' and V_B' and averaging over time (or, equivalently, using a low-pass filter), we get their correlation function as a function of τ'.

$$R_{AB}(\tau') \propto \cos[(\omega - \omega_0)\tau' - \omega\tau] . \tag{20.22}$$

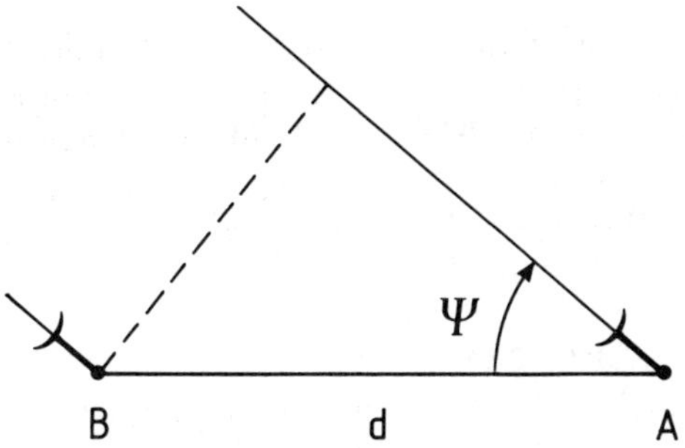

Fig. 20.3. The geometry of VLBI.

The constant part of the phase $\varphi = \omega\tau = 2\pi(d/\lambda)\cos\psi$ of the correlation function determines the delay τ. From eq. (20.21') we obtain the errors in the distance d and the angle ψ for a given error in the phase:

$$\sigma_d = \lambda\sigma_\varphi/(2\pi\cos\psi), \qquad \sigma_\psi = \lambda\sigma_\psi/(2\pi d\sin\psi). \qquad (20.23)$$

For example, in S-band (λ = 13 cm) and with a 10,000 km baseline, assuming just fringe detection ($\sigma_\varphi = 2\pi$), we obtain σ_d = 13 cm/cosψ, σ_ψ = 2.6 mas/sinψ. However, the phase is determined to within a multiple m of 2π:

$$\varphi = 2\pi(d/\lambda)\cos\psi - 2m\pi = \omega\tau + 2m\pi. \qquad (20.24)$$

We need at least two frequencies ω_1, ω_2 to obtain the quantity we need:

$$\tau = \frac{\varphi_1 - \varphi_2}{\omega_1 - \omega_2}. \qquad (20.25)$$

In practice, the method of *bandwidth synthesis* is used: several adjacent, narrow-band channels within the total bandwidth of the radiotelescope are independently measured; the tapes of the corresponding signals from the two stations are correlated with a computer and the relevant unknown parameters, in particular the delay τ, are estimated. The phase uncertainty depends upon the power of the source, its distance, the bandwidth and the integration time. The total bandwidth and the number of channels is the result of a compromise.

Since the delay τ is modulated by the rotation of the earth, VLBI allows also a very precise determination of polar motion. For example, if the baseline and the source lie in the equatorial plane, $\psi = 2\pi\tau$/LOD and the determination of the Length Of the Day (LOD) amounts to the determination of the period of τ. If the baseline lies in a meridian plane, making an angle α with the polar axis, and the source has polar coordinates (θ, φ), we have:

$$\cos\psi = \sin\alpha \sin\theta \cos(2\pi t/LOD - \varphi) + \cos\alpha \cos\theta \ . \quad (20.26)$$

This allows a determination of the angle α with roughly the same accuracy as ψ. This method is currently used to provide polar motion data every few days with milliarcsec accuracy.

The use of VLBI ground antennas has several limitations: the additional and partially unpredictable delay due to the troposphere and the ionosphere; the lack of coverage of the sky due to the opacity of the earth and the difficulty of making observations at low elevation; the limit to the length of the baseline. The tropospheric effect is due both to the dry part and the wet part of the atmosphere. It can be shown (Problem 20.10) that the former can be evaluated by measuring the ground pressure; the water vapour contribution, instead, can be measured only *in situ* or by monitoring the intensity of a microwave line on the ground with *water vapour radiometers*. All these limitations could be relieved by deploying a VLBI antenna in space and correlating it with a ground antenna.

20.6. TESTING RELATIVITY IN SPACE

The theory of general relativity was created by Albert Einstein mainly on the basis of criteria of mathematical simplicity and the principle of equivalence; hence it is important to subject it to tests as stringent as possible. Indeed, if the conventional simplicity requirements are relaxed, many other relativistic theories of gravitation can be constructed, all satisfying the principle of equivalence. The solar system, where very precise measurements of distance and relative velocity are possible, is truly a laboratory to test theories of gravitation. It should be stressed, however, that there the gravitational field is very weak (the metric differs very little from the Minkowskian form) and the velocities are much smaller than c: an extrapolation is needed for the important applications in relativistic astrophysics, where the field is strong and bodies may move fast.

We are going to discuss separately relativistic effects in interplanetary space and in the vicinity of the earth.

a. Relativistic effects in interplanetary space.

The relativistic dynamical corrections due to a body with gravitational radius m (see eq. (1.15)) are characterized by the same length m: it gives the order of magnitude of the displacement in the orbit of a test body in a time span of the order of the period. In addition, the only relativistic secular effect displaces the periastron by an angle of order m/a in an orbit.

The main relativistic effects on the motion of planets and satellites are due to the sun, for which m = 1.5 km. Optical observations from the ground, with an accuracy of order 1 arcsec = $5 \cdot 10^{-6}$ rad, correspond, at 1 AU, to a transversal displacement of 750 km. Only the secular accumulation of the displacement of the periastron over many orbits is accessible optically; even in this case a part of the error is given by the uncertainty in the classical advance of its perihelion (see later).

Things changed drastically with the implementation of radio communications in space. For example, a fractional frequency accuracy of 10^{-13} – currently available – corresponds to a relative velocity 0.003 cm/s and, with a time scale of 10^7 s, to a change in the radial distance of 300 m. By means of modulated radio pulses (see Sec. 19.3), one can measure the distance to a spacecraft with an accuracy of a few meters. At these very high levels of accuracy the requirements on the calculation of the classical forces (both gravitational and non-gravitational) are very demanding; indeed, the main present limitation to relativistic experiments with ordinary spacecraft, especially for long-period effects, are due to our poor knowledge of non-gravitational forces. For example, the estimates given in Secs. 15.4 and 18.1 of the effect of solar radiation pressure, with a cross-section of 1 m^2 and a mass of 1 ton, give a displacement of the order of 100 km, which can be reduced to, say, 5 km by modelling. Drag-free bodies or systems, therefore, are important for precise tests of general relativity with spacecraft (Sec. 18.1).

A way to solve the problem of non-gravitational forces is to use bodies with a very small area-to-mass ratio, like planets and natural satellites. An extensive set of measurements of distance to Venus and Mercury was carried out in the 60's in the United States and the Soviet Union using high-power radars at frequencies below 1 GHz. The return signal is not only affected by the orbital motion of the centre of gravity of the planet, but is also spread out in time and frequency by the shape, the topography and the rotation of the planet. After suitable modelling of these effects the distance of the centre of gravity is estimated and compared with the predictions of relativistic celestial mechanics; residuals of a few km have been obtained, allowing a good determination of the perihelion advance.

A much better use of planets was made with the two Viking landers on Mars. Their radio transmitters provided from 1976 to 1982 1136 measurements of distance with an error of about 7 m. These measurements made it possible not only to make very accurate estimates of the relativistic parameters β and γ (see Ch. 17), but also to set an excellent upper limit to the cosmological change in the gravitational constant G. It has been conjectured (by Dirac and other people) that as the universe expands, gravitation gets weaker, so that G decreases appreciably with time in a Hubble time $1/H \approx 10^{10}$ y. The effect of such a slow change on the Kepler motion can be deduced from the adiabatic constancy of the action (see Sec. 9.1). In our case the action is proportional to the ratio between kinetic energy and mean motion; for a small eccentricity this is, in turn, proportional to $\sqrt{(Ga)}$, where a is the semimajor axis. (This result can be obtained also by noting that, even when G is a slowly varying function of time, we still have a radial force which conserves the angular momentum $\propto \sqrt{(Ga)}$.) As a result of the cosmological change of G, therefore, the semimajor axis would increase by a part in 10^{10} in a year, to wit, in 7 years by 100 m for the earth and by 160 m for Mars. Using this effect, the Viking measurements gave the result

$$dG/(GHdt) = 0.02\cdot(1 \pm 2) \ , \tag{20.27}$$

thus practically ruling out this conjecture.

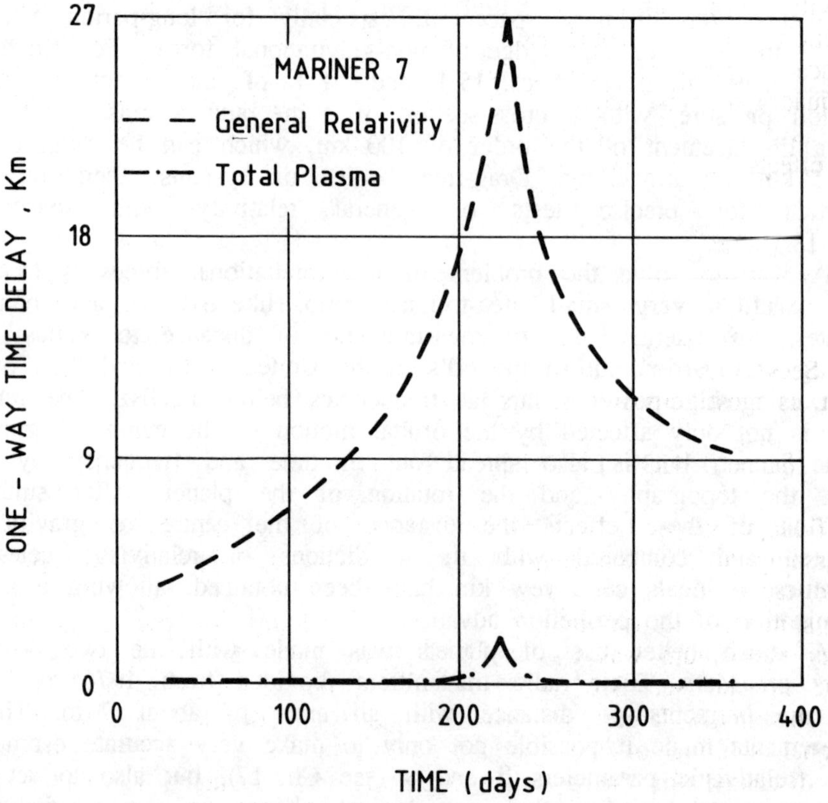

Fig. 20.4 The predicted relativity delay and total plasma delay for Mariner 7 in the near occultation of August 10, 1969, with a ray impact parameter of 4.9 $R_{\odot}$.

Precise tracking of interplanetary spacecraft by range and velocity measurements have provided other excellent tests of theories of gravitation. The parameter γ, which controls the gravitational deflection of an electromagnetic beam (eq. (17.46)), can be estimated rather accurately by measuring the gravitational delay of a radio signal coming from a spacecraft on the opposite side of the sun. The integration of the equation of a null geodesic shows that, as the solar impact parameter of the ray diminishes, the delay has a peculiar spike, easily distinguishable from the slower change due to the orbital motion of the earth and the spacecraft (see Fig. 20.4). There is of course the competing effect of the solar corona, which produces a delay of a similar shape; it can be eliminated using two frequencies (Sec. 19.5).

Today relativistic corrections are an essential part of the laws of celestial mechanics and of the appropriate software used to describe spacecraft trajectories. With space navigation very good estimates of the dimensionless parameters γ and β have been obtained by different methods; they have been found to agree with the values of general relativity, i. e. unity, to within about 0.1%. However, an intrinsic limitation to the combination $(2 - \beta + 2\gamma)$, which determines the advance of periastron (eq. (17.44)), comes from our poor knowledge of the quadrupole coefficient J_2 of the sun. One can easily show that the total advance of the perihelion longitude in one orbital period, due to both effects, is

$$\delta\omega = \frac{2\pi m_\odot}{a(1 - e^2)}(2 - \beta + 2\gamma) + J_2\frac{3\pi R_\odot^2}{a^2(1 - e^2)^2} . \qquad (20.28)$$

(see Secs. 11.6 and 17.6). For Mercury (which, being the innermost planet, is most sensitive to both) this gives the numerical value

$$\delta\omega = 0.4298\cdot[1/3(2 + 2\gamma - \beta) + 10^{-4}(J_2/10^{-7})]''/\mathrm{y}.$$

The fiducial value $J_2 \approx 10^{-7}$ has been used because it corresponds to the assumption that the angular velocity of the sun does not change with the distance from the axis (see eq. (13.3)). This certainly is an underestimate, because during the early evolution of the solar nebula angular momentum was transferred from the outer layers of the proto-sun to the nebula (see Sec. 16.1). In fact, people have speculated that J_2 could be much larger; recently, however, the detailed analysis of the rotational splitting of the oscillatory eigen-modes of the sun (similar to those discussed for the earth in Sec. 6.1) have favoured a value a few times 10^{-7}. In any case, an uncertainty δJ_2 in the solar quadrupole coefficient gives rise to a corresponding uncertainty in the relativistic combination $(2 + 2\gamma - \beta)$. An independent measurement of the solar quadrupole coefficient would be important not only for solar physics, but also for gravitation theory.

We mention also an independent way to measure the gravitational deflection (17.46) by *differential VLBI*. This method uses the difference in the delays of the electromagnetic signals coming from two nearby radio sources which happen to be near the sun. Since the gravitational contribution to the travel time increases with the solar impact parameter, such a difference measures the parameter ($\gamma + 1$) independently of the absolute position of the source in the sky. Finally, by precise Doppler tracking of interplanetary spacecraft the detection of *low frequency gravitational waves* (in the mHz band) will be attempted in the future.

b. Relativistic effects near the earth.

The smallness of the gravitational radius of the earth (0.5 cm) makes it rather difficult to observe dynamical relativistic corrections for earth satellites. Even with the best tracked satellite – LAGEOS – the secular relativistic advance of the perigee could not be observed because its very small eccentricity makes the position of the perigee very uncertain. It is possible, however, that in the future LAGEOS will provide a test of the Lense-Thirring gravito-magnetic effect, namely, the orbital precession due to the angular momentum of the earth.

An important exception, however, is provided by the verification of the principle of equivalence by means of the lunar laser ranging. In other words, we would like to set an upper limit to the quantity

$$\delta = (m_i - m_g)/m_i \,. \tag{20.29}$$

To see the consequence of this violation, consider the newtonian equation of relative motion of the moon with respect to the earth, obtained by taking the difference between their accelerations. Denoting with 2 and 1 the moon and the earth, respectively, we get

$$\ddot{\mathbf{r}}_2 - \ddot{\mathbf{r}}_1 = (1 - \delta_2)\mathbf{g}_2 - (1 - \delta_1)\mathbf{g}_1 \quad .$$

Here $\mathbf{g}$ is the ratio between the gravitational force and the (passive) gravitational mass m_g. When there are no other perturbations, because of the equality of action and reaction, $\mathbf{g}_1$ and $\mathbf{g}_2$ are proportional and the change induced by δ in the relative acceleration is a trivial coefficient of proportionality very near unity. To see a true effect we need a third body, like the sun. Its acceleration fields at the moon and the earth are very nearly equal; in newtonian mechanics what counts is their small difference, proportional to the separation vector $\mathbf{r}$ between the two bodies: this is the tidal force. When the principle of equivalence is violated, however, we have an extra term for the moon

$$(\delta_2 - \delta_1)GM_\odot\mathbf{R}/R^3 = -\nabla U_\delta \;, \tag{20.30}$$

where **R** is the vector from the sun to the earth. The corresponding potential energy for the relative motion

$$U_\delta = -(\delta_2 - \delta_1)GM_\odot \mathbf{R} \cdot \mathbf{r}/R^3 \qquad (20.30')$$

is a linear function of the relative coordinate **r** and depends upon time through the change of the angle between **R** and **r**. Neglecting the eccentricity of the earth, the perturbation Hamiltonian is proportional to cosD, where D is the *sinodic longitude*, i. e., the moon-earth-sun angle. The effect of this perturbation on the earth-to-moon distance r is also a sinusoidal function of D; its amplitude (see also the Problem 20.8) can be roughly estimated by taking the ratio of the disturbing energy to the main potential energy:

$$\frac{\delta r}{r} = (\delta_2 - \delta_1)\frac{M_\odot}{R^2}\frac{r^2}{M_\oplus} = 2(\delta_2 - \delta_1) \ . \qquad (20.31)$$

With an accuracy of 5 cm an uncertainty in δ of order 10^{-10} can be achieved; but one can do much better by using the peculiar signature of the effect. The test of the principle of equivalence for the earth and the moon is particularly important because their gravitational binding energy contributes appreciably ($\approx 10^{-10}$) to the total mass.

In the theory of general relativity the electromagnetic transfer of frequency, and therefore the synchronization of clocks is governed by eq. (17.33) and its particular cases (17.37) and (17.20). The relativistic corrections are $O(v^2/c^2)$ for bodies moving with velocity v and $O(\Delta U/c^2)$ for a gravitational potential difference ΔU. Note that for a satellite at an altitude comparable to, or greater than, the earth radius, they are of the same order of magnitude $\approx 10^{-10}$. Indeed, both effects are just contingent aspects of a single phenomenon and their separation depends on the frame of reference.

A change in the clock rate of a part in 10^{10} is readily observable with the frequency standards available today; indeed, relativistic effects are an integral part of the time and frequency measurements. It was thus possible to test experimentally the *twin paradox*, according to which two clocks in general show a different time when they come back together, after having travelled apart. Mathematically, this is an obvious and general consequence of the fact that the proper time interval ds is not an exact differential: its integral between two events depends of the world line. When the gravitational field can be neglected (i. e., when both clocks stay at the same distance from the centre of the earth) the elapsed proper time is given by the integral

$$\int dt\sqrt{(1 - v^2(t)/c^2)} = \int dt(1 - v^2(t)/2c^2 + \dots) \ .$$

For example, in a day a clock at rest on the equator falls behind a clock at rest in an inertial frame by $\frac{1}{2}\omega^2 R^2 \cdot LOD \approx 0.1\ \mu s$. An effect of this kind has been measured with a clock carried around the earth on an airplane; since its speed is comparable to the earth's equatorial velocity, the time lag is of the same order.

In 1976 a maser clock, with an accuracy of about a part in 10^{14}, was flown for about 2 hours on a rocket up to an altitude of about 10,000 km. It controlled the frequency of a radio signal, whose Doppler shift was measured on the ground, as a function of time. In this way it was possible to test the relativistic Doppler shift (eq. (17.33)) to about 1 part in 10^4 (R. Vessot).

PROBLEMS

20.1. Using a parametric representation of the ellipses (2.16), prove eq. (20.16').

20.2. Show that the error in determining the coefficient c_n of the best power law $y = c_n t^n$ that fits N ($\gg 1$) independent and equally spaced observations y_i is proportional to the power $-1/2$ of N.

*20.3. A problem of *aliasing*. Fit a record of independent observations y_i to the sum of a straight line and a sinusoid of known frequency ω and evaluate the large correlation arising when the length of the record is of the order of (or less than) $1/\omega$. Hint: fit the data y_i with the expression

$$a t_i + b \cdot \sin \omega t_i$$

with respect to the slope a and the amplitude b.

20.4. Compute the round trip light time from a station to LAGEOS as a function of the elevation. Neglect the rotation of the earth, atmospheric effects and assume that the satellite passes at the zenith.

20.5. Calculate, in order of magnitude, the phase noise at 1 GHz of a quasar at 10^8 light-years with a power 10^{45} erg/sec. Assume an effective temperature of 50 K and a dish of diameter 30 m.

*20.6. Find the expression of the VLBI delay for a generic position of the pole and the source and compute the formal accuracies in the coordinates of the pole for a given accuracy in the delay.

20.7. Prove eq. (20.28).

*20.8. Study the effect of a violation of the principle of equivalence on the motion of the moon. Neglect the eccentricities of the earth and the moon and the inclination of the moon on the ecliptic plane.

20.9. Evaluate the difference in elapsed time between two clocks when they meet again at the equator, one on the ground and one flying around the world westward or eastward.

20.10. The phase change of a radio beam is determined by the integral over the ray of the refractive index, which in a dry troposphere is proportional to the air density. Show that in a parallely stratified and static atmosphere the phase change is determined by the ground pressure.

20.11. How does the angular accuracy of a stellar interferometer depend upon the apparent luminosity of the source?

FURTHER READINGS

There are many good textbooks on data analysis, a marginal topic in this book; we quote only one which includes an introduction to Kalman filtering: A.H. Jazwinski, *Stochastic Processes and Filtering Theory*, Academic Press (1970). On LAGEOS, see the special issue of the J. Geophys. Res. of September 30, 1985; the result about the change in J_2 is described in C.F. Yoder *et al.*, Nature, **303**, 757 (1983).

The relativistic experiments are described in detail in the book by Will (see Ch. 17); for a more recent update, see the contribution by C.M. Will, *Experimental Gravitation from Newton's* Principia *to Einstein's General Relativity*, in the volume *Three Hundred Years of Gravitation* (S.W. Hawking and W. Israel, eds.), Cambridge University Press (1987).

SUBJECT INDEX